S. R. Gib
D1743742

Small Particles Technology

Small Particles Technology

Jan-Erik Otterstedt

Chalmers University of Technology
Gothenburg, Sweden

and

Dale A. Brandreth

Widener University
Chester, Pennsylvania

Plenum Press • New York and London

Library of Congress Cataloging in Publication Data

Otterstad, Jan Erik.
  Small particles technology / Jan-Erik Otterstedt and Dale A. Brandreth.
    p.   cm.
  Includes bibliographical references and index.
  ISBN 0-306-45935-3
  1. Particles. 2. Metallic oxides. 3. Silica. 4. Surface chemistry. 5. Colloids. 6. Catalysts. I.
Brandreth, Dale A. II. Title.
TA418.78.O88   1998   98-41315
620′.43—DC21   CIP

ISBN 0-306-45935-3

© 1998 Plenum Press, New York
A Division of Plenum Publishing Corporation
233 Spring Street, New York, N.Y. 10013

http://www.plenum.com

10 9 8 7 6 5 4 3 2 1

All rights reserved

No part of this book may be reproduced, stored in a data retrieval system, or
transmitted in any form or by any means, electronic, mechanical, photocopying,
microfilming, recording, or otherwise, without written permission from the Publisher

Printed in the United States of America

To the memory of
Ralph Iler

Preface

It is difficult to imagine modern technology without small particles, 1-1000 nm in size, because virtually every industry depends in some way on the use of such materials. Catalysts, printing inks, paper, dyes and pigments, many medicinal products, adsorbents, thickening agents, some adhesives, clays, and hundreds of other diverse products are based on or involve small particles in a very fundamental way. In some cases finely divided materials occur naturally or are merely a convenient form for using a material.

In most cases small particles play a special role in technology because in effect they constitute a different state of matter because of the basic fact that the surface of a material is different from the interior by virtue of the unsaturated bonding interactions of the outermost layers of atoms at the surface of a solid. Whereas in a macroscale particle these differences are often insignificant, as the surface area per unit mass becomes larger by a factor of as much as 10^9, physical and chemical effects such as adsorption become so pronounced as to make the finely divided form of the bulk material into essentially a different material— usually one that has no macroscale counterpart.

The creation of finely divided forms of different materials depends a great deal on the material itself. Grinding will produce a certain fraction of very small particles, but economic, technical, and safety factors frequently make that primeval method of little use, particularly when the desired size is less than about 10 μm (10,000 nm). In addition in some materials such as catalysts, the specific surface is greatly increased by the presence of pores, which are not created by grinding.

For the past twelve years one of the authors has given a course in Small Particles Technology to senior engineering students and graduate students at Chalmers University of Technology in Gothenburg, Sweden, where this subject has also been a very active and successful area of research in the Department of Engineering Chemistry at that university.

The authors, who both graduated from the Department of Chemical Engineering at the University of Toronto in Canada, and who were also colleagues at E. I. Du Pont & Company in Wilmington, Delaware, agreed that it would be worthwhile to expand the course in Small Particles Technology to a book that would seek to be a useful compendium of facts and ideas to those working in research, development, and manufacturing involving aspects of small particles technology.

The goal of our book is, therefore, to present the chemical, and also to some extent, the physical principles of methods for making small particles of various materials, especially metal oxides, and to describe the particle surface and methods for modifying it. We also want to demonstrate how small particle technology can be used in various technical applications and how to make technically important materials.

In particular, we want to bring out the possibility of tailoring materials to meet special needs and standards of performance that small particles technology offers. Pigments, for instance, are small particles whose performance depends on their size and shape. In many cases pigments are made by grinding or milling coarser particles to a finer size—procedures that do not guarantee uniformity of particle size or shape.

Ceramics are made of small particles that also have been obtained by grinding coarser particles. The high performance of ceramics, especially high-tech ceramics in demanding applications, depends critically on the quality of the small particles, e.g., their size and shape, that make up the starting material. Small particles with carefully controlled size, shape, and surface characteristics can be made by the methods of small particles technology as described in this book.

Thus we have devoted attention to descriptions of typical preparation methods for the most important types of small particles. As it turns out, particles in the colloidal range of, say, 10 to 1000 nm, are not easy to handle in the dry state and in many cases are not easy to prepare in the dry state, so that in this book the emphasis is on preparations in the wet state, which frequently is a convenient state in which to ship and use the product, for instance, pigments.

The majority of important heterogeneous catalysts have a porous structure, the surface of which is either catalytically active or has been made so by depositing catalytically active substances on it. Small particles, 5-1000 nm, in different packing configurations, make up the structure. Catalytically active materials, e.g., noble metals, deposited on the porous structure are present in the form of very small particles, 1-10 nm. We believe that if one can make small particles of various materials, e.g., zeolites, silica, alumina, titania, zirconia, and other metal oxides, in different sizes and shapes, and can chemically modify the surface of such particles, one has an excellent tool for developing not only new and improved catalysts, but also porous materials for other technically important applications.

In this book we have sought to examine small particles technology by dividing the topic into a consideration of the principal types of materials of importance to industry. Thus, in the first part of this book, silica in its various forms, alumina and other metal oxides, carbon, and various metals in finely divided forms, are considered with regard to their special properties and types, after which, in the second part, applications in specific industries such as papermaking are considered.

In so doing, of course, we inevitably neglect some materials and applications that may well deserve more consideration were it practical to double the book's size. As with all such choices, the experiences, knowledge, and interests of the authors are key factors in determining what is included. On the other hand, we believe that many of the methods and techniques described in particular applications can be generalized to include neglected or new uses.

Although some more basic, theoretical material is used, we have not sought to make this a partial text in surface chemistry and physics, for there are many excellent texts available to fill that need. We will mention Adamson: *Physical Chemistry of Surfaces* (1983); Fridrikhsberg: *A Course in Colloid Chemistry* (1986); Prutton: *Introduction to Surface Physics* (1994); Hiemenz: *Principles of Colloid and Surface Chemistry* (1977); and Evans and Wennerström: *The Colloidal Domain* (1994).

Special mention should also be made of the two books by Ralph Iler on the chemistry of silica: *The Chemistry of Silica* (1979), *The Colloid Chemistry of Silica and Silicates* (1955), and the book by Brinker and Scherer: *Sol-Gel Science* (1990).

In his earlier book Iler assembled and correlated information on the behavior of silica and silicates in the colloidal state so as to present a coherent picture of such phenomena as solubility, formation, and behavior of colloidal particles of silica and surface chemistry of the solid phase.

In Iler's second book (1979), which he first intended to be an updated edition of his first book but which became the definitive book on silica chemistry, he presented a complete and coherent account of the chemistry of amorphous silica, including soluble silica and silicate precursors of soluble silica, polymerization of polysilicic acids, colloidal sols and gels, and the surface chemistry of silica.

In the last three areas there is some overlap among Ralph Iler's two books and our book, but we have reviewed, condensed, and updated material in those areas, and also included results of our own work. Furthermore, we describe fewer technical applications, but cover each application in considerably more depth and detail.

In their book *Sol-Gel Science*, Brinker and Scherer state that their goal is to present the physical and chemical principles of the sol-gel process in a manner suitable for graduate students and practitioners in that field. They define sol-gel rather broadly as the preparation of ceramic materials by preparation of a sol, gelation of the sol, and removal of the solvent. The sol may be produced from inorganic or organic precursors and may consist of dense oxide particles or polymeric clusters. Their emphasis is on the science rather than the technology. There is some overlap between Brinker and Scherer's book and this book's Chapter 3 dealing with hydrolysis of metal ions, but we have tried to minimize it.

In addition to stating what this book is about, perhaps it is worthwhile stating some of the topics which are not included, despite their obvious importance. We have not included biomedical, safety, toxicological, natural phenomena, and environmental aspects of fine particles except in passing mention for three main reasons: 1) we interpret the word "technology" in the title to mean industrial technology aimed at creating small particles, and using them in industrial processes; 2) each of these topics is highly specialized and very diverse with a profuse literature of its own, generally outside areas where we can claim much experience; 3) we have restricted ourselves to *solid* particles, because aerosols with liquid droplets are an essentially different field.

Conventional photography too, in terms of both its enormous commercial and general significance, has its roots in small particles technology, but we consider it to be too specialized, beyond fundamental considerations treated in several chapters here, to be able to do justice to it.

This is not to say that there are not areas of common interest and overlap. In many applications of small particles one may be concerned with the behavior of such particles in settling, or in other transport modes, i.e., the mechanics of small particles. That subject is a much studied area of vast importance in atmospheric science and biomedical applications, but it already has been very thoroughly covered in the literature, most especially in a survey by the Russian scientist, N. A. Fuchs, in a classic work, *The Mechanics of Aerosols* (1989).

Chemical engineers have long been interested in the transport properties of gases and liquids in pores such as those found in solid catalysts and have developed a very wide literature covering both the theoretical and experimental sides of that topic. Those areas too we neglect here as being too tangential.

Currently, with the impressive number of new analytical tools for exploring surfaces, considerable progress is being made in understanding the details of processes such as catalysis occurring at solid surfaces. Here we have occasionally made brief mention of some results based on these modern methods, but again the constraints of space make it necessary for interested readers to look at some of the references for more details.

It is the future of small particles technology that we find the most intriguing. Exciting developments abound. For example, nanotechnology—the creation of devices on the super-microscale—is rapidly emerging as an area with enormous implications in medical science, computers, communications, and control devices, to name but a few possibilities. Small particles technology must inevitably play a role in the materials used and as a limitation of the sort already accepted in microchip computer manufacturing, where miniaturization depends on controlling the size of small particle contamination. Other new areas, e.g., in catalysts and zeolites, are mentioned throughout this book.

As Kingery pointed out, high-tech ceramics have an enormous leverage derived from their criticality to the effective functioning of a whole range of devices to which our world's culture is dedicated. *"Companies or nations that do not remain at the forefront of ceramic developments will find it impossible to remain at the forefront of the device developments that will play a dominant role in the future of our materials culture."*—Kingery (1988). As we show in Chapter 10, high-tech ceramics are an integral part of small particles technology.

We believe that our approach to small particles technology has some unique features and hope that this book will stimulate further interest in the field and will contribute to the understanding of properties of materials and the development of new materials needed to enhance the quality of our lives in maintaining a high standard of living while protecting our environment.

Jan-Erik Otterstedt, Bohus, Sweden

Dale Brandreth, Hockessin, Delaware, USA

References

Adamson, A., *Physical Chemistry of Surfaces*, 4th ed., John Wiley, NY (1983).

Brinker, C. J., and Scherer, G. W.: *Sol-Gel Science*, Academic Press, NY (1990).

Evans, D. F., and Wennerström, H.: *The Colloidal Domain*, VCH, NY (1994).

Fridrikhsberg, D. A., A Course in Colloid Chemistry, Mir Publishers, Moscow (1986).

Fuchs, N. A., *The Mechanics of Aerosols,* Dover Publications, NY, (1989).

Hiemenz, P. C., *Principles of Colloid and Surface Chemistry* , Marcel Dekker, NY (1977).

Iler, R. K., *The Colloid Chemistry of Silica and Silicates*, Cornell University Press, NY (1955).

Iler, R. K., *The Chemistry of Silica*, John Wiley, NY (1979).

Kingery, W. D., The Materials Revolution, ed. T. Forester, The MIT Press, Cambridge, 315 (1988)

Prutton, M.,*Introduction to Surface Physics* , Oxford University Press, (1994).

Acknowledgment

The authors acknowledge their indebtedness to the many people who made the publication of this book possible.

We appreciate the helpful discussions with Dr. Horacio Bergna, Dr. William O. Roberts, and Dr. Paul C. Yates, our former colleagues at E. I. Du Pont & Company, and also with Dr. Randol Carroll and his staff at The PQ Corporation in Valley Forge, PA. We are particularly grateful to Dr. John Bugosh, who not only followed the progress of the book, but also provided the information on hydrolysis and formation of fibrous particles of alumina in Chapter 3.

The contributions from the work of IngMari Axelsson, Lars Eriksson, Lennart Evaldsson, Börje Gevert, Peter Greenwood, Lars Löwendahl, Anders Persson, Brian Schoeman, Magnus Skoglundh, Johan Sterte, Anders Törncrona, Zhong-Shu Ying, and Yan-Min Zhu, former graduate students and associates of one of the authors in the Department of Engineering Chemistry at Chalmers University of Technology, have been crucial and are gratefully acknowledged.

Our very great appreciation goes to Mary Mattsson, who did many of the drawings and performed many experiments on colloidal silica and silicates, the results of which have heightened the value of important parts of the information in this book.

We are in the greatest debt to Elisabeth Hawami, who patiently and efficiently sent out, collected, and organized the several hundred copyright permission requests. This book would not have been finished without the help and support of Elisabeth.

We also gratefully acknowledge the help of Amelia McNamara, Mary Curioli, Meri Zeltser, and their staffs at Plenum Publishing Corporation for their help, patience, and forebearance in what has turned out to be a far greater task than the authors ever could have envisioned.

Table of Contents

CHAPTER 1.   INTRODUCTION TO SMALL PARTICLES

1.   Small Particles and Materials

One of several ways to classify materials is to divide them into the following groups:

1. Metals
2. Ceramics
3. Polymers-Plastics and Elastomers (Rubber)
4. Cellulose-Wood-Paper

Figure 1-1 shows these groups of materials and some typical important uses of them. Many materials also contain different kinds of additives and auxiliary chemicals: e.g., fillers in polymers-plastics, rubber, paper. A closer examination will show that many products of different materials, and many additives as well, are made of or consist of small particles. Ceramics are thus made of metal oxides, e.g. oxides of aluminum, silicon, zirconium, or nitrides of, for instance, carbon. Fillers in plastics and paper may consist of small particles, e.g., various types of clays. Porous structures, which form the basis of such structures as catalysts, adsorbents, and membranes, are frequently made of small particles. Inorganic binders may consist of small particles of silica or alumina.

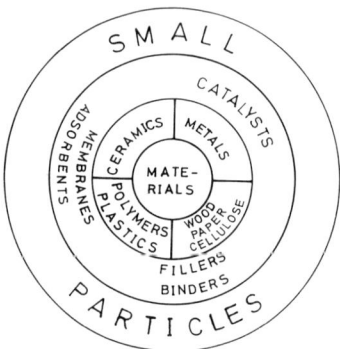

Fig. 1-1. Small particles play an important role in the material world.

A fundamental question, which has fascinated physicists and theorists for a long time, is what happens when one, two, tens, hundreds, thousands, or millions of atoms bind together to larger systems such as ultrafine particles, powders or solids. Today we have good knowledge of the first building block, the atom, and of the final result, macro-scale solids, whereas our knowledge of intermediate systems is relatively modest. Many properties of atoms and solids can be found in handbooks with tables and diagrams. The periodic table is one of the most comprehensive general compilations wherein the position of an element gives information about some of its special properties. However, handbook values almost always refer to the properties of the solids in bulk form, i.e. without reference to specialized forms such as finely divided particulates, spongy forms, etc., each of which may have vastly different properties than the bulk form.

Atoms are often characterized by such properties as electron configuration, ionization energies, oxidation states, and electron affinity. For solids the discrete energy levels of the atom have been broadened to energy bands and the first ionization energy corresponds to the work function. Moreover, solids are often characterized by a certain crystal structure and can be classified as conductors, semi-conductors, or insulators, with different electrical and optical properties; Fig. 1-2. There is today a great deal of interest in fundamental as well as applied research to investigate how properties change when atoms, one by one, successively bind together and grow to molecules and building blocks and finally transform to condensed matter.

The build-up of clusters and small particles from single atoms and molecules can be viewed as a growth process where atoms are successively added. If the total energy of the new system is lower than the sum of the energies of the single atoms or molecules, a cluster-small particle is formed with certain geometry and equilibrium distances between atoms-molecules, which can be described by potential energy diagrams. Compared with a diatomic molecule, a molecule made of several atoms has more degrees of freedom. However, various spectroscopic methods can be used to study transitions between electronic, vibrational, and rotational levels in molecules and clusters-small particles. The results of such studies give information about structures as well as the chemical bonding between atoms.

Just as for atoms and solids, the properties of clusters and small particles are determined by the interaction of the outer valence electrons of the single atoms as they grow to larger structures. As it becomes possible to make clusters and very small particles in a controlled manner and to do calculations on them, several interesting questions arise. Compared with the band gap of 1.12 ev for bulk silicon, how does the band gap between the highest occupied and lowest

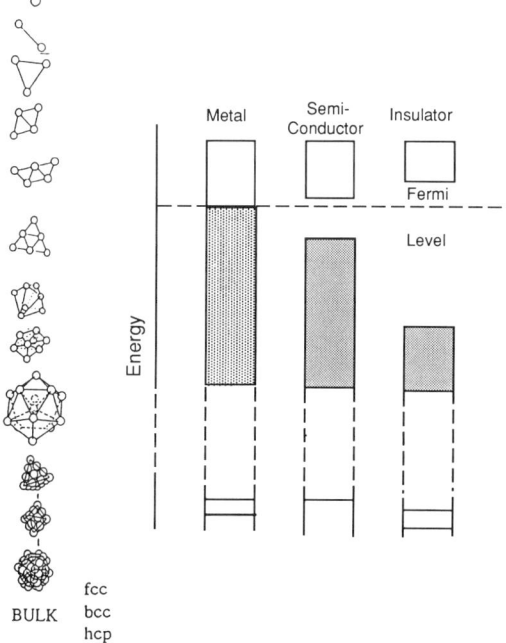

Fig. 1-2. Atoms bind together to molecules, clusters and larger systems to form small particles and solids. On the right are shown schematic energy level diagrams for a metal, a semi-conductor, and an insulator. Rosén (1993). Courtesy Swedish Science Press.

unoccupied levels change in clusters with 2, 5, 10, or 100 silicon atoms? How many metal atoms are required before a cluster shows metallic properties? How do the optical properties and the color change with the size of the cluster compared to what is known for the corresponding solid material? It is known, for instance, that color filters obtain their color from the size of the particles dispersed in the binder. Similarly, the color of Bohemian crystal or the glass in some church windows depends on the size of the particles present in the glass.

2. How Many Atoms/Molecules Are There in Small Particles?

In order to approach the questions in the previous section and also to gain greater familiarity with small particles, it will be helpful to derive a relationship between the numbers of atoms and molecules in a small particle and its size. We will do this for the case of colloidal silica and assume that similar relationships exist for small particles of other materials, the shapes of which are spherical or nearly so.

The surface is always an important region of a material, and this is especially the case for small particles because the surface-to-volume ratio increases with decreasing particle size. Feigl (1949) discussed the relationship between the surface and the interior of a small particle and concluded that every molecule in the interior of the solid is surrounded on all sides by like molecules that are arranged in orderly fashion in crystals and in disorderly fashion in amorphous materials. Such molecules, which are hedged in on all sides, are fully saturated with respect to their binding power. In contrast, the molecules at the free surface are unsaturated in this respect, and consequentially can exert binding forces directed outward. Carman (1940) discussed the structure of a single

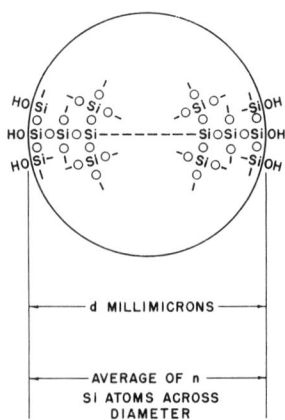

Fig. 1-3. Particle of colloidal silica, according to Carman (1940). Courtesy The Royal Society of Chemistry.

particle of colloidal SiO_2 in an aqueous dispersion and visualized that in a particle made up of amorphous SiO_2, in which every oxygen atom lies between two silicon atoms, a peculiar situation exists at the surface. The surface silicon ions will strive for complete tetrahedral coordination and in contact with moisture, water will add to this unsatisfied surface, with the hydrogen going to the oxygen atom on the surface and the hydroxyl to the silicon on the surface (Fig. 1-3). Using Carman's picture, Iler (1955) calculated the relation of particle composition to particle size purely from geometry and densities of the components. With the following notation:

n_t = total number of silicon atoms in a particle

n_s = number of silicon atoms at the particle surface

d = diameter of particle on anhydrous basis (nm)

d_n = diameter of hydroxylated particle (nm)

x = ratio of SiOH groups to total Si atoms

$= n_s/n_t$, assuming one OH per surface silicon

w = weight of one anhydrous SiO_2 particle (grams)

w_h = weight of one surface hydroxylated particle

p = average number of silicon atoms across the diameter of a particle

and using the relationship:

$$n_t = \frac{\pi}{6}p^3 \tag{1-1}$$

and the following two equations for the weight of an anhydrous SiO_2 particle

$$w = \frac{60n_t}{6.03 \times 10^{23}} = 2.2\frac{p}{6}d^3 \times 10^{-21} \tag{1-2}$$

he derived equations for the total number of silicon atoms in a particle, the average number of silicon atoms across the diameter of a particle , and the number of silicon atoms at the particle surface:

$$n_t = 11.5d^3 = 0.524p^3 \tag{1-3}$$

$$p = 2.80d = 1.24(n_t)^{1/3} \tag{1-4}$$

$$n_s = \frac{\pi}{6}[p^3 - (p-2)^3] \tag{1-5}$$

The ratio of SiOH groups to total Si atoms is equal to n_s/n_t assuming one OH per surface silicon atom:

$$n_s/n_t = x = 2.14d^{-1} - 1.53d^{-2} + 0.36d^{-3} \qquad (1\text{-}6)$$

The hydroxyl content, expressed as wt. % water, is :

$$\%H_2O = 32d^{-1} - 23d^{-2} + 5.45d^{-3} \qquad (1\text{-}7)$$

The composition of the particle may be expressed as

$$Si_nO_a(OH)_b$$

where $2a - b = 4n_t$. Then in terms of anhydrous particle diameter (nanometers) the composition is given by

$$n_t = 11.5d^3 \qquad (1\text{-}8)$$

$$a = 23d^3 - 12.3d^2 + 8.8d - 2.09 \qquad (1\text{-}9)$$

$$b = 24.6d^2 - 17.6d + 4.18 \qquad (1\text{-}10)$$

The various properties of the silica particles that can be calculated by the equations above are summarized in Table 1-1.

Table 1-1. Relation of particle composition to particle size for silica particles, Iler (1979). Reprinted by permission of John Wiley & Sons, Inc.

n_t	d	d_h	OH:Si	O:Si	OH nm^{-2}
8	0.89	1.0	0.99	1.51	2.9
40	1.52	1.7	0.85	1.57	4.3
100	2.06	2.2	0.72	1.64	5.0
311	3.0	3.2	0.55	1.72	5.7
1438	5.0	5.2	0.37	1.82	6.5
11500	10.0	10.2	0.20	1.90	7.2
•	•	•	0	2.0	7.8

The ratio n_s/n_t is of particular importance in catalytic processes using noble metals such as Pt, Rh, or Pd as catalysts deposited as small particles on a support—see Chapter 7. The ratio for the small particles of noble metals, in this case called the degree of dispersion, then indicates the fraction of the noble metal atoms present as surface atoms, and therefore catalytically active atoms. Naturally, it is desirable that the degree of dispersion of the noble metals is as high as possible; i.e. they are present as very small particles. In the case of platinum as a catalyst, the molecular weight and the density of platinum can be inserted in Eqn. 1-2 and, assuming that the platinum particles are spherical in shape, the degree of dispersion can be calculated from the equation:

$$n_s/n_t = 1 - (1 - 2/p)^3 = 1 - [1-2/(4.04d)]^3 \qquad (1\text{-}11)$$

Figure 1-4 shows that the degree of dispersion of platinum decreases from 83% to 28% when the particle size grows from 1 nm to 5 nm.

That the properties of a material depend on particle size, can be readily seen from a simple consideration. The total energy of a system, E_{tot}, is made up of two contributions, namely the internal energy, E_i, and the surface energy, E_s.

$$E_{tot} = E_i + E_s = e_i V + gA \qquad (1\text{-}12)$$

where e_i is the internal energy per unit volume and g is the interfacial energy per unit surface area. V and A are the total volume and surface area, respectively, of dispersed material.

The contributions from the two terms may vary with the conditions, and how they relate to each other depends on the geometric factor. The total energy in Eqn. 12 expressed per unit volume will be:

$$(E_{tot})_v = e_i + g(A/V) \qquad (1\text{-}13)$$

Since e_i and g are properties characteristic of a given material, A/V is the only variable in Eqn. 1-13. If this ratio is large, the second term will be significant; e.g. A/V ~ 10^4 to 10^7 cm^{-1}. Properties of materials may thus be strongly affected by the contribution of surface energy to the total energy of the system. High values of A/V can be obtained by allowing one, two, or three of the dimensions to be small, resulting in films, fibers, and small particles.

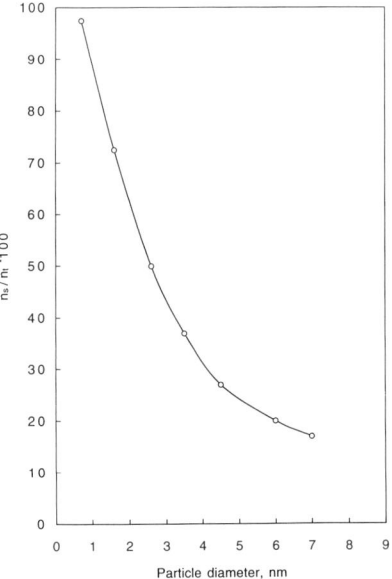

Fig. 1-4. Degree of dispersion of Pt particles vs. particle size.

3. How Small Is a "Colloid" Particle?

The term colloid comes from the Greek word for "glue", which no doubt was made from boiling hides and bones to obtain gelatin. Thomas Graham in 1861 in London was the first to recognize what he called the "colloidal" state of matter—a substance so finely divided it does not settle out of suspension but is not so finely divided as to pass through a parchment membrane or to depress the freezing point of water as small molecules do. Graham defined a colloid as a substance having a slow diffusion rate in solution.

Hackh's Chemical Dictionary says that Brownian motion is a characteristic of a colloidal particle and that the size range is 1×10^{-7} to 1×10^{-5} cm, or one nanometer (10Å) to 100 nm (1000Å). Others have defined colloidal particles as being in the range from 5 nm to 500 nm. Conventionally, a colloidal particle need have only one dimension in the colloidal range to be a colloid, since even that one dimension confers extraordinary properties to the material as a result of falling into that range of influence of forces governing the behavior of that material in a particular medium, as noted in the introduction.

Taking silica as an example, with a density of 2.2 g/cm^3, particles down to 5 nm diameter can form stable sols, but smaller particles are increasingly unstable because they become increasingly soluble—see Ch. 2. Particles above 500 nm in size settle from water in a matter of days, but if they are less than 70 nm, they do not settle under gravity because Brownian motion keeps them in suspension.

All of the foregoing is true if the particles are prevented from colliding by ionic forces or surface films of stabilizers to prevent direct silica-silica contact during collision. This stabilization requires carefully controlled conditions of concentration and pH, or stabilizers.

Most commonly colloidal particles collide and stick together either slowly or rapidly and thus aggregates form which settle as a precipitate or thicken the liquid to a gel.

What do you *see* in silica sols of different sized particles? If particles are 5-10 nm in size, even concentrated 30% sols are essentially clear. With particles 10 to 30 nm, the sols become increasingly "opalescent"—still transparent but whitish in color. From 50 to 100 nm the sols are white, like milk, due to light scattering as the particles are approaching the wavelength of visible light, 400-700 nm.

It is difficult to appreciate or visualize colloidal dimensions. Colloidal particles are beyond our scale of perception of reality. How do you visualize a typical colloidal particle, say 25 nm in diameter, which is twenty-five times larger than the smallest colloids and twenty-five times smaller than the largest colloidal particles?

Such particles can be magnified by the electron microscope, typically 50,000 times, and are then 12.5 mm in diameter. Under the optical microscope at 1000-fold magnification, the largest colloidal particle of 500 nm is barely visible at half a millimeter in size.

4. What Materials Form Colloids?

Except for our rigid skeletons, we ourselves are essentially an organic colloid full of water. Here we limit ourselves to inorganic colloids.

What is the most obvious characteristic of colloidal particles in a "sol", which has long been ambiguously called a "colloidal solution"?
It is the insolubility of the particles in the liquid in which they are suspended. Thus sodium chloride could never form a colloidal suspension in water, but it might do so in gasoline.

Water is the most important dispersing medium, and we shall first consider the elements as the dispersed phase. It is apparent that we cannot have a sol of an element that reacts at once with water. This immediately eliminates many elements such as sodium and potassium. The element must be a solid that does not vaporize easily and is insoluble in water.

Thus we have the precious metals gold, platinum, etc. Then there are the elements selenium and especially carbon, but in these cases there is a monolayer of the oxide which wets the particle by water. This illustrates another requirement—namely that the particle must be wettable by water. Just as oil requires a wetting and dispersing agent at the surface of droplets in an oil-in-water emulsion, so do most elements require a surface film of an oxide.

With regard to salts and compounds of metals we realize that in almost all cases the colloids are very insoluble in water. Thus oxides of the heavy metals are insoluble and can be made in colloidal form, e.g., oxides of titanium, zirconium, vanadium, chromium, manganese, iron—or more generally the oxides of the elements of the IIIB to VIIIB of the periodic table as well as IB and IIB.

Then there are the oxides of aluminum and homologs and of silicon and homologs germanium, tin, and lead. Besides these are the oxides of all the rare earths as well as thorium and uranium. The colloidal properties of many of these, except thoria, are not very well known.

But of all these oxides, amorphous SiO_2 presents an anomaly. It is soluble in water at about 0.01 % whereas all other inorganic colloids are generally less soluble than 10^{-4} to $10^{-7}\%$. Also, of all the inorganic colloids it remains amorphous regardless of particle size whereas the other particles are generally crystalline.

The reason why silica systems remain amorphous seems to be that the Si-O-Si bond angle is somewhat variable, so that irregular networks can be formed. In a way it may be similar to the variable bond angle between carbon atoms and the possibility of forming so many amorphous polymer structures.

5. Small Particles by Polymerization of Monomers

Figure 1-5 shows that the following starting materials can be used to prepare inorganic as well as organic small particles by polymerization.

5.1. Organic Monomers

Small particles of different polymers, suspended in e.g., water, can be prepared by emulsion or suspension polymerization of different monomers. Latex is the common name for such suspensions of small particles in water. The particle size of the latex depends on the amount of dispersing agent present during the polymerization reaction, but will usually fall into the range from 100

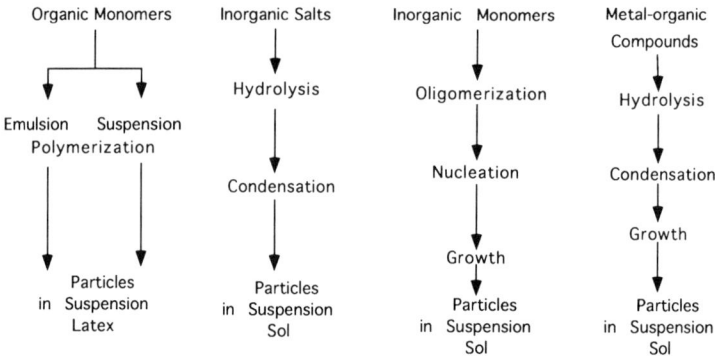

Fig. 1-5. Small particles by polymerization of monomers. Courtesy Swedish Science Press.

to 1000 nm. Polymer latices are used as binders in paint, lacquers, coating compositions for paper, etc.; see Chapter 10.

5.2. Inorganic Salts

Small particles of oxides-hydroxides of many transition metals, e.g. aluminum, titanium, zirconium, etc., can be made from aqeous solutions of salts of such metals by hydrolysis and condensation reactions—see Ch. 3.

5.3. Polymerization of Silica

Small particles of silica can be prepared by polymerization of monomeric silicic acid as will be described in Ch. 2.

5.4. Metal-organic Compounds

Small particles of silica having special properties can be prepared by hydrolysis and condensation of silicon alkoxides—see Ch. 2. Small particles of other metal oxidess can also be prepared by similar methods—see Ch. 3.

6. Clusters — "Soccer Balls"

This book is about small particles, including very small particles, but it does not deal specifically with clusters. It would be inappropriate, however, to not say anything about clusters. In particular, as will be described in Chapter 7, supported noble metals, as, for instance, platinum and rhodium in catalytic converters for automobiles, may be present as particles of such minute size that they can be called clusters.

Clusters may be characterized as an agglomeration of atoms or molecules that normally do not exist in the gas phase. Sulfur, which in the gas phase can exist as a cyclic molecule of eight atoms, and phosphorus, which can exist as a tetrahedron of four atoms, are, strictly speaking, therefore not clusters. A cluster may consist of atoms of the same as well as of different elements. Clusters thus consist of tiny aggregates comprising from two to several hundred atoms. This poses questions that lie at the heart of solid-state physics and chemistry, which we have already touched upon: how small must an aggregate of atoms become before the character of the substance they once formed is lost? If the substance is a metal, how small must its cluster of atoms be to avoid the characteristic sharing of free electrons that underlies conductivity? Do growing clusters proceed gradually from one stable structure to another, largely through the simple addition of atoms, or do they undergo radical transformations as they grow?

Many cluster properties are determined by the fact that a cluster is mostly surface. A closely packed cluster of 20 atoms has only one atom in the interior; a cluster made up of 100 atoms may have only 20; compare Table 1-1 and the discussion on the degree of dispersion in § 2 above. Other properties stem from clusters' unfilled electronic bonding capability, which leaves them "naked" and hence extremely reactive. This reactivity makes them effective tools for the study of the solid state and, potentially, for such industrial processes as the growing of crystals, selective chemical catalysts, and the creation of entirely new materials with made-to-order electronic, magnetic, and optical properties.

Small clusters of atoms also offer new clues about melting and freezing points—namely, that they are not so easy to pinpoint. These clusters can coexist as solids and liquids over a finite range of temperatures and have distinctly differing melting and freezing points. It is the unique character of clusters that make them suitable to probe the secrets of their freezing and melting points. Clusters are bigger than individual molecules yet smaller than bulk matter— which consists of so many atoms that the number can be treated as infinite—and so they exhibit the properties of both. Because of their intermediate size, clusters can be studied in almost as precise detail as the atoms or molecules constituting them while simultaneously illustrating some properties of bulk matter.

Of special interest and very appealing for their mathematical beauty are the structures that have been proposed for cluster species of carbon by Smalley et al (1985) in *Nature* . They advanced the idea that the cluster of 60 carbon atoms could only be explained by a structure in which all bonds are saturated and there are no edge or border atoms. The structure reminds one of a soccer ball made of 12 pentagons and 20 hexagons with a carbon atom at each of the 60 corners—see Fig. 1-6. The diameter of the soccer ball is estimated at 7Å. Smalley called the structure "Buckminsterfullerene" in honor of the inventor of the geodesic dome, Richard Buckminster Fuller.

Krätschmer et al. (1990) succeeded in preparing relatively large quantities, a few grams, of C_{60} and related fullerenes, C_{70}, etc., by burning carbon in the form of graphite electrodes in an electric arc in a helium atmosphere. During the discharge atoms and ions of carbon are formed, which form soot by collision with the helium gas. When they shook the soot with benzene in a flask, they obtained a reddish solution after the soot particles settled to the bottom. By evaporating the solution crystals of C_{60} were obtained.

The availability of relatively large quantities of buckminsterfullerene, C_{60}, and the related fullerenes, C_{70} etc., has generated an explosion of research activity by synthetic and physical chemists. It is now evident that fullerene-derived solids are especially exciting from both the fundamental and potential technological viewpoints, and that these solids possess many novel properties of interest to physicists and material scientists.

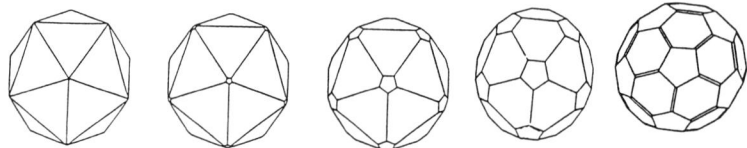

Fig. 1-6 Structure of soccerball C_{60} may be obtained by successively truncating the 12 corners of an icosohedron. A structure with single and double bonds is obtained by making common edges of hexagons with two lines. Rosén (1995). Courtesy Swedish Science Press.

7. Other General Methods for Making Small Particles

7.1 Grinding

This method of comminution has long been the principal method of making fine particles in a wide range of industries from ceramics and minerals to dyestuffs and cement. However, from energy and handling considerations it is usually not practical to grind to particles sizes less than about 1 μm, and for our purposes in this book, where we are generally concerned with particles in the colloidal range from 1 to 500 nm, only brief mention will be made here. There are special cases where other methods for making small particles below 1μm are not known or are impractical, so that grinding is used.

A vast literature devoted to crushing and grinding exists as evidenced by the bibliography by Bickle (1960) with over 2800 references to all aspects of this subject, including energy consumption, equipment and safety. Most such references are on methods for making particles in the range larger than a few microns because energy consumption for making particles much less than a few microns is enormous and quickly becomes much too expensive.

A more modern method of fine grinding involving the use of a Szego mill—a planetary ring-roller mill with radially mobile helically grooved rollers—has been studied in depth by Trass and coworkers (1986, 1990, 1995) for grinding particles in dry form, and in pastes and slurries. This particular type of mill allows some determination of resultant particle shape, e.g., flakes versus granular, depending on operating conditions. Such grinding is also more appropriate where the resulting form, e.g., a slurry, is something like a coal slurry in a combustible fluid, i.e., where the grinding medium is a part of the final product.

7.2. Pyrolytic Methods

In certain cases, as noted in Section 6 above for preparing fullerenes, pyrolysis is a specialized method for preparing small particle forms of some materials. The largest-scale use of the pyrolytic method is in the production of carbon black. In the oil furnace process a heavy aromatic fraction of petroleum distillate is atomized into a preheated, closed furnace followed by cooling and collecting the formed carbon particles. The process is efficient and permits a high degree of control of carbon black properties. These particles consist of layer planes, similar to graphite, but not as precisely oriented and aligned. Electron microscopy reveals that the carbon black is composed of clusters of fused prime particles which are called primary aggregates, Cabot (1990).

Overall, due to the surprising results in making buckminsterfullerene, this will likely become a well explored research area for trying to make other novel materials.

8. Characterization of Small Particles

8.1. Introduction

As noted in the introduction, small particles may often be thought of as totally different materials as compared to the bulk solids of which they are composed largely because the specific surface areas of the finely divided form can be on the order of a million to several billion times greater than that of the bulk form of that substance. This enhanced specific surface area can have effects that are quite dramatic from the standpoints of safety—very violent large-scale explosions of grain storage facilities are well known—to catalytic effects, efficient adsorption of toxic compounds, as with gas masks, and the opacity of paint pigments, to mention but a few examples.

The most important characteristics of small particles are:

• chemical analysis: elemental analysis, total solids, stabilizing agents, soluble salts.

• particle characteristics: size and size distribution, porosity, degree of aggregation, specific surface area.

• physical characteristics: density, refractive index, and in a liquid medium: pH, viscosity, turbidity.

Characterization of structures of small particles, e.g. gels and powders,

must take into account the following variables:
1. The size and shape of the primary particles
2. The spatial distribution of the particles, including the order and density of packing
3. The strength of the bond between particles (coalescence).

These three topics are covered in various sections of this book, but are not covered here specifically.

It is not our purpose to go into the details of characterizing small particles, but rather to describe briefly some of the more general and widely used characterization methods. For a more extensive coverage of small particle characterization, especially of catalysts, we refer the reader to the book by Anderson and Pratt (1985).

8.2. General Methods

8.2.1. Particle Size Measurement This characteristic is of primary importance because it is most closely related to the *raison d'être* for small particle applications. Two critical points must be made about small particle size measurements: 1) without great care in obtaining a representative sample to test, results may be grossly incorrect; 2) colloidal particles dispersed in a liquid usually have to be measured in that medium, or if first dried, must be treated in such a way as to maintain the individual integrity of the original particles, i.e., not be agglomerated.

8.2.1.1. Sampling Inasmuch as a given quantity of small particles almost always covers a range of particle sizes, and because the sampling method itself or preceding treatments may introduce a bias in the selected size range, the first consideration must be given to obtaining a representative sample of the material to be analyzed. Several pitfalls await the unwary:

1. In transporting a large mass of particles in drums or bags via road or rail, frequent vibration generally causes a non-uniform distribution of the particles with respect to size or type (for a mixture).

2. Pouring pellets or powders usually causes a loss of some of the finest particles into the surrounding atmosphere. Since the smallest particles may have most of the total surface area, this can be significant even when a small weight fraction is lost.

3. A sample submitted by a manufacturer may already not be representative for the above reasons.

Every case has its own peculiarities, but two rules of thumb for sampling can be identified which also relate to obtaining a representative sample:
1. A sample is best taken when the material is in motion—as in a pipeline.

2. It is better to sample the entire stream for a short time than part of the stream for a longer time.

Many techniques for representative sampling have been developed and we refer the reader to Allen (1992) for a good review of the methods.

Although representative sampling may be of greatest importance in particle size measurement, clearly other properties such as bulk density and surface area will also vary with particle size and particle size distribution; thus sampling is usually important no matter what the characterization being made.

8.2.1.2. Size Parameters For a full treatment of particle size measurement methods we refer the reader to the book by Allen (1992). Here we shall only point out some of the considerations involved along with a few methods for measuring the size of small particles.

Most particles being irregular in shape, it is apparent that there is some difficulty in defining exactly what "size" means, and this in turn depends somewhat on the use to be made of the measurements. Flake materials such as mica may have length and width dimensions many times the thickness dimension, so that even if the size distribution were very narrow, a given average "size" may mean something quite different than the size for spherical beads.

If estimates of the surface area of non-porous particles are to be made from particle size measurements, then clearly one must have knowledge of the particle shape. For this reason it is useful to examine a representative sample of the material with optical microscopy or by electron microscopy, in both cases preferably obtaining photomicrographs to facilitate making further measurements of particular dimensions at later times. Such examination also gives additional important information such as surface texture, degree of agglomeration, impurities, and possibly porosity. As a rule of thumb, optical microscopy can be used for particles in the range from 0.5 to 500 µm, while electron microscopy is applicable below 10 µm.

For irregular particles one can substitute an equivalent particle of defined geometrical shape such as a plate, a sphere, or a cube, appropriate to the material. For convenience it is common to reduce a particle to an equivalent sphere because there is then only a single parameter—the diameter.   Commonly the extreme limits of a particle in a fixed direction is taken as the diameter.   To do this one needs some visual information as noted above.

The *nominal* diameter of any particle is defined as the diameter of a sphere having the same volume as the particle. Thus, if there are N particles per gram, and if the particle density is ρ, then the nominal particle diameter, d_n, is

$$d_n = \sqrt[3]{\frac{6}{\pi \rho N}} \qquad (1\text{-}14)$$

An indirect method for obtaining a measure of the effective particle diameter, d_e, is by using Stokes' law for a falling body achieving a terminal velocity V_t

$$V_t = \frac{gd_e^2(\rho - \sigma)}{18\mu} \tag{1-15}$$

where   g= the gravitational acceleration
d_e= the "effective" diameter as calculated using this relationship without taking secondary factors into account
μ = viscosity of the surrounding fluid
ρ = particle density
σ = fluid density

This is, however, a simplistic approach inasmuch as other factors such as particle shape, the condition of the particle surface, and its size and density play a role. Two particles having the same nominal diameter might have different effective diameters.

Various types of average diameters are sometimes used: arithmetic average, geometric mean, harmonic mean, and median diameter.

The arithmetic average of various particle diameters is the sum of the diameters of the individual particles divided by the total number of particles.

$$d_{av} = \frac{\sum_i (n_i d_i)}{\sum_i n_i} \tag{1-16}$$

The geometric mean, d_g, is the nth root of the product of the n particles measured (one of each diameter)

$$d_g = \sqrt[n]{(d_1 d_2 d_3 \cdots d_n)} \tag{1-17}$$

or, when weighted for n_i particles of diameter d_i

$$d_g = \exp\left(\frac{\sum_i (n_i d_i)}{\sum_i n_i}\right) \tag{1-18}$$

The harmonic mean is the reciprocal of the diameters measured. When there are n_1, n_2, ..n_n particles with diameters d_1, d_2, ...d_n, then

$$\frac{1}{d_h} = \left(\frac{\sum_i (\frac{n_i}{d_i})}{\sum_i n_i} \right) \qquad (1\text{-}19)$$

The geometric mean is less than the arithmetic average and the harmonic mean is less than the geometric mean. The harmonic mean is related to the specific surface area.

The median diameter is the diameter for which 50% of the particles measured are less than the stated size.

The usefulness of these various averages depends on the nature of the problem at hand. For example, if the surface area of a collection of particles is of interest, more weight should be given to fine particles because the surface of a unit-weight of material increases as the particle size decreases. For this purpose a mean diameter can be defined as

$$d_s = \frac{\sum_i (n_i d_i^2)}{\sum_i (n_i d_i)} \qquad (1\text{-}20)$$

This gives a mean based on the observed surface where the summation of the surface areas is divided by the summation of the diameters.

However, if the specific surface area(surface area per unit weight, usually expressed as m^2/g) is important, then a mean diameter can be defined as follows:

$$d_u = \frac{\sum_i (n_i d_i^4)}{\sum_i (n_i d_i^3)} \qquad (1\text{-}21)$$

Comparing d_s and d_u for a hypothetical distribution with many small particles and a few much larger particles, a sample calculation shows that d_s is heavily weighted by the effect of the small particles while d_u takes into account the weight contribution of the larger particles.

In addition to the previously mentioned problems of obtaining a representative sample of the material to be sized and defining a satisfactory measure of particle size, there are problems associated with instrument bias for any method used and of distinguishing loose clumps of particles from the fundamental particles. For example, if size distribution estimates are made by counting various size particles by direct visual observation, even if a representative sample has been selected,   much scanning must be carried out to ensure that a representative viewed area is chosen. Furthermore, optical microscopy has such a short depth of focus at high magnification that it is difficult to estimate particle thickness, given that plate-like particles tend to lie flat.

Finely divided solids usually exhibit skewed size distributions depending on their method of manufacture. Figure 1-7 shows a typical skewed distribution compared to a normal distribution. Since most particle size distributions are skewed, the normal distribution does not apply, but frequently the asymmetrical distribution curves can be made symmetrical if the logarithms of the sizes are substituted for the sizes, Hatch and Choate (1929).

In this book we have emphasized the importance of small particles in the colloidal range, and because many of the commonly used methods of particle size determination such as light microscopy and electron microscopy are sorely limited by slowness, sample preparation, and very high cost, we will briefly discuss two modern methods well suited to measuring the particle size distributions (PSD's) of colloidal range particles using modern instrumentation which greatly facilitates obtaining PSD's rapidly and with far less effort than many older methods. As with so many measurement techniques, there is no universal answer: different problems usually have different best solutions, so that consideration should also be given to other methods—see Allen (1992).

Methods for particle size determination can be grouped into two categories: fractionation and non-fractionation—see reviews by Barth and Sun (1985, 1991). Well-known non-fractionation methods include microscopy, neutron scattering, quasi-elastic light scattering, and tubidimetry. Fractionation methods include sedimentation, capillary hydrodynamic fractionation (CHDF), disc centrifugation, size exclusion chromatography (SEC), and field-flow fractionation (FFF).

In the non-fractionation methods, all particles in the sample are analyzed simultaneously. Thus electron microscopy, for all its power to give additional information, looks at all the particles within the chosen field of view.

In the fractionation methods, particles are initially sorted according to size and/or density after which further characterization such as turbidimetry, infrared analysis, or multi-angle laser light scattering can be carried out.

Using CHDF, colloidal particles are fractionated according to size as they flow in a capillary tube, Dos Ramos and Silebi (1990) and Dos Ramos, Jenkins, and Silebi (1991). A fluid, the eluant, usually a surfactant solution, is pumped continuously into the capillary tube while a high-pressure liquid chromatography injection valve is used to inject the small particles into the flowing fluid. The parabolic laminar flow profile with its maximum fluid velocity at the tube center and zero velocity at the wall effects a separation of particles according to size because the larger particles travel faster. As a result, the particles exit the capillary in order of decreasing size. As the particles exit, they are detected by a UV light detector by both their scattering and absorption. The particle average velocity in the capillary is determined from its residence time in the capillary. The diameter of the particle is found from a calibration curve for the capillary. The calibration can be carried out with a monodisperse latex standards. Using the

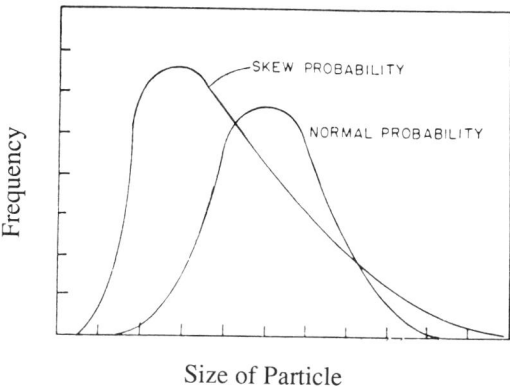

Size of Particle

Fig. 1-7. Normal and skewed particle size distributions.

Mie theory of light scattering, the relative amounts in the sample can be calculated from the detection signal amplitude. The final result is a complete particle size distribution for the sample, Dos Ramos and Silebi (1989). Particle size measurements of a wide variety of polymer colloids, carbon black, and organic pigments are made routinely with commercial CHDF instruments available from Matec Applied Sciences (Hopkinton, MA).

Another modern method utilizing an instrument called an Acoustosizer® (Matec Applied Sciences) measures both the particle size and the zeta potential for samples in the 1 - 40 vol. % solids range without dilution. Both sedimentation and light scattering have severe limitations for even mildly concentrated suspensions. In this method the instrument applies an alternating electric field, at megahertz frequencies, to an aqueous colloidal suspension. This produces a sound wave and by measuring the magnitude and phase angle of that wave with respect to the applied field, both the effective charge (zeta potential) of the particles and their size can be determined. The instrument requires a sample of about 350 ml in volume at concentrations of at least 0.5% by weight, Oja et al. (1985). However, unless the sample is first fractionated with respect to size, the instrument gives an average particle size.

8.2.2. Surface Area Measurement The importance of the surface area of small particles is noted frequently throughout this book. Here we shall simply mention two common surface area measurement methods: 1) the BET method for the total accessible surface area; and 2) chemisorption for specific adsorption on metallic surfaces to determine a particular partial surface area. References such as Anderson and Pratt (1985) cover these techniques in much greater detail.

The Brunauer, Emmett, and Teller (B.E.T.) method for total surface measurement depends on the non-specific physical adsorption of simple non-polar gases such as nitrogen or argon wherein the amount of gas adsorbed can be related to surface area when an area per adsorbed molecule in forming a unimolecular layer over the surface of the particles is known at a particular adsorption temperature. The reader is referred to the sample calculation of surface area for the B.E.T. equation as given in Masel (1996).

Determination of a metal surface area in a supported metallic catalyst requires specific adsorption (chemisorption) solely on the fraction of the surface area that is metallic. Commonly used gases are hydrogen, carbon monoxide, and oxygen under conditions such that chemisorption is so well defined as to allow the number of surface metal ions—and thus the metal surface area—to be estimated. Here too experience is often needed to properly interpret the data.

8.2.3. Porosity Measurement Many particulate materials are porous, some naturally occurring that way, and others by design. The true surface area, i.e. that area accessible to adsorption—in contrast to the "superficial" surface area of the particle considered as a pore-free geometric volume—is usually relevant in small particle technology. For that reason it is necessary to be able to measure porosity. In some cases sectioning particles, aided by staining with appropriate substances, allows microscopic examination to determine roughly the size of pores and their shape.

For quantitative determinations of pore sizes and pore size distributions, three methods are commonly used, each with its range of pore size application and each with its pros and cons with respect to ease of measurement and interpretation, accuracy, and requirements. In general it is not easy to attain unambiguous answers in terms of pore sizes and their distributions because of the complications which arise in each method, but at the same time if one considers these methods as comparative *characterization* methods for porous particles in terms of looking at differences among batches of ostensibly the same material, then the value of the information in terms of quality control and understanding of the material in at least a semi-quantitative way is considerable.

In most porous materials—charcoal being one example—the adsorption accessible area due to porosity is far greater than the superficial surface area. Most industrial catalysts are porous pellets for just that reason and can have specific surface areas of hundreds of square meters per gram.

In particulate catalysts the pore structure is a determinant in controlling reactant accessibility and product selectivity, and in that sense there is great interest in transport phenomena of gases in pores, Satterfield and Sherwood (1963). The phenomena involved are quite complex both in their realistic description (different types of pores in terms of size, shape and tortuosity and their three-dimensional representation) and the mathematical description of multicomponent mixtures diffusing into and out of the pore structure in time varying thermal fields. That subject is beyond the purview of this book. Here we shall mention the basics of pore structure and its determination. Basic references are Gregg and Sing (1967), Everett and Stone (1958), and Anderson and Pratt (1985).

Three common methods for the measurement of pore size distribution are: 1) small-angle x-ray scattering (pores about 1-100 nm); 2) mercury porosimetry (4 nm-150 μm); and 3) physical adsorption isotherm analysis (2-20 nm).

The x-ray method, though non-destructive, is limited in its application because it cannot distinguish between discontinuities due to pores and chemical composition variations, or pores which are the same size as particles in the structure.

In mercury porosimetry the pressure to force a measured quantity of liquid mercury into the porous solid is monitored and the trace is interpreted to yield the pore diameter distribution. However, the mercury contact angle with the solid must be known and a realistic model of the pore shape (round or slit shaped) must be adopted in order to make the calculations. The sample must also be physically strong enough to withstand the pressure without fracturing. Samples with fissures or cracks will give spurious results. Applied pressures are in the range from 1 - 350 atm. (for small pores about 4 nm). Errors can be appreciable using this method, largely due to the variation of contact angle with pressure, Drake and Ritter (1945).

Determining pore structure by physical adsorption can be carried out by determining adsorption isotherms for gases such as nitrogen and then analyzing the results in terms of the filling of the pores in the various isotherm regions as the fraction of the surface covered by adsorbed molecules increases. However, some knowledge of pore shape is needed in advance of the isotherm analysis. Additional information on pore shape can be obtained from the shape of the hysteresis loop on the plot of relative pressure vs. adsorbate amount, de Boer (1958). Considerable experience is needed to interpret results with this method, especially for materials not examined previously. Its advantage is that it is not subject to as many other limitations as the other methods described here.

8.2.4. Bulk Properties With respect to most of the subjects discussed in this book, the bulk properties of small particles are seldom directly involved unless the end use is one such as insulation where the thermal conductivity is paramount.

However, the bulk density is almost always of utilitarian importance because any material used in quantities from bag or drum lots to railway hopper cars must involve some estimate of the volume required for a given weight of material.

The bulk density unlike the true "crystal density", or density of a non-agglomerated, non-porous sample of the material, depends on the particle size, shape, and particle size distribution as well as the pressure in the container. For a particular grade of material poured into some standard container and shaken down, a reasonably reproducible result can generally be obtained.

The true density of a porous solid can usually be measured accurately using helium as a displacement fluid because helium is able to penetrate into very small pores and does not adsorb significantly at room temperature on most materials, Anderson and Pratt (1985).

For large catalyst beds the mechanical strength of porous catalysts is important in preventing crushing and undesirable changes in the bed. In addition to the mechanical strength *per se*, other factors such as pore size distribution, pore shape, and particle size distribution determine the ultimate strength, Hasselman, (1963, 1969).

Although not chosen as a topic for more than brief mention in this book, there is a growing technical importance of small particles in powder insulation. The relative demise of high R-value plastic foam insulation such as isocyanates and isocyanurates in appliances and building construction due to the phase-out of R-11 (CCl$_3$F) as the low thermal conductivity blowing agent gas of choice in those products led to increased research activity and significant progress in making insulation panels with low thermal conductivity particulate matter as the filler. For this reason as well as the importance of thermal conductivity in catalyst beds, the transport property, thermal conductivity, is worthy of note as a bulk property of particulate materials.

The properties abrasion resistance and thermal shock resistance are also of interest in catalyst applications.

8.2.5. Surface Characterization This subject was treated briefly in terms of adsorption phenomena (surface area measurement) in Section 8.2.2 and more extensively in a general way in Chapter 6. Many modern spectroscopic techniques have been developed in the past forty years or so which can be applied to surfaces in regard to the special properties conferred upon them by the unsaturated bonds exposed at the surface. The books by Minachev and Shpiro (1990) and Campbell (1988) give many details of such techniques particularly related to catalytic activity of particular materials.

In addition various forms of specialized microscopy: field ion microscopy (FIM), and field electron microscopy (FEM), both due to Müller (1951); low energy electron diffraction (LEED), Farnsworth (1950), Jona (1970); scanning tunneling microscopy (STM), Binning and Rohrer (1986); transmission electron microscopy (TEM); and scanning electron microscopy (SEM), can be used to obtain images showing the structure of surfaces. These methods are expensive and are usually of interest only where specialized information is required relevant to the problem at hand.

8.2.6. Chemical Characterization In addition to the overall chemical composition of the small particles, which we assume can be treated by conventional analytical techniques both for the principal constituents and trace materials, small particles, due to their very large specific surface area, present special features such as the many unsaturated bonds which are of special importance in catalysis and adsorption. Because there are so many possibilities for various reactions and so many methods for studying the chemical nature of surfaces involving various spectroscopic techniques as well as acid/base titrimetry and differential thermal analysis, only the briefest mention is made here, though several specific cases are noted in Chapter 6.

In most industrially important applications the relevance of chemical characterization is related to the particular function of the small particles. Thus, for a certain small particle catalyst a determination of its surface acidity using titration with a base and an indicator may yield pertinent information as to the activity of that catalyst for a particular reaction. Anderson and Pratt (1985) give an extensive treatment for catalytic applications.

As an ultra-practical consideration, the careful storage of these materials in water impermeable containers along with monitoring water content before use is often quite important.

In general for many applications of small particles, various treatments such as thermal treatment or exposure to reactive materials can be carried out without affecting the physical characteristics such as particle size, density, etc., so that such physical characterizations may not be helpful where the use has a chemical basis, as in catalysis. The key to chemical characterization of small particles is usually to find a test related to the chemical function the particles fulfill.

8.3. Special Methods

For colloidal silica special methods have been developed for determining the particle size and specific surface area, since they are inversely related.

8.3.1. Sears Method
The Sears method is a rapid method for determining the specific surface area of non-porous silica particles in the size range 5 to several hundred nm which involves a titration of the silica surface with sodium hydroxide, in a medium of 20% aqueous sodium chloride between pH 4 and 9, Sears (1956). At pH 9, 1.26 hydroxyl ions are adsorbed per square nanometer of surface (see Ch. 2). The titer, therefore, is a measure of the total surface present and can be related by the following empirical equation to the specific surface of the colloidal particles, or gel powder, as determined by nitrogen adsorption:

$$A = 26.4(V_t - V_b) \qquad (1\text{-}22)$$

where A is square meters per gram determined by the BET method, V_t the milliliters 0.1M NaOH required for 1.50 g SiO_2 and V_b, the titration blank in the absence of silica, usually about 0.3 ml.

The sol or powder sample containing 1.50 g SiO_2 is diluted with CO_2-free water at 25°C to a concentration of 2-3 % SiO_2, acidified with HCl to about pH 3, and diluted to 135 ml volume. Then 30 g pure crystalline NaCl is added and the mixture is stirred rapidly. As soon as the salt is dissolved, the pH is adjusted to 4.00 with a 0.1M NaOH and the volume is noted after the pH has remained at 9.00 ± 0.05 for about 1 minute.

The method is mainly of value for comparing relative surface areas of particle sizes in a given system which can be standardized. Under these conditions results are reproducible within ±5%. Where there are variations in the types of silica, differences from BET values may range up to ±10%.

The particle size can be related to the surface area by the equation

$$S = 6/\rho d \qquad (1\text{-}23)$$

where S is the specific surface area, d is the particle size, and ρ is the density of silica. Inserting the value of 2.2 g/cm^3 for the density of silica and measuring the particles size in nanometers yields

$$d = 2727/S \qquad\qquad (1\text{-}24)$$

The Sears method can be used to determine the sizes of particles smaller than 5 nm. However, in sols of such small particle size there is an appreciable concentration of monomeric silica at equilibrium. Since monomer reacts with base at pH 9, it is therefore necessary to correct the titration for the effect of soluble silica in order to obtain a reliable value for the specific area or the particle size, Iler (1955).

For colloids of silica of extremely small size it is preferable to use the molybdate method to determine the particle size.

8.3.2. Silico-Molybdate Method   Until fairly recently, workers investigating the course of polymerization of silica stopped the reaction at a certain point and converted all the polymer species to trimethyl silyl derivatives. Trimethyl silanol was added to the polymer mixture and all silanol groups were condensed with trimethyl silanol, which is a method resembling end-capping polymers in an organic system. Each polymer and oligomer were converted to a stable unit surrounded by trimethyl silyl groups. These were then separated and characterized by chemical analysis, molecular weight, etc. The method is slow and so Höbbel et al. (1973-1977) used the reaction of molybdic acid with monomeric silicic acid to develop a colorimetric method for determining the size of polymeric silicic acid. A detailed account of the silico-molybdate method is given by Iler (1979).

The method depends on the fact that only $Si(OH)_4$, but not polymers thereof, can react directly with acidified ammonium heptamolybdate to form the yellow silico-molybdic acid, since the latter molecule contains only one silicon atom:

$$7\ Si(OH)_4 = 12\ H_6Mo_7O_{24}{\cdot}4H_2O + 174\ H_2O = 7\ H_8Si(Mo_2O_7)_6{\cdot}28\ H_2O$$

$$7\ SiO_2 + 84\ MoO_3 = 7\ SiO_2(MoO_3)_{12}$$

The method, which can also be used to determine the concentration of monomeric silica and silicate ions in a sample, is quite sensitive due to the fact that 1 g SiO_2 consumes 35.3 g of the ammonium molybdate. Höbbel et al. (1973-1977) started with pure monomer in dilute solution of low pH and, as it aged, withdrew samples on which two types of tests were run.

First, they used the trimethyl silylation method and separated the species present and characterized them by analysis and molecular weight by osmometry. Then they measured the rate of reaction of the aqueous samples with the molybdic acid at each stage of the aging.

By assuming that the depolymerization of a particular species of silicic acid, and hence the reaction of $Si(OH)_4$ with molybdic acid, is a first-order reaction, they were able to derive the following relation between the first-order rate constant, k, and the molecular weight of the silicic acid species, W_h:

$$\log k = 2.505 - 1.161 \log W_h \tag{1-25}$$

Alexander (1953, 1954) developed a procedure to measure the rate of reaction of specific silicic acids with molybdic acid. The ratio of $Mo: H^+$ in the solution is important, especially when the silica has polymerized to particles which depolymerize only slowly over a period of about one hour. The silico-molybdate is first generated in the yellow beta-form, which fades to a less intense yellow alpha-form. This fading is slower at lower pH, but on the other hand, so is also the rate of dissolution of silica polymers and formation of the yellow color. A pH of about 1.2 is a compromise so that the absorbance of a standard sample decreases less than 5% per hour.

Iler (1980), in his study of the polymerization of silicic acid to colloidal silica prepared a stable stock solution of ammonium molybdate containing 100 g/l $(NH_4)_6MoO_{24}•H_2O$ and 47 g/l concentrated ammonium hydroxide. The molybdic acid reagent was made up with 100 ml stock solution, 500 ml distilled water, and 200 ml 0.75 M H_2SO_4, so that concentrations are 0.0707 M $MoO\backslash S(-2,4)$ 0.188 M SO_4^{-2}, and 0.148 M NH_4^+, with a pH of about 2.

To 20 ml of this reagent was added no more than 5 ml of the sample solution containing from about 0.01 to 1.0 mg SiO_2 with dilution of the total volume to 25 ml with distilled water. Thus there was at least a sevenfold excess of molybdate ion over that required to convert all the silica to silico-molybdic acid. Adsorption was measured at a wavelength of about 410 nm.

Molybdic acid reacts rapidly with monomer and progressively more slowly with polymers of increasing molecular weight. This is because the polymers have to dissolve or depolymerize before reaction can occur. At least up to the point where 50% of the smaller polymers or 20% of the larger polymers have reacted, the reaction rate appears to be of the first order with respect to silicic concentration thus:

$$-\ln (p_2/p_1) = k(t_2-t_1) \tag{1-26}$$

where k is the rate constant, p_1 is the percent of the silica *not* reacted at time t_1 (minutes) and p_2 at time t_2.

A typical first-order plot is shown in Fig. 1-8 where on the ordinate is plotted the logarithm of the percent of the silica that has not reacted in the number of minutes shown on the abscissa. The fact that after the molybdic acid has reacted for ten minutes the line is still straight, indicates that the latter stages of the reaction are first order and a reaction rate constant can be calculated from the slope. The line, which has a slope of about 45°, extrapolates to the vertical dotted line at 2 minutes, showing that about 20% of the silica was present as monomer at t = 0. Extrapolation to the ordinate shows that about 45% of the silica was originally present as polymers. The balance, about 35%, of the silica was therefore present as oligomers.

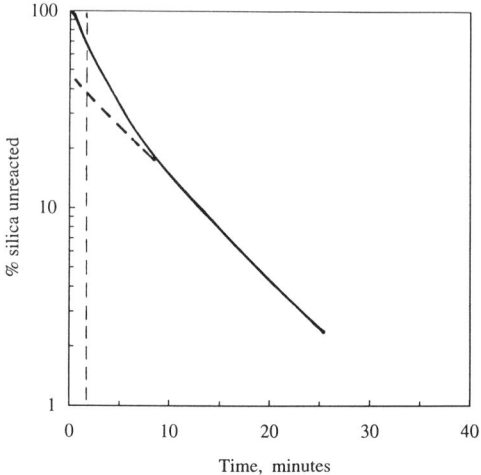

Fig. 1-8. Reaction of molybdic acid with a solution of silicic acid in a certain stage of polymerization.

Thus it is observed that when molybdic acid reacts with polysilicic acid, i.e. colloidal particles in the 1-5 nm size range, the rate of reaction in fact depends on the rate at which monomer is released by dissolution of particles.

However, it is not immediately apparent how a first-order reaction rate is observed under these conditions, namely, that the rate of formation of the yellow silicomolybdic acid is proportional to the total concentration of remaining silica.

An explanation is based on the fact that when a particle in this size range begins to dissolve and diminish in diameter, it then dissolves at an ever increasing rate and disappears, providing that there is very little soluble silica in solution. The molybdic acid reacts with 99% of monomer in 2-3 minutes so that the concentration of monomer remains very low.

The second consideration is the fact that in a sol of particles of this size, the particles soon establish a narrow size distribution(see Ch. 2), because because of the Ostwald-Freundlich equation by which small particles become much more soluble and thus disappear much faster with decreasing particle diameter. As a result, in any sample of small particles, the small particles, being more soluble, dissolve more rapidly and disappear.

Thus, in effect the rate of reaction depends directly on the *number* of particles present, since all are dissolving, but the amount of monomer available for forming silicomolybdate color depends simply on the *number* of particles that remain. This gives a first-order reaction rate.

Equation 1-27 expresses the relation between first-order rate constant k and the molecular weight of polymeric silica, W_h. Iler (1980) calculated the degree of polymerization, that is the number of silicon atoms per polymer unit, n, from the following equation:

$$n = 1.176 \frac{W_h}{60} \tag{1-27}$$

and then used Eqn. 1-28 to obtain the following relation between the first-order rate constant and the particle diameter, d (nm):

$$\log d = 0.250 - 0.287 \log k \tag{1-28}$$

References

Alexander, G. B., J. Am. Chem. Soc., **75**, 2887 (1953).

Alexander, G. B., J. Am. Chem. Soc., **75**, 2094 (1954).

Allen, T., *Particle Size Measurement*, Chapman Hall, 4th ed., London (1992).

Anderson, J. R., and Pratt, K. C., *Introduction to Characterization and Testing of Catalysts*, Academic Press, Sydney (1985).

Barth, H. G., and Sun, S. T., Anal. Chem., **57**, 151 (1985).

Barth, H. G., and Sun, S. T., Anal. Chem., **63**, 1R (1991).

Binning, G., and Rohrer, H., IBM J. Res. Dev., **30**, 355 (1986).

Cabot Technical Report S-134, *Special Blacks for Plastics* (1990).

Campbell, I. M., *Catalysis at Surfaces*, Chapman and Hall, London (1988).

Carman, P. C. Trans. Faraday Soc., **36**, 964 (1940).

de Boer, J.H., *The Structure and Properties of Porous Materials*, eds. D.H. Everett and F.S. Stone, Butterworths, London (1958).

Dos Ramos, J. G., and Silebi, C. A., J. Colloid Intefac. Sci., **135**, 165 (1990).

Dos Ramos, J. G., Jenkins, R. D., and Silebi, C. A., Particle Size Distribution Assessment and Characterization II, ed. T. Provder. ACS Symposium Series No. 472, American Chemical Society, Washington, D.C., 279 (1991).

Drake, L. C., and Ritter, H. L., Ind. Eng. Chem., Anal. Ed., **17**, 787 (1945).

Everett, D. H., and Stone, F. S., eds., *The Structure and Properties of Porous Materials*, Butterworths, London (1958).

Farnsworth, H. T., Rev. Sci. Instrum., **21**, 102 (1950).

Feigl, F. *Chemistry of Specific, Selective, Sensitive Reactions*, Academic Press, NY, 63 (1949).

Gregg, S. J., and Sing, K. S. W., *Adsorption, Surface Area and Porosity*, Academic Press, London (1967).

Hasselman, D. P. H., J. Amer. Ceram. Soc., **46**, 564 (1963); **52**, 457 (1969).

Hatch, T., and Choate, S., J. Franklin Inst., **207**, 369 (1929).

Höbbel, D., Wieker, W., and Stade, H., Allg. Chem., **366**, 139 (1974); **400**, 148 (1973); **405**, 163 (1974); **428**, 43 (1977).

Höbbel, D., Wieker, W., and Stade, H., J. Chromatogr., **119**, 173 (1976).

Iler, R. K., *The Colloid Chemistry of Silica and Silicates*, Cornell University Press, NY, 99 (1955).

Iler, R. K., *The Chemistry of Silica*, John Wiley & Sons, NY, 7 (1979).

ibid. 354. 205, 95-100, 138-140, and 195-202.

Iler, R. K., J. Colloid Interfac. Sci., **75**, 138 (1980).

Jona, F., IBM J. Res., **14**, 445 (1970).

Krätschmer, W., Lamb, L. D., Fostiropoulos, K., Huffman, D. R., Nature, **347**, 354 (1990).

Masel, R. I., Principles of Adsorption and Reaction on Solid Surfaces, John Wiley & Sons, NY (1996).

Minachev, Kh. M., and Shpiro, E. S., *Catalyst Surface: Physical Methods of Studying*, CRC Press, Boca Raton (1990).

Müller, E. W., Z. Phys., **131**, 136 (1951).

Oja, T., Petersen, G., and Cannon, D., U.S. Patent 4,497,207 (1985).

Rosén, A., Kosmos, 119 (1993).

Satterfield, C. W., and Sherwood, T. K., *The Role of Diffusion in Catalysis*, Addison-Wesley Publishing Co., Reading (1963).

Sears, G. W., Anal. Chem. 28, 1981 (1956).

Smalley, R. E., Nature, **318**, 162 (1985).

Trass, O., and Koka, V. R., Particulate Sci. Tech., **4**, 379 (1986).

Trass, O., and Gandolfi, E. A. J., Powder Technology, **60**, 273 (1990).

Trass, O., Papachristodoulou, G. L., and Gandolfi, E. A. J., Coal Preparation, **16**, 179 (1995).

Trass, O., Koka, V. R., and Papachristodoulou, G. L., Particles and Particle Syst. Charact., **12**, 158 (1995).

CHAPTER 2. SMALL PARTICLES OF SILICA

CHAPTER 2. SMALL PARTICLES OF SILICA

Except for water, silica is the most abundant substance on the face of the earth. It is the major ingredient of our rocks, sand, and soil.

Silica has played an important role in the evolution of life. Today it is important not only in microscopic forms of marine life, but also in many plants, and perhaps, surprisingly, in many animals. Each of us contains about half a gram of silica, without which our bones could not have been formed, and probably also, not our brains.

Silica has played key roles in the development of civilization. Flint was used for the earliest improvements in tools and weapons. Sand and clay were the basis for pottery, food storage, cooling vessels, and brewing alcohol.

A particular volcanic ash, which is pure amorphous silica powder, was used by the Romans for their remarkably strong, permanent cement. Today this lost art is being revived in the form of fine amorphous silica powder from electro-metallurgical furnaces, which, added to our present cement, greatly improves its strength.

Silica chemistry has been the basis for the catalysts essential in refining crude oil to gasoline and other fuels. Colloidal silica is used in making the molds for casting the super alloys in jet engines and for polishing silicon wafers in the electronics industry. Pure silica is needed for the glass in fiber optics. Many of us probably have a quartz crystal on our wrist.

The word silica has a very broad connotation. It includes silicon dioxide in all its forms—crystalline, amorphous, soluble, or chemically combined. This includes silicon in any chemically combined form in which the silicon atom is surrounded by four or six oxygen atoms. Thus, we can speak of the silica content of clay, or of sodium silicate, in terms of SiO_2.

Silicon is the most abundant metal in the crust of the earth, but, originally, on the earth's surface there was very little free silica. As the earth cooled and water condensed, it was the water that dissolved silica from the rocks and permitted the silica to crystallize as quartz and sand. In a few places it later came out of solution in the amorphous form of opal, which is made of uniformly-sized spherical particles of silica in the size range from 200-400 nm packed into regular arrangements.

A. Solubility and Rate of Dissolution of Silica

1. Solubility of Silica

The oxidation (Z) and coordination (N) numbers of silicon in naturally occurring compounds are most often the same and equal to 4. In general, silicon is less electropositive than aluminum and the transition metals to be discussed in Chapter 3. Thus, the partial positive charge on silicon in $Si(OEt)_4$ is +0.32 whereas it is +0.63 for Ti and +0.65 for Zr in $Ti(OEt)_4$ and $Zr(OEt)_4$, respectively, Livage, Henry, and Sanchez (1988). The smaller partial charge on silicon makes silicon less susceptible to nucleophilic attack, and spontaneous

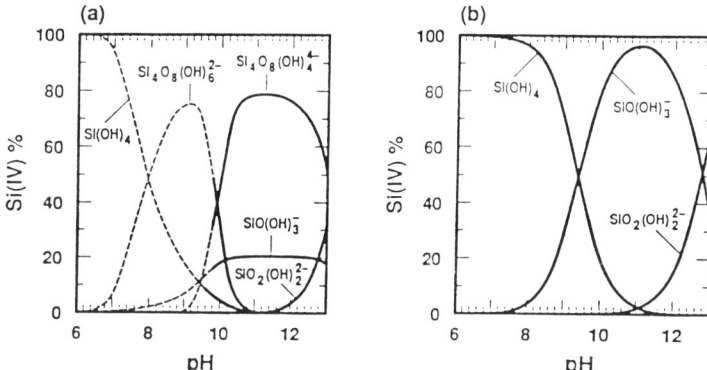

Fig. 2-1. Distribution of aqueous silicate species at 25°C in (a) 0.01M Si(IV) and (b) 10^{-5}M Si(IV). Ionic strength, I=3M. Baes & Mesmer (1976). Reprinted by permission of John Wiley & Sons, Inc.

expansion of the coordination with nucleophilic reagents does not occur since N = Z. The kinetics of hydrolysis and condensation of silicon compounds is therefore considerably slower than for the corresponding compounds of transition metals and group III metals. At pH below 7 the predominant hydrolysis product of silicon is Si(OH)$_4$, which is a weak acid. At higher pH it will ionize first to Si(OH)$_2^-$ and then to SiO$_2$(OH)$_2^-$ at pH above 12—see Fig. 2.1.

The solubility of silica is a function of temperature, pressure, structure, particle size, and pH. Figure 2-2 shows that the solubility of amorphous silica at pH below 8 is in the range 100-150 ppm at 20°C and increases with temperature. Quartz, which is a crystalline form of silica, is about ten times less soluble than amorphous silica. The dissolution of silica is a hydration reaction by which monomeric silicic acid or soluble silica is formed.

This is quite different from sugar dissolving in water, where the sugar molecules in the crystal disperse into the water unchanged. Fournier and Rowe (1977) measured the solubility of amorphous silica under hydrothermal conditions in an autoclave. Their data are expressed by the equation

$$\log C = -\frac{731}{T} + 4.52 \tag{2.1}$$

where C = ppm SiO$_2$ and T is the absolute temperature. According to this equation, the solubilities at 25°C and 100°C are 117 ppm and 364 ppm, respectively.

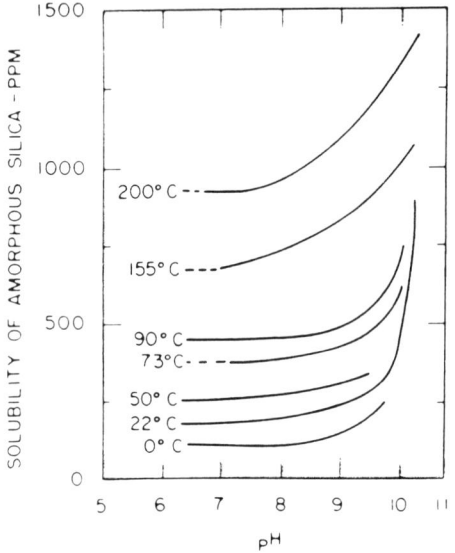

Fig. 2-2. Solubility of amorphous silica versus pH at different temperatures. Goto (1955). Courtesy Chemical Society of Japan.

2. Solubility and Particle Size

The surface of solids tends to decrease in area so as to decrease surface area to a minimum. This is the driving force in sintering at high temperatures and in the growth of small particles into larger ones.

The relation between the solubility, S_r, of a particle and its radius, r, which is a consequence of this general principle, is given by the Ostwald-Freundlich equation:

$$S_r = S_0 \exp[\tfrac{2\gamma_{SL}V_m}{RT_r}] \qquad (2.2)$$

where S_0 = solubility of a flat surface or particle or infinite radius

V_m = molar volume (27.2 cm^3 for amorphous silica)
γ_{SL} = interfacial surface free energy (ergs/cm^2)

R = gas constant (8.31 x 10^7 erg/mole deg)

The general principle, as expressed by the Ostwald-Freundlich equation, controls some seldom-mentioned basic aspects of silica chemistry.

At the top of Fig. 2.3 is shown the cross-section of the surface of amorphous silica. On the right the surface has projections or bumps, Iler (1979). At the left there are depressions or cracks or crevices. On the right, the projections have a radius of curvature corresponding to the radius of the spherical particles just above. On the left, the depressions have a surface with a negative radius of curvature. Just above, the two particles in contact have a connecting surface with a negative radius of curvature.

When applied to amorphous silica, the Ostwald-Freundlich equation, which simply expresses the tendency of a surface to minimize its area so as to minimize its surface energy, gives some useful and interesting information about properties and behavior of such chemical products as silica sols and gels.

Consider again the right side of Fig. 2.3. Along the horizontal axis, starting from the center, is radius of curvature up to 10 nm. At the top are the rounded bumps with corresponding radii of curvature. Above are spherical particles with the same radii as the bumps below them. Near the bottom is a dotted line showing the solubility of a flat surface—which corresponds to an infinite radius of curvature—77 ppm. Just above, the solubility curve at a radius of 10 nm is about 100 ppm, which is what one usually finds for silica powders and sols. However, as the radius becomes smaller than 5 nm, as shown by the particles at the top, the solubility rises rapidly; See also Fig.2-9. The result of the very strong dependence of the solubility on the curvature at radii smaller than 5 nm is that, in water, projections with sharp points or small radii of curvature dissolve and the silica is deposited on the flat surface. The surface tends to become smooth. Small

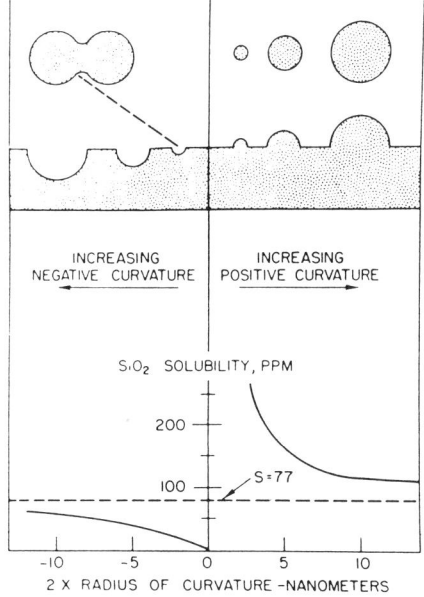

Fig. 2-3. Variation in solubility of silica with radius of curvature of surface. The positive radii of curvature are shown in cross-section as particles and projections from a silica surface. Negative radii are shown as depressions or holes in the silica surface, and in the crevice between two particles. Iler (1979). Reprinted by permission of John Wiley & Sons, Inc.

particles in suspension dissolve and the silica is deposited on larger particles. In particular, if only small particles are in suspension in contact with a flat surface, the particles dissolve and are deposited as a film on the flat surface. This is actually the way relatively large particles of pigments, e.g. TiO2, are coated with a thin film of dense amorphous silica.

On the left-hand side of the ordinate a depression has a negative radius of curvature. The equation shows that the smaller the negative radius, the less soluble the silica. For a very fine crevice the solubility at the bottom is almost zero. This means that the silica will dissolve from the flat surface to fill the depressions. Again, the surface becomes smoother and the surface area moves toward a minimum. Also, when two particles touch, the area of contact has a negative radius of curvature. Silica will dissolve from the areas of positive radius of curvature to form a neck between the particles making them grow larger.

The different processes we have described of moving silica from one place to places of larger radii of curvature or negative radii of curvature have been given the common name *Ostwald ripening*. Ostwald ripening can be harnessed to play a very useful role in many important technical processes involving silica, e.g., the manufacture of silica sols, strengthening of gels, coating particles with a layer of dense amorphous silica, etc.

3. Sources of Monomeric Silica

Studies on the polymerization of silica require access to supersaturated solutions of monomeric silicic acid. Such solutions can be made in several ways. First, the simplest one is to suspend silica gel in water and stir it for several hours at the boiling point and filter it hot. It then contains about 400 ppm SiO_2, or four-fold saturation at 20°C, which, if adjusted to pH 2, will remain monomeric for several hours.

Another way is to start with a cold dilute solution of a strong acid such as hydrochloric or sulfuric acid stirred rapidly while adding pulverized sodium metasilicate hydrate until the pH rises to 1.8. The amounts should be such as to give a concentration of about 1000 ppm SiO_2.

In a third method, the ethyl ester of silicic acid, tetraethyl-orthosilicate, can be stirred into water, adjusted to pH 2, until the silica concentration reaches about 1% SiO_2. This solution begins to polymerize at once, but less rapidly if diluted.

4. Rate of Dissolution of Silica

We have already mentioned that amorphous silica dissolves in water to form a saturated solution.   The rate at which silica dissolves in water depends on

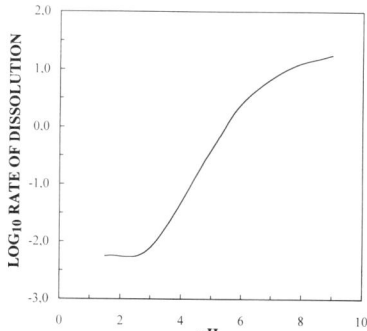

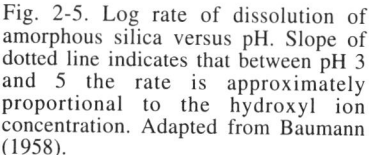

Fig. 2-4. Proposed mechanism of dissolution of silica in water in the presence of hydroxyl ions. The dotted line represents the interface between silica on the left and water on the right. Iler (1979). Reprinted by permission of John Wiley & Sons, Inc.

many factors such as temperature, pH, particle size, etc. The dissolution of the silica surface, shown in Fig. 2-4 in cross-section, Iler (1979), requires the hydroxyl ion, that is alkali, as a catalyst, forming monomeric silica on the right. Notice that this reaction is completely reversible and that silica is deposited on the surface by the same reaction. Actually, the dissolution in water is a hydrolytic de-polymerization and the solubility of silica is the concentration of monomeric silica reached at a steady state in the de-polymerization-polymerization equilibrium. Figure 2-4 also shows that, as a catalyst, the hydroxyl ion is chemisorbed on the surface and increases the coordination number of a silicon atom to more than four, hence weakening the silicon-oxygen bonds in the surface.

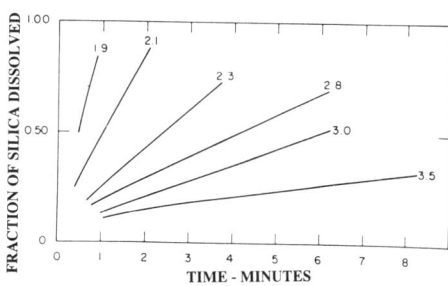

Fig. 2-5. Log rate of dissolution of amorphous silica versus pH. Slope of dotted line indicates that between pH 3 and 5 the rate is approximately proportional to the hydroxyl ion concentration. Adapted from Baumann (1958).

Fig. 2-6. Fraction of silica dissolved versus time for particles of indicated initial diameters in nanometers. Iler (1979). Reprinted by permission of John Wiley & Sons, Inc.

After the adsorption of OH⁻ ion, a silicon atom goes into solution as a silicate ion. If the pH is much below 11, the silicate ion hydrolyzes to monomeric silicic acid, $Si(OH)_4$, and OH⁻ ions.

In Fig. 2-5 the log_{10} rate of dissolution is plotted against pH, Iler (1979). Below pH 3 the rate is very slow because the concentration of catalyst, that is, OH⁻ ions, is very low. Between pH 3 and 6 the rate is proportional to the hydroxyl ion concentration. The declining rate of dissolution at pH above 6 may be due to a slower rate of desorption of silicic acid from the surface and the fact that OH⁻ ions may find it increasingly difficult to chemisorb on the surface as its negative charge increases with pH.

Iler (1979) showed that the rate of dissolution increases exponentially with decreasing particle size. Fig. 2-6 shows the rapid increase of the experimentally determined rate of dissolution with decreasing particle size and that the rate of dissolution of particles of a given size remains reasonably constant while 10-70% of the silica is dissolving.

B. Polymerization of Silica

1. Polymerization of Monomeric Silica

Freundlich made monomeric silicic acid by acidifying solutions of soluble silicates (see Section 2C above), and noticed that the silicic acid thus formed readily diffused through parchment membranes. Determining the molecular weight by freezing point depression, he concluded that the silicic acid was present in monomeric form. With time the molecular species grew larger and passed more slowly through the membrane. These observations can be explained either by the monomers or other small primary particles forming aggregates or individual particles increasing in size and decreasing in number, Iler (1979).

Since solutions of silicic acid slowly become more viscous and finally gel—superficially they behave like organic gels—many researchers believed for a long time that $Si(OH)_4$ polymerized into siloxane chains, which then were branched and cross-linked as in many organic polymers. Iler, however, makes it clear that there is no relation or analogy between silicic acid polymerizing in an aqueous solution and condensation-type organic polymers. According to Iler (1979), the polymerization of silica takes place in these steps:

1. polymerization of monomers to particles
2. growth of particles
3. particles linking together into chains, and then networks, extending throughout the liquid medium, thickening it to a gel.

In the last ten to fifteen years the polymerization of silica in aqueous solutions has been studied by ^{29}Si NMR. In general, these studies support Iler's view that the condensation occurs in such a way that the number of Si-O-Si bonds will be maximized and the number of terminal hydroxyl groups will be minimized by internal condensation. Thus rings will form rapidly, to which monomers will be added forming three-dimensional small particles. These particles will be compacted by internal condensation so that OH⁻ groups are present only on the surface of the particles. Fig. 2-7a is a picture of models of monomers and polymers. The circles represent oxygen atoms. The black dots are hydrogens. Silicon atoms are so small they are hidden in the center of four oxygen molecules forming a tetrahedron. The figure shows monomer, dimer, trimer, cyclic trimer and, at the lower right, cyclic tetramer. If two cyclic tetramers are linked face to face, one obtains the cubic octamer A—see Fig. 2-7b—which also shows theoretical colloidal particles, B and C.

In a polymerizing system of silica these three-dimensional particles, e.g. B and C in Fig. 2-7b, serve as nuclei for further growth, which can occur through Ostwald ripening. Particles grow in size and diminish in number in such a way that highly soluble very small particles dissolve to monomer, which re-deposits on larger, less soluble nuclei. Growth will cease when the difference in solubility between the largest and the smallest particles is reduced to a few ppm. At higher

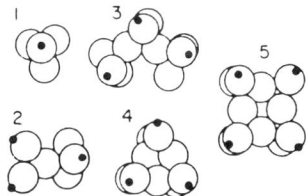

Fig. 2-7a. Molecular models of silicic acids. Spheres represent oxygen atoms; black dots, hydrogen atoms. Silicon atoms within oxygen tetrahedra are not visible. Not all hydrogen and oxygen atoms are visible: (1) Si(OH)$_4$; (2) (HO)$_3$OSiOSi(OH)$_3$; (3) (OH)$_3$SiOSi(OH)$_2$OSi(OH)$_3$; (4) [(OH)$_2$SiO]$_3$; (5) [(OH)$_2$SiO]$_4$. The existence of cyclic trimer is questionable. Iler (1977). Courtesy of Plenum Publishing Corporation.

Fig. 2-7b. Models of (A) cubic octasilicic acids, and (B) and (C), the corresponding theoretical colloidal particles formed by condensing monomers to form closed rings until the original species is surrounded by one layer of deposited silica bearing silanol groups. When formed above pH 7, the inner silica contains few silanol groups. Different kinds of incompletely condensed oligomers could form the cores of colloidal particles. There is no evidence that A and B are specifically involved. Spheres are oxygen atoms, black dots, hydrogen atoms. Silicon atoms are not visible. Iler (1979). Courtesy of Plenum Publishing Corporation.

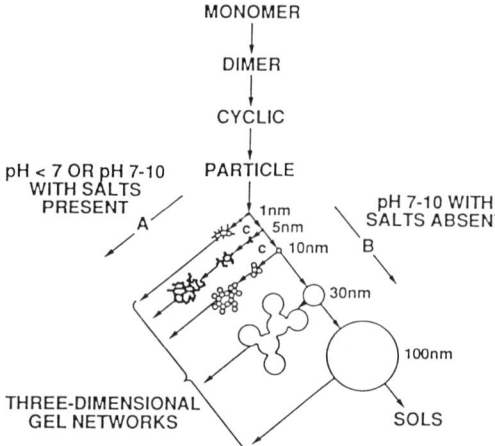

Fig. 2-8. Polymerization behavior of aqueous silica. In basic solution (B) particles grow in size with decrease in number; in acid solution or in the presence of flocculating salts (A), particles aggregate into three-dimensional networks and form gels. Iler (1979). Reprinted by permission of John Wiley & Sons, Inc.

temperatures, the increased solubility will cause the particles to grow still larger, particularly at pH above 7. Fig. 2-8 gives an overview of the polymerization of silica in aqueous solution.

There are really two ways in which the molecular weight of the silica can increase. On the right, in salt-free alkaline solution, particles can be grown to almost any size or molecular weight, e.g. by letting the polymerization take place at high temperatures and pH above 8. On the left, at any stage, the sol of discrete particles can be linked together into aggregates. At any stage of aggregation the aggregation can be stopped by raising the pH to 8 or 9, giving a sol of porous particles, which can be called microgel sol.

Iler (1979) considered the polymerization of silica in the three approximate pH ranges: pH < 1.5-2(14); pH 2-7 (15); and pH > 7 (16). The pH range 1.5-2 can be considered a natural boundary since the point of zero charge, i.e., where the surface charge is zero, and the isoelectric point, i.e., where the electrical mobility of the silica particles is zero, are both around pH 2. Another natural boundary occurs at pH 7 since both the solubility and rate of dissolution are maximized at or above pH 7. Also, the silica particles are noticeably ionized, i.e., negatively charged, at pH above 7, so that particle growth can occur without aggregation or gelling—see Fig. 2-8.

1.1. Polymerization At pH 1.5-7

The most complete and recent study of the polymerization of silica in this pH range was done by Iler (1979). The purpose of his investigation was to study the formation of nuclei and their growth to colloidal particles under such conditions that the polymerization is so slow that several different methods could be used to characterize the system. For this reason Iler started with an aqueous solution of primarily monomeric silicic acid containing 6000 ppm SiO_2 at pH

1.75, prepared by acidifying a solution of sodium metasilicate, to which NaOH had been added to ensure that only monosilicate species were present. Sample solutions were aged for different lengths of time and characterized by the silicomolybdate method so as to determine the particle size—see Ch. 1—by separating polymeric particles from monomer and oligomers by extraction with tetrahydrofuran and using ultrafiltration to confirm the particle size determined by the silicomolybdate method.

1.1.1. Nucleation The polymerization of monomer is characterized by an induction period, which is really the time required for some three-dimensional nuclei to form on which silica then condenses rapidly. Figure 2-9 shows that a change in the polymerization behavior occurs after about two hours.

The concentration of monomer starts to decrease more rapidly and the amount of silica extractable with tetrahydrofuran becomes considerable, about 10%; see Ch. 6, § C. In addition, rate constants of 0.2 to 0.1 min^{-1} for the reaction with molybdic acid reveal the presence of polymers as well as oligomers since an appreciable fraction of the silica needs more than 10 minutes to react with molybdic acid. Line K in Fig. 2-9 shows rate constants for polymers in the aqueous solution as well as for polymers isolated as THF-complex.

As shown in the section on particle growth below, it is reasonable to conclude that nucleation occurs after the solution has been aged for 2-3 hours, after which time no new particles form, but those present grow by condensation of monomer on the particle surface while oligomers dissolve when the monomer concentration decreases. When the oligomers disappear, the average particle size increases while the number of particles decreases by Ostwald ripening.

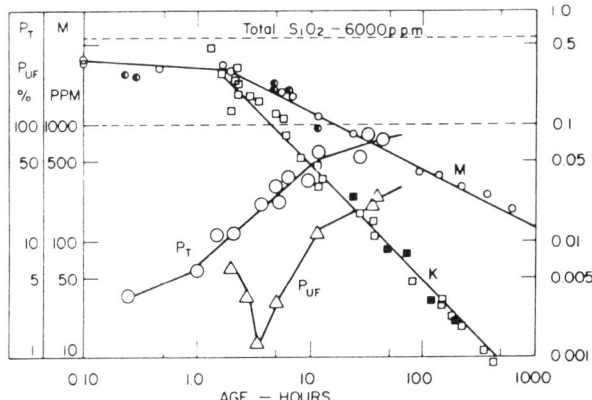

Fig. 2-9. Data during aging of silicic acid solution at pH 1.75, 25°C, 0.6% SiO$_2$. Ordinates: k, minutes^{-1}, first-order rate constant of reaction of polymeric particles with molybdic acid. (■) data from Hoebbel et al (1974); M, concentration of monomer, Si(OH)$_4$, as parts per million of SiO$_2$.(●) in ultrafiltrate; P$_T$, percent of total silica extracted with tetrahydrofuran; P$_{UF}$, percentage of total silica retained by an ultrafilter with pores 2.4 nm in diameter. Abscissa: hours after preparation of solution as monomer-dimer. Iler (1980). Courtesy of Academic Press.

Table 2-1. Distribution of SiO_2 for Monomers, Oligomers, Nuclei or Particles at Various Aging Times, Iler (1980). Courtesy of Academic Press.		
	Distribution of $SiO_2(\%)$ Time	
	2h	4h
Monomer	50	11
Oligomer	40	14
Nuclei or particles	10	75

At the nucleation stage, Iler estimated that 50% of the silica still is present as monomer, 40% consists of oligomers whereas 10% may be considered to be nuclei which can be extracted as THF-complex; see Table 2-1. The nuclei, or polymer, however, has a surprisingly large rate constant, 0.3 min^{-1}. According to Hoebbel et al (1974,1973,1977,1976), this corresponds to a molecular weight for the polymer, $SiO_2.xH_2O$, of 500-1000. This would be a particle with 10-20 silicon atoms or $(SiO_2.xH_2O)_{10-20}$, corresponding to an anhydrous particle of SiO_2 about 0.9 nm in diameter. This is the smallest three-dimensional particle that can be formed with maximum condensation of silanol groups to form siloxane bonds, $\equiv Si - O - Si \equiv$. Perhaps it is similar to the cubic octamer in Fig. 2-10 in which the silicon atoms occupy the eight corners of a cube, but has a less regular structure and contains additional silicon atoms.

1.1.2. Particle Growth Table 2-1, Iler (1980), shows that as the solution ages, the nuclei will grow to larger particles at the expense of monomer and oligomers.

Assuming that no new nuclei or particles are formed after two hours, these results imply that the average particle diameter has increased by a factor of $(75/10)^{1/3}$. If the nuclei were 0.9 nm in diameter, the diameter of the particles would thus be 1.8 nm after 40 h, assuming that the particles are anhydrous. In reality they contain water roughly corresponding to $Si_{65}O_{108}(OH)_{52}$ —see Eqns. 1-8 to 1-10. From the densities of water and silica (2.2 g/cm^3), one can estimate that the average diameter of the hydrated particle is about 2 nm. This is consistent with the observation that at this stage, i.e., at 2 hours, about 80% of the silica species are still small enough to pass through a membrane with an average pore size of 2.4 nm.

Extrapolation of line K in Fig. 2-9 to an age of 500 h indicates that the particles then would have grown to a diameter of 4 nm. At pH 1.75, the sol, containing 600 ppm SiO_2, will remain fluid for several months, but at pH below 1.5 or above 2.0, the particles will aggregate and the viscosity of the solution will increase until gelling occurs.

1.1.3. Mechanism Since the gel time steadily decreases between pH 2 and 6, it is generally assumed that the rate of condensation above the isoelectric point is

proportional to [OH⁻] according to the following reaction scheme, Brinker and
Scherer (1990);

$$\equiv Si\!-\!OH + OH^{-} \xrightarrow{\text{fast}} \equiv Si\!-\!O^{-} \tag{2-3}$$

$$\equiv Si\!-\!O^{-} + HO\!-\!Si \equiv \xrightarrow{\text{slow}} \equiv Si\!-\!O\!-\!Si \equiv\ + H_2O \tag{2-4}$$

Thus, the first step in the polymerization of silica is the ionization of a
silanol group, the extent of which, of course depends on the acid strength, or
pK_a. For the monomer with four hydroxyl groups attached to silicon, the pK_a is
9.8, which makes it a very weak acid. Silicate ions, therefore, appear in
substantial amounts only above pH about 10. However, on the surface of silica
particles—see Fig. 2-7—each silica atom bears only one hydroxyl ion and the
pK_a of this is about 7. For oligomeric silica species the pK_a will fall between 7
and 9.8.

Figures 2-7 and 2-8 show that the first molecular species of the
polymerization are dimers and so-called oligomers, consisting of linear or cyclic
trimers, tetramers, pentamers etc. For a given distribution of silicate species, the
most acidic silanol groups, which are also the easiest to de-protonate according to
Eqn. 2-3, are present in the most highly condensed molecular species.
Condensation according to Eqn. 2-4, therefore, takes place preferably between
more highly condensed ionic species and less condensed neutral species. This
implies that the rate of dimerization is slow, but when dimers once have been
formed, they react preferably in ionized form with neutral monomers to form
trimers, which in turn ionize and react with monomers to form tetramers. At this
stage of the polymerization ring closure is fast since the chain ends are close to
one another and the monomer population has been depleted. The cubic octamer in
Fig. 2-10 is the first predominant polymer where each of the eight silicon atoms
bear only one hydroxyl group. Therefore, these hydroxyl groups are more acidic
than those of lower species, several of which can be attracted to one silicon atom.

Further growth occurs by continued addition of species of lower molecular
weight to more highly condensed species and aggregations of condensed species,
so that chains and networks are formed. Near the isoelectric point, where there is
no electrostatic repulsion between particles, growth and aggregation occur
simultaneously, and it may be impossible to distinguish between the two
processes. Fig. 2-8 shows that at pH above 7, growth primarily takes place by
adding smaller species to more highly condensed species. Electrostatic repulsion
between the particles prevents aggregation in this pH range providing the salt
level is below the critical coagulation concentration.

1.2. Polymerization at pH Above 7

We have just described how monosilicic acid at low pH first polymerizes to
about 1-nm nuclei, which then slowly grow to about 4-nm particles by Ostwald
ripening. These systems are unstable and will gel in time—see Fig. 2-8. At pH
above 7 the polymerization takes place in the same way, i.e., by the same
nucleophilic mechanism expressed by Eqns. 2-3 and 2-4,   but the process is much

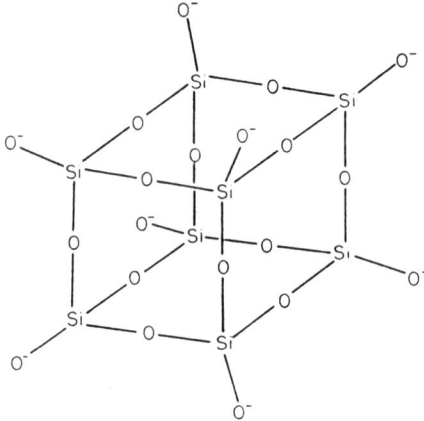

Fig. 2-10. The double-four ring or cubic octasilicate ion. Iler (1979). Reprinted by permission of John Wiley & Sons, Inc.

faster. Since all condensed species are more highly ionized at pH above 7 and therefore repel one another more strongly than at lower pH, particle growth occurs primarily by addition of monomers to larger species, namely particles, than by particle aggregation. Particles with diameters of 1-2 nm will form very rapidly and then grow by Ostwald ripening to a size that mainly depends on the temperature. Ostwald ripening implies that the particle growth takes place by dissolution of smaller particles and deposition of soluble silica on larger particles. The rate of growth will therefore depend on the particle size distribution; the wider the distribution, the faster the particle growth.

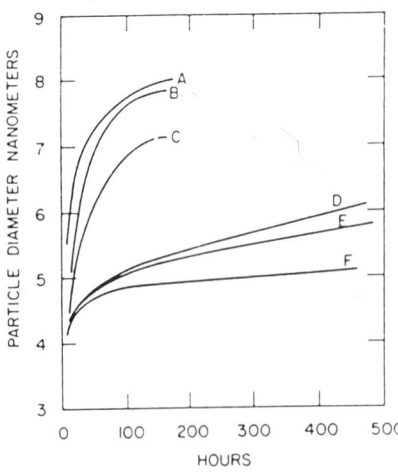

Curve	Temperature °C	$SiO_2:Na_2O$	pH
A	90	97	8.8-9.7
B	90	186	7.9-9.7
C	90	470	7.1-8.1
D	50	97	8.8-9.0
E	50	186	7.9-8.4
F	50	470	7.1-7.5

Fig. 2-11. Rate of growth of silica particles at 50°C and 90°C and different pH values, Iler (1979). Reprinted by permission of John Wiley & Sons, Inc.

The growth curves in Fig. 2-11 for particles made at 50° C and 90° C and at three different pH's demonstrate that the final size of the particles is predominantly determined by the temperature, Iler (1979). The rate of growth depends, of course, also on pH, and, at a given temperature it is possible to modify the particle size to a certain degree by varying the pH.

Commercial silica sols of different particle size are made by polymerization and particle growth at pH above 7, as will be described in Section D below.

1.3. Polymerization at pH Below 1.5-2

For silica particles the pH range 1.5-2 represents a metastable region where gel times are rather long (on the order of several hours). Below this range the rate of polymerization is proportional to $[H^+]$. Iler (1979) and others proposed that the acid- catalyzed mechanism of polymerization utilizes a siliconium ion, Si^+, as an intermediary species:

$$\equiv Si\text{-}OH + H_3O^+ \longrightarrow \equiv Si^+ + H_2O \qquad (2\text{-}5)$$

$$\equiv Si^+ + HO\text{-}Si \equiv \longrightarrow \equiv Si\text{-}O\text{-}Si \equiv \ + H^+ \qquad (2\text{-}6)$$

Based on recent results, obtained from polymerization of alkoxides, Brinker and Scherer (1990), however, have proposed alternatives to this mechanism.

In the absence of fluoride ions the solubility of silica at pH below 2 is rather low (except at pH much below 0) and the silicate species in solution are not strongly ionized. For these reasons it is therefore likely that the formation and aggregation of primary particles occur simultaneously and that Ostwald ripening does not contribute much to the particle growth once the particles have reached a diameter of about 2 nm. Gel networks, made of extremely small particles, will form—see Fig. 2-8. Trace amounts of F^- or addition of small amounts of HF will reduce the gel times and form gels similar to those formed at pH above 2. Fluoride and hydroxyl ions have about the same size and influence the polymerization behavior in about the same way.

2.   Hydrolysis and Condensation of Silicon Alkoxides

Sols and gels of metal oxides with very special properties can be prepared by hydrolysis and condensation of metal alkoxides according to the sol-gel method, which is a very important technique for making so-called high-tech ceramics. Monodisperse spheres of silica can be prepared by this method.

Silica sols and gels can be synthesized by hydrolysis of monomeric tetrafunctional alkoxide precursors using a base, e.g. NH_3, in the case of sols, or an acid, e.g. HCl, in the case of gels, as catalyst. Using the functional groups, the sol-gel method is generally described by the following three equations, Brinker and Scherer (1990):

$$\equiv\text{Si-OR} + H_2O \quad \underset{\text{esterification}}{\overset{\text{hydrolysis}}{\rightleftarrows}} \quad \equiv\text{Si-OH} + \text{ROH} \tag{2-7}$$

$$\equiv\text{Si-OR} + \text{HO-Si}\equiv \quad \underset{\text{alcoholysis}}{\overset{\text{alcohol condensation}}{\rightleftarrows}} \quad \equiv\text{Si-O-Si}\equiv \; + \; \text{ROH} \tag{2-8}$$

$$\text{Si-OH} + \text{HO-Si} \quad \underset{\text{hydrolysis}}{\overset{\text{water condensation}}{\rightleftarrows}} \quad \equiv\text{Si-O-Si}\equiv + H_2O \tag{2-9}$$

where R is an alkyl group, C_nH_{2n+1}. The hydrolysis reaction, Eqn. 2-7, substitutes hydroxyl groups, OH, for alkoxide groups, OR. Subsequent condensation reactions, comprising silanol groups, form siloxane bonds, Si-O-Si, and the byproducts, alcohol, ROH, or water, according to Eqns. 2-8 or 2-9, respectively. Under most conditions, condensation (Eqns. 2-8 and 2-9) will begin before hydrolysis (Eqn. 2-7) has been completed. Since water and alkoxysilanes are immiscible—see Fig. 2-12 for the system tetraethylorthosilicate and water—a co-solvent such as an alcohol is commonly used as an homogenizing substance. The alcohol, however, is not only a solvent. According to Eqns. 2-7 and 2-8, it can participate in the esterification and alcoholysis reactions.

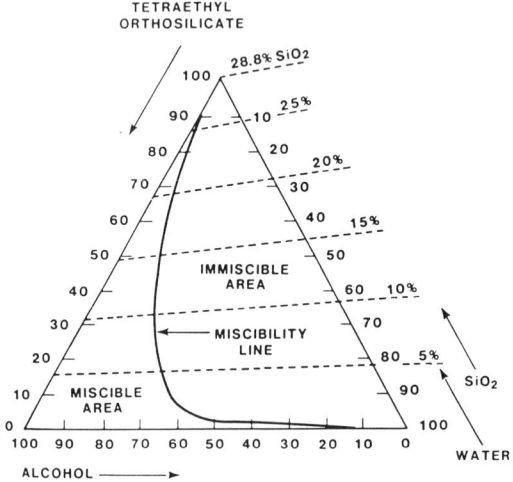

Fig. 2-12. TEOS, H_2O, Synasol (95% EtOH, 5% water) ternary phase diagram at 25°C. For pure ethanol the miscibility line is shifted slightly to the right. Cogan and Setterstrom (1947). Courtesy American Chemical Society.

In studies of the hydrolysis and condensation of silicon alkoxides, depending on the particular final product, the H_2O:Si molar ratio has been varied from less than one to above 50, and the concentrations of acids or bases have been varied from less than 0.01M to 7M, Coenen and De Kruif (1988), and it is possible to discern some consistent trends.

Acid-catalyzed hydrolysis at low H_2O:Si ratios yields weakly-branched "polymeric" sols, whereas base-catalyzed hydrolysis at high H_2O:Si ratios gives highly condensed "particulate" sols. Intermediate conditions produce structures between these two extremes. Monodisperse silica sols can be prepared by using a base, e.g., NH_3, as catalyst and H_2O:Si ratio of about 50.

2.1. Preparation of Monodisperse Silica Spheres

Stöber, Fink, and Bohn, "SFB", (1968) developed a widely-used method for preparing monodisperse silica spheres. They hydrolyzed a tetraalkylorthosilicate, e.g. tetraethylorthosilicate, TEOS, in a basic solution of water and alcohol, e.g. ethyl alcohol. The reactions for hydrolysis and condensation are:

$$Si(OC_2H_5)_4 + 4H_2O \xrightarrow{\text{hydrolysis}} Si(OH)_4 + C_2H_5OH \tag{2-10}$$

$$Si(OH)_4 \xrightarrow{\text{condensation}} SiO_2 + 2H_2O \tag{2-11}$$

We have already pointed out that hydrolysis and condensation occur simultaneously and would like to bring attention to the fact that the stoichiometric amount of water needed for hydrolysis is four moles of water for each mole of silicon alkoxide, or two moles if the condensation goes to completion. In order to prepare silicon particles, the ratio of water to TEOS, however, is typically higher than 20:1 and pH is very high. These two factors promote condensation, which aids the formation of compact structures rather than widely-branched networks. According to the SFB-method, alcohol, ammonia, and water are first mixed. The TEOS is then added and a visible opalescence can be observed in less than ten minutes. The resulting particles are very monodisperse; typically less than 5% of the particles deviate by more than 8% from the average particle size.

Figure 2-13 shows that the particle size depends on the composition of the reaction mixture. With TEOS as the silica source the largest average particle size with a narrow particle size distribution is about 0.7mm. The width of the distribution increases and the particles become less spherical as the concentration of TEOS exceeds about 0.2M. Stöber et al also studied alkoxides with smaller and larger alkoxy groups than TEOS and also used mixtures of alcohols as solvents. In general, lower alcohols as solvents yielded smaller particles and narrower particle size distributions. Smaller alkoxides gave smaller particles and faster reactions. Particles with diameters up to 2 mm can be prepared by using tetrapentylorthosilicate, or by either adding more alkoxide after the particles have been formed, or by carrying out the reaction at low temperatures, Hsu, Yu, and Matejevic (1993).

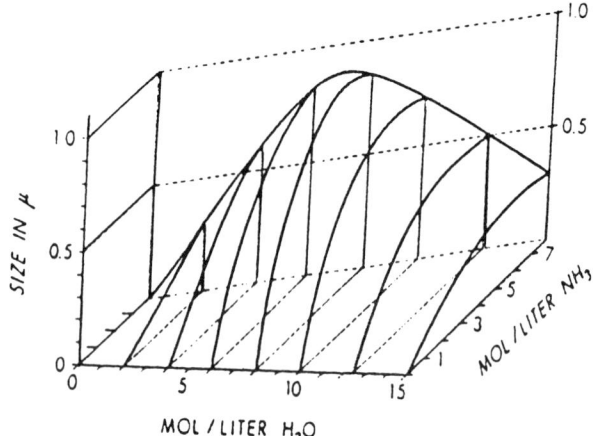

Fig. 2-13. Final particle sizes as obtained by reacting 0.28 M TEOS with various concentrations of water and ammonia in ethanol. Stöber, Fink, and Bohn (1968). Courtesy Academic Press.

Smaller particles, with diameters 20-35 nm, of uniform size can be prepared by using conventional silica sols, e.g. Ludox® sols—see Section 2-D—as nuclei, Coenen and De Kruif (1988).

Monodisperse spheres of many other metal oxides, e.g. TiO_2, ZrO_2, Al_2O_3, etc., can also be prepared by hydrolysis and condensation of the corresponding metal alkoxides (see Ch. 3).

2.2. Structure of SFB Particles

Monodisperse silica spheres, prepared by hydrolysis and condensation of alkoxides, have several common, micro-structural characteristics, Fegley and Barringer (1984):

1.   The skeleton density, as measured by helium pycnometry, is less than 80% of the density of amorphous silica (2.20).

2.   The BET specific surface area is much larger than the geometrical surface area that can be calculated from the particle diameter, as determined by transmission electron microscopy.

3.   The weight loss on heating at 110 C is about 8-20%, most of which is water. Organic residue constitutes less than 1% of the weight of the dried particles.

The low density of the silica skeleton and the high specific surface area suggest that the particles are porous. From nitrogen adsorption and desorption isotherms it has been possible to show that the particles actually are porous, LeCloux et al. (1986); the smaller the SFB-spheres, the more porous they are. The contribution of micropores, i.e., pores smaller than 2 nm in diameter, increased from about 2% of the total pore volume of 200-nm particles to about 50% for the smallest particles, about 8 nm in diameter.

Light scattering has been used to study the refractive index of 110-nm SFB-spheres, Van Helden and Vrij (1980). The refractive index varied strongly with the radius, from about 1.40 at the center of the particles to about 1.46 near the surface, implying that the density of the particles increases while they grow. In the beginning of the growth process the particles are highly porous, but as they grow larger, at least a shell under the surface becomes more dense.

2.3. Growth Mechanism

La Mer and Dinegar (1950) developed a theory for the formation of monodisperse particles. According to this theory the formation of the dispersed material should be arranged in such a way that all nucleation takes place in a very short period and additional material can be supplied so slowly that it can find its way to the nuclei without the supersaturation reaching a level at which further nucleation could occur.

Figure 2-14 illustrates this situation in a simple diagram relating to the preparation of sulfur sols from acidified thiosulfate. Sulfur is formed by the chemical reaction, its concentration increases, and rises above the saturation concentration until nucleation occurs. This condition results in the formation of many nuclei in a short burst. They grow rapidly, which lowers the concentration to a value below the nucleation concentration, but high enough to allow particle growth to occur at a rate that just consumes all further sulfur generated.

If nucleation thus occurs as one single, explosive outburst which exhausts the excess of dissolved matter, one single particle size may be obtained. However, if nuclei are formed during the growth period, a range of particle sizes will result.

If growth takes place by diffusion of molecules to the surface of the spheres, as suggested by the La Mer-Dinegar theory, it would be expected that the particles would be non-porous. Thus it appears that the porous structure and the low density of SFB-spheres cannot be explained by their theory.

Instead, as has been proposed by Bailey and Mecartney (1992), the most likely mechanism for the formation of monodisperse SFB-spheres is that small particles of silica are first formed in some way and then aggregate to larger particles. The results of Bailey and Mecartney's work on the formation of colloidal silica from alkoxides suggest the mechanism shown in Fig. 2-15.

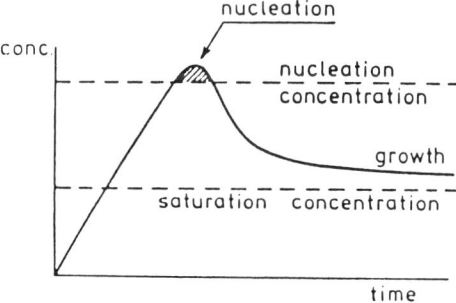

Fig. 2-14. Formation of a monodisperse system by controlled nucleation and growth. LaMer and Dinegar (1950). Courtesy American Chemical Society.

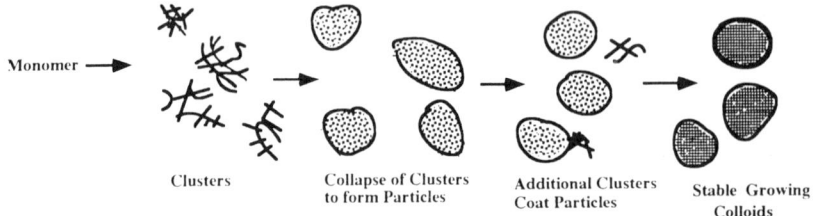

Fig. 2-15. Growth mechanism for the formation of monodisperse colloidal silica particles from alkoxides, Bailey and Mecartney (1992). Reprinted from *Colloids and Surfaces* with kind permission from Elsevier-Science-NL, Sara Burgerhartstraat 25, 1055 KV Amsterdam, The Netherlands.

Hydrolyzed tetraethylorthosilicate polymerizes and forms large clusters of microgel, which attain such a size, about 25 nm, and degree of cross-linking, that they become insoluble and collapse. After, or perhaps under, the collapse, the microgel will undergo further condensation and form colloidally stable nuclei of high density, i.e., relative to the microgel. Monomers are continuously still being formed by hydrolysis and will condense to polymers, which either grow to a size large enough that they collapse or they collide with another particle and attach themselves to its surface. The attached polymers may collapse onto the surface, thus allowing the particles to retain a spherical shape.

2.4. Factors Affecting the Size of SFB particles

Iler (1984) prepared particles according to the Stöber, Fink, and Bohn method, SFB particles, by mixing dry alcohol, concentrated ammonia solution (28% NH_3) and water and heating the mixture to the appropriate temperature in a thermostated water bath. The tetraethylorthosilicate, TEOS, was heated to the same temperature before it was added rapidly to the thermostated mixture under vigorous stirring. A few seconds after the addition of TEOS, the mixture turned milky. The stirring was continued for ten seconds. After 24 hours at room temperature, ammonia and alcohol were removed by evaporation under vacuum

Table 2-2 Stöber Sols made by Iler (1984)										
	Alcohol			Water		Ammonia (2%)		Tetra- ethylortho- silicate TEOS		Temperature ±0.1° C	Diameter nm
	Type	ml	per mole TEOS	ml	per mole TEOS	ml	per mole TEOS	ml	no. of moles		
1	EtOH	1000	25.5	210	17.5	240	5.35	150	0.670	50	320
2	MeOH	700	38.8	150	18.7	150	5.04	100	0.446	50	200
	"	"	"	"	"	"	"	"	"	40	330
	"	"	"	"	"	"	"	"	"	35	480
3	"	250	23.1	150	31.1	35	1.95	60	0.268	38	400
4	"	1050	38.8	225	18.7	225	5.01	150	0.670	50	200
5	"	70	38.8	15	18.7	15	5.04	10	0.045	50	240
6	"	1050	38.8	255	21.2	225	5.01	150	0.670	45	200
7	"	1000	37.0	190	15.8	260	5.81	150	0.670	40	440

Table 2-3 Stöber Sols made with Methanol as Cosolvent. Mattsson and Otterstedt (1993)						
	Mole Ratio			TEOS		Temperature	Diameter
	MeOH: TEOS	H2O: TEOS	NH3: TEOS	ml	moles	± 0.1 C	nm
1	38.8	18.7	5.04	100	0.446	34.0	438
2	"	"	"	"	"	35.0	386
3	"	"	"	"	"	37.0	376
4	"	"	"	800	3.57	39.5	360
5	"	"	"	"	"	39.6	334
6	"	"	"	"	"	40.0	314
7	"	"	"	"	"	40.0	327
8	"	"	"	"	"	40.2	317
9	"	"	"	"	"	45.0	250
10	"	"	"	"	"	46.2	212
11	"	"	"	"	"	46.2	245

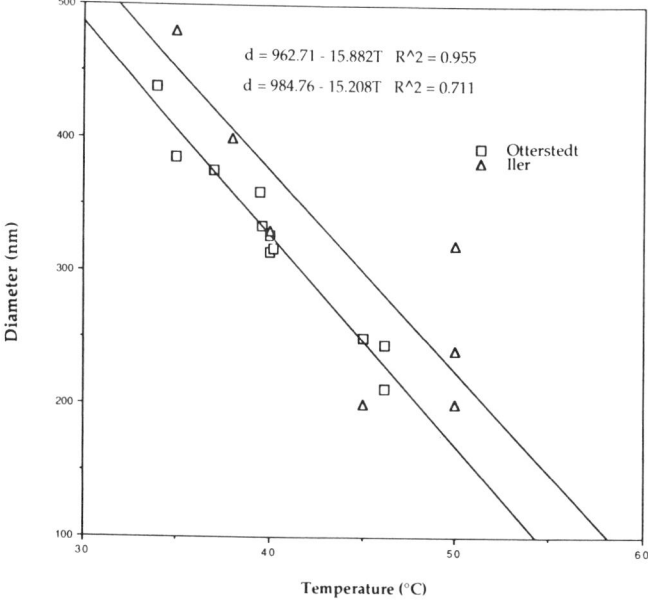

$d = 962.71 - 15.882T \quad R^2 = 0.955$

$d = 984.76 - 15.208T \quad R^2 = 0.711$

□ Otterstedt
△ Iler

Fig. 2-16. Particle diameter of Stöber sols vs. temperature made by Iler (1984) and Mattsson and Otterstedt et al. (1993).

at 35-40°C at constant volume, which was maintained by adding deionized water to the evaporator. The silica content was determined by the molybdate method and the particle size by scanning electron microscopy—see Chapter 1. The data from Iler's experiments are summarized in Table 2-2.

Fig. 2-16, in which the particle sizes of the experiments in group 2 of Table 2-2 are plotted against temperature, shows that there is a strong inverse dependence of the particle size on temperature. The results of experimental groups 1 and 5 and those of experimental groups 2 and 7 indicate that ethanol and more ammonia, respectively, yield larger particles.

Similar experiments   carried out at the Department of Engineering Chemistry, Chalmers University of Technology, Mattsson and Otterstedt (1993), yielded the results summarized in Table 2-3 which are also plotted in Fig. 2-16. The agreement between these results and Iler's is quite close. Also, the results show that, for a given batch composition, the size of the batches can be varied over a wide range, by factors of 10 and 8 for Iler's and Chalmers' experiments, respectively, without affecting the size of the Stöber particles.

C. Alkali Silicate Solutions

Jean Baptiste Van Helmont, a physician in Brussels, discovered in the first half of the 17th century that a water soluble material could be obtained by fusing pebbles with an excess of potash. About 300 years ago Stradivarius of Cremona impregnated the wood for his superior violins with a solution of a glass made by heating sand with the ashes of grape vines, which contained potassium carbonate.

However, it was not until about 1825 when Johann Nepomunk von Fuchs in Munich started a production of water soluble potassium and sodium silicate, which he called "water glasses", that alkali silicate solutions became commercial products for various applications.

Aqueous solutions of alkali silicates have thus been used for many years in numerous applications, but it is only fairly recently that their constitution has become known. The soluble commercial silicates have the general formula: $M_2O \cdot mSiO_2 \cdot nH_2O$ where M is an alkali metal and m and n are the number of moles of SiO_2 and H_2O relative to one mole of M_2O; m is called the ratio of the silicate. Sodium silicates are the most common silicates but potassium and lithium silicates are produced to a limited extent for special applications, with 3.3-ratio sodium silicate being the most common form of soluble silicate.

What is actually the difference between a small particle, sodium hydroxide stabilized silica sol and a high ratio, say 3.3, sodium silicate solution? As will be described in this section and in section D of this chapter, most of the silica in a concentrated 3.3-ratio sodium silicate solution is present as about 2-nm silica particles, the surface of which is covered with Na^+OH^- ion pairs while a 50 ratio, sodium hydroxide stabilized silica sol may consist of negatively charged 5-nm silica particles. Historically, sodium silicate solutions and sodium hydroxide stabilized silica sols have been the most thoroughly studied systems and, commercially, they are the most important materials. It turns out that it is very difficult to make sodium silicate solutions of a ratio higher than four and stable, sodium hydroxide stabilized, 4- to 5-nm silica sols of ratio lower than, say, 25. These two materials therefore appear to be separated by a large ratio gap. However, as will be described in this section, cesium, lithium, and potassium silicate solutions can be made with ratios up to 25 and in such solutions, the silica is present as colloidal particles, the size of which increases with ratio. The difference between high-ratio alkali silicate solutions and low-ratio, small particle silica sols may therefore mainly be a question of semantics.

1. Constitution of Solutions of Alkali Silicates

There have been many investigations of alkali, in particular sodium, silicate solutions over the last 50 years, identifying and measuring the component polysilicates as exemplified by the work of Ray and Plaisted (1983) and the studies of Dent-Glasser and associates (1982). In almost all cases, the solutions of sodium silicate of different ratios of SiO_2/Na_2O were diluted to one mole SiO_2 dm^{-3} or less before being examined.

Iler (1982) investigated the constitution of concentrated sodium silicate solution by using similar methods used to study the polymerization of silica (see section 2-B). The silicate solutions were suddenly injected into cold dilute solutions of H_2SO_4 with violent stirring to obtain the corresponding silicic acids of pH 1.8 at a concentration of 0.1M SiO_2 (6 g dm-3). A portion of the solution was mixed at once with tetrahydrofuran (THF) and then saturated with NaCl to salt out the organic liquid phase containing colloidal particles of silica but not oligomers or monomer; see Ch. 6, § C. Samples of the sol, as made, and of the THF extract were reacted with molybdic acid to obtain reaction rate curves from which were determined k, the first order reaction rate constant, and the amount of silica recovered as colloid. Figures 2-17 and 2-18 show silico-molybdate

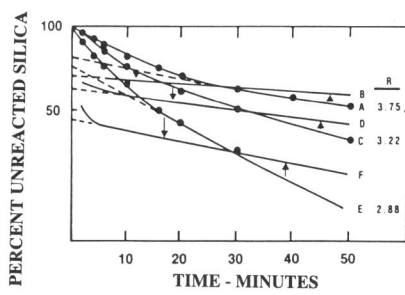

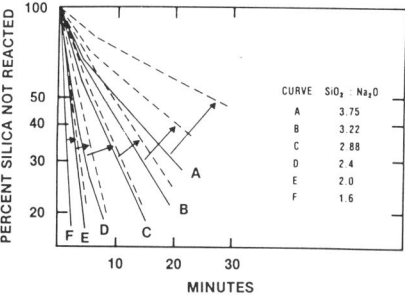

Fig. 2-17. Silicomolybdate reaction data for silicic acids, 16.6 mM SiO_2; pH 1.7; 25°C, made from made from sodium silicates, 5 M SiO_2; ratios (R) indicated; A, C, and E sols aged 1 min; and B, D, and F aged 10 days at 25°C. Iler (1982). Courtesy American Chemical Society.

Fig. 2-18. Silicomolybdate reaction data for silicic acid sols silicate solutions. Silicic acids of 16.6 mM SiO_2; pH 1.7; 25°C; aged less than 15 min. Silicate solutions diluted from 5-1 M SiO_2, aged 3 months at 25°C. Dashed lines are for polymeric silica extracted into THF within 15 min. Iler (1982). Courtesy American Chemical Society.

Table 2-4. Silicic acids from concentrated and diluted solutions. Iler (1982). Courtesy American Chemical Society.

$SiO_2:Na_2O$ Ratio	Silica from conc. silicate (5 M SiO_2)			Silica from diluted silicate (1.0 M SiO_2)		
w/w	% colloid	k of colloid	d (nm) calc.	% colloid	k of colloid	d (nm) calc
3.75	76±2	0.007	2.3	83±2	0.050	1.3
3.22	77	0.014	1.9	85	0.071	1.25
2.88	76	0.025	1.7	85	0.095	1.2

Table 2-5. Comparison between properties of colloidal silica particles in aqueous solutions and extracted into THF. Iler (1982). Courtesy American Chemical Society.

$SiO_2:Na_2O$ Ratio	SiO_2 extracted in THF			SiO_2 colloid in hydrosol		
w/w	% colloid	k of colloid	d (nm) calc.	% colloid	k of colloid	d (nm) calc
3.75	92	0.008	2.2	92	0.012	2.0
3.22	87	0.016	1.8	87	0.020	1.7
2.88	83	0.032	1.5	77	0.037	1.4
2.4	75	0.042	1.4	48	0.048	1.3
2.0	37	0.080	1.2			

reaction data for silicic acids made from 5M SiO_2 sodium silicates of different ratios and from sodium silicates, also of different ratios, diluted from 5M to 1M SiO_2 and aged from three months at 25°C, respectively.

From k, the nominal particle diameter of the colloid can be calculated from the equation (Eqn. 20 in Chapter 1): $\log_{10} d = -0.250 - 0.287 \log_{10} k$.

In Table 2-4, the results for concentrated and diluted sodium silicate solutions are compared. The fraction of colloidal silica present does not vary with ratio for either type of sol but it is larger in sols made from diluted solutions than from concentrated solutions. The effect of dilution may be to re-distribute silica between different species so that the fraction of colloids increases but the particle size of the colloidal silica particles decrease.

Table 2-5 shows that for higher ratios the amount of colloids that can be extracted from hydrosols made from 5 M SiO_2 silicate solutions is the same as the fraction of colloids in the hydrosols. Iler obtained the fraction by extrapolating the linear parts of the curves in Fig 2-17. At lower ratios, this extrapolation excludes smaller particles, which can be extracted into THF. The value of 48% colloids in solutions of ratio 2.4 is therefore too low.

From these results, Iler concluded that in concentrated solution of sodium silicate of ratios from 2 to 4, about 75% of the silica is present in the form of particles with diameters ranging from 0.8 nm to about 2.0 nm. Dilution of such solutions brings about a redistribution of silica between different species so that the fraction of colloidal particles increases but their size decreases.

In an investigation of the constitution of aqueous silicate solutions Ray and Plaisted (1983) measured the amounts of monomeric silica and of polymers (anions with charges of 10 or more) in solutions that had been diluted to a series of concentrations ranging from 0.01 to 3.0 molar SiO_2, but not adjusted to pH 1.8 before the measurements were made, and found that most of the silica was present as monomer (which includes both $Si(OH)_4$ and $(HO)_3 SiO^-$) and as polymer, which was defined above. They plotted, Fig. 2-19, the percentage of the total silica present as monomer and as polymer versus the logarithm of c, the molar concentration of SiO_2, and could express their data in terms of equations:

$$\% \text{ Monomer} = -1.48-15.28 \ln c + 6.35 c$$

$$\% \text{ Polymer} = 88.26+14.17 \ln c + 0.84 c$$

These equations represent the values shown in Fig. 2-19 for values of up to 1.0.

The results from the work of Ray and Plaisted do not describe the constitution of concentrated sodium silicate solutions, and indeed other alkali silicate solutions, since, as was pointed out by Iler (1982), and will be discussed further below, dilution of such solutions before adjusting pH to 1.8 will lead to extensive redistribution of silica between various species.

McCormick, Bell, and Radke (1987) identified 19 different silicate species in sodium silicate solutions of ratio 1-3 going from monomer to cyclic trimer to prismatic hexamer to double hexamer, etc. and determined how the concentration of some of these species varied with ratio (Fig. 2-20). The fraction of more complex species increases with ratio.

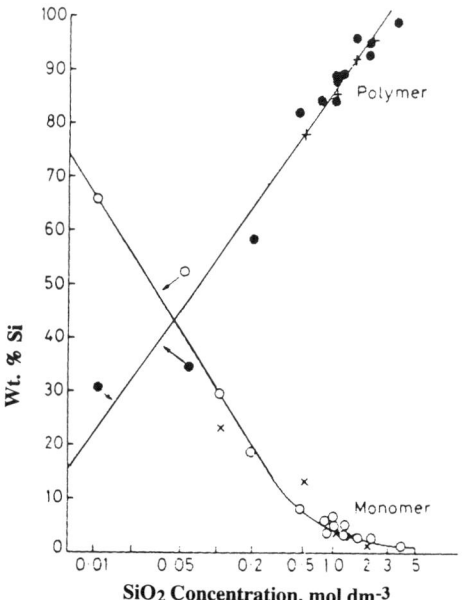

Fig. 2-19. Distribution of anions in sodium silicate solutions (SiO_2-Na_2O = 2.8-3.6) as a function of SiO_2 concentration, Ray and Plaisted (1983), (•, O) and Dent-Glasser (1982) (x, +). Courtesy of The Royal Society of Chemistry.

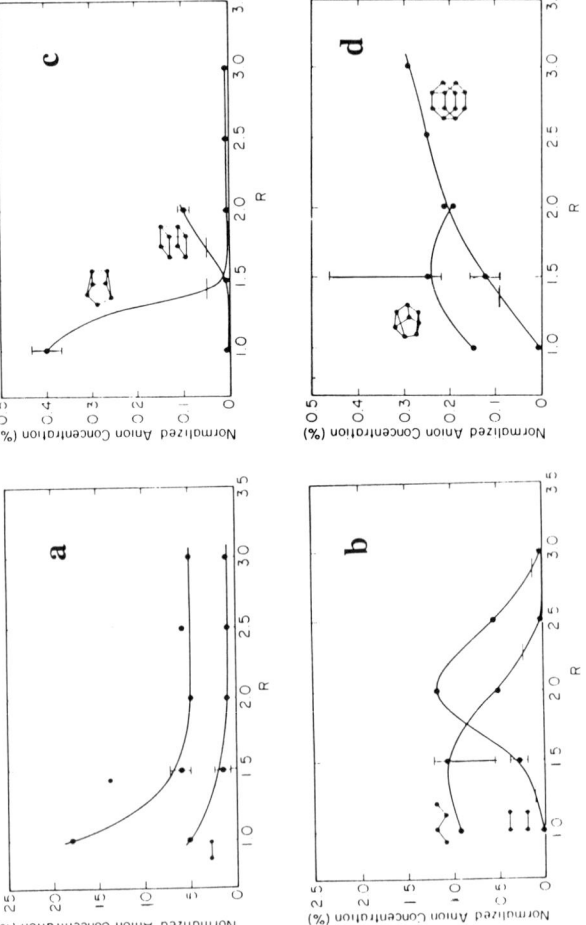

Fig. 2-20. Normalized concentrations of various silicaceous species as a function of silicate ratio. a) Normalized concentration of monomer and dimer as a function of silicate ratio: 3 mol% SiO_2. b) Normalized concentration of linear and cyclic tetramer as a function of silicate ratio: 3 mol% SiO_2. c) Normalized concentration of pentacyclic heptamer and cubic octamer as a function of silicate ratio: 3 mol% SiO_2. d) Normalized concentration of hexacyclic octamer and double hexamer as a function of silicate ratio: 3 mol% SiO_2. McCormick, Bell, and Radke (1987). Reprinted from *Zeolites* with kind permission from Elsevier-Science-NL, Sara Burgerhartstraat 25, 1055 KV Amsterdam, The Netherlands.

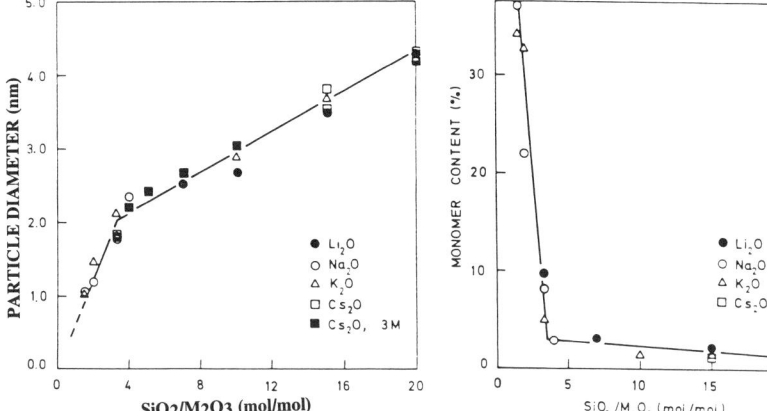

Fig. 2-21. Particle diameter in 5M SiO2 alkali-silicate solutions versus SiO2/M2O molar ratio, as measured by the molybdic acid reaction.Otterstedt et al. (1987). Courtesy Academic Press.

Fig. 2-22. Fraction SiO2 in monomeric form in 5 M SiO2 alkali-silicate solutions versus SiO2/M2O molar ratio, as estimated from the molybdic acid reaction. Otterstedt et al. (1987). Courtesy Academic Press.

Otterstedt et al. (1987) used the method of Iler to determine the particle size of colloidal silica in solutions of alkali silicates of different ratios. Figure 2-21 shows the particle size of colloidal silica, as determined by the reaction of silica with molybdic acid, for 5 M SiO_2 solutions of lithium, sodium, and potassium silicates and for 3 M SiO_2 cesium silicates of different ratios in the range 1.5-20. The curve for potassium silicate solutions was interpolated in the ratio range 3.3-10.0 where no liquid solutions could be made. The particle size of the colloidal particles does not appear to depend on the type of cation, but increases with ratio for all four alkali silicates and is about 1.8 nm for ratio 3.3 and 4.3 nm for ratio 20. At ratio 4 there appears to be a break-point in the relationship between ratio and particle size. For ratios below 4, the relation is linear and is still linear at ratios above 4, but with significantly smaller slope.

Another observation related to the breakpoint is that the monomer concentration is low and almost constant at ratios above 4, but increases rapidly below ratio 4; see Fig. 2-22.

2. Dilution of Concentrated Alkali Silicate Solutions

Iler (1979) noted that the particle size decreased when 5 M SiO_2 3.3-ratio solutions of Na-silicate were diluted to 1 M SiO_2; (Figs. 2-17 and 2-18). McGarry and Hazel (1965), however, found an increase in particle size when 3.45-ratio potassium silicate solutions were diluted from 10% SiO_2 (~1.8 M) to 0.3% SiO_2 (~0.05 M).

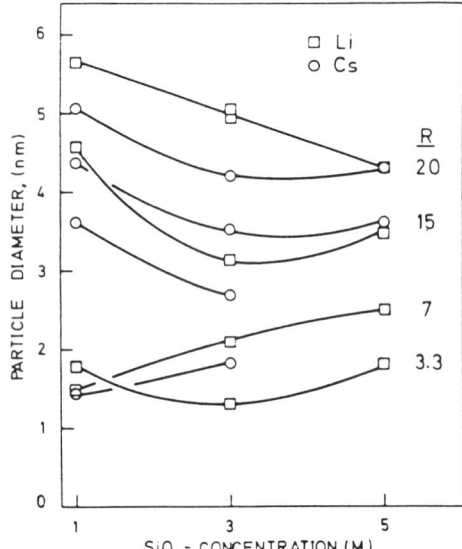

Fig. 2-23. Particle diameter in Li- and Cs-silicate solutions of SiO_2/M_2O ratios in the range 3.3-20 versus SiO_2 concentration, as measured by the molybdic acid reaction. Otterstedt, Ghuzel, and Sterte (1987). Courtesy Academic Press.

Otterstedt et al. (1987) studied the effect of dilution on the size of colloidal components in alkali silicates by diluting Si- and Cs-silicate solutions after aging them for one week at room temperature before particle sizes were determined by the molybdic acid reaction. In the first set of experiments, 5 M SiO_2 solutions of SiO_2/M_2O molar ratios of 3.3, 7.0, 15, and 20 were diluted to 3 and 1 M SiO_2. Figure 2-23 shows that the particle size increased for the higher ratios. For the 20-ratio solutions, it grew continuously from 4.3 nm in the 5 M SiO_2 solutions to 5.6 nm in the 1 M SiO_2 solutions whereas for the 15-ratio solutions there appeared to be a slight decrease down to 3 M SiO_2 followed by an increase on further dilution to 1 M SiO_2.

These results can be explained by assuming that the particle size is not completely uniform but distributed around a mean value and that the surface of the particles are saturated with M^+ ions (see below). Dilution will lower the concentration of M^+ ions, both on the surface of the particles and in the solution, causing the smaller particles to dissolve and deposit on the larger particles. These will grow to a size that gives a total surface area, saturated with M^+ ions, consistent with the new concentration of M^+ ions.

For the 7- and 3.3-ratio solutions, dilution from 5 M to 1 M SiO_2 caused the particle size to decrease. To explain this behavior, 3.3-ratio Li-silicate solutions were further diluted in a series of steps to 0.1 M. The particle size and pH were measured after each step and after aging the solutions for one week at room temperature. Figure 2-24 shows that the particle size decreased from 1.8 nm to a minimum of 1.3 nm at 3 M SiO_2 and then increased to 2.2 nm at 0.1 M

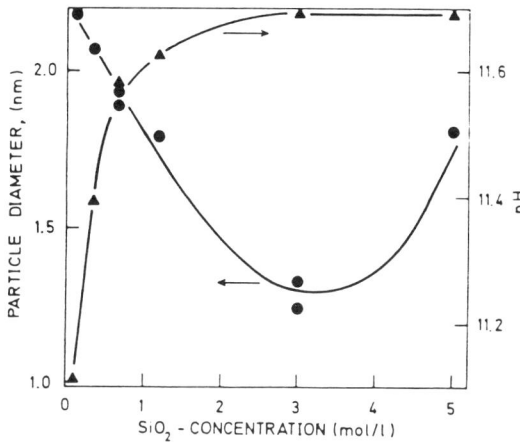

Fig. 2-24. Particle diameter (circles) and pH (triangles) of 3.3 ratio Li-silicate solutions versus SiO_2 concentration. Otterstedt, Ghuzel, and Sterte (1987). Courtesy Academic Press.

SiO_2. The pH decreased slowly at first but faster as the SiO_2 concentration decreased below 3 M SiO_2.

In the 5 M SiO_2 solution, pH was about 11.7, which is also the pK_a of H_3SiO_4, Scherban (1967). The concentration of $H_2SiO_4^{2-}$ is therefore appreciable and each $H_2SiO_4^{2-}$ ion is associated with 2 M^+ ions. The slight decrease in pH when diluting the 5 M SiO_2 solution to 3 M SiO_2 will cause a large decrease in $H_2SiO_4^{2-}$ with a simultaneous release of M^+ ions. The colloidal particles will increase in number, but decrease to a size that gives a total surface area consistent with the increased concentration of M^+ ions free to associate with the particle surface. Further dilution will cause the particles to decrease in number but increase in size to adjust to the decreasing concentration of M^+ ions.

Si(OH)$_4$, at high pH as in concentrated silicate solutions, is present almost entirely as (HO)$_3$SiO$^-$, as can be deduced from the pK_a. At high concentrations of sodium ions these ions are held on the surface of the colloid and/or incorporated into the surface through the sodium ions. This seems to be the explanation of why very little monomer is present in the corresponding acidified solutions. However, if the silicate solution is diluted ten- or twenty-fold only for a few seconds before being added to the acid, high monomer and oligomer content is found.

3. Association of Sodium Ions with Silicate Ions

Iler (1979) summarized past investigations of silicate solutions and the equilibrium of the monomer, Si(OH)$_4$, monosilicate ion, (HO)$_3$SiO$^-$, and polymers, but the data all relate to dilute solution. However, little attention has been given to the possibility that in concentrated solutions, the Na^+ ion and the SiO^- ion may not be present as usually depicted: $SiO^- + Na(H_2O)_6^+$.

Instead, Iler proposed that essentially all the ions are present as ion pairs and that on the surface of the silica particles the silanol groups may replace one or more of the water molecules of the hydrated sodium ion:

This possibility is suggested by several related observations:

1. Lawrence and Vivian (1961) found that at high concentrations (3 M) the lithium ion is strongly adsorbed on the surface of silica so that the latter does not dissolve in alkali. Since the silica dissolves in dilute LiOH solution, lithium silicate is readily soluble at ordinary temperature and thus the failure to dissolve is not due to a coating of insoluble lithium silicate. Instead, it is suggested that the lithium ion forms an ion pair on the surface and this blocks dissolution.

2. A sufficiently high concentration of sodium ions (3-4 M) prevents the dissolution of amorphous silica at pH 12, at which point the surface adsorbs 3.5 OH^- ions per square nanometer, Heston, Iler and Sears (1960).

3. Particles of colloidal silica in an alkali-stabilized sol (see Section 2-D) are coagulated when the concentration of sodium ion (added salt) exceeds a critical value. This has usually been explained by the double-layer theory, in which the counter ions move in closer to the surface, reducing the neutral charge repulsion so that the particles are drawn together by the assumed "Van der Waals forces". However, Iler (1979) suggested that at a sufficiently high concentration of sodium ions, ion pairs are formed on the surface and that a colliding particle with its SiOH surface can become attached by coordination with the sodium ion, which thus acts as a link between the particles. It is significant that when coagulation of silica occurs at only weakly alkaline pH, the aggregates do not become dispersed when the solution is diluted. However, if the pH is at 9-10.5, where the surface is highly charged, the addition of salt causes precipitation, but if the suspension is diluted to lower concentrations of sodium ions, the precipitate redisperses to the original sol state. This indicates that the particles were held together through sodium ions and not by SiO-Si bonds, which are irreversible under these conditions.

4. The formulation of ion pairs on the surface was proposed by Yates et al (1974) who called it the "site-binding theory" and the idea was supported by Iler (1979).

The "ion pair" theory was not applied to the soluble silicate system until Iler suggested that in concentrated solutions, much of the silica was present as colloidal particles coated with $SiO^- Na^+$ ion pairs, Iler (1979).

4. Acidity of Silanol Groups

The acidity of silanol groups in silica species and its importance to the polymerization behavior of silica was discussed in section B-2. Here, some of the salient features of the acidity relevant to the constitution of alkali silicate solutions will be emphasized.

The acid strength of the OH groups on silicon, like those on phosphoric acid, increases with fewer OH groups on the central atom. Thus for $Si(OH)_4$ the first pK_a is about 9.6 while for the $(OSiO)_3$ SiOH it is about 6.8, Iler (1979). This has several consequences:

1. The cubic octamer, $Si_8O_{12}(OH)_8$ (see Fig. 2-10) is a stronger acid, i.e., more highly ionized, than monomer, branched chain or cyclic polymers.

2. Since polymerization involves reaction of $Si(OH)_4$ with SiO^-, the $Si(OH)_4$ reacts more rapidly with octamer than with itself or lower oligomers.

3. Likewise, the $(OSiO)_3$ SiOH groups on the surface of the colloidal particles react more rapidly with monomers than do the small oligomers.

4. When the cubic octamer first adds more monomers, the latter cannot fully condense all the OH groups so, at an intermediate stage, there is a polymer that is less reactive with monomers than either the cubic octamer or the fully covered and condensed surface of a larger colloidal particle. Based on a molecular model, Iler concluded that at this latter stage there is present a total of about 40 Si atoms in the polymer.

5. At high concentrations of Na^+ ions, only the more strongly acidic or ionized $(OSiO)_3$ SiO^- sites can form ion pairs.

Thus it is only the cubic octamer (or some very similar closely related, fully condensed species) and the colloidal particles that have stabilized surfaces of SiO^-Na^+. The oligomers with $Si(OH)_3$ or $Si(OH)_2$ are too weakly acidic to be stabilized and disappear when the system reaches equilibrium.

5. Calculation of Particle Size

From his experimental data, Iler (1983a) calculated the particle size of the colloidal particles in sodium silicate solutions of different ratio. From k, the reaction rate constant with molybdic acid, the particle size and amount of polymer can be estimated. Knowing the amount of alkali associated with the polymer and the amount of polymer present, one can calculate the specific surface area based on the fact that the fully charged surface of a colloidal particle has about 3.5 anionic charges per square nanometer, Heston, Iler, and Sears (1960). The particle diameter can be calculated from the specific surface area assuming spherical particles (see Ch. 1).

The amount of alkali adsorbed on the particles was expressed in terms of the normality in the system, N_a. The total alkali in the system is N_t and thus the difference is the alkali associated with the oligomer, N_o. Thus $N_o = N_t - N_a$.

From N_o and the amount of silica present as oligomer, the number of ionic charges per silicon atom (in oligomers) were calculated. For solutions having

Table 2-6. Calculation of ionic charge per silicon atom in oligomer. Iler (1983)

Ratio of silicate SiO_2/Na_2O	% Silica recovered in THF	SiO_2 in the oligomer %	M_O	k 1st-order reaction rate constant of colloid with molybdic acid	N_t normality of total alkali in the system	d - diameter of colloidal particle calculated from the equation below (nm)	A- area colloid (= 2750/d) (m²/g)	CA total area of colloid (=Ax(grams colloid per cm³)	N_c normality of alkali associated with surface of colloid	N_O normality of alkali associated with oligomer (=N_t-N_c)	Oligomer ions per Si atom
3.75	92	8	0.008	0.008	0.052	2.29	1200	6624	0.037	0.015	1.9
3.22	87	13	0.013	0.016	0.060	1.64	1495	7804	0.043	0.017	1.3
2.88	83	17	0.017	0.032	0.067	1.512	1818	9053	0.050	0.017	1.0
data from smoothed curve				0.025		1.62	1698	8456	0.047	0.021	1.2
2.4	75	25	0.025	0.042	0.081	1.40	1964	8838	0.049	0.032	1.3
2.0	37	63	0.063	0.080	0.097	1.17	2350	5217	0.029	0.068	1.1

SiO_2/Na_2O ratios of from 2.0 to 3.2, this value averaged about 1.2 (in the case of the 3.75 ratio silicate the amount of oligomer was small and could have been in error by 50%). It may be noted that the ionic charge on the cubic octamer is 1.0 per silicon atom.

The details of Iler's calculations are shown in Table 2-6 (1983) using the equation: $\log_{10} d = -0.250 - 0.287 \log_{10} k$

6. Manufacture of Soluble Alkali Silicates

Various manufacturing routes for commercial sodium silicates are shown in Fig. 2-25. Similar routes are being used for the manufacture of potassium silicates.

Mixtures of sand and sodium carbonate in appropriate proportions are continuously fed to an open-hearth furnace, kept at above 1300°C. A 3.3-ratio glass is formed by the following reaction:

$$3.3\ SiO_2 + Na_2CO_3 \dashrightarrow CO_2 + 3.3\ SiO_2\ Na_2O$$

The sand should be of high purity and contain no more than 300 ppm of iron. The particle size of the quartz sand should be in the range of 0.1-0.5 mm. The sodium carbonate, or potassium carbonate, must also be of high purity.

The glass from the furnace can be drawn into a rotary dissolver. Cold glass lumps can also be dissolved in a pressure dissolver at pressures of about 5 bar and temperatures of 150°C or higher.

Sodium metasilicate pentahydrate is produced by dissolving finely ground sand or amorphous silica in sodium hydroxide in an autoclave with metasilicate feed liquid, which after evaporation crystallizes to the pentahydrate.

Drying the metasilicate feed liquor in a drum dryer yields anhydrous metasilicate, Na_2SiO_3.

Sodium orthosilicate, Na_2SiO_4, is produced by blending anhydrous metasilicate and NaOH beads.

Potassium silicates are manufactured by similar routes. Since lithium silicate glasses are insoluble, lithium silicate solutions are normally prepared by dissolving silica gel in a LiOH solution or mixing a silica sol with LiOH.

Tables 2-7a and 2-7b show the composition of typical American commercial sodium and potassium silicate solutions.

Table 2-7a. Typical Properties and Composition of PQ Standard Potassium Silicate Solutions								
Product Name	Wt. Ratio SiO_2/K_2O	Mole Ratio SiO_2/K_2O	K_2O %	SiO_2 %	Density @ 20°C g/cm^3	pH	Viscosity cp	Characteristics
Kasil® #1	2.50	3.92	8.30	20.8	1.26	11.3	40	clear liquid
Kasil® #33	2.10	3.29	11.6	24.4	1.34	11.7	430	clear liquid
Kasil® #6	2.10	3.29	12.65	26.5	1.38	11.7	1050	clear liquid

Courtesy PQ Corporation

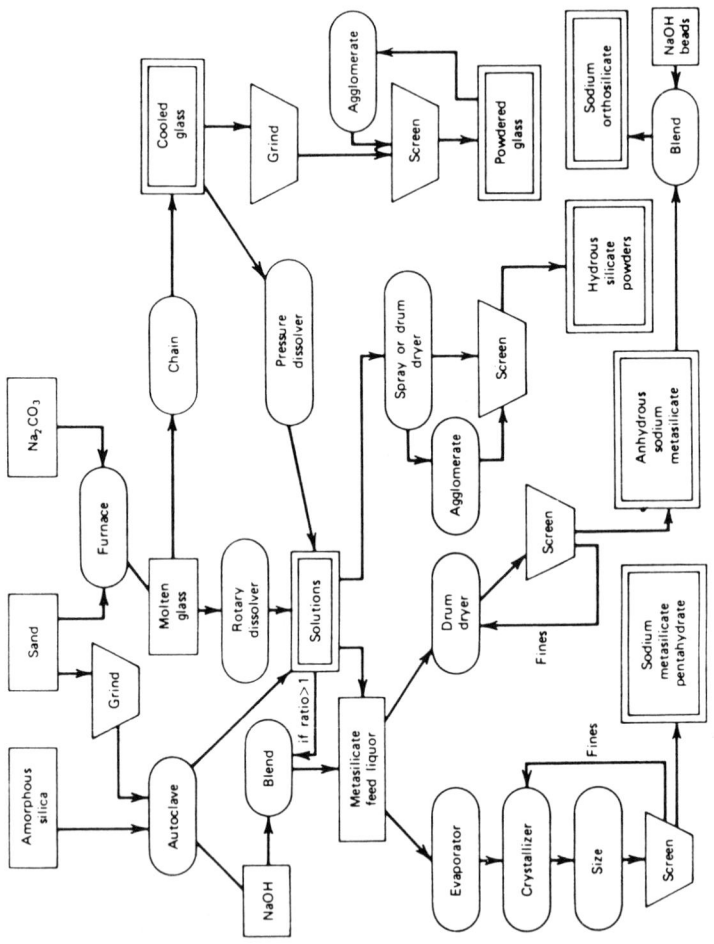

Fig. 2-25. Various manufacturing routes for commercial sodium silicates. Courtesy of PQ Corporation.

Table 2-7b. Typical Properties and Composition of PQ Sodium Silicate Solutions

Product Name	Wt. Ratio SiO_2/Na_2O	Na_2O %	SiO_2 %	Density @ 20° C g/cm^3	pH	Viscosity cp	Characteristics
STIXSO® RR	3.25	9.22	30.0	1.41	11.3	830	syrupy liquid
N® & N® Clear	3.22	8.90	28.7	1.38	11.3	180	syrupy liquid
E®	3.22	8.60	27.7	1.37	11.3	100	specially clarified
O®	3.22	9.15	29.5	1.41	11.3	400	more conc. than N®
K®	2.88	11.0	31.7	1.47	11.5	960	sticky, heavy silicate
M®	2.58	12.45	32.1	1.50	11.8	780	syrupy liquid
STAR®	2.50	10.60	26.5	1.40	11.9	60	brilliantly clear, stable solution
RU®	2.40	13.85	33.2	1.55	12.0	2100	heavy syrup
D®	2.00	14.70	29.4	1.53	12.7	400	syrupy alkaline liquid
C®	2.00	18.00	36.0	1.68	12.7	70,000	heavy alkaline liquid
STARSO®	1.80	13.40	24.1	1.43	12.9	60	specially clarified
B-W®50	1.60	16.35	26.2	1.53	13.4	280	high alkalinity; syrupy liquid

Courtesy PQ Corporation

7. Uses of Soluble Alkali Silicates

Figure 2-26 shows the uses of sodium silicates in the United States. The key uses will be described briefly.

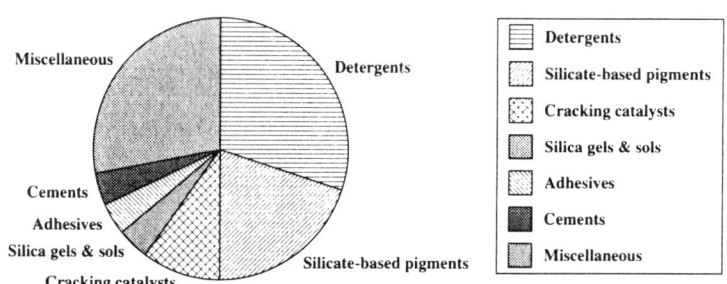

Fig. 2-26. Use of sodium silicates in the United States.

Detergency The major single use for soluble silicates as a functional additive is in soaps and detergents. Detergency, or soil removal, from surfaces immersed in liquids is a complex process consisting of several steps, most of which are aided very significantly by soluble alkali silicates, especially the most available and cheapest one, sodium silicate. Among these are:

Wetting: Soil removal is made easier by the reduction in surface tension of the wash water by small amounts of sodium silicate.

Deflocculation: An important part of soil removal is the breaking up of larger clumps into smaller particles which are easier to remove from surfaces and easier to suspend in the bulk of the liquid so as to allow removal by rinsing.

Redeposition prevention: Removal of soil particles is important, but can readily have its effectiveness diminished by redeposition of the dirt particles.

Emulsification: Since many common soils are composed of both oily compounds along with particulates, it is vital that oily soil be dispersed into fine droplets that can be suspended in the wash solution.

Buffering: Neutralizing acidic soil, emulsifying fats and oils, and dispersing and/or dissolving proteins are all aided by the alkalinity of sodium silicates, whose buffering capacity—maintaining a relatively constant high pH on dilution or in the presence of acidic components—is stronger than most alkaline salts such as sodium tripolyphosphate, sodium carbonate, and sodium tetraphosphate.

Stabilization: Sodium silicates stabilize chlorine and oxygen bleaches during the cleaning cycle.

Combating water hardness: Sodium silicates minimize the negative effects of Mg^{+2} hardness in water and therefore work well when combined with the Ca^{+2}-ion-activity reducers, soda ash and zeolites.

Corrosion inhibition: Silicate polyanions act to form a physical barrier to alkali attack by adsorbing at oxide surfaces thereby providing protection for sensitive glazed dishware, glass, and metal surfaces.

Phosphate replacement: Sodium silicates find wide use as partial phosphate replacements in phosphate-free formulations.

Spray-drying detergents: In spray-dried detergent powder formulations their glassy nature confers a crisp quality to the dehydrated detergent granules.

Adhesives and Binders In general sodium silicate adhesives have several outstanding properties for these applications: good spreading and penetration; good bond formation that results in a strong bond that resists heat and to some extent water; and setting rates that are controllable over wide limits. One of the best known applications is as an adhesive in corrugated box manufacture where one of the advantages over starch is the dislike by rodents for chewing through sodium silicate. Other common similar uses are in spiral tube winding, lamination of metal foil to paper, and fiber drum manufacture. Silicates are used extensively in manufacturing refractory and acid-resistant mortars and cements with potassium silicate being the choice for higher temperature materials. Another common use for potassium silicate is as a binder for phosphors for television screens. Silicates also find use in the pelletizing, granulating, and briquetting of fine particles. Binder and coating applications requiring high water insensitivity are roofing granules, welding-rod coatings, and corrosion-resistant paints for iron and steel. Another binder application is in sand molds used in foundries, where the use of sodium silicate inorganic binder systems promotes cleaner foundry environments because these binder systems are nontoxic, odor-free, and are easy to use. Reclamation of silicate-bound sands is widely practiced. The spent sands from these sodium silicate castings are low in residual organics, thus permitting easier, less expensive disposal.

Potassium silicate is available in electronic grades with low levels of iron, copper, and sodium for use as an efficient phosphor binder and settling agent for CRTs used in computer monitors, oscilloscopes, and radar screens.

Protective and Decorative Coatings Potassium silicate is favored over sodium silicate in coating welding rods because it yields a smooth, quiet burning arc in AC welding and for stainless steel. Applied to stone, concrete, stucco, brick, and metal surfaces, films of potassium silicate remain free of frosty bloom because they do not form carbonates. In decorative applications the colorless, nearly transparent potassium silicate does not hide the pigments. In addition, masonry treated with a dilute solution of potassium silicate can be made non-porous and moisture resistant.

Potassium silicate can be used to make zinc-rich protective coatings. Zinc dust is mixed with the silicate prior to application. When the coating dries, a curing system is applied. The resulting coating is resistant to water, rain, and corrosion.

Building and Construction Silicates are used in soil stabilization and as sealants in concrete pipes by direct impregnation. After a concrete mix has hardened, sodium silicate can be surface-applied to penetrate the concrete and increase resistance to wear, water, oil, and acid. Lime and other ingredients react slowly with the silicate solution to form an insoluble gel within the concrete pores. Sodium silicates combine chemically with cement ingredients to form masses which have strong bonding properties. These cements can be dry or liquid. Sodium silicates are also used to modify the physical properties of portland cement. Adding silicate to cement reduces permeability by increasing the total number of bonds formed between aggregate particles. Silicates can be used to modify the set time in cold weather or adverse conditions.

Chemical grouting, or soil solidification, is used to: strengthen soil formations that are not strong enough to carry the required load, such as underwalls and footings of structures; to make impermeable the porous soils that otherwise would allow the flooding of mines and tunnels; to prevent water loss from cracks in dams and other containment structures; and to encapsulate soils contaminated with hazardous materials and prevent toxic components from migrating into groundwater. Sodium silicate is used with one or more chemicals to form a grout-gel bond. In the Joosten process, for example, sodium silicate and calcium chloride are injected separately into the ground to form an instant grout.

Enhanced Oil Recovery In some oil recovery processes the presence of silicates enhances the surfactant's effectiveness in attaining higher oil recovery where chemical flooding is used. In the very complex surface chemistry processes which occur, silicates are thought to play several important functions in changing surfactant adsorption and retention. This newer application offers great promise when oil prices are high enough—perhaps in the $50/bbl range—because of the importance of high recovery efficiency and attaining commercial viability in otherwise depleted fields.

Pulp Bleaching Used increasingly for pulp bleaching, hydrogen peroxide offers an alternative to chlorine bleaching processes. Sodium silicates are an important component of hydrogen peroxide bleach liquors where they act to stabilize the peroxide and produce a brighter pulp at a lower cost. In addition the sodium silicate deactivates metals such as iron, copper, and manganese, which catalyze the decomposition of hydrogen peroxide. An added benefit is that the silicates buffer the bleach liquor at the pH at which the peroxide is most effective.

Mineral Beneficiation and Wet-ore Grinding The mineral beneficiation process of froth flotation is run on a vast scale in recovering many metals such as copper and zinc. Sodium silicates are used extensively as strong, selective agents for increasing the hydrophilic nature of specific particles which therefore sink, yet at low concentrations may aid the flotation of some calcium minerals. Soluble silicates are also excellent dispersants. Variable rates of settling can be achieved by using soluble silicates at low concentrations to adsorb preferentially on specific mineral surfaces. At higher concentrations soluble silicates cause non-selective depression of all minerals.

Silicates also find use in cost reduction in wet-ore grinding where they aid in reducing wear on grinding balls and rods.

Water Treatment The principal uses of silicates in water treatment are as a coagulant aid in the alum coagulation of suspended matter in raw or waste water streams; for corrosion protection of metal surfaces in contact with water; for boiler water treatment to reduce scale formation; and as a stabilizer in reducing iron and manganese in water supplies. Generally these applications depend on the alkalinity of silicates to form complexes with metal ions in solution and to adsorb at charged interfaces. Indirectly, sodium silicates also function in another way to yield better quality water for drinking by reducing concentrations of lead, copper and other heavy metals by forming a microscopic film on the inside of water supply pipes, preventing the leaching of lead solders and other metals throughout the system.

Other Uses Other large-scale applications include, slurry thinning, and as a source of building-block anions for manufacturing silica-based products such as synthetic zeolites, silica gels and sols, pigment-grade silicas, and heavy-metal silicates. Hundreds of other industrial uses could be cited. Many of these are minor in terms of tonnage, but are often important in terms of cost effectiveness or being unique.

Derivative Products The single largest use of soluble silicates is as a source of silicate anions in the manufacture of silica-based products such as silica sols and gels, zeolites (see Ch. 5) and clays and pigment-grade silicas.

D. Silica Sols and Colloidal Silica

A silica sol is a stable dispersion of discrete, colloid-size particles of amorphous silica in an aqueous phase. Concentrated sols are stabilized against gelling, brought about by interparticle siloxane bonding, by an ionic, usually anionic, charge on the particles. Commercial sols do not gel or settle even after several years of storage. They may contain up to 50 wt. % silica and particle sizes up to several hundred nanometers, although particles larger than 70 nm slowly settle. This definition excludes solutions of polysilicic acid, in which the polymers or particles are so small that they will either slowly grow in size or aggregate. Although silica sol particles of about 5 nm will slowly grow and therefore fall outside the conventional definition of silica sols, they are of commercial interest, e.g. as retention aids in paper making, and will be treated in this text.

1. Formation of Sol Particles

The stepwise formation of small silica particles from monomeric silicic acid at pH 1.7-2.0 was described in Section B-2. At that pH, the polymerization takes place so slowly that the particle growth can be followed by various analytical techniques.

Commercial production of silica sols occurs in the pH range 8-10 and at elevated temperatures so as to rapidly attain the desired particle size. Furthermore, solutions of alkali silicates, often 3.3 ratio sodium silicate solution, and not monomeric silicic acid, are used as raw materials. The basic mechanism for particle growth, however, is the same as at pH 2, but the rate of growth is several orders of magnitude faster.

Figure 2-27 shows that the first step in the manufacture of commercial silica sols is to dilute a 3.3 ratio sodium silicate solution to less than 5% SiO_2, often to 2-3% SiO_2. At higher SiO_2 concentrations gel can form and clog the cation exchange column. After filtration, the diluted Na-silicate solution is passed through a column of cation exchange resin in the H^+-form, usually a strong acid resin containing sulfonic acid groups, where H^+ ions are exchanged for Na^+ ions and the pH falls from about 11.5 to 2-3. It is often useful to make high solids silica sols from which it is then necessary to remove destabilizing anions, e.g. sulfate ions, from the decationized waterglass solution by passing it through a column with anion exchange resin, usually a strong base resin in the OH^--form.

The effluent from the cation exchanger contains the colloidal particles, about 2 nm in diameter, originally present in the waterglass solution, but the sodium ions, present on the particle surface as Na^+OH^- pairs, have been replaced by H^+ ions and the pH of the sol of extremely fine silica particles is typically in the range 2-3. If an anion exchanger is also used to remove destabilizing anions, the pH of the effluent is in the range 4-6.

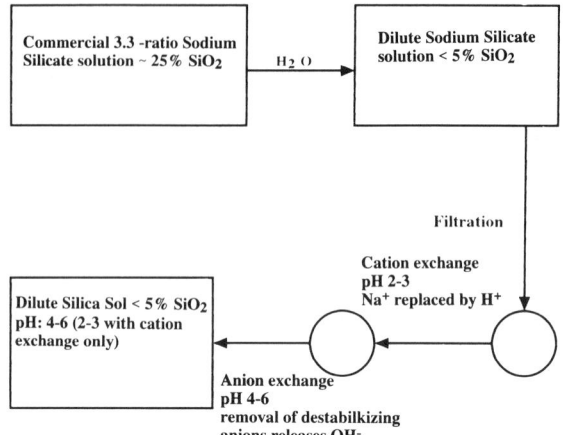

Fig. 2-27. Schematic diagram of manufacture of commercial silica sols.

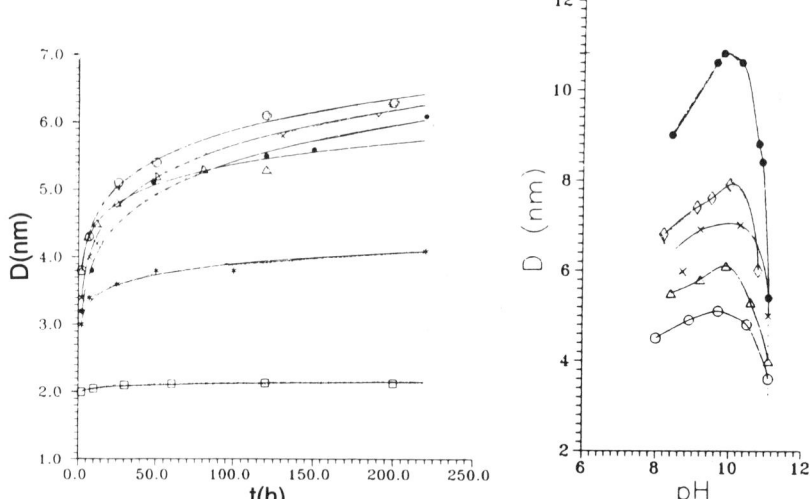

Fig. 2-28. Particle diameter D vs. time of growth (T=343 K) at various values of the mole ratio SiO₂/Na₂O: •, 283; x, 120; o, 65; Δ, 26; *, 9.0; □, 4.5. Liu et al (1993). Reprinted from Colloids and Surfaces **74**, 8 (1993) with kind permission from Elsevier Science - NL, Sara Burgerhartstraat 25, 1055 KV Amsterdam, The Netherlands.

Fig. 2-29. Particle diameter D vs. pH at various temperatures, T(K), and times, t(h); •, T=443, t=11.5; ◊, T=403, t=11.5; x, T=363, t=75; Δ, T=343, t=120; o, T=343, t=25. Liu et al (1993). Reprinted from Colloids and Surfaces **74**, 8 (1993) with kind permission from Elsevier Science - NL, Sara Burgerhartstraat 25, 1055 KV Amsterdam, The Netherlands.

Figure 2-28 shows that the growth rate at 70°C rapidly increases with ratios of about 26 and then levels off or even decreases at the highest ratios, Liu et al. (1993). Figure 2-29 indicates that at a given temperature and reaction time, the particle size increases with pH to a maximum at pH 10. Beyond pH 10 the particle size decreases because the particles will dissolve with formation of soluble silicate ions. A cross-plot of the curves in Fig. 2-28 with temperature in the pH range 8-10 would show that particle size increases rapidly with temperature in the pH range 8-10; compare Fig. 2-11.

2. Processes for Making Silica Sols

The development of commercial silica sols has been a process of gradual evolution. In § D-1 it was stated that 3.3 ratio sodium silicate solutions are the most common raw materials for making silica sols. Neutralizing such a silicate solution by mixing it into a dilute solution of an acid, e.g., HCl, one obtains acidic or basic silica sols depending on how much acid is used. The particle size will be quite small, although it is larger in basic than in acidic sols. Due to the small particle size, the sols gel when one tries to increase the silica concentration. Furthermore, it is observed that gelling is accelerated when the temperature is increased. It appeared logical at one time to make sols with low concentrations of silica at ordinary or low temperatures.

P. C. Bird (1941) therefore created something of a sensation when he showed that by removing sodium ions from a 3.3 ratio Na-silicate solution, but maintaining the pH on the alkaline side, he could raise the silica content by a factor of two by vacuum evaporation; from 2-3% in the deionized sodium silicate solution to 6-7% in the final product.

In the next two decades, DuPont and other companies made improvements in Bird's basic process so that today commercial silica sols having a wide range of particle sizes are available in concentrations up to 50% SiO_2.

The main types of processes or methods for making silica sols are:

1. *Column methods*   dilute sodium silicate solution is passed through a column bed of ion exchange resin.

2. *Concentrated sol or consol method*   concentrated sodium silicate solution and ion exchange resin are added to a heel of water.

3. *Autoclave method*   a small-particle silica sol is heated in an autoclave to give a large-particle sol.

4. *Aggregation method - Stöber sols* porous silica particles of uniform size are made by hydrolyzing silicon alkoxides.

In these processes control of pH and rate of addition of different components are very important.

2.1. Column Processes

Bird process Bird (1941) passed a solution of 3.3 ratio sodium silicate containing approximately 3% SiO_2 through a bed of acid regenerated cation exchange resin (see Fig.2-30). The total effluent from the resin column consists of a solution of silica containing about 3% SiO_2 and has an alkalinity which Bird does not specify in terms of pH, but as "60 to 75 grains per gallon calculated as $CaCO_3$." This effluent is concentrated by evaporation to abvout half its original volume, but he does not specify the temperature and rate of evaporation. A typical analysis of the concentrated effluent is about 6.5% SiO_2 and 0.13% Na_2O, which has a ratio of 50:1.

In contrast to later processes, Bird did not decationize the Na-silicate solution to pH 2-3, but operated all the time on the alkaline side, although he did not specify the pH change he was working in. The small, ~2-nm, particles originally present in the 3.3-ratio silicate solution, will, after decationization, rapidly grow at pH above 8 even at room temperature. Since it takes a certain time for the silicate solution to pass through the bed of ion exchange resin and form the total amount of effluent, some of the particles will be older than others and have time to grow larger than younger particles. The silica sol from the Bird process will therefore have a wider particle size distribution than products from later processes. Furthermore, the sodium content of the sol is somewhat undefined and difficult to control. It is true that the Bird process meant a breakthough in the tecnology for making commercial silica sols, but the products were characterized by a solids content that still was low, wide particle size distribution, and relatively high sodium content. The Bird process can also be used to make sols of vanadia, titania, zirconia, and oxides of the rare earth metals.

Bechtold - Snyder Process Bechtold and Snyder (1951) developed Bird's process so that the process could yield sols with particle sizes in the range 10-130 nm and solids contents of up to 30% SiO_2. In their process, as in Bird's process, a 3.3 ratio sodium silicate solution containing about 3% SiO_2 was passed through a bed of a strong-acid ion exchange resin in the hydrogen form to remove enough Na^+ ions to yield an $SiO_2:Na_2O$ ratio in the range 60:1 to 130:1. Thus far the steps are the same as in Bird's process. However, instead of directly evaporating water from the ion exchange bed effluent, as in Bird's process, part of the effluent, a heel, was heated at a temperature above 60°C, often 100°C, for about 1-2 hours. The small particles, about 2-3 nm, present in the effluent, then grow to a particle size in the range 6-9 nm. Then the rest of the alkaline effluent is gradually added to the hot heel at such a rate that all the added silica, in the form of about 2 to 3-nm particles, is deposited on the original particles in the heel by the process of Ostwald ripening. At the same time water is evaporated so that the number of particles per volume remains constant, but they grow in size as the sol becomes concentrated—see Fig. 2-31.

Alexander Process Originally, commercial silica sols were made to have a particle size of at least 7-8 nm so that costs could be reduced by concentrating the

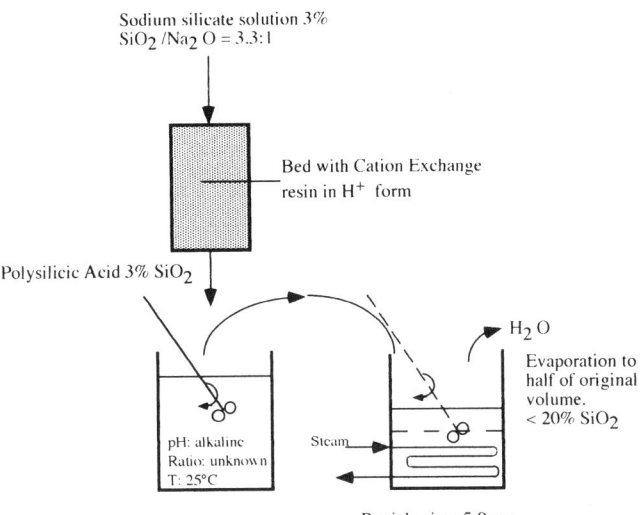

Fig. 2-30. The Bird Process, Bird (1941).

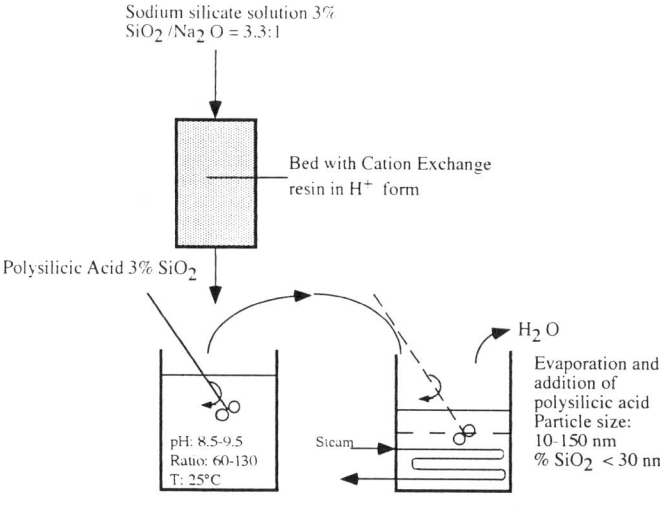

Fig. 2-31. The Bechtold-Snyder Process. Bechtold and Snyder (1951).

sols to at least 30% SiO_2. Such sols are very stable with regard to further particle growth at ordinary temperatures—see Fig. 2-11. When it was realized that in some applications, e.g. anti-soiling agents and retention aids in paper making, smaller-particle sols would be advantageous, such sols were developed. Sols of particles smaller than 8 nm, and certainly much smaller than 5 nm, will spontaneously grow in storage with corresponding change in properties. If concentrated to, say 15% SiO_2, sols of a particle size of about 5 nm may aggregate to some extent, leading to further changes in properties.

Alexander (1956) showed how to make silica sols with a particle size in the range 5-8 nm and in concentrations of at least 15% SiO_2. Figure 2-32 shows that, as in the Bird and Bechtold-Snyder processes, he decationized a 3.3 ratio sodium silicate solution containing 3% SiO_2 by passing it through a bed of a strong-acid resin in the hydrogen form, but unlike the other two processes, decationization proceeded down to pH 2-3. The effluent is alkalized to a ratio of about 100 and is kept at temperature and time conditions that will give the desired particle size. In order to determine these conditions, one must study what combinations of temperature and time will give particle sizes in the range 5-8 nm.

Once the desired particle size has been achieved, the sol is deionized in a bed consisting of a mixture of anion and cation exchange resin to remove destabilizing anions and cations, to which small particles are particularly sensitive. Then the ratio is restored to its initial value of about 100 by adding alkali and the sol is evaporated to a concentration which, depending on the particle size, is in the range 15-30% SiO_2 under conditions that do not bring about further particle growth.

Large particles by the buildup process In the buildup process the particle size is increased by deposition of soluble silica, e.g., in the form of decationized 3.3 ratio sodium silicate, also called polysilicic acid, on the particles of a heel under carefully controlled conditions of pH, temperature, and addition rate.

Bechtold and Snyder's process is a buildup process, but it is difficult to make sols of high concentrations because anions such as sulfate and chloride ions, which are already present in sodium silicate solutions, destabilize the sol and will cause aggregation or gelling. Rule (1951) modified the Bechtold-Snyder process by introducing a bed of anion exchange resin in the hydroxyl form after the bed of cation exchange resin so as to remove destabilizing anions—see Fig. 2-33. Part of the salt-free acid effluent from the anion exchange bed is used to make a heel by adding enough sodium hydroxide to give a ratio of about 150 and reflux the bed for about 30 minutes. Salt-free acid effluent is added to this heel, containing 7-8 nm particles, at a carefully controlled addition rate together with sodium hydroxide solution to maintain the ratio at 150.

Albrecht (1969) patented the optimal rate of addition of polysilic acid in the Rule process for making silica particles 45-100 nm in diameter and up to 60% SiO_2 in concentration.

Iler (1979) defined the "buildup ratio", B_r, as

$$B_r = W_a/W_n \qquad (2\text{-}12)$$

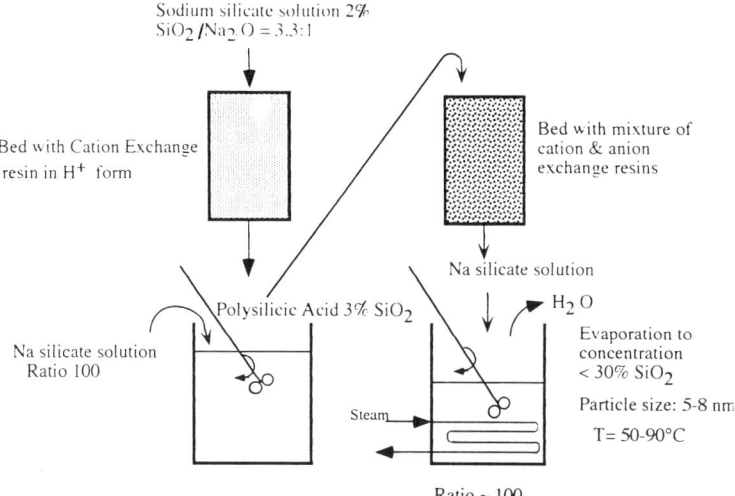

Fig. 2-32. The Alexander Process. Alexander (1956).

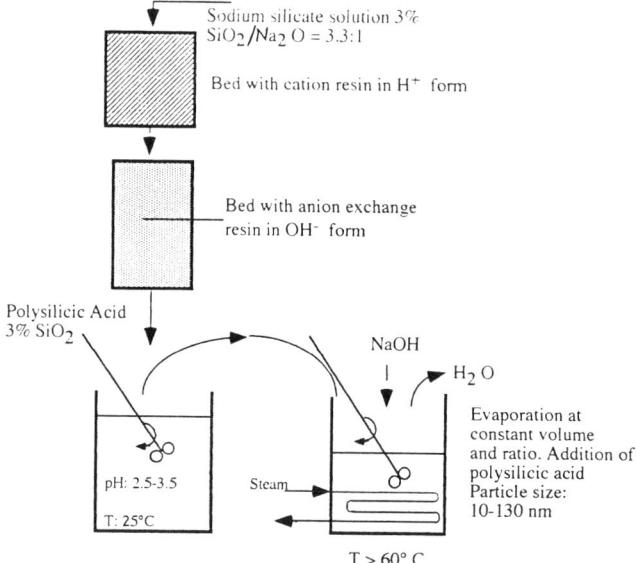

Fig. 2-33. The Rule Process. Rule (1951).

where W_a is the amount of active silica added to the particles and W_n is the weight of silica initially present as nuclei in the heel. Obviously, the particle diameter will be increased from the initial size d_i to the final size d_f in accordance with the following equation:

$$(d_f/d_i)^3 = (W_n + W_a)/W_n = 1 + B_r \tag{2-13}$$

which means that, in order to double the particle diameter, seven times the amount of silica initially present must be added to the heel.

Albrecht (1969) derived the following equation for the maximal feed rate, F_t, at time t of the polysilicic acid in grams of silica per milliliter of alkaline sol per hour:

$$F_t = k(C_t)^{0.67} (C_0)^{0.33} S_0 \tag{2-14}$$

k is a constant with a value of 0.005 at the boiling point of the aqueous sol and pH of the starting sol from 9.5 to 11.5, C_t is the silica concentration in grams of silica per ml of the alkaline sol at any time, C_0 is the initial concentration of the alkaline sol, and S_0 is the initial specific surface area of the alkaline sol in square meters per gram of SiO_2. The initial feed rate is given by:

$$F_0 = kC_0S_0 \tag{2-15}$$

Otterstedt, Neald and Sterte (1993) studied the buildup of silica particles by the Albrecht process at T= 95 -100°C and pH = 9.5-10.0 for k-values in the range 0.001-0.01. In a typical run the buildup reactor, Fig. 2-34, was first charged with 4 l of the starting sol, the heel, which was Ludox®TM, a 22-nm sol from DuPont, in all the experiments. The amount of silica in the heel was chosen to allow an initial feed ratio of at least 3.5 ml/min. If necessary, the pH of the heel was adjusted to 9.5-10.0 by addition of 0.10 M NaOH. The feed solution was an active silica solution, which was obtained by decationizing a 3.3 ratio sodium silicate solution containing 6% SiO_2. It could be stored without gelling for 4-5 days in a refrigerator. After addition of the amount of active silica corresponding to the desired final particle size according to Eqn. 2-13, the product sol was cooled, taken out of the reactor, and characterized by measuring particle size by light scattering, Sears titration (see Ch. 1) and BET surface area.

Assuming ideal particle growth, i.e. all acid active silica feed will accrete on particles already present, Albrecht (1969) derived the following equation for the particle diameter, d_i, at time t during the buildup process:

$$(d_t/d_0) = (C_t/C_0)^{0.33} \tag{2-16}$$

where d_0 is the particle diameter at t = 0 and C_t and C_0 are as before. The logarithmic form of this equation is more useful in data analysis:

$$\log d_t = 1/3 \log (C_t/C_0) + \log d_0 \tag{2-17}$$

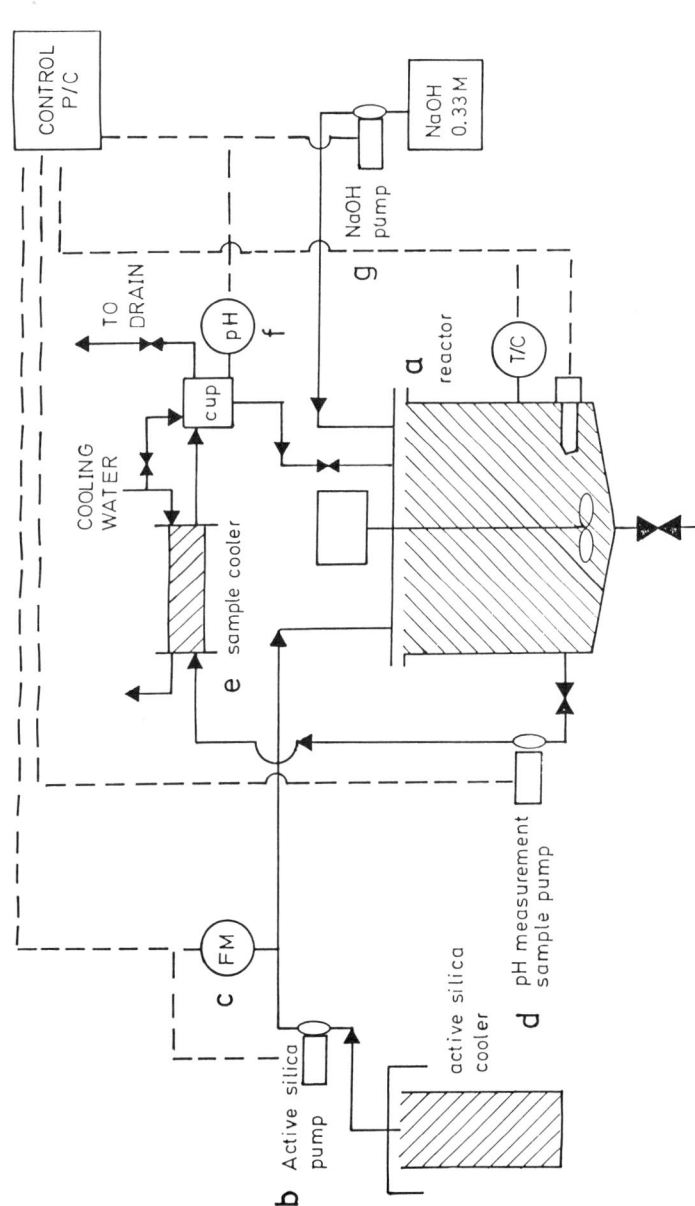

Fig. 2-34. Reactor set-up for the buildup of silica particles by the Albrecht process.

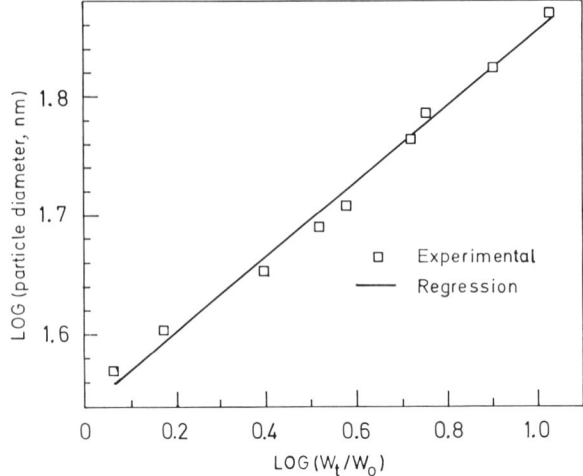

Fig. 2-35. Log particle diameter as a function of log (W_t/W_o). W_t is the amount of active silica added to the reactor. Otterstedt et al. (1993).

Figure 2-35 is a logarithmic growth curve for k=0.002, but is also typical of the other k-values studied. Table 2-8 shows the parameters when regressed in the form of Eqn. 2-17.

Below a k of 0.003, they indicate a rate of increase of particle size which is in concurrence with Eqn. 2-14. Albrecht (1969) states that the maximum rate at which particles can be built up without nucleating new particles occurs at a k-value of 0.005. At k-values of 0.007 and above, the particle growth rate is significantly less than the ideal value of 1/3 in Eqn. 2-17, which indicates that all of the silica added to the sol is not being deposited on the heel particles, but some of it is being used to form new particles. Figure 2-36, which shows photomicrographs from reflux boiled samples of two sols manufactured at k-values of 0.002 and 0.007, respectively, bear out these conclusions. These shows that secondary particles of a significant size are present at k=0.007, whereas this is not the case at k=0.002. The maximum buildup rate at which a monodisperse sol can be produced thus lies between k-0.004 and 0.007, which is in agreement with the results of Albrecht.

Table 2-8. Regressed data from buildup experiment. Otterstedt et al.(1993)			
k	Line slope	antilog (y-intercept)	r^2
0.001	0.30	35	0.989
0.002	0.32	35	0.989
0.004	0.34	33	0.996
0.0066	0.28	38	0.994
0.008	0.26	36	0.996
0.01	0.26	36	0.997

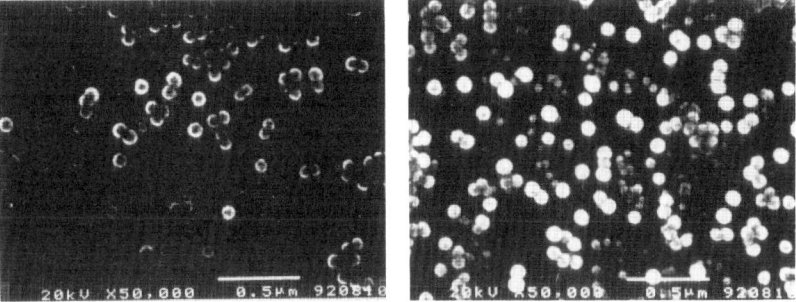

Fig. 2-36. Particle buildup with k=0.002 (left) and k=0.007 (right). Otterstedt et al. (1993).

2.2. Concentrated Sol "Consol" Process

Iler and Wolter (1953) developed the concentrated sol-consol process, by which concentrated (up to 12% SiO_2) sols can be made directly by ion exchange, compared with about 3% SiO_2 for column processes. Relatively concentrated solutions of 3.3 ratio sodium silicate, 10-20% SiO_2, and cation exchange resin in the hydrogen form, are added simultaneously to a heel of water or dilute sodium silicate, which is kept at a temperature in the range of 60-100°C, at a rate which maintains pH between 8.5 and 10.5. Alternatively, the exchange resin may be suspended in water and the sodium silicate solution may then be added at such a rate as to maintain pH within the range above. This is particularly effective where the cation exchanger withdraws sodium ions sufficiently slowly that pH control is no great problem, which is the case with the weakly acidic cation exchangers such as the carboxylic exchange resins. If sodium silicate solution is used in the heel it should be dilute enough that the sodium ion concentration does not exceed 0.35M after the $SiO_2:Na_2O$ ratio reaches about 10:1. In order to prevent gelling or aggregation, it is important that the sodium ion concentration be kept at less than 0.35 M while the pH is maintained between 8.5 and 10.5 by adding cation exchanger. As in column processes, the particle size can be controlled by temperature. Example 2 in Iler and Wolter's patent (1953) shows that a 7-nm sol containing 12% SiO_2 can be made by carrying out the process at 80°C. At 60°C and with careful control of pH and rates of addition of resin and silicate solutions 4-nm particles are obtained; see example 8 in that patent.

In the Consol process, when 3.3 ratio sodium silicate solution and cation exchanger in the hydrogen form are added simultaneously to a heel of water, Na^+ ions are removed from those silicate species—primarily about 2-nm particles—which make up 3.3 ratio sodium silicate solutions. At the pH and temperature of the process, monomeric silicic acid, which in part is present initially in the sodium silicate solution, but primarily is formed by dissolution of the smaller polymers, is deposited on the large particles; that is particle growth occurs by

Ostwald ripening. If the addition of sodium silicate solution and cation exchanger is fairly rapid, in about 30 minutes as in patent example 2, 4 to10-nm particles will form, with the size depending on the temperature and to a lesser extent on the pH—see Fig. 2-11. If, on the other hand, the heel contains silica particles, the Consol process can be used to build them up to about 150 nm. Of course, the feed rate of 3.3 ratio sodium silicate solution must not exceed the maximum value specified by Albrecht in Eqn. 2-14. Unlike the buildup processes described earlier, which use active silica prepared in a separate operation, the Consol process produces active silica *in situ* by neutralizing the sodium silicate solution with cation exchanger in the heel.

The Consol process has the advantage over column processes to directly produce stable sols with a solids content of at least 10% SiO_2. Furthermore, providing the sodium concentration is kept less than 0.35 M in the process, no microgel will form since the sol is never exposed to pH lower than 8.5, where gelling may occur. Conversely, the Consol process yields sols containing more sodium than do modern column processes since the sols have grown in a sodium-rich environment with the result that sodium may be occluded in the particles and be difficult to remove.

2.3   Autoclave processes

The rate of particle growth by the process of Ostwald ripening becomes very slow when the difference in solubility between the smallest and the largest particles, i.e., the driving force for particle growth, becomes small. Otterstedt and Brandreth (1995) carried out a qualitative study to show how this driving force of solubility difference depends on temperature for particles differing in size by ±10 to ±20%.

Equation 2-2, adapted to the system of amorphous silica, expresses the dependence of the solubility of a particle on its diameter d:

$$\log S_d/S_o = 5.7 \ E/(Td) \tag{2-18}$$

Equation 2-1 shows how the solubility of a flat surface of silica, S_o, varies with temperature

$$\log S_o = -731/T + 4.52$$

Combining these two equations yields

$$S_d = 10 \ \exp(-731/T + 4.52) \bullet 10\exp(5.7E/(Td)$$

$$\text{or, } S_d = 10\exp((1/T)(5.7E/d - 731) + 4.52)$$

The difference in solubility, ΔS, between particles of diameters d_1 and d_2, where $d_1 < d_2$ is:

$\Delta S = S_{d1} - S_{d2} = 10\exp((1/T)(5.7E/d_1 - 731) + 4.52) - 10\exp((1/T)(5.7E/d_2 - 731) + 4.52)$

Differentiation with respect to T yields:

$d(DS)/dT = (\ln 10)/T^2\{(5.7E/d_2 - 731)\ 10\ \exp((5.7E/d_2 - 731)/T + 4.52) - (5.7E/d_1 - 731)\ 10\ \exp(5.7E/d_1 - 731)/T + 4.52\}$

Changing 10-based exponentials to e-based ones, and using $e^x \sim 1+x$ for $x < 1$, the expression in brackets can be simplified to :

$$5.7E\ (d_2 - d_1)/(d_1 d_2)\ [2.303\ (731 - 5.7E/d_1)/T - 1] = \sigma$$

For $E < 100$ and $d_1 > 1.5$, $\sigma > 0$ for $T < 600°K$. Thus $d(\Delta S)/dT > 0$ for temperatures up to at least 300°C. The rate of particle growth in a system of silica particles, the sizes of which are distributed about a mean value by Ostwald ripening, will thence be promoted and accelerated by using higher temperatures.

When heating an alkali-stabilized sol of amorphous silica particles to over 300°C, quartz will form unless the sodium concentration is below 1% Na_2O. Broge and Iler (1954) and Rule (1961) described methods for making large-particle silica by autoclaving procedures. Rule found that if a sol containing particles between 8 and 15 nm in diameter is first thoroughly deionized by cation exchange and then is heated at 300-350°C under autogenous pressure, the sodium concentration is insufficient for the formation of quartz, but enough to raise the pH and stabilize the sol by the release of occluded sodium ions during the growth process. In this manner stable sols have been made with particle sizes up to 150 nm with solids content of over 60% SiO_2. Broge and Iler pointed out that if the original sol consists of discrete particles, they will grow as discrete spheres, but if the sol contains aggregates, the final sol will contain particles of irregular shape.

Table 2-9, which combines data from Alexander and McWhorter (1958) and Broge and Iler (1954), shows particle growth in deionized solution of 3.3 ratio sodium silicate solutions containing 4% SiO_2 under different conditions of time and temperature.

It is worthwhile to note that in processes using autoclaving to achieve particle growth, the initial distribution of particle sizes about a mean value will have marked effect on the final particle size. Silica sols made by autoclaving are characterized, providing necessary precautions have been taken to avoid particle aggregation, by very uniform particle size.

2.4. Aggregation Methods - Stöber Sols

The preparation, properties, and structure of silica sols formed by hydrolysis of tetraalkyl orthosilicates are described   in subsection 2 of Section B of this chapter.

Table 2-9. Growth of silica particles by heating a 4% sol of silicic acid at pH 8-10. Iler (1979). Reprinted by permission of John Wiley & Sons, Inc.

Mole Ratio SiO_2/Na_2O	Time	Temperature °C	Specific Surface Area (m^2/g)	Estimated Particle Diameter $(m\mu)$
100	1 h	80	600	5
64	6 h	85	510	6
100	5 h	95	420	7
78	6h	98	406	7
80	30 min	100	350	8
85	3 h	160	200	15
85	3.25 min	270	200	15
85	0.9 min	250	225	15
90	3.1 min	200	271	10
85	10 min	200	228	12
85	10 min	295	78	36
85	30 min	295	--	64
very high	3 h	340	--	88
very high	6 h	340	--	105
very high	3 h	350	20	150

3. Commercial Silica Sols and Their Uses

Iler (1979) lists the following uses of colloidal silica, some of which will be discussed in more detail in Chapter 7 and 10.

1. Catalyst bases and adsorbents. Here the colloidal silica provides a good inorganic binder which is inert to most chemical systems, is stable at normal reaction temperatures, and has a high specific surface area to enhance the activity of the support catalyst.
2. Stiffening and binding agents for fibrous and granular materials as in precision casting molds, molded refractory products, and high-temperature insulating materials.
3. Increasing the friction of surfaces as in textile fibers, railway tracks, and waxed floors. An example is as an antislip agent to avoid seam slippage during fabric cutting and sewing. Another is for paper frictionizing applications.
4. Anti-sticking, anti-blocking, and anti-static effects on polymer films.
5. Anti-soiling agents which provide an ultrasmooth, oleophobic surface on porous materials by filling micropores to exclude dirt particles, as in textiles, paper, and painted surfaces.
6. Agents which provide hydrophilic, oleophobic surfaces by virtue of the highly polar SiOH groups of the silica surface, e.g., lithographic plates. Another example is to improve the wetting of plastic pieces in dishwashers, etc.
7. Agents which reduce or increase adhesion between surfaces, depending on the nature of the substrate and the method of application, e.g. surfaces of polymer films, glass, and metals. An example is to increase the adhesion of emulsion paints and coatings on paper, wood, and metal foil.

8. As a component of refractory electrically insulating films on conductive surfaces, as in laminations in transformer cores; or in conducting films on insulating materials, e.g. graphite coatings on paper.

9. Agents for cross-linking, stiffening, and reinforcing polymers, e.g. leather, latex-foam products, elastomers. For instance, to increase the compression resistance of foams.

10. Polishing agents for silicon wafers.

11. Agents for modifying surfactant properties: flocculating, coagulating, dispersing, stabilizing, emulsifying, suspensions, anti-foam properties. Activated silica sol offers efficient coagulation at low temperatures and can also act as a filter aid. Many industries achieve a clear effluent by using activated silica to gather finely divided impurities into a fast-forming floc. The floc is then separated from the water by sedimentation and/or filtration.

12. Thickening agents (increasing viscosity) for a wide variety of liquid or paste-like products.

13. Adsorbed coatings on surfaces; optical effects such as anti-sheen applications on textiles (delustering).

14. Components of multilayer film systems.

15. Use in biological research; culture media; centrifuging medium.

16. Source of chemically reactive silica. Used to provide a pure silica reactant to increase the yield of desired crystal forms.

A relatively new use of small particle silica sols is as a retention aid in papermaking; see Ch.9. Silica sols, often in combination with gelation, are used for clarifying wine and beer—see Ch.10.

Silica sols are principally manufactured by the following companies: Bayer, Akzo Nobel (EKA Nobel), E. I. DuPont, Nalco, Nissan, among others.

Tables 2-10 and 2-11 illustrate the properties of the types of silica sols available from AKZO Nobel and DuPont, respectively.

Table 2-10. Typical Properties of representative Nyacol® Silica from AKZO NOBEL

	\multicolumn{8}{c}{Grades}							
	215	830	1440	2050	5050	9950	2040 NH_4	2034D1
Stabilizing Counter Ion	Na	Na	Na	Na	Na	Na	NH_4	chloride
Particle charge	-	-	-	-	-	-	-	+
Avg. Particle diameter nm	4	10	14	20	50	100	20	20
Specific Surface Area m^2/g	545	273	195	136	55	27	136	136
Silica (as SiO_2) wt. %	15	30	40	50	50	50	40	34
pH (25°C)	11	10.5	10.4	10.0	9.3	9.0	9.0	3
Titratable alkali (as Na_2O) wt. %	0.80	0.55	0.50	0.50	0.15	0.12	a	b

Courtesy AKZO Nobel

a   Alkalinity as % NH_4 is 0.2%

b   Nyacol® 2034D1 is an acid stable colloidal silica

Table 2-11. Typical Properties of representative Ludox® Colloidal Silicas

	\multicolumn{6}{c}{Conventional Grades}						\multicolumn{4}{c}{New Grades}			
	HS-40	TM	SM	AMa	AS	LS	SK	TMA	CL	CL-P
Stabilizing Counter Ion	Na	Na	Na	Na	NH_4	Na	c	c	Cl-	Cl-
Particle Charge	-	-	-	-	-	-	-	-	+	+
Avg. Particle Diameter nm	12	22	7	12	22	12	12	22	12	22
Specific Surface Area m^2/g	220	140	345	220	135	215	230	140	230	140
Silica (as SiO_2) wt %	40	50	30	30	40	30	25	34	30	40
pH (25°C)	9.7	9.0	10.0	8.9	9.1	8.2	4-7	4-7	4.5	3.5-4.5
Titratable alkali (as Na_2O) wt %	0.41	0.21	0.56	0.24	b	0.10	b	b	-	-

Courtesy E. I DuPont de Nemours Company

a   Surface modified with aluminate ions

b   Sol contains less than 0.1% Na_2O (occluded inside the particles)

c   Deionized sol

References

Albrecht, W. L., U. S. Patent 3,440, 174 (1979).

Alexander, G. B., U. S. Patent 2,750,345 (1956).

Alexander, G. B., and McWhorter, J. R., U.S. Patent 2,833,724 (1958).

Anderson, K. R., Dent-Glasser, L. S., and Smith, D. H., *Soluble Silicates*, ed. Falcone, J. S., American Chemical Society Symposium Series 194, Washington, D. C., 115, 1982.

Baes, C. F., and Mesmer, R.E., *The Hydrolysis of Cations*, John Wiley & Sons, NY (1976).

Baumann, H., Beitr. Silkose-Forsch., **37**, 47 (1955).

Bailey, J. K., and Mecartney, M. L., Colloids and Surfaces. **63**, 151 (1992).

Bechtold, M. F., and Snyder, O. E., U. S. Patent 2,574,902 (1951).

Bird, P. C., U.S. Patent 2,244,325 (1941).

Brinker, C. J., and Scherer, G. W., *Sol-Gel Science.*, Academic Press, NY (1990).

Broge, E. C., and Iler, R. K., U. S. Patent 2,680,721 (1954).

Coenen, S., and De Kruif, C. G., J. Colloid Interfac. Sci., **124**, 104 (1988).

Cogan, H. D., and Setterstrom, C. A., Chem. and Eng. News, **24**, 2499 (1947).

Fegley, J. B, and Barringer, E. A., Mater. Res. Soc. Symp. Proc., **32**, 187, eds. C. J. Brinker, D. E. Clark, and D. R. Ulrich.. North Holland Publishing Co., NY (1984).

Fournier, R. O., and Rowe, J. J., Amer. Mineral. **62**, 1052 (1977).

Goto, K., J. Chem. Soc. Japan, Pure Chem. Sect., **76**, 1364 (1955).

Heston Jr., W. M., Iler, R. K., and Sears Jr., G. W., J. Phys. Chem. **64**, 147 (1960).

Hoebbel, D., Wieker, W., and Stade, H., Z. Allg. Chem. **366**, 139 (1974). **400**, 148 (1973). **405**, 163 (1974). **428**, 43 (1977). J. Chromatogr. **119**, 173 (1976).

Hsu, W. P., Yu, R., and Matijevice, E., J. Colloid Interfac. Sci., **156**, 56-65 (1993).

Iler, R. K., J. Colloid Interfac. Sci., **75**, 138 (1980).

Iler, R. K., Am. Chemical Soc. Symposium Series 194, Washington, D. C., (1982).

Iler, R. K., private communication (1984).

Iler, R. K., private communication.(1983).

Iler, R. K., private communication.(1983a).

Iler, R. K., *Biochemistry of Silicon and Related Problems*, eds. G. Bendz and I. Lindquist, Plenum Publishing Corp., NY (1977).

Iler, R. K., *The Chemistry of Silica*, John Wiley & Sons, NY (1979).

Iler, R.K., *Soluble Silicates*, ed. J. S. Falcone, American Chemical Society Symposium Series #194, 95 (1982).

Iler, R. K., and Wolter, F. J., U.S. Patent 2,631,134 (1953).

Kirk-Othmer. *Encyclopedia of Chemical Technology*, Vol. 20, 868, John Wiley and Sons, NY (1982).

La Mer, V.K., and Dinegar, R. H., J. Amer. Chem. Soc., **72**, 4847 (1950).

Lawrence, M., and Vivian, H. E., Aust. J. Appl. Sci. **12**, 96, (1961).

Le Cloux, A. J., Bronckart, J., Noville, F., Dodet, C., Marchot, P., and Pirard, J. P., Colloids and Surfaces, **19**, 359 (1986).

Liu, H. C., Wang, J. X., Mao, Y., and Chen, R. S., Colloids and Surfaces A: Physicochemical and Engineering Aspects, **74**, 8 (1993).

Livage, J., Henry, M., and C. Sanchez. Sol-Gel Chemistry of Transition Metal Oxides, *Progress in Solid State Chemistry*, Plenum Publishing Corp. (1989).

Mattsson , and Otterstedt, J-E. A., unpublished work (1993).

McCormick, A. V., Bell, A. T., and Radke, C. J., Zeolites, **7**, 183 (1987).

McGarry, M. A., and Hazel, J. F., J. Colloid Sci., **20**, 72 (1965).

Otterstedt, J-E. A., unpublished results (1993).

Otterstedt, J-E. A., Ghuzel, M., and Sterte, J. P., J. Colloid Interfac. Sci., **115**, 95 (1987).

Otterstedt, J-E., A., Neald, S., and Sterte, J. P., unpublished work.

Otterstedt, J-E. A., and Brandreth, D. A., unpublished results (1995).

Ray, N. H., and Plaisted, J., J. Chem. Soc. Delfon Trans., 475 (1983).

Rule, J. M., U.S. Patent 2,577,484 (1951).

Rule, J. M., U.S. Patent 3,012,972 (1961).

Scherban, J. D., Dokl. Akad. Nauk SSSR, **177**, 1200 (1967).

Stöber, W., Fink, A., and Bohn E., J. Colloid Interfac. Sci., **26**, 62 (1968).

van Helden, A. K. and Vrij, A., J. Colloid Interfac. Sci., **76**, 418 (1980).

Yates, D. E., Levine, S., and Healy, T.W., J. Chem Soc. Faraday Trans., **70**, 1807 (1974).

CHAPTER 3. SMALL PARTICLES OF HYDROUS METAL OXIDES

Introduction

A. HYDROLYSIS

1. Fundamental Concepts

2. Hydrolysis of Metal Cations

3. Mononuclear Complexes

4. Polynuclear Complexes

5. Different kinds of Complexes

6. Formation of Small Particles

7. Effect of the Anion

B. HYDROLYSIS OF SOME METAL SALTS

1. Hydrolysis of Al^{3+}
   1.1. Aluminum Species in Aqueous Solutions
   1.2. Colloidal Boehmite
      1.2.1. Basic Aluminum Salt Solutions
      1.2.2. Conversion of Basic Aluminum Salts to Colloidal Aluminas
      1.2.3. Effect of Variations in Aluminum and Chloride Ion Concentrations on Morphology of Colloidal Particles
      1.2.4. Effect of Anion Type on Morphology of Colloidal Alumina Particles
      1.2.5. Effect of Time-temperature on Morphology of Colloidal Alumina Particles
      1.2.6. Constitution of Alumina Sols and Mechanism of Particle Growth
      1.2.7. Importance of "Unpolymerized Alumina"
      1.2.8. Alkaline Stable Alumina Sols
      1.2.9. Some Properties of Fibrous Boehmite Sols

2. Hydrolysis of Fe^{3+}
   2.1. Polymerization and Formation of Colloidal Particles

3. Hydrolysis of Ti^{4+}
   3.1. Polymerization and Formation of Colloidal Particles

Introduction

As noted in the main introduction, this book focuses generally on a smaller range of particle sizes than older literature on this subject. In the past there have been limitations on producing very fine particles by the conventional method of attrition due to the large amounts of time and mechanical energy required and the often important concomitant problem of contamination by the grinding medium.

Thus newer methods involving controlled particle size by means of production in situ via chemical reactions assume a considerable importance because it is often advantageous to use ultra-fine particles to attain particular objectives. In this chapter we lay the fundamental background for making controlled particle sizes in solutions. In some cases the product dispersed in that liquid medium may be used in that same medium or admixed with other components to directly make the desired final product without the added (often difficult) separation of the fine particles from the liquid medium and possible redispersion in another medium.

A. HYDROLYSIS

As shown in Fig. 1-5, small particles can be formed by the hydrolysis of metal salts or metal-organic compounds. However, before covering those two routes to making small particles, we will discuss the common and important, yet somewhat obscure concept of hydrolysis.

The term hydrolysis is applied to reactions in which a substance is split or decomposed by water, or to a reaction involving the splitting of water into its ions and the formation of a weak acid or base or both. In inorganic chemistry, hydrolysis is usually applied to solutions of salts and the reactions by which they are converted to new ionic species or to precipitates—oxides, hydroxides, or basic salts. The hydrolysis of salts can involve either the cation, the anion, or both, Baes and Mesmer (1976).

1. Fundamental Concepts

In aqueous solutions metal cations (and anions) are hydrolyzed:

$$M^{z+} + n\,H_2O \longrightarrow M(H_2O)_n^{z+} \tag{3-1}$$

Due to hydrolytic reactions, aqueous solutions of hydrated metal cations are often acidic:

$$M(H_2O)_n^{z+} + H_2O \longrightarrow M(H_2O)_{n-1}(OH)^{(z-1)+} + H_3O^+ \tag{3-2}$$

In this context hydrolysis thus means that the metal cations act as acids by donating protons from coordinated water molecules in the solvation shell around the cations. Metal cations of higher valency are more readily hydrolyzed, i.e., the hydrolysis equilibrium in Eqn. 3-2. is shifted more to the right, than cations of lower valency. The hydrolysis may proceed in several steps and not necessarily stop after one step as in Eqn. 3-2. In many cases polynuclear complexes are formed during the course of hydrolysis and such complexes are precursors of small particles, which may eventually form as a result of various condensation reactions between the precursors.

A general hydrolysis reaction can be written as follows:

$$xM^{z+} + yH_2O = M_x(OH)_y^{(xz-y)+} + yH^+ \tag{3-3}$$

with the formation quotient

$$Q_{xy} = \frac{[M_x(OH)_y^{(xz-y)}][H^+]^y}{[M^{z+}]^x} \tag{3-4}$$

and the formation constant

$$K_{xy} = Q_{xy} \frac{g_{xy}(g_{H^+})^y}{(g_{M^{z+}})^x (a_{H_2O})^y} \tag{3-5}$$

where K_{xy} equals the equilibrium constant Q_{xy} multiplied by the activity coefficients of the species in Eqn. 3-3.

In many investigations of hydrolysis reactions and products the concentrations of metal ions have been so low that the deviation of the activity coefficient from unity can be neglected with the result that $Q_{xy} = K_{xy}$.

2. Hydrolysis of Metal Cations

Hydrolysis of metal cations is commonplace and is an important step towards the formation of small particles. The basis of these hydrolysis reactions is that most metal ions tend to form strong bonds with oxygen and hydroxyl ion ligands, which are always present in water, albeit in concentrations that depend on pH.

Identification of the products from a hydrolysis reaction, e.g., as represented in general terms by Eqn. 3-3, has proven to be a difficult task for several reasons. Not only can polynuclear complexes form, i.e., complexes containing two or more metal cations, but the same number of metal cations can make up several different complexes. A further complication is that the hydrolysis products usually form in a narrow pH range. When this range is exceeded, the hydrolysis products precipitate as metal oxides, hydroxides, or hydrous oxides.

In a dilute aqueous system with low total concentration of metal ions, hydrolysis product can exist over a wider pH range. Such systems are of special interest in geochemistry and mineralogy since hydrolysis products can adsorb on the surface of mineral or soil particles, or other naturally occurring colloidal particles and cause them to coagulate. The hydrolysis of Fe^{+3} is of particular interest in this context and will be described in detail, as will the hydrolysis of Al^{3+}, Ti^{4+}, Zr^{4+}, Cr^{3+}, Y^{3+}, Ce^{4+}, Zn^{2+}, and Ga^{3+} below.

3. Mononulear Complexes

At the beginning of this century scientists realized that metal cations can undergo hydrolysis and form larger entities, so-called polycations, Bjerrum (1907). Werner (1907) and Pfeiffer (1907) suggested the following mechanism, in which metal aquaions release protons stepwise, for formation of hydrous metal oxides.

$$M(H_2O)_6^{3+} \longrightarrow M(H_2O)_5(OH)^{2+} \longrightarrow M(H_2O)_4(OH)_2^+ + \longrightarrow \dots \tag{3-6}$$

It is true that this mechanism for explaining the hydrolysis of metal cations is too oversimplified, but it describes quite well how the first mononuclear complexes are formed. The key to gaining better understanding of the mechanism of hydrolysis is accurate pH measurement. After the glass electrode was developed around 1950, it became possible to calculate the formation quotient Q_{11} for the first hydrolysis step for a given metal cation from accurate pH measurements of solutions of different metal cation concentrations, to which different amounts of base had been added.

$$Q_{11} = \frac{[M(OH)^{z-1}][H^+]}{[M^{z+}]} \tag{3-7}$$

The degree of hydrolysis of metal cations depends on the total metal ion concentration. Below a certain concentration, which depends on the type of cation, only mononuclear complexes will form, whereas at higher concentrations bi- and tri-nuclear complexes often form. Most metal cations form polynuclear complexes with up to four metal ions, but in certain cases still larger complexes may form, e.g., $Bi_6(OH)_{12}^{6+}$, $Th_6(OH)_{15}^{9+}$, and $Al_{13}(OH)_{32}^{7+}$.

Information about the products formed in a hydrolysis reaction is given by the molar ratio, $\bar{n} = OH/M$, the hydroxyl ligand ratio, which represents the average number of hydroxyl ions coordinated to each metal ion in a given system:

$$\bar{n} = \frac{\sum y[M_x(OH)_y^{(xz-y)+}]}{m_M} \tag{3-8}$$

where

$$m_M = \sum x[M_x(OH)_y^{(xz-y)+}] \tag{3-9}$$

For instance, if in an aqueous solution of Fe^{3+} ions, OH/Fe = 2.5, the system is close to precipitating $Fe(OH)_3$ and the solution contains polynuclear complexes and polycations. Using Eqn. 3-4, $\bar{n}$ can be expressed as:

$$\bar{n} = \frac{\sum y Q_{xy}[M^{z+}]^x / [H^+]^y}{m_M} \tag{3-10}$$

$$m_M = \sum x Q_{xy}[M^{z+}]^x / [H^+]^y \qquad (3\text{-}11)$$

For systems containing only mononuclear complexes Eqn. 3-10 takes the form

$$\bar{n} = \frac{\sum y Q_{ly}[M^{z+}]/[H^+]^y}{m_M} \qquad (3\text{-}12)$$

where

$$m_M = \sum Q_{ly}[M^{z+}]/[H^+]^y \qquad (3\text{-}13)$$

which can be simplified to

$$\bar{n} = \frac{\sum y Q_{ly}/[H^+]^y}{\sum Q_{ly}/[H^+]^y} \qquad (3\text{-}14)$$

Equation 3-14 shows that when only mononuclear complexes are present, $\bar{n}$ does not depend on the total concentration of metal ions, but only on H^+, i.e., pH. In the rare cases where only mononuclear complexes are formed, seldom more than two such complexes exist simultaneously. If only one type of mononuclear complex is present, Eqn. 3-14 is reduced to

$$\frac{Q_{ly}}{[H^+]^y} = \frac{\bar{n}}{y - \bar{n}} \qquad (3\text{-}15)$$

Q_{ly} can be calculated from the pH where $\bar{n} = y/2$

$$-\log Q_{ly} = y \cdot (pH)_{\bar{n}=y/2} \qquad (3\text{-}16)$$

Figure 3-1 shows how $\bar{n}$ varies with pH for the complex MOH and for the mixture of the two complexes MOH and $M(OH)_2$.

4. Polynuclear Complexes

Hydrolysis of metal cations usually results in formation of polynuclear complexes containing 2-4 metal atoms. However, as mentioned above,

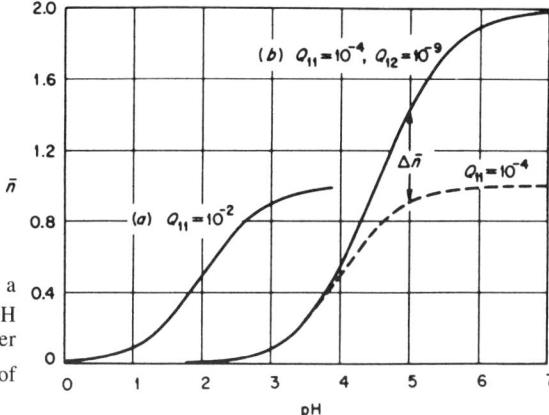

Fig. 3-1. Hydroxyl:ligand ratio as a function of pH. (a) MOH; (b) MOH and $M(OH)_2$. Baes & Mesmer (1976). Reprinted by permission of John Wiley and Sons.

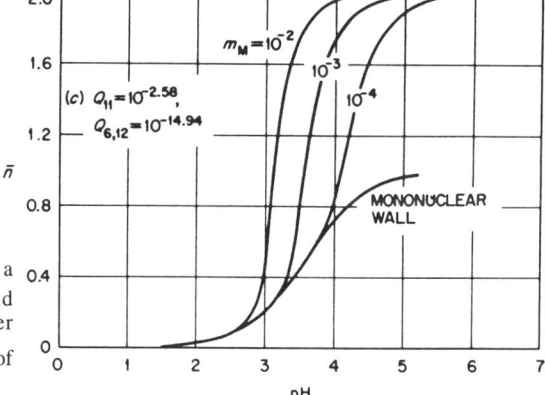

Fig. 3-2. Hydroxyl:ligand ratio as a function of pH. MOH and $M_6(OH)_{12}$. Baes & Mesmer (1976). Reprinted by permission of John Wiley and Sons.

considerably larger complexes, e.g., $Al_{13}(OH)_{32}^{7+}$, may also form in some cases. The mathematical treatment of experimental data, primarily pH measurements, for solutions containing polynuclear complexes is considerably more complicated than the simple treatment of mononuclear complexes. The basic principle for establishing whether polynuclear complexes are formed in a hydrolysis reaction is to investigate how the hydroxyl:ligand number, n, varies with pH at a different total concentration of metal ions. If the n-pH relationship depends on the concentration of metal ions, polynuclear species are present in the system; see Fig. 3-2.

The mathematical treatment of hydrolysis systems involving polynuclear complexes has been described by Baes and Mesmer (1976) and Sillén and Martell (1964).

It is of great interest for understanding the formation of small particles in a hydrolysis system to know which hydrolysis products are present since polynuclear complexes, polycations (polyanions), are precursors of small particles.

5. Different Kinds of Complexes

A metal ion can form oxo, hydroxo, or aquo complexes, or combinations of these complexes. The valency of an ion strongly influences the types of complexes the ion can form. High valencies, e.g., as in Cr (VI), promote formation of oxo-hydroxo complexes, whereas lower valencies, as in Cr(III), promote formation of aquo, aquo-hydroxo, or hydroxo complexes, but never oxo complexes, in aqueous solutions; see Table 3-1.

Table 3-1. Complexes of (a) chrome (VI) and (b) chrome (III). Livage et al. (1988). Here h is defined as the molar ratio of hydrolysis and is equal to 0 when the complex is an "aquo-ion", e.g. $[Cr(OH_2)_6]^{3+}$, and 2N when the complex is an "oxo-ion" $[MO_N]^{(2N-Z)-}$ formed by hydrolysis of the "aquo-ion" $[M(OH_2)_N]^{z+}$, e.g. $[CrO_2]^{2-}$ formed by hydrolysis of $[Cr(OH_2)_2]^{6+}$. Reprinted from Livage, J., Henry, M., and Sanchez, C., Prog. Solid State Chem. **18**, 259 (1988), with kind permission of Elsevier Science Ltd., The Boulevard, Langford Lane, Kidlington, OX5 1GB, UK.

(a)	$[CrO_2(OH)_2]^0$	h=6
	$[CrO_3(OH)]^-$	h=7
	$[CrO_4]^{2-}$	h=8
(b)	$[Cr(OH_2)_6]^{3+}$	h=0
	$[Cr(OH)(OH_2)_5]^{2+}$	h=1
	$[Cr(OH)_2(OH_2)_4]^+$	h=2
	$[Cr(OH)_3(OH_2)_3]^0$	h=3
	$[Cr(OH)_4]^-$	h=4

Figure 3-3 shows in a qualitative manner that formation of oxo and hydroxo complexes is promoted by high pH and requires metal ion valences of at least +3. The qualitative nature of the figure can be realized from the fact that it indicates that +3-valent metal ions would not form hydroxo complexes at pH lower than about 7. It is, however, well known that Fe^{3+} precipitates as $Fe(OH)_3$ at pH values much below 7.

Livage et al (1958) developed a partial charge model so as to explain in a more quantitative manner the dependency of the types of complexes formed in aqueous solutions on pH and the charge on the metal ion.

Hydroxo bridges between metal ions in hydroxo complexes can form by different condensation processes, so-called *olation* mechanisms; see Fig. 3-4. The condensations proceed by nucleophilic substitution, in which the hydroxy group is the nucleophile and H_2O is the leaving group.

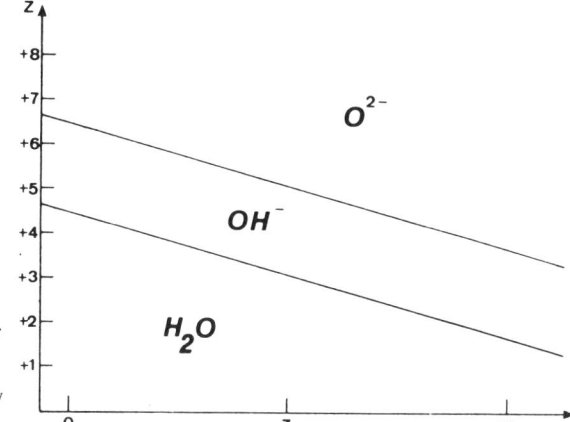

Fig. 3-3. Dependency of type of complex on pH and metal ion charge. Kepert (1972). Courtesy Academic Press, London.

The different types of hydroxo bridges in the figure are defined by the symbol $_x(OH)_y$, in which x indicates the number of metal ions bridged by a single OH-group and y the number of bridges formed between these x metal ions.

Oxo bridges between metal ions in oxo complexes are also formed by different condensation processes, so-called oxolation mechanisms. It is possible to distinguish between two main types of oxolation reactions. In the first type oxo

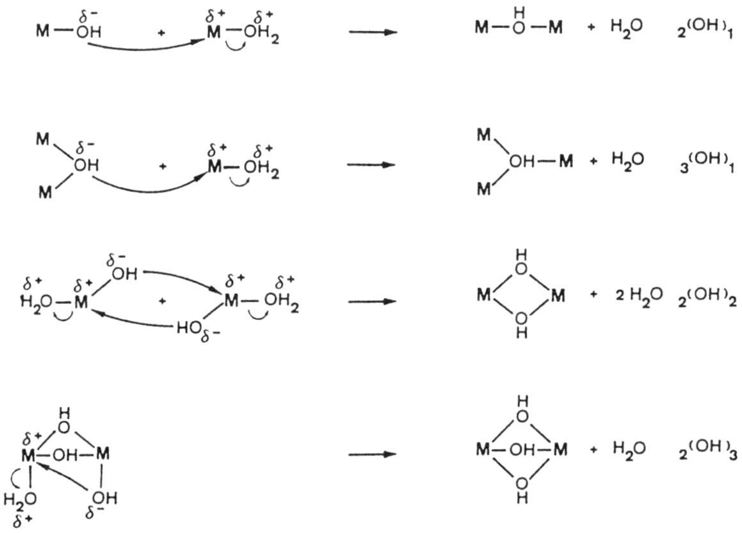

Fig. 3-4. Hydroxo-bridges by different olation mechanisms. Reprinted from Livage, J., Henry, M., and Sanchez, C., Prog. Solid State Chem. **18**, 259 (1988). with kind permission of Elsevier Science Ltd., The Boulevard, Langford Lane, Kidlington, OX5 1GB, UK.

bridges are formed between coordinately unsaturated metal ions and oxolation takes place by nucleophilic addition with rapid kinetics resulting in polyhedra, in which edges and faces are shared, Livage et al. (1988):

$$- M\overset{O}{<} + M\overset{O}{<} - \longrightarrow - M\overset{O}{<}\underset{O}{>}M - \qquad 2^{(O)_2} \qquad (3\text{-}17)$$

$$- M\overset{O}{<}_{O} + O-M- \longrightarrow -M\overset{O}{<}\text{-O-}M\overset{}{>} - \qquad 2^{(O)_3} \qquad (3\text{-}18)$$

Polyanions of the type $M_4O_{12}(OH)_4^{-4}$, where e.g., M=W or Mo, are examples of complexes formed by this mechanism from the mononuclear complexes $MO_3(OH)^-$.

In the second type of oxolation reaction oxo bridges are formed between coordinately saturated metal ions and oxolation occurs by a two-step nucleophilic substitution reaction involving nucleophilic addition, Eqn. 3-19, followed by elimination of H_2O, so-called β-elimination, Eqn. 3-20, Livage et al. (1988).

$$\overset{\delta^-}{M}\text{-}\overset{\delta^+}{OH} + M\text{-}OH \longrightarrow M\text{-}\overset{H}{\overset{|}{O}}\text{-}M\text{-}OH \qquad (3\text{-}19)$$

unstable OH-bridge

$$\overset{H}{\underset{|}{}}\overset{OH}{\underset{|}{}}\text{-M-O-M-} \xrightarrow{\beta\text{-elimination}} \text{-M-O-M-} + H_2O \qquad (3\text{-}20)$$

Referring to Fig. 3-3, oxolation can take place over a wider pH range than olation. The following rule-of-thumb applies to the two types of condensation reactions leading to metal ion complexes: metal ions with oxidation numbers between 2+ and 4+ form polycations by olation at low to moderate pH, whereas metal ions with oxidation numbers higher than 4+ form polyanions by oxolation.

6. Formation of Small Particles

If polycations in the form of aquo-hydroxo complexes combine to react by the olation mechanism, small particles of metal hydroxides, $M(OH)_z$, will eventually form. In some cases the hydroxy bridges are unstable and transform into oxo bridges by giving off water, resulting in particles of composition $MO_{x/2}(OH)_{z-x}$. In other cases all hydroxo bridges may convert to oxo bridges and the composition of the particles will then be $MO_{z/2}$.

Polyanions, on the other hand, often continue to react by the oxolation mechanism so as to form small particles of metal oxide.

The polymerization reaction will continue until a thermodynamically stable state is reached, usually represented by precipitation of the metal hydroxide or oxide. However, this may in many cases take a very long time and metastable states can form, in which polymerization stops and particle growth seems to cease. Thus, small particles of $(Fe_4O_3(OH)_4)_4^{2n+}$ with a molecular weight of about 10^4 and a size of 2-4 nm may form in solutions of $FeCl_3$. The subcolloidal particles will give the solution a reddish-brown color.

Particle growth can be accelerated by forced hydrolysis, which can be accomplished by raising the temperature or adding base, i.e. OH^- ions. Hydrolysis reactions, e.g., represented by Eqns. 3-3 and 3-6, may be forced to the right by adding compounds such as urea or formamide, which release hydroxyl ions when they are made to decompose by, e.g., raising the solution temperature. Thus, urea will decompose above 70°C. This method of adding OH^- ions brings about a much more uniform increase of solution pH compared to adding a base, e.g., NaOH, which may cause local precipitations where the individual drops of NaOH enter the solution of metal ions undergoing hydrolysis. Since oxolation is more favored by higher temperatures than olation, it must be realized, however, that if a higher temperature is required to release OH^- ions, the resulting particles may have a different composition from those particles formed by forced hydrolysis accomplished by adding NaOH solution at a lower temperature.

7. Effect of the Anion

The anions accompanying the metal ions in solutions of metal salts will, although they may not participate in the hydrolysis reactions leading to particle formation, usually affect the shape and size of the particles. There are no simple explanations why a given metal ion sometimes gives rise to spherical particles and sometimes to fibrillar particles. Based on the present state of knowledge, it is not possible to make reliable predictions of particle shapes formed by hydrolysis in a solution of a given metal salt. In some cases the anions coordinate strongly to the metal cations and they may then be incorporated into the particles. In other cases they are just counter ions to the particles and can readily be removed by washing. It is, however, not always the case that strongly coordinating anions become part of the particles. Thus in solutions of $Ce(SO_4)_2$ and dilute H_2SO_4 the particle composition depends on the molar ratio between Ce^{4+} and $(SO_4)^{2-}$ as well as on the concentration of Ce^{4+}, Hsu et al. (1988) and Bottero and Fiessinger (1989). Figure 3-5 shows that at low Ce:SO$_4$ ratios and Ce^{4+} concentrations spherical particles of CeO_2 are formed, whereas at higher ratios and concentrations rod- or peg-like particles of $CeOSO_4$ are formed, see Fig. 3-6.

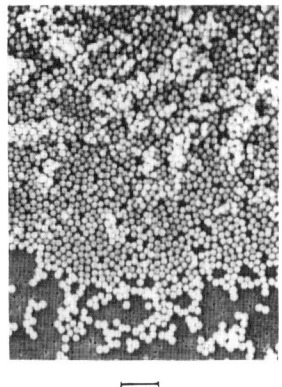

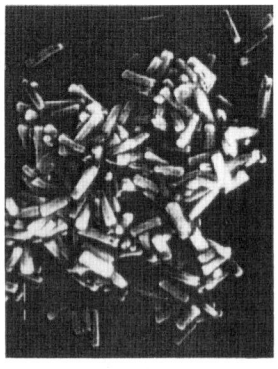

2 μm

Fig. 3-5. CeO$_2$ particles made at
90°C (48 h) from a solution of 1.2
mM Ce(SO$_4$)$_2$ and 80 mM H$_2$SO$_4$.
Hsu et al. (1988). Courtesy American
Chemical Society

5 μm

Fig. 3-6. CeOSO$_4$ particles made at
90°C (48 h) from a solution of 2.5
mM Ce(SO$_4$)$_2$ and 10 mM H$_2$SO$_4$.
Hsu et al. (1988). Courtesy American
Chemical Society

B. HYDROLYSIS OF SOME METAL SALTS

It is generally recognized that the formation of colloidal hydrous oxides from
solutions of metal salts by hydrolysis involves polymerization of polynuclear
cations, but it is perhaps not widely appreciated that the size, shape, and degree of
aggregation of the resulting colloidal particles in many cases can be accurately
controlled. In this section methods for preparing colloidal hydrous oxide
particles by hydrolysis of aqueous solutions of Al^{3+}, Fe^{3+}, Ti^{4+}, Zr^{4+}, Cr^{3+},
Y^{3+}, Ce^{4+}, Zn^{2+}, and Ga^{3+} will be described.

1.   Hydrolysis of Al^{3+}

The formation of colloidal particles of boehmite, AlOOH or Al$_2$O$_3$•H$_2$O, by
hydrolysis of Al^{3+}-solutions involves a stepwise formation of complexes of
increasing size and complexity resulting eventually in the polycation
[Al$_{12}$^{VI}(OH)$_{24}$Al^{IV}O$_4$(H$_2$O)$_{12}$]$^{7+}$, which is considered to be the precursor of
colloidal boehmite.

1.1.   Aluminum Species in Aqueous Solutions
Bottero and Fiessinger (1989) used small-angle x-ray scattering, ^{27}Al high-

resolution liquid and solid NMR, microelectrophoresis, and potentiometric titration to study the hydrolysis products formed when base, NaOH, was added intermittently to a solution of aluminum chloride. The 0.1M aluminum chloride solution was kept in a reactor at 20°C and NaOH solutions of different concentrations were added under vigorous stirring, which was continued for at least one hour after completed base addition before any measurements were done.

Figure 3-7 shows that two ^{27}Al NMR peaks at 17 and 80 ppm from the reference, four-coordinated Al in $Al(OH)_4^-$, varied with the molar ratio, r, between NaOH and $AlCl_3$.

The signal at 80ppm from the reference corresponds to mononuclear species such as $Al(H_2O)_6^{3+}$, $Al(H_2O)_5OH^{2+}$ and $Al(H_2O)_4(OH)_2^+$ and decreases with r. Bottero attributed the peak at 17ppm from the reference, which increases with r, to the tetrahedral Al-ion in the polycation $Al_{12}{}^{VI}(OH)_{24}Al^{IV}O_4(H_2O)_{12}^{7+}$, the so-called Al_{13}^{7+}-ion identified and described by Johansson (1960); see Fig. 3-8.

Using the results from the small-angle x-ray scattering measurements, and assuming the shape to be spherical, Johansson calculated the radius of the

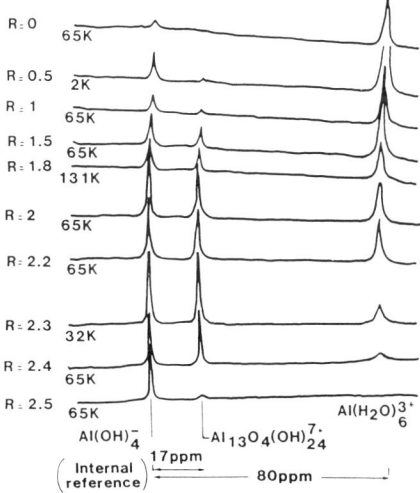

Fig. 3-7. Typical ^{27}Al NMR spectra of liquid samples as a function of the ratio r between added molar amounts of NaOH and $AlCl_3$. Bottero & Fiessinger (1989). Courtesy Arbor Publications.

Al_{13}^{7+}-ion to be 1.26 nm. From this value of the radius he calculated that the specific surface area of the ion was 2380 m^2/g, see Eqn. 1-15.

Figure 3-9 shows the distribution of aluminum between monomers and polymers as a function of r (and pH).

The concentration of Al_{13}^{7+}-ions increased to a maximum at r≈2,   at which r-value almost 100% of the aluminum was present in the form of these polycations. The turbidity of the solutions remained constant up to an r-value between 2 and 2.2 (pH 4.06 to 4.28), but then increased rapidly with r, indicating that further hydrolysis made the Al_{13}^{7+}-ions condense and form larger aggregates. Bottero used an aggregation model in combination with small-angle x-ray data to evaluate the structure of these larger colloids, see Fig. 3-10.

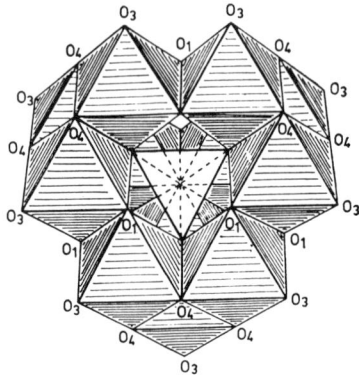

Fig. 3-8 Schematic structure of the Al_{13} polycation $[Al_{13}O_4(OH)_{24}(H_2O)_{12}]^{7+}$-ion. The twelve AlO_6-octahedra are joined together by means of common edges. The tetrahedron of oxygen (O_2) atoms in the center of the group contains the 4-coordinated Al-atom. Johansson (1960). Courtesy Acta Chemica Scandinavica.

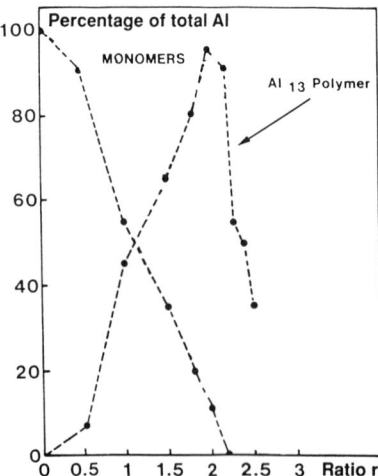

Fig. 3-9. Distribution of aluminum in aqueous solutions between monomers (Al^{3+}) and polymers (Al_{13}^{7+}) vs. the molar ratio, r, between NaOH and $AlCl_3$. Bottero and Fiesinger (1989). Courtesy Arbor Publications.

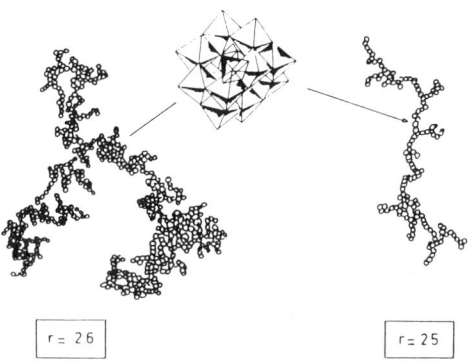

Fig. 3-10. Schematic representation of aggregates of Al_{13}^{7+}-polycations in aqueous solutions of r=2.5 and 2.6. Bottero and Fiesinger (1989). Courtesy Arbor Publications.

1.2. Colloidal Boehmite

Bugosh, in a series of investigations, studied the formation of colloidal boehmite from solutions of basic aluminum salt solutions which were heated under carefully controlled conditions of pH, temperature, and time, Bugosh (1959, 1961). Dr. Bugosh graciously allowed us the use here of extensive and important work carried out at E.I. DuPont Co. in 1961, which was officially approved for publication by the DuPont Company but was never published.

1.2.1. Basic Aluminum Salt Solutions   Basic aluminum salts can be characterized in several ways. A given basic aluminum chloride solution can be described as having an atomic ratio of Al to Cl of 2:1 or a basicity, which is the molar ratio of hydroxyl groups to aluminum (called "r" in §1.1 this chapter), of 2.5. Formulae such as $Al_2(OH)_5Cl$ are also used, with the understanding that they refer only to stoichiometric relationships and not to structures of the salts in solution.

In the laboratory, basic aluminum chloride solutions can be made by adding carefully controlled amounts of base, e.g., NaOH, to solutions of aluminum chloride. In industry, however, basic aluminum chloride solutions, and also basic aluminum nitrate solutions, are produced by adding aluminum metal powder with good agitation to solutions of aluminum chloride, Huehn and Haufe. (1940) and aluminum nitrate, Denk and Alt (1952). By varying the amount of aluminum metal added, the basicity of the fluid product can easily be controlled.

The storage stability of concentrated basic aluminum chloride and nitrate solutions differs markedly. Concentrated basic aluminum chloride solutions of varying ratios aging at room temperature develop only a very slight amount of precipitated alumina, whereas with basic aluminum nitrate solutions a very voluminous white precipitate develops after about a month.

1.2.2. Conversion of Basic Aluminum Salt Solutions to Colloidal Aluminas Bugosh (1961) heated solutions of basic aluminum chloride in unagitated autoclaves at temperatures from 100°C to 360°C and for times ranging from 0.5 to 85 h. The samples were cooled slowly to a temperature of about 50°C before removing them from the autoclave. The alumina particles were characterized by x-ray and electron diffraction, electron microscopy, and BET surface area. Unpolymerized alumina, defined as the amount of aluminum ions, normal and polycations, plus very readily acid-soluble alumina which is present in an alumina sol, were also determined.

1.2.3. Effect of Variations in Aluminum and Chloride Concentration on Morphology of Colloidal Particles In Fig. 3-11 domains corresponding to various products obtained by heating basic aluminum chloride solutions of different alumina and chloride ion concentrations at 160°C for times between 4 and 16 h are defined. Bugosh drew the straight lines going through the origin for reference purposes, and their slopes are the Al/Cl ratios of the basic aluminum chloride solutions that were autoclaved.

   a. fibrillar lath-like particles (<0.10 moles Al/liter). At very low alumina and chloride concentrations broad lath-like particles as well as platelets of various types were obtained as shown in Fig. 3-12-a, -b, -c, -d, and -e.

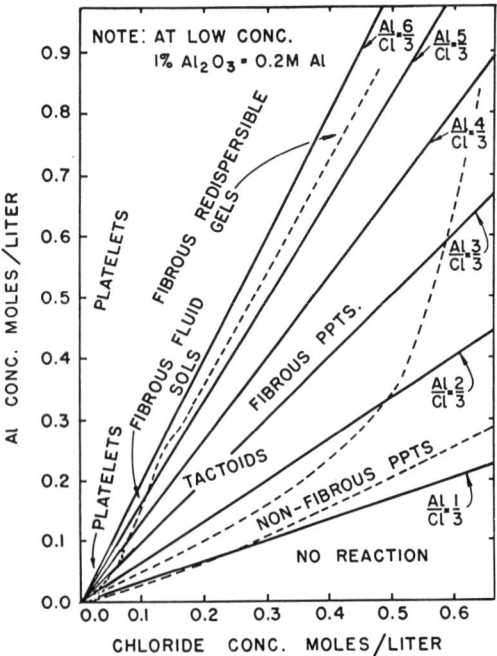

Fig. 3-11. Products obtained by heating basic aluminum chloride solution at 160°C for 4-16 h at various concentrations. Bugosh (1961).

The first particles to form appeared to be fibrillar as in Fig. 3-12a. These particles then aggregated to form platelets such as shown in Fig. 3-12b. Platelets or parts of platelets of this type could then aggregate to form the very unusual structures shown in Fig. 3-12c and at higher magnification in Fig. 3-12d. Closer examination of the electron micrographs showed that these latter particles were composed of elongated platelets which are aggregated with the long edge of each platelet common to the aggregate.

b. Fibrous particles - tactoids (0.1-1.0 M Al; 0.1-0.4 M Cl). The term *tactoid* refers to a structure containing fibers aggregated with their long axes predominantly parallel. Figs. 3-12a and 3-12b show typical tactoid structures.

1. Tactoidal area. In the region of Fig. 3-11 centered at 0.2 M Al and 0.2 M Cl, formation of a very peculiar tactoid-type of particle was observed. Figure 3-13 shows photomicrographs of typical products composed of oriented fine hair-like particles. Tactoids composed of coarser particles are shown in Fig. 3-14. The definite tendency these tactoids have to aggregate in an end-to-end manner is especially noticeable in Fig. 3-13a. X-ray diffraction of the powder produced by the dried tactoid powder showed that boehmite was the only crystalline phase present.

If the chloride ion concentration was reduced from the tactoid region in Fig. 3-11 to about 0.1 M, well dispersed stable sols of the fibrous form of boehmite were produced, e.g., at Al/Cl = 6/3 and 0.2 M Al, which showed intense streaming birefringence when viewed between crossed polarizing elements. By increasing the aluminum concentration above about 0.3 M Al, highly viscous and thixotropic "gels" were produced which also contained fibrous boehmite particles; see Fig. 3-12h and -i. These highly thixotropic "gels" were reversible and would revert to a stable sol upon dilution. They should thus be considered as concentrated thixotropic sols rather than gels, even though the external appearance was one of a typical gel structure.

A true alumina gel could be formed from such a concentrated sol by the addition of alkali to a pH of 8. This gel would no longer redisperse to a stable sol upon dilution nor is it birefringent.

2. Boehmite Platelet Area. As the ratio of Al/Cl was increased above 6/3, the colloid particles became broader until hexagonal boehmite platelets were produced. These platelets are typical of the relatively low surface area products made by heating pure alumina gels, alkalized alumina gels, or those containing only a small amount of anion. Typical hexagonal boehmite platelets made by heating an alkaline alumina gel are shown in Fig. 3-12f.

Bugosh noted that it was not possible to prepare clear, homogeneous basic aluminum chloride solutions with Al/Cl ratios above about 6/3. Compositions above this ratio were therefore usually prepared by adding an acid to various types of reactive alumina gels.

3. Nonfibrous Precipitate Area. The variation in alumina particle morphology when heating a basic alumina chloride solution with Al/Cl = 2/3 is llustrated in Fig. 3-14a, -b, -c, and -d. Pine cones, fibrils, tactoids and non-

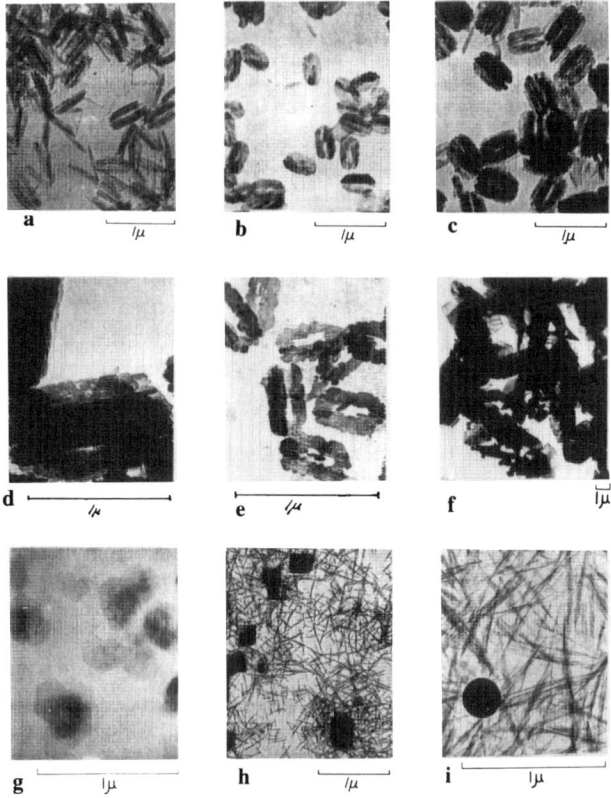

Fig. 3-12. Colloidal alumina particles obtained by heating basic aluminum chlorides. (a) Heated basic aluminum chloride (Al/Cl = 5/3; 0.10 M Al) 160°C, 16 h. (b) Heated basic aluminum chloride (Al/Cl = 4/3; 0.02 M Al) 160° C, 16 h. (c) Heated basic aluminum chloride (Al/Cl = 3/3; 0.033 M Al) 160°C, 16 h. (d) Heated basic aluminum chloride (Al/Cl = 1/3; 0.01 M Al) 160°C, 85 h. (e) Heated basic aluminum chloride (Al/Cl = 5/3; 0.05 M Al) 220°C, 1 h. (f) Normal aluminum chloride (0.6 M Al adjusted to 22.3/6 = OH/Al) pH = 10.6 after heating at 160°C for 4 h.160°C, 16 h. (g) Basic aluminum chloride (Al/Cl = 6/3; 0.10 M Al) aged at room temperature 7 months. After one more year aging particles become much denser with distinct spiral aggregation. (h) Alumina sol prepared according to U.S. 2,763,620 using sodium aluminate and hydrochloric acid; 0.2 M sol then autoclaved for 16 h at 160°C. The dense cubical crystals are NaCl which disappear upon deionization. (i) Heated basic aluminum chloride (Al/Cl = 6/3; 0.4 M Al) 160°C- 4 h. The large sphere is a Dow latex polystyrene calibration sphere, 280 mμ in diameter. Bugosh (1961).

fibrous particles were obtained. As can be seen from Fig. 3-11, above a chloride molarity of about 0.5-0.6, the fibrous nature of the products was reduced. Fibrous particles were obtained at much higher chloride ion concentration for basic aluminum chlorides with ratios of Al/Cl =5/3 and 6/3 than for those with lower ratios.

With normal aluminum chloride only a slight amount of alumina was produced at these temperatures, and that only at high dilution.

c. Unoriented aggregates; occlusion of chloride ions (> 1.0 M Al, > 0.6 M Cl). As the chloride ion concentration became greater than about 0.8 M, the fibrous products became more aggregated and shorter until finally nonfibrous precipitates were formed. With compositions containing Al/Cl = 6/3 or 5/3, however, fibrous particles were produced even at as high as 5-6 M Al, but at these high concentrations the particles were aggregated into unordered particles which did not redisperse upon dilution. A typical product of this type is shown in Fig. 3-14e. From x-ray diffraction data Bugosh concluded that at high concentrations chloride ions were probably occluded in the boehmite lattice.

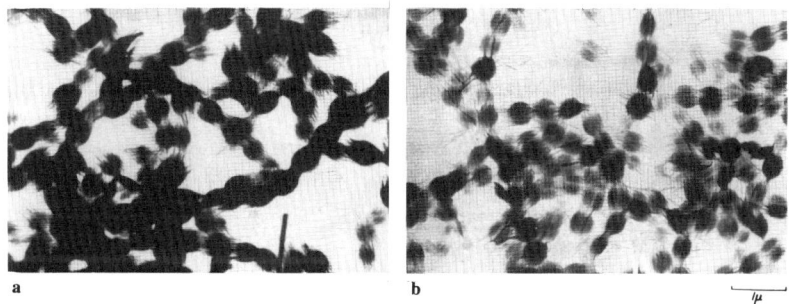

Fig. 3-13. Products obtained by heating basic aluminum chloride. (a) Al/Cl = 4/3; 0.2 M Al; heated 5 h at 160°C. (b) Al/Cl = 3/3; 0.20 M Al; heated 5 h at 160°C. Bugosh (1961).

Small amounts of other additives present during the polymerization of basic aluminum chloride to fibrous boehmite markedly affected the course of the reaction. For example, in a basic aluminum chloride solution with Al/Cl = 6/3 heated at 140°C for 8 h, Bugosh found that as little as 0.01 mole percent Mo (based on Al in the form of ammonium molybdate) completely inhibited fiber formation and markedly inhibited formation of any kind.

1.2.4. Effect of Anion Type on Morphology of Colloidal Alumina Particles

a. Strong monobasic acids. Comparison of products made from basic aluminum chlorides and nitrates at temperatures of 160-220°C showed that at comparable polymerizing conditions, comparable products are obtained. At 360°C chloride, bromide and nitrate are equivalent, giving similar types of lath-like products as shown in Figs. 3-14f and -g. Fluoride did not work, since it produced aluminum hydroxy fluoride hydrate.

At high temperatures (e.g., 360°C) the anion could be added either as free acid or the normal salt (to gelatinous $Al(OH)_3$) and comparable products resulted under similar polymerization conditions.

The criticality of the Al/anion ratio was most strikingly demonstrated at high temperatures. For example, with chloride and bromide at 360°C, 8 h and 0.51% Al_2O_3 compared at Al/anion of 100/1, 20/1, 5/1, and 1/1, long laths were produced in both cases.

At lower temperatures similar results were obtained, but the transition was not quite as sharp and it occurred at a slightly lower Al/anion ratio.

b. Strong polybasic acids. Bugosh found that at temperatures as high as 220°C no fibrous or other well-defined particles were produced by heating basic aluminum sulfates. On the other hand, with basic aluminum sulfate at 360°C very long fibers were obtained at at least twice the Al/anion ratio at which laths form in the chloride system. The long asbestiform particles containing sulfate were described by Arthur (1960).

With Na_2SO_4 as the anion source and at the same level, no fibers at all were formed. This was a further indication that not only did the anion type and level have to be correct, but also the acidity, which probably controlled the charge on the polymerizing ions.

As the ratio of Al/SO_4 was decreased, an unidentified "alunite-type" phase appeared. This alunite-type phase was first observed by Arthur, and it does not correspond to any of the numerous known crystalline basic aluminum sulfates, Bassett and Goodwin (1949).

1.2.5. Effect of Time-Temperature on Morphology of Colloidal Alumina Particles

a. Aging basic aluminum chloride solutions at room temperature (25-30°C). A basic aluminum chloride with Al/Cl = 6/3 and 0.10 M Al aging at room temperature for two years gradually became turbid and exhibited strong streaming birefringence. X-ray analysis of a dried residue of the aged sol shows gibbsite with four unidentified lines. An electron photomicrograph of these platelets is shown in Fig. 3-12g.

When aging a group of dilute basic aluminum chloride solutions for about 6 months, all of the solutions which had a chloride molarity below about 0.07 developed a noticeable turbidity. Only in the case of the 0.1 M Al Al/Cl = 6/3 solution was an opalescent sol obtained. Normal aluminum chloride solution was not turbid at a chloride molarity of 0.03.

b. Heating basic aluminum salt solutions at 100°C. The rate of the polymerization reaction to colloidal alumina at 100°C was studied by Bugosh with the results given in Fig. 3-15, where the data are plotted as for a first-order reaction.

The top curve shows that the reaction is first-order with $k_{100°C} = 1.23 \times 10^{-4}$ min^{-1} and approaching an equilibrium at about 9% unpolymerized alumina. The bottom curve is an indication of the rate of particle growth from a turbidity measurement. There appear to be three distinct types of particle growth

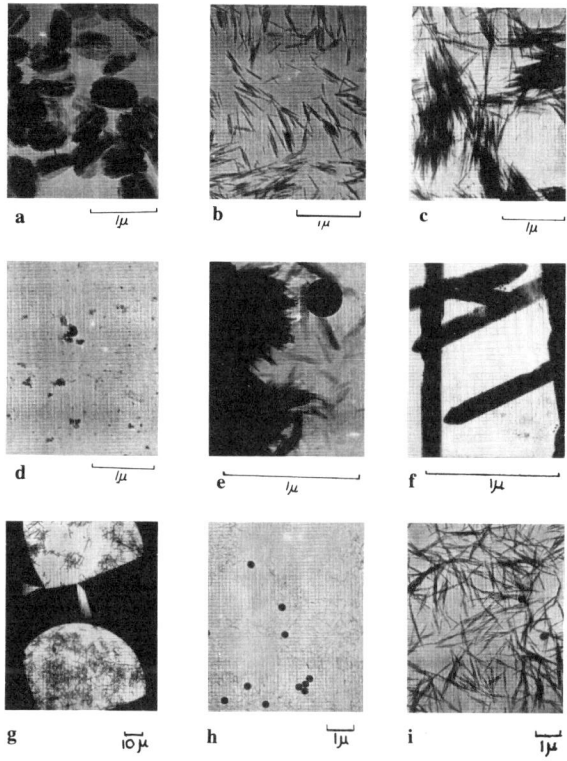

Fig. 3-14. Products from heating basic aluminum chloride. Effect of variation in concentration of Al and Cl at constant Al/Cl ratio. (a) Basic aluminum chloride (0.04 M Al; Al/Cl = 2/3) heated 16 h at 160°C. (b) Basic aluminum chloride (0.1 M Al; Al/Cl = 2/3) heated 16 h at 160°C. (c) Basic aluminum chloride (0.2 M Al; Al/Cl = 2/3) heated 16 h at 160°C. (d) Basic aluminum chloride (1.5 M Al; Al/Cl = 2/3) heated 16 h at 160°C. (e) Basic aluminum chloride (5.0 M Al; Al/Cl = 6/3) heated 16 h at 160°C. (f) Basic aluminum chloride (0.2 M Al; Al/Cl = 6/3) heated 16 h at 160°C. (g) Basic aluminum chloride (0.2 M Al; Al/Cl = 6/3) heated 11 h at 350°C in sealed platinum tube. (h) Basic aluminum chloride (0.6 M Al; Al/Cl = 6/3) heated 1 h at 160°C. (i) Basic aluminum chloride (0.6 M Al; Al/Cl = 6/3) heated 16 h at 160°C. Bugosh (1961)

occurring, i.e., from 0-125 h; 125-200 h; and finally >200 h. The last portion of the curve is probably also associated with the formation of the large gibbsite prisms which are observed in the final product. The dilute sols made at 100°C by heating, e.g., for 385 h, have a tendency to stratify and settle upon standing.

From plots of data of the time necessary to reach a given pH, at temperatures of 100°C, 160°C, and 220°C, an activation energy between 20-30 kcal/mole was calculated.

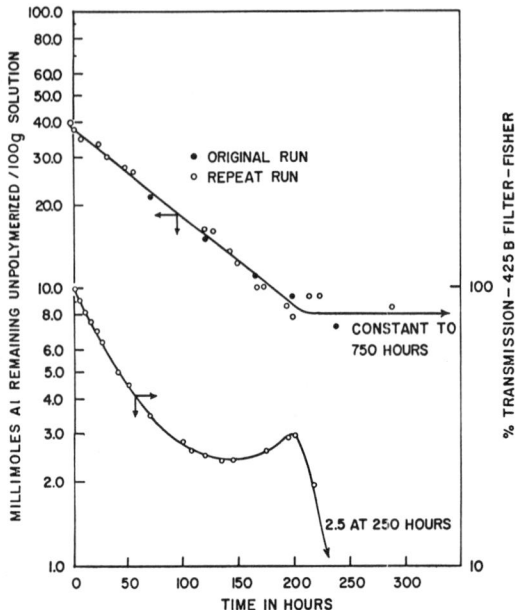

Fig. 3-15. First-order reaction plot for basic aluminum chloride: Al/Cl $= 2/1$, 2% Al_2O_3; 100°C; $k_{100°C} = 1.24 \times 10^{-4}$ min^{-1} from slope. Bugosh (1961)

c. Heating basic aluminum salt solutions at temperatures greater than 140°C - Formation of stable boehmite fibrils. As the heating time of a dilute basic aluminum salt solution was increased from 1 to 85 h, the sols produced became more turbid as the time increased until finally a precipitated product was formed. These changes were associated with an increase in particle length and thickness. Figure 3-16 shows that as the duration of heating at 160°C increased, the particle length increased. Typical fibrous products made by heating a basic aluminum chloride at 160°C for one hour and 16 h are compared in Fig. 3-14h and -i. Figures 3-17a, -b, and-c shows products made at 160°C for 4 h and 85 h. That some of the fibers were probably end-to-end aggregates was inferred from the fact that vigorous agitation in a Waring Blender ("whizzing") for 30 minutes reduced the apparent particle length.

Not only did the particle length increase, but the thickness did as well. In Fig. 3-18 the specific surface area of the fibrous particles is plotted vs. the duration of heating at 160°C. The coarsening of the tactoid-like particles is readily apparent in Fig. 3-17h and -i. At a constant time of 4 h, Fig. 3-19 shows how rapidly the specific surface area falls off as the temperature increases. The products in Fig. 3-17b, -e, -f, and -g show that as the temperature was raised, especially above 200°C, the fibers became wider (more lath-like) as well as thicker (lower specific surface area). The width dimension appeared to increase just slightly faster than the thickness dimensions, since at 300 m^2/g the width to thickness ratio was 2 and at 60 m^2/g it was about 3.

1.2.6. Constitution of Alumina Sols and Mechanism of Particle Growth A schematic representation of the chemical changes as well as changes in particle morphology when heating a basic aluminum chloride is shown in Fig. 3-20. Initially the basic aluminum chloride solution contained essentially no normal aluminum ions—compare with Fig. 3-9. Complex basic aluminum cations containing at least six aluminum atoms were present as well as a small number of nuclei. These nuclei dissolved in hydrochloric acid at a rate far slower than the complex basic aluminum ions. The presence of some nuclei was suspected since the basic aluminum chloride solutions showed a faint Tyndall cone. That these nuclei were either platelets or fibrillar could be inferred from Fig. 3-21 where viscosity of the basic aluminum chloride is related to the flow birefringent phase angle at a constant gradient. The phase angle is directly proportional to the volume fraction of the material producing the flow birefringence. Such anisometric nuclei only appeared in freshly made solutions. Particles such as these or the more crystalline particles formed from them after some heating might serve to nucleate the fibrous boehmite phase.

When a dilute basic aluminum chloride solution was heated for one hour at

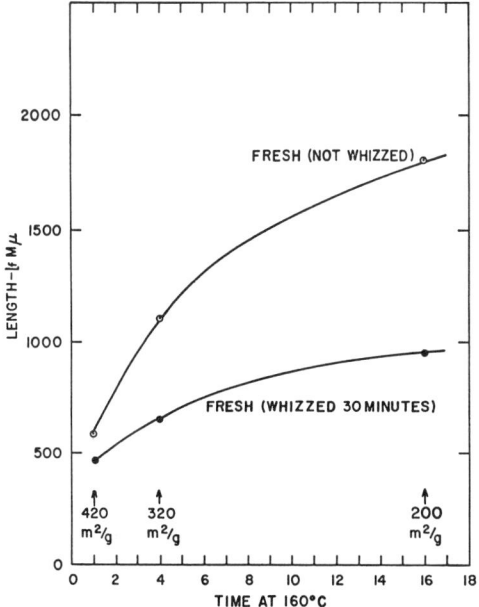

Fig. 3-16. Variation of length of particles of polymerized basic aluminum chloride (3% Al$_2$O$_3$; Al/Cl = 6/3; 160°C) vs. time at 160°C. Bugosh (1961).

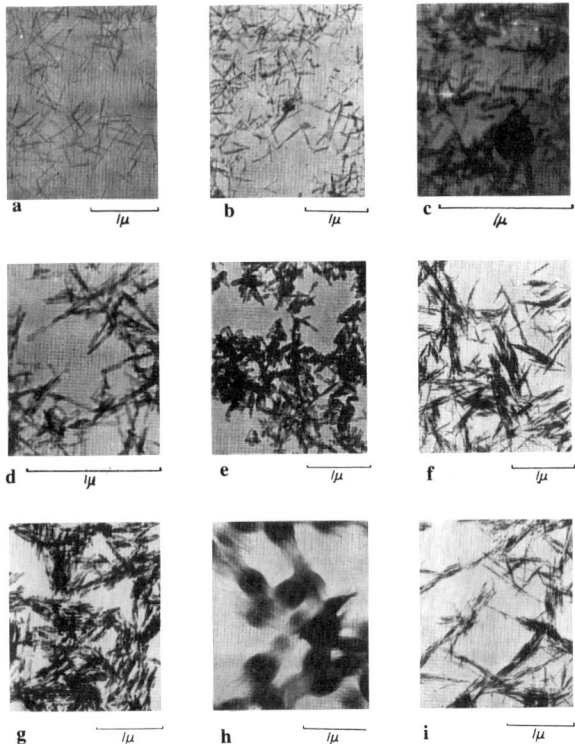

Fig. 3-17. Product resulting from heating basic aluminum chlorides. (a) Basic aluminum chloride (0.6 M Al; Al/Cl = 6/3) heated 160°C for 4 h. (b) Basic aluminum chloride (0.1 M Al; Al/Cl = 6/3) heated 160°C for 16 h. (c) Basic aluminum chloride (0.1 M Al; Al/Cl = 6/3) heated 160°C for 85 h. (d) Basic aluminum chloride (0.2 M Al; Al/Cl = 6/3) heated 160°C for 1 h. (e) Basic aluminum chloride (0.1M Al; Al/Cl = 6/3) heated 220°C for 16 h. (f) Basic aluminum chloride (0.2 M Al; Al/Cl = 3/3) heated 160°C for 16 h. g) Basic aluminum chloride (0.2 M Al; Al/Cl = 3/3) heated 220°C for 16 h. h) Basic aluminum chloride (0.2 M Al; Al/Cl = 3/3) heated 160°C for 6 h. i) Basic aluminum chloride (0.2 M Al; Al/Cl = 3/3) heated 160°C for 16 h. Bugosh (1961).

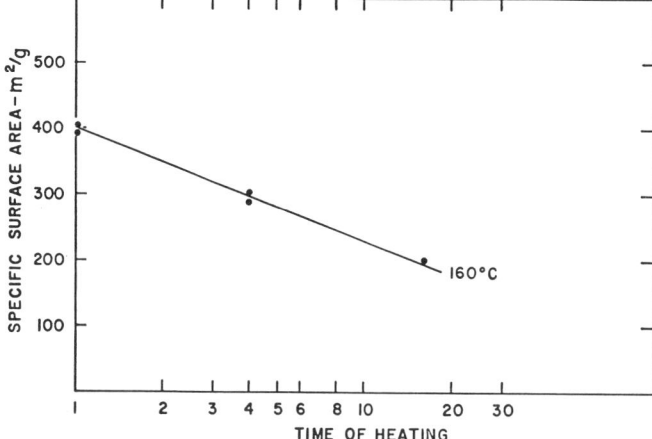

Fig. 3-18. Specific surface area of fibrous boehmite vs. autoclave time at 160°C. (BAC; 0.4 M Al; Al/Cl = 2/1) Bugosh (1961).

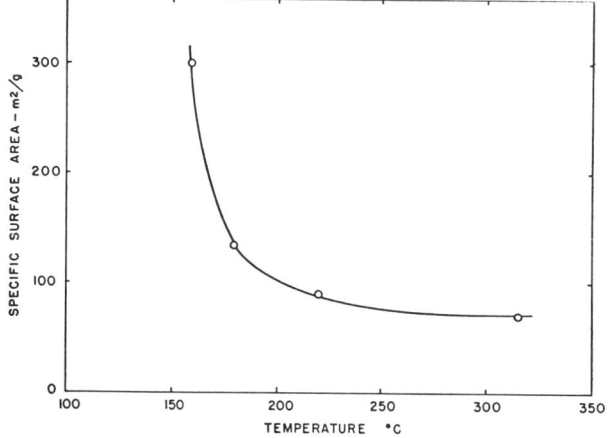

Fig. 3-19. Variation of specific surface area with temperature at which 3% Al_2O_3; Al/Cl = 2/1 basic aluminum chloride is polymerized at a time of 4 h. Bugosh (1961).

160°C, the complex basic aluminum cations disproportionated to some normal aluminum ions and higher molecular weight species.

It was believed that the fibrillar particles in the intermediate fraction aggregated in an end-to-end manner to produce longer fibers and more normal aluminum ions (equivalent to the chloride ion present) upon continued heating, as shown in Fig. 3-20c. End-to-end aggregation of rodlets to form longer fibers of V_2O_5 was shown by Brumberg et al. (1951). Figure 3-17h and -i show the increase in particle length when the heating time was increased from 1 to 16 h. Any further heating beyond the point of stoichiometric disproportionation yielded broader, lower surface-area fibers. Since normal aluminum ions would not hydrolyze sufficiently under these conditions (see Fig. 3-11), the additional alumina required to broaden the fibers probably came from the dissolution of the smallest particles or a recrystallization of individual particles already present.

Disproportionation of normal and basic aluminum chloride solutions of different ratios is shown in Fig. 3-22. The six experimental points of curve A fall on the line drawn according to the reactions I, II, and III given in the figure.

At temperatures considerably higher than 160°C, e.g., 330°C, normal aluminum ions hydrolyzed and less unpolymerized alumina resulted than expected from reactions I, II and III , but the particles were broad low surface-area laths (see Fig. 3-22, curve B).

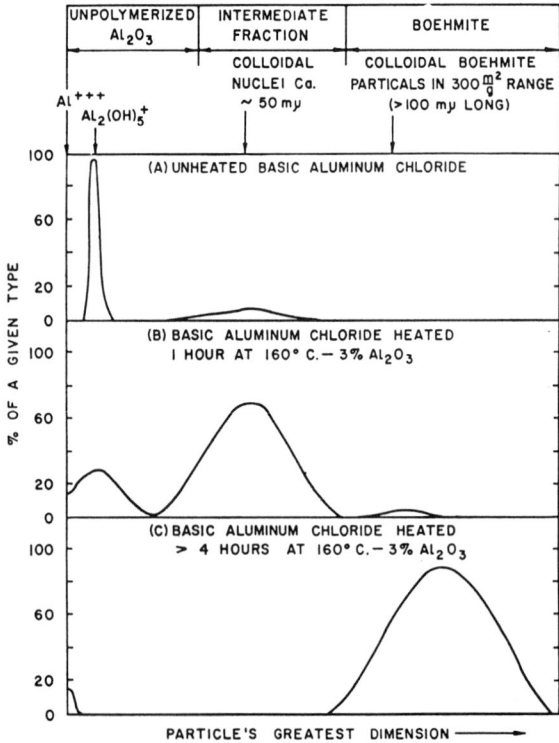

Fig. 3-20. Schematic representation of the polymerization of basic aluminum chloride to fibrous boehmite. Bugosh (1961).

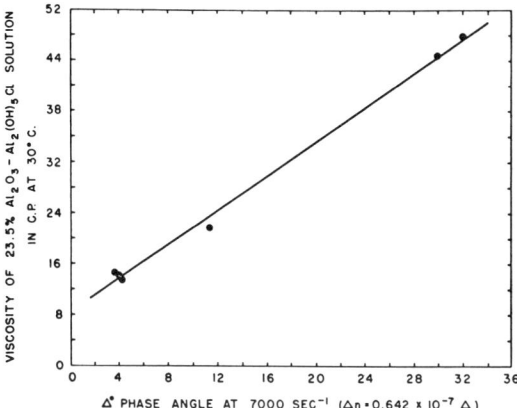

Fig. 3-21. Viscosity of unpolymerized aluminum chloride vs. phase angle at 7000 sec^{-1}. Bugosh (1961).

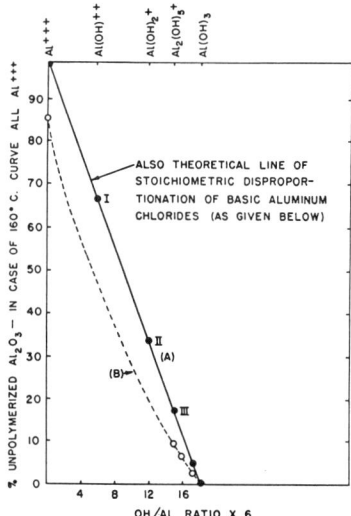

Fig. 3-22. Polymerization of basic aluminum chloride at different OH/Al levels at: 160°C—4 h (A); 330°C—4 h (B). Bugosh (1961).

$$I—3 \, Al(OH)Cl_2 \longrightarrow AlOOH + 2 \, AlCl_3 + HOH$$

$$33.3\% \qquad 66.6\%$$

$$II. \, 3 \, Al(OH)_2Cl \longrightarrow 2 \, AlOOH + AlCl_3 + 2 \, HOH$$

$$66.6\% \qquad 33.3\%$$

$$III. \, 3 \, Al(OH)_5Cl \longrightarrow 5 \, AlOOH + AlCl_3 + 10 \, HOH$$

$$84.4\% \qquad 16.6\%$$

Bugosh (1961) suggested that the effect that variables such as anion concentration, anion type and heating temperature have on the morphology of boehmite alumina could be understood using the following working hypothesis: from the layer crystal structure of boehmite one would normally expect to get platelet-type particles and, indeed, such particles are formed in nature, Bohnstedt-Kupletskaya and Vlodavetz (1945); and synthetically, Ervin and Osborn (1951). He proposed that the anions adsorbed from solution onto the growing boehmite crystal and thus modified its crystal habit. A proposed orientation of the crystallographic axes of boehmite with respect to the fiber axis determined from electron and x-ray diffraction studies is shown in Fig. 3-23. Sasvari and Zalai (1957) give a diagram of the unit cell of boehmite which can be related to this figure.

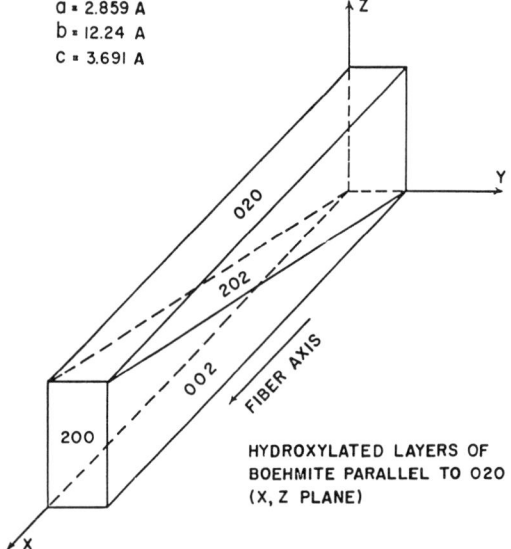

Fig. 3-23. Proposed orientation of boehmite crystal planes with respect to fiber axis. Bugosh (1961).

To account for his results, Bugosh further postulated that at a given anion concentration and a given temperature there was an equilibrium amount of anion adsorbed on each of the three different types of crystalline faces. For unknown reasons, chloride, bromide, and nitrate apparently adsorbed primarily on the sides of the crystallites, thus allowing the crystals to grow in the lengthwise direction. Particle growth might involve a combination of end-to-end aggregation of nuclei already present in solution and deposition of the basic aluminum chloride cations on the ends of the growing fiber. Although the chloride and nitrate ions were not too well adsorbed on the ends of the growing fiber, apparently molybdate ion was, since it was able to inhibit fiber growth at extremely low concentrations.

Bugosh also showed that the aluminum to anion ratio was related to the fraction of the surface of boehmite blocked, f, the surface chloride per surface aluminum, n, and the specific surface area in square meters per gram, A, as follows:

$$Al/anion = 1000/(nfA) \qquad\qquad (3\text{-}21)$$

Thus, with a specific surface area of about 300 m^2/g, assuming one chloride for each aluminum and f=1, the aluminum to anion ratio comes out to be about 3. This was about the ratio found experimentally above which the fibers began to broaden and get shorter. Referring again to Fig. 3-11, at about 0.6 M Al, Al/Cl = 6/3 and 160°C, the surface adsorption equilibrium was such that very little adsorption occurred on the ends of the fibers, thus promoting growth in this direction.

Again from Fig. 3-11, at Al/Cl = 6/3 and high chloride concentrations, the fibers tended to become shorter and more aggregated. Some chloride appeared to be occluded in the lattice under these conditions and a large amount of short intermediate fraction was formed. At very high chloride levels the chloride ions began to be adsorbed on the less favorable end faces, thus inhibiting fiber growth. Very low chloride levels favored desorption and therefore platelet formation, as actually observed. As the temperature was raised, desorption of the ions occurred first from the least favorable face and this face grew at the expense of the other faces. Either broader fibers, laths, or platelets formed, depending on the specific conditions.

Although chloride and bromide appeared to be about equivalent in their lattice blocking effectiveness, sulfate ions appeared to be much more strongly adsorbed. This was expected since sulfate forms a wide variety of insoluble basic aluminum sulfates and previous work with various alumina gels showed sulfate to be more strongly adsorbed than the halogen ions, Weiser and Middleton (1920).

At temperatures of about 160°C sulfate ion is so strongly adsorbed on the basic aluminum cations that basic aluminum sulfates tend to form instead of boehmite, Denk and Alt (1952). As noted before, at temperatures in the region of 360°C, basic aluminum chloride solutions yielded laths. These laths were noticeably longer and very much thinner if a small amount of sulfate was added as a modifying agent.

In general, in order to obtain fibrous boehmite particles, the pH of the starting solution should be between 3.5 and 5. From the standpoint of ease of polymerizability, the higher the basicity of the basic aluminum salt (i.e., the higher its pH), the more readily it polymerized due to the lower unit charge on the cation. On the other hand, the concentration of cations capable of charging up the alumina surface (basic aluminum ions, hydrogen ions) must be high enough (and the pH low enough) to prevent the growing particles from crosslinking or flocculating.

1.2.7. Importance of "Unpolymerized Alumina" Bugosh concluded that the formation of a well dispersed, high specific surface area (i.e., greater than about $200m^2/g$) fibrous boehmite involved a delicate balance of alumina and anion concentration, anion type, as well as time and temperature of heating.

Even though a potentially stable high surface-area fibrous boehmite sol could be prepared according to the principles outlined above, it does not necessarily mean that the sol would be permanently stable. When using strong acid basic salts there was always present at the end of the reaction in addition to the boehmite, some "unpolymerized alumina" which usually consisted of either normal or basic aluminum salts or a mixture of the two. If only normal aluminum ions and fibrous boehmite were present in the sol, it would remain stable with respect to particle shape and crystal type for several years—unlike the unstable pseudo-boehmite, Papee (1958)—as long as the pH was not raised (to convert the normal aluminum ions to basic aluminum ions). As shown above, dilute basic aluminum salts readily aged at room temperature to crystalline aluminas. Were such salts present in a fibrous boehmite sol, they would also revert to a non-boehmite crystalline alumina upon aging; and depending on specific conditions, either bayerite or gibbsite would eventually form. Therefore, in order to insure that the final sol does not develop unwanted new phases on standing, it is desirable to remove all the free ionic aluminum species. As these are removed, the anions associated with them will also be removed. The anion remaining will then be only that closely associated with the boehmite particle surface, Bugosh (1961).

In addition to its importance in the aging stability of the alumina sols, the unpolymerized alumina is also important in obtaining alumina sols which are stable toward settling and which exhibit streaming birefringence under alkaline conditions. Alkaline stable alumina sols prepared without the use of organic protective colloid were previously unreported, Bugosh (1961).

Removal of unpolymerized alumina, in general, is also desirable since its properties are typical of a water soluble ionic compound, and, if present in considerable amounts, it can mask the properties of the colloidal alumina fraction in the sol.

In addition to the slow, conventional method of dialysis, several other methods of deionization could be used:
- deionization by gelling and washing
- deionization with specially prepared ion exchange resins (followed by heat treatment)
- deionization by vapor-phase removal of chloride, Bugosh (1961).

1.2.8. Alkaline Stable Colloidal Alumina The sols made by heating a basic aluminum chloride were normally slightly acidic, positively charged, and were compatible with cationic organic latices and other positive colloids. An alkaline stable alumina can be made, however, which is presumably negatively charged since it is compatible with such negative colloids as colloidal silica and organic lattices, Bugosh (1961).

It appears that in order to make an alkaline stable alumina sol, it is necessary to produce on the surface of the particles an aluminum hydroxide which can form an aluminate ion. Such an aluminum hydroxide cannot be produced by alkali precipitation from a basic aluminum chloride solution or by *slowly* adding a base to a normal aluminum chloride solution, Treadwell and Zurcher (1932). A suitable monomeric aluminum hydroxide can, however, be formed under conditions which avoid the formation of polynuclear basic aluminum cations.

1.2.9. Some Properties of Fibrous Boehmite Sols Unusual properties of these sols were streaming birefringence, paracrystallization and unusual rheological behavior.

When a dilute fibrous boehmite sol was made by heating a basic aluminum salt (e.g., 0.6 M Al, Al/Cl = 6/3; 4 h; 160°C), a viscous thixotropic sol resulted. Deionization of this sol reduced both the apparent viscosity and the thixotropy. A high degree of thixotropy could again be obtained with the sol if the pH was adjusted to 5-6.5. Raising the pH to 7-8 yielded a gel; see also Figs. 3-10-3-15. This gel had the appearance of a paste stable toward hardening rather than a "ringing gel".

Although it was generally desirable to prepare well-dispersed fibrous boehmite particles at a relatively low concentration (1-6% Al_2O_3), the sols once made could be concentrated if they were first deionized and the thixotropic viscosity reduced. The extent to which the sols could be concentrated depended to a large extent on the specific surface area of the particles. Particles of about 200 m^2/g could be concentrated to about 15% Al_2O_3 and a stable, opaque, and highly thixotropic sol resulted. Sols containing particles with a specific surface area of about 400 m^2/g could be concentrated to about 10% Al_2O_3, and a viscous, opalescent sol resulted.

When fibrous boehmite alumina dispersions were dried on a surface, a coherent film of boehmite alumina formed. This film-forming ability was a direct consequence of the fibrous nature of the ultimate particles and the fact that there was probably a considerable amount of inter-particle association through the hydroxyl groups on the surface. It was observed that the maximum film thickness attainable with a given sol depends on the fiber length and specific surface area of the particles. With particles at least 300 nm long and in the 200-300 m^2/g range, self-supporting films 6 x 6 in. square and several mils thick were readily prepared by casting from a dilute solution onto a glass surface coated with a parting agent. Translucent to transparent films were made which remained coherent and transparent to 1000°C.

This result was unexpected since the boehmite originally in the film was first transferred to an almost anhydrous γ-alumina at about 400°C and to θ-alumina at about 950°C. In separate experiments, however, fibrous boehmite alumina powder was heated at these same temperatures and reexamined with the electron microscope. Both of the heated samples had retained the fibrous character of the original fibrous boehmite.

Fibrous boehmite films, when formed by air drying were, in general, very water sensitive, i.e., water markedly softened the film, but usually did not repeptize it. Under these conditions there were apparently very few Al-O-Al bonds formed between the particles to yield insolubility (as in the case of a silica sol; see §B-1, Ch. 2). Highly insoluble, harder, as well as chemically nonreactive films resulted, however, if they were given a heat treatment above 400-500°C. Such films dissolved only very slowly in warm concentrated hydrochloric acid, Bugosh (1961).

Table 3-2. Fibrous Boehmite Sol Properties. Bugosh (1961).

	% Al$_2$O$_3$	% Cl	pH	Unpolymerized Alumina	Specific Surface Area m^2/g
1	2.14	0.65	3.78	26.0	392
2	2.25	0.72	3.02	16.1	252
3	2.18	0.71	2.46	15.6	194
4	1.86	0.06	5.51	20.6	388
5	1.74	0.01	5.84	9.9	274
6	1.96	0.006	5.30	12.1	202
7	----	----	4.12	<2.0	274

Table 3-3. Typical Film Properties. Bugosh (1961)

	Levelling & Uniformity	Water Resistance	Hardness	Powderiness	Birefringence*	Clarity	Cracking & Crazing
1	Fair	Fair	Poor	Good	None	Good	None
2	Poor	Poor	Poor	Poor	Slight	Fair	None
3	Very Poor	Very Poor	Poor	Poor	Slight	Poor	None
4	Excellent	Very Good	Excellent	Excellent	Upon Tilting	Excellent	Few cracks, not crazed
5	Fair	Poor	Poor	Poor	Good	Fair	None
6	Fair	Very Poor	Poor	Poor	Intense	Fair	None
7	Fair	Poor	Poor	Poor	Slight	Good	None

* Birefringence between crossed polaroids

A series of fibrous boehmite sols was prepared (Table 3-2) containing a range of particle lengths, specific surface areas, anion contents, and unpolymerized aluminas and was then compared in film-forming ability. These sols were made by heating a basic aluminum chloride (Al/Cl = 6/3; 0.4 M Al) for one hr, 4 h, and 16 h at 160°C. These sols contained fibrous particles which by quantitative streaming birefringence measurements had most frequent particle lengths of 400 nm, 800 nm, and 1600 nm, respectively.

Sols 1, 2, and 3 were the original sols; sols 4, 5, and 6 correspond to 1, 2, and 3 except that they were deionized with an ion-exchange resin; sol 7 is sol 4 reheated to reduce the amount of unpolymerized alumina present. Enough of each of the sols was used to give a film thickness corresponding to 2 mg Al_2O_3/ cm^2, by air drying the sols on glass. Typical film properties obtained from these sols are given in Table 3-3. The differences in leveling and uniformity were primarily a result of the differences in viscosity and thixotropy, i.e., the most viscous sols would not flow evenly over the glass plate and so appeared somewhat non-uniform. In each case there was a noticeable improvement in water resistance as the ionic content of the sol was reduced. The very good water resistance of sample 4 was apparently due to the greater amount of interparticle reaction produced because of the much higher surface area of these particles. Since such films (e.g., from sample 4) were highly and uniformly birefringent, the particles present in the films were probably not arranged in a random, disordered manner (as in an alumina gel), but rather oriented with their long axes parallel to the substrate, Bugosh (1961).

2. Hydrolysis of Fe^{+3}

Iron is commonly present and plays an important role in many areas such as geochemistry, environmental chemistry, and, of course, corrosion; and the hydrolysis of Fe(III)⁻ salts has therefore received a great deal of attention in the last three decades. Fe^{3+} is readily hydrolyzed in aqueous solutions and forms dimers and trimers before particles of different shape and composition are formed, Baes and Mesmer (1976). Common crystalline phases are α-FeOOH (goethite), β-FeOOH (akaganeite), α-Fe_2O_3 (hematite), γ-FeOOH (lepidocrocite), and γ-Fe_2O_3 (maghemite). When a base, e.g., NaOH solution, is rapidly added to a Fe^{+3} solution so that the ratio OH:Fe = 3, an amorphous precipitate will form instantly, which will in time turn crystalline. Table 3-4 gives formation quotients for products formed by hydrolysis of Fe^{3+} ions in aqueous solutions, and also of Fe^{2+} ions, although this ion, compared with Fe^{3+}, undergoes only slight hydrolysis, Baes and Mesmer (1976). In all of these hydrolysis products Fe^{3+} is octahedrally coordinated to O^{2-}, OH^-, or H_2O. It is only on the alkaline side Fe^{3+} assumes a tetrahedral coordination, as in $Fe(OH)_4^-$.

The formation quotient of $Fe_2(OH)_2^{4+}$ indicates that considerable amounts of this complex will be formed even at low concentrations of Fe^{3+}. The low value of the formation constant of the ferrate ion, $Fe(OH)_4^-$, log K_4 = -21.6, means that it does not form at all at low pH. It will, however, form at pH above 9, as is shown by Fig. 3-24, which also shows the distribution of the different hydrolysis products.

Figure 3-25a shows that only mononuclear complexes are formed at low Fe^{3+} concentrations, 10^{-5} M Fe, whereas polynuclear complexes will form at higher concentrations, 0.1M Fe, Fig. 3-25b.

Table3-4. Hydrolysis products of Fe^{2+} and Fe^{3+}.
$$\log Q_{xy} = \log K_{xy} + aI^{0.5}/(1 + I^{0.5}) + bmX \qquad \text{Baes and Mesmer (1976)}$$

Species or phase	log K_{xy}	a	b mX=0.1	mX=1.0	mX=3.0	σ(log Q_{xy})
			Fe^{2+}			
$FeOH^+$	-9.5	-1.022	0.4	0.2	0.1	±0.1
$Fe(OH)_2$	-20.6	-1.022	0.30	0.05	-0.04	±1.0
$Fe(OH)_3^-$	-31	0	-0.05	-0.21	-0.26	±1.5
$Fe(OH)_4^{2-}$	-46	2.044	-0.34	-0.34	-0.34	±0.3
$Fe(OH)_2$ (act) (log Q_{s10})	12.85	1.022	-0.30	-0.05	-0.04	±0.2
			Fe^{3+}			
$Fe(OH)^{2+}$	-2.19	-2.044	1.18	0.44	0.18	±0.02
$Fe(OH)_2^+$	-5.67	-3.066	2.02	0.78	0.25	±0.1
$Fe(OH)_3$	< -12	-3.066	2.40	0.86	0.34	
$Fe(OH)_4^-$	-21.6	-2.044	2.08	0.67	0.13	±0.2
$Fe_2(OH)_2^{4+}$	-2.95	0	1.07	0.32	0.06	±0.05
$Fe_3(OH)_4^{5+}$	-6.3	1.022		(0)		±0.1
α-FeO(OH)(log Q_{s10})	0.05	3.066	-2.40	-0.86	-0.34	±0.8
FeO(OH)(am)	2.5					±0.1

Reprinted by permission of John Wiley & Sons, Inc.

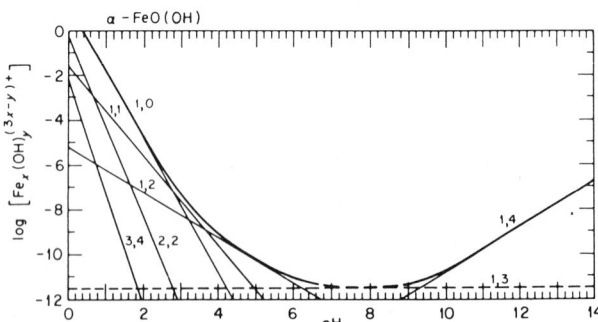

Fig. 3-24 Distribution of hydrolysis products (x,y) at ionic strength = 1M and 25°C in solutions saturated with α-FeOOH as a function of pH. The heavy curve is the total concentration of Fe(III). Baes and Mesmer (1976). Reprinted by permission of John Wiley & Sons, Inc.

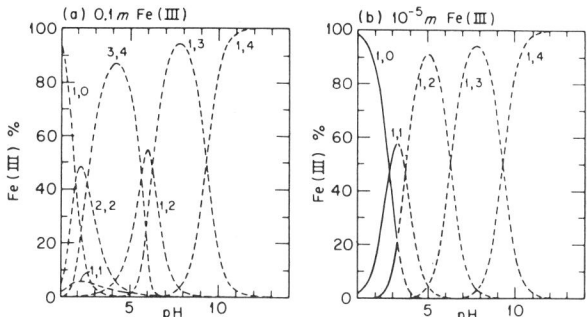

Fig. 3-25. Distribution of hydrolysis products (x,y) at ionic strength = 1M and 25°C in: (a) 0.1M Fe(III), and (b) 10^{-5} M Fe(III). Baes and Mesmer (1976). Reprinted by permission of John Wiley & Sons, Inc.

Table 3-5 shows the solubility products of particles of the various oxides and hydrous oxides that can be formed by hydrolysis of aqueous solutions of Fe^{3+}.

Table 3-5. Solubility products of various oxides and hydrous oxides of iron. $(1/2)Fe_2O_3(s)+ (3/2)H_2O = Fe^{3+} +3OH^-$ $FeO(OH)(s) + H_2O = Fe^{3+} + 3OH^-$ Flynn (1984)	
Solid phase	$-\log K_{s0}$[a]
$(1/2)\alpha$-Fe_2O_3 (hematite)	41.7[b] 42.7 ≤41.9±0.4
$(1/2)\gamma$-Fe_2O_3 (maghemite)	≥38.8±0.5
α–$FeO(OH)$ (goethite)	41.7[b] 41.5 ≤41.2±0.4
β-$FeO(OH)$ (akaganeite)	36 + x[c]
γ-$FeO(OH)$ (lepidocrocite)	≥38.7±0.5
"$Fe(OH)_3$" (amorphous)	37.1-39.0[d]

[a]Uncertainties are ±0.1 to 0.2 in most cases. [b] Based on $\Delta G^{\circ}_{f\ 298}$ [Fe^{3+}(aq)] = -17 kJ/mol (see text) and CODATA values for H_2O(l) and OH^-(aq). [c] Estimated from data given for $Fe(OH)_{2.7}Cl_{0.3}$ in references, with correction for chloride complexation from data of Smith and Martell (1976), and estimation of a correction of ~1 log K unit to convert from ionic strength 0.5 to 0. The term x is log K for the reaction β-$FeO(OH)$(s) + H_3O + $0.31Cl^-$ = β-$Fe(OH)_{2.7}Cl_{0.3}$(s) + $0.3OH^-$. [d] Range for fresh to aged precipitate.
Courtesy American Chemical Society

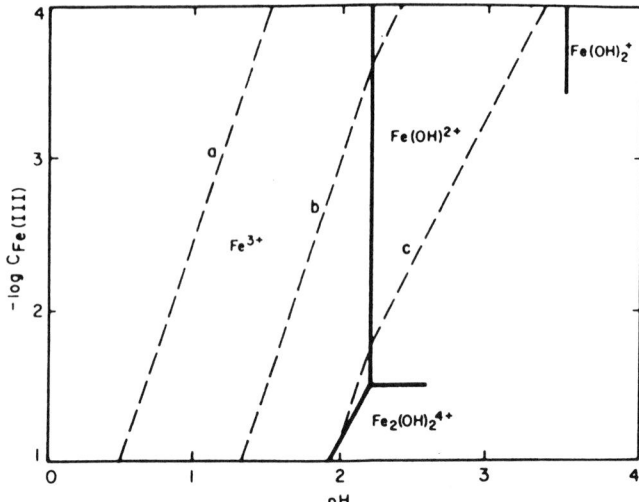

Fig. 3-26 Approximate predominance-area pH vs. log $C_{Fe(III)}$ (M) diagram, 25 °C, ionic strength 0. The dashed lines labeled a, b, and c are the saturation lines for goethite, aged amorphous hydrous oxide, and fresh amorphous hydrous oxide, respectively. Flynn (1984). Courtesy American Chemical Society.

The saturation lines of three different Fe(III) hydroxides and hydrous oxides, i.e., $[Fe^{3+}]$ in equilibrium with the precipitated materials, are shown in Fig. 3-26.

2.1. Polymerization and Formation of Colloidal Particles

An intermediate step in the formation of α-FeOOH or α-Fe$_2$O$_3$ is the development of cationic, subcolloidal particles in the early stages of the hydrolysis. If a base is added to solutions of Fe(III)-chloride, -nitrate, or -perchlorate in amounts insufficient to cause precipitation, the yellowish color of the solution turns reddish brown due to the formation of small particles, about 2 nm, of hydrous iron oxide. If the base is added dropwise, local precipitation may occur, but if the amount of base added is such that the ratio OH:Fe is less than 2.5, the precipitate will always redissolve. The molecular weight, as determined by gel permeation chromatography, is of the order 10^4, Murphy et al. (1976), and the number of iron atoms per particle can be estimated to be about 100 in the form of monomers of the average composition Fe(OH)$_{2.5}^{0.5+}$, see §2 of Ch. 1.

If the solutions of subcolloidal particles are left standing, pH will gradually decrease, indicating that processes leading to the formation of larger particles occur. Protons are split off when condensation reactions take place by the formation of hydrogen bridges—olation; or oxo bridges—oxolation. Olation is considered to be much slower than oxolation.

$$\text{(oxo)} \quad [Fe(OH)_2^+]_n \quad \longrightarrow \quad [Fe(OH)]_n + nH^+$$

$$\text{(hydroxo)} \quad [Fe(OH)_2^+]_n + Fe^{+3} + 2H_2O \longrightarrow [Fe(OH)]_{n+1} + 2H^+$$

The anion in the Fe(III) system plays an important role in the kinetics of particle formation for the composition of the particles, Flynn (1984). The chloride ion gives preference to β-FeOOH whereas α-FeOOH is formed in systems where ClO_4^- or NO_3^- are the anions. The SO_4^{2-}-ion, in contrast to monovalent anions, come into being at the beginning of the particle formation process as an iron sulfate complex, in which SO_4^{2-} is strongly coordinated to the central atom. Particles form faster and at lower ratios of OH to Fe, compared to systems containing monovalent anions, and the preferred compositions are α-Fe_2O_3 or α-FeOOH. When the particle formation is slow it appears that all the sulfate ions leave the structure during the course of particle formation. Figure 3-27 summarizes schematically the spontaneous polymerization of Fe^{3+} in aqueous solutions to colloidal and larger particles.

The slow formation of colloidal particles by spontaneous hydrolysis can be speeded up by raising the temperature and/or adding a base, i.e., by so-called forced hydrolysis. Matijevic and co-workers have done pioneering work in studying particle formation by forced hydrolysis in solutions of Fe^{3+} ions, Matijevic et al. (1975, 1978), Ozaki et al. (1984), Sapieszko et al. (1972), Hamada and Matijevic (1981). They found that many factors such as temperature, time, total concentration of metal ions, type of anion, molar ratio between anion and metal cation, and the composition of the solvent affected the size, morphology, and composition of the particles. The particles grow with increasing temperature, reaction time, and total metal ion concentration, but it is more difficult to predict how the other factors affect the particle formation. This is demonstrated by Fig. 3-28, which shows how particle shape and composition vary with the concentration of $FeCl_3$ and HCl, Matijevic and Scheiner (1978). Spherical, ellipsoidal, rod-like, irregular, and double ellipsoidal shapes have been observed in this system depending on the molar ratio between $FeCl_3$ and HCl.

Matijevic and co-workers also studied the effect of other anions such as NO_3^-, ClO_4^-, $H_2PO_4^-$, $H_2PO_3^-$, and HSO_4^- on particle formation, and Fig. 3-29 shows micrographs of particles prepared with different anions.

The micrographs demonstrate clearly that the shape and morphology depend strongly on the anion present in the system. Different ratios between Fe^{3+} and anion, Cl^-, and temperature resulted in not only different shapes, but also in different compositions (Figs. 3-29a and -b). With 50% ethanol in the system α-Fe_2O_3, particles of more pronounced cubical shape were formed (Fig. 3-29-d) than in Fig.3-29-b. In contrast to other anions, HSO_4^- is incorporated into the

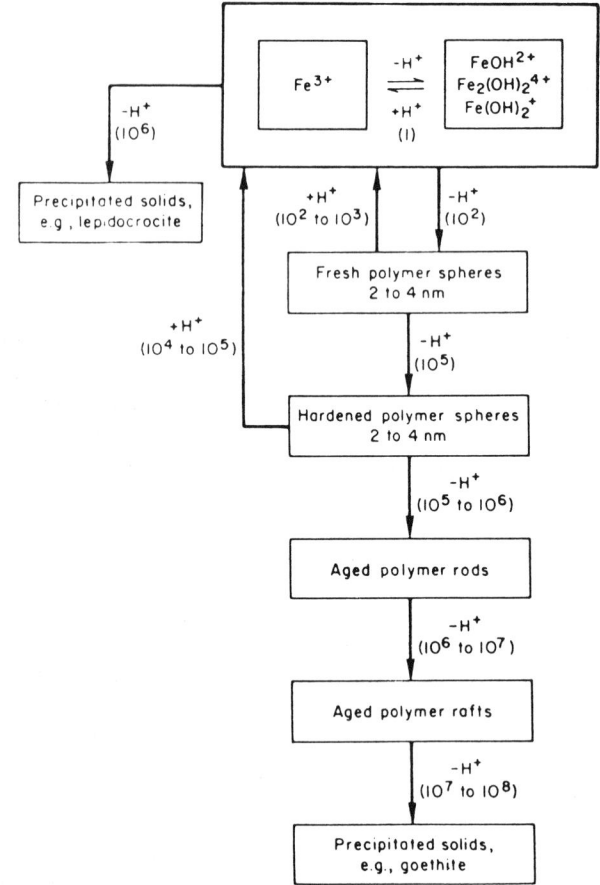

Fig. 3-27. Hydrolysis processes in Fe(III) solutions. Numbers in parentheses give reaction time(s) at 25°C. Flynn (1984). Courtesy American Chemical Society.

final product (Fig. 3-29h), which may be due to the fact that sulfate ions may form from HSO_4^- by deprotonation and this ion coordinates much more strongly to Fe^{3+} than OH^-.

The theory of hydrolysis is not yet sufficiently advanced to allow reliable predictions as to the effect of the type of anion on particle shape and composition.

Reaction kinetics is probably very important and polymerization mechanisms leading to particles of a given shape and composition may be favored in certain domains of temperature and ratio of anion to Fe^{3+}. While awaiting an adequate theory, empiricism may still provide useful guidance as to what experimental conditions will give particles of desired shape and composition.

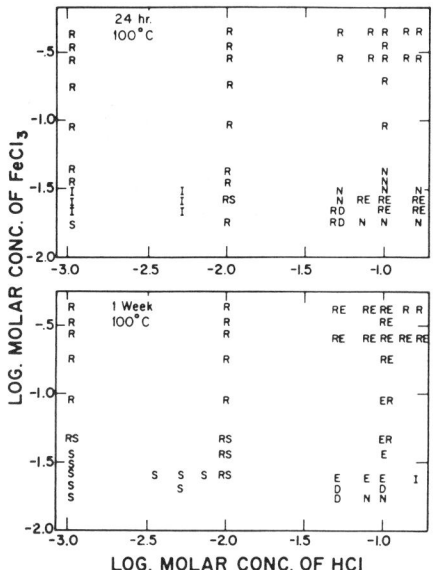

Fig. 3-28 Concentration domains of solutions containing $FeCl_3$ and HCl aged at 100°C for 24 h (upper) and for 1 week (lower). Symbols: N, no particle formation. Particle shapes: D, double ellipsoid; E, ellipsoidal; I, irregular of varying sizes; R, rodlike; S, spherical. Pairing of symbols indicate a mixture of corresponding particles in the suspension. Particle composition: R, β-FeOOH; all other particles α-Fe_2O_3. Matijevic and Scheiner (1978). Courtesy Academic Press.

The anions may play a dual role in the particle formation. At the beginning of the hydrolysis they coordinate to Fe^{3+} with a strength that depends on the type of anion. However, after the subcolloidal cationic particles have been formed, anions may act as coagulants and cause them to aggregate to larger particles. From TEM micrographs it is sometimes possible to decide whether colloidal particles of iron oxides and hydrous oxides, and indeed other metal oxides and hydrous oxides, are aggregates of subcolloidal particles or have been formed by a growth mechanism involving adding monomeric species to the growing particles.

3. Hydrolysis of Ti^{4+}

The Ti^{4+} cation readily undergoes hydrolysis in aqueous solutions at low pH to hydroxo complexes such as $Ti(OH)_2^{2+}$ and $Ti(OH)_3^+$. Figure 3-30 shows that these complexes exist only at pH below 3 and that the solubility of $Ti(OH)_4$ is very low above pH 2. Even in very acid solution, e.g., 1.5 M $HClO_4$, it is completely hydrolyzed to $Ti(OH)_2^{2+}$. There has been speculation as to the existence of a titanyl ion, TiO^{2+}, Baes and Mesmer (1976), but it probably has the composition $Ti(OH)_2^{2+}$.

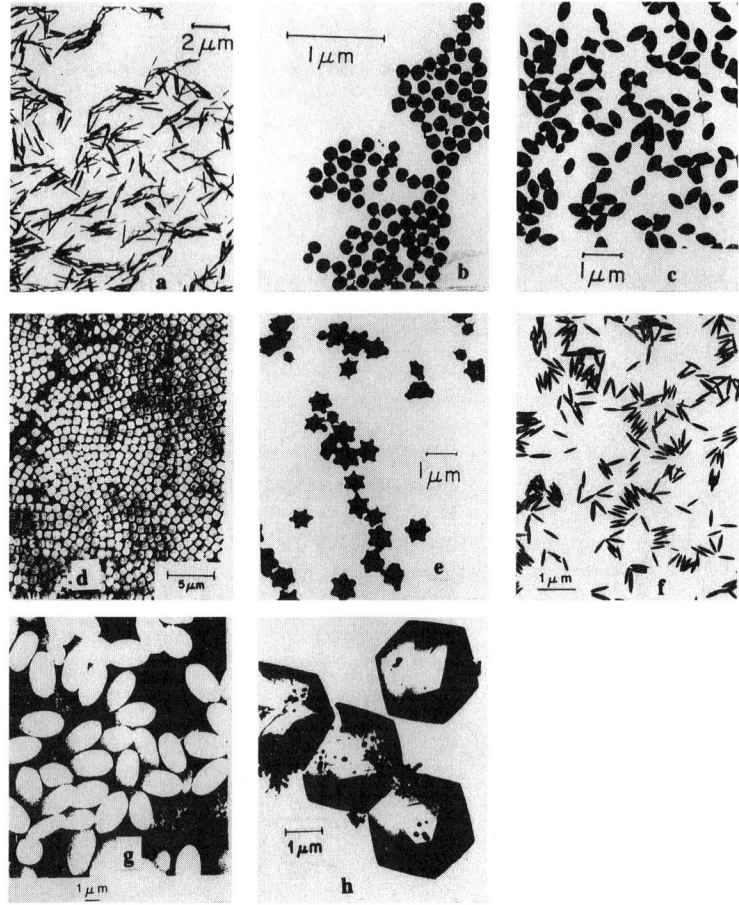

Fig. 3-29. Particles of Fe(III) oxides and hydrous oxides prepared with different anions. (a) β-FeOOH obtained by aging at 100°C 0.45 M FeCl₃ + 0.01 HCl for 1 week. (b) Aging a solution of 0.018 M FeCl₃ and 0.001 M HCl for 24 h at 100°C. (c))Aging a solution .018M in Fe(NO₃)₃ and 0.05 M HNO₃ for 24 h at 100°C. (d) α-Fe₂O₃ separated in system of β-FeOOH particles precipitated by aging 50% v/v ethanol/water solution of 0.019 M FeCl₃ and 0.0012 M HCl for 2 days at 100°C. (e) α-Fe₂O₃ sol prepared by aging 0.018 M Fe(ClO₄)₃ and 0.05 M HClO₄ for 3 days at 100°C. (f) Spindle-type hematite particles prepared in presence of phosphate ion. (g) Hematite particles prepared in presence of hypophosphite ion (1.2x10⁻⁴ M) obtained in water/ethanol mixture (50%). (h) Basic ferric sulfate sol particles prepared by aging a solution 0.088 M in Fe₂(SO₄)₂ and 0.17 M urea at 98±5°C for 3 h. Matijevic and Scheiner (1978): (a), (b), (c), and (e). Hamada and Matijevic (1981): (d). Ozaki et al. (1984): (f), (g). Courtesy Academic Press.

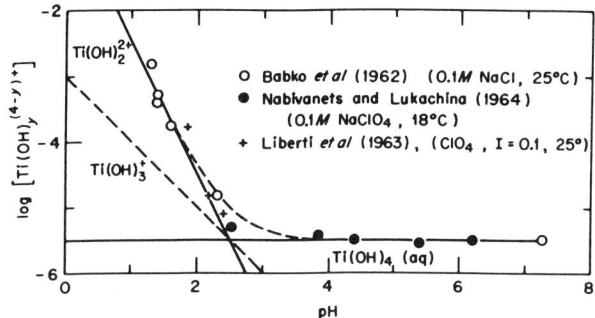

Fig. 3-30 Solubility behavior of hydrous TiO_2. Lines indicate the estimated contributions of $Ti(OH)_2^{2+}$, $Ti(OH)_3^+$, and $Ti(OH)_4(aq)$ to the solubility, Baes and Mesmer (1976). Reprinted by permission of John Wiley & Sons, Inc..

Polycations of Ti^{4+} have not been found, which is probably due to the fact that it is very difficult to identify such species in strongly acidic solutions. Table 3-6 shows the hydrolysis products, and corresponding formation constants, formed in aqueous solutions of Ti^{3+} and Ti^{4+}.

Table 3-6. Hydrolysis products of Ti^{3+} and Ti^{4+} in aqueous solutions at 25°C. Baes and Mesmer (1976)

$$\log Q_{xy} = \log K_{xy} + aI^{0.5}/(1+I^{0.5}) + bm_x$$

Ti^{3+}				
Species or phase	$\log K_{xy}$	a	b	$\sigma(\log Q_{xy})$
$TiOH^{2+}$	-2.2	-2.044	(0.4)	± 0.3
$Ti_2(OH)_2^{4+}$	-3.6	0	(0)	±0.5
$Ti(OH)_2^{2+}$				
$Ti(OH)_3^+$	< -2.3[a]	-1.022[a]	(0.1)[a]	
$Ti(OH)_4(aq)$	-4.8[a]	-1.022[a]	(0)[a]	±0.3
$TiO_2(c)$ ($\log Q_{s12}$)	~ -4.8[a]	-1.022[a]	(0)[a]	?

[a] Defined for reactions of the type

$$Ti(OH)_2^{2+} + (y-2)H_2O = Ti(OH)_y^{(4-y)+} + (y-2)H^+$$

Reprinted by permission of John Wiley & Sons, Inc.

3.1. Polymerization and Formation of Colloidal Particles

Since small particles of TiO_2 are widely used, especially as pigments (see Chapters 8 and 9), but also as catalyst supports, their formation under various conditions has been extensively studied. One important commercial process for producing pigment-grade TiO_2 particles involves leaching ilmenite, $FeTiO_3$, with H_2SO_4 and boiling the leachate, which results in small particles of irregular size and shape. Small particles of TiO_2 of uniform size and shape have been made by heating acid solutions of $TiCl_4$ containing Na_2SO_4, Matijevic et al. (1977), and

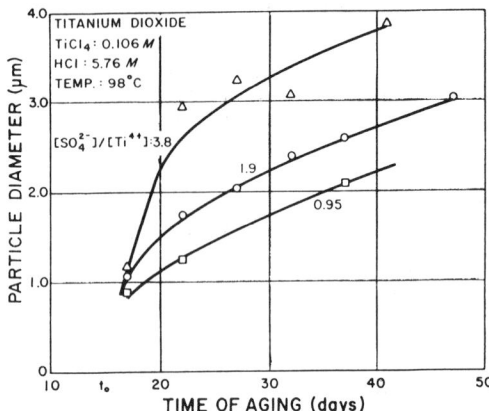

Fig. 3-31 Plot of the modal particle diameter of titanium dioxide sols as a function of days of aging at 98°C of solutions of 0.106 M in $TiCl_4$ in 5.76 M HCl which contained different concentrations of Na_2SO_4 to give $[SO_4^{2-}]:[Ti^{4+}]$ equal to 0.95($\square$), 1.9 (O), and 3.8($\triangle$), respectively. Matijevic et al. (1977). Courtesy Academic Press.

Fig. 3-31 shows that the rate of particle growth depends strongly on the ratio $[SO_4^{2-}]:[Ti^{4+}]$.

In contrast to the system Fe^{3+}-SO_4^{2-} (see § 2.1 this chapter), SO_4^{2-} was not found in the final product. This is probably due to the fact that SO_4^{2-} does form a sulfate complex with Ti^{4+}, but it will slowly decompose on heating and give rise to hydrolysis species of titanium, which will grow to small particles. In terms of the language of the La Mer theory of particle formation and growth (see Ch. 2), the sulfate ions can be said to affect the number of nuclei in the system, which in turn will affect the final particle size.

Matijevic et al. (1977) proposed the following mechanism for the formation of TiO_2 particles by hydrolysis of aqueous solutions of Ti^{4+}. The hydroxy ligands in reaction (a) are provided by water, which first coordinates to Ti^{4+}, but then immediately becomes deprotonated. Reaction (b), which involves polymerization by formation of hydroxo bridges, olation, is considered to be the

$$Ti^{4+} + y\,OH^- \rightleftarrows Ti(OH)_y^{(4-y)+} \qquad (a)$$

$$2Ti(OH)_y^{(4-y)+} \rightleftarrows (2y-2)\,OH^- + \;\underset{H}{\overset{H}{Ti\diagup O \diagdown Ti}} \quad (\equiv E\;) \quad (b)$$

$$E \rightleftarrows H_2O + \;\rangle Ti - Ti\langle \; (\equiv F) \quad (c)$$

$$E + F \longrightarrow TiO_2 \;\text{(crystals)}$$

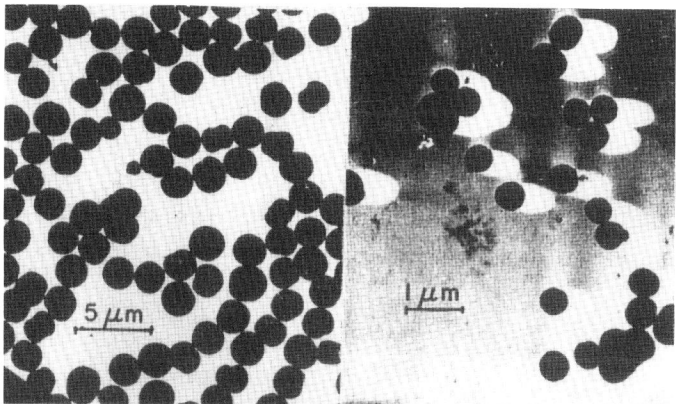

Fig. 3-32 *Left:* Electron micrograph of titanium dioxide sol particles obtained by aging for 37 days at 98°C a solution of 0.106 M TiCl$_4$ in 5.76 M HCl containing Na$_2$SO$_4$ to give [SO$_4^{2-}$]:[Ti^{4+}] = 1.9. *Right.* Electron micrograph of chromium-shadowed titanium dioxide sol particles obtained by aging for 16 days at 98°C a solution of 0.1M TiCl$_4$ in 4.9 M HCl which contained Na2SO4 to give [SO$_4^{2-}$]:[Ti^{4+}] = 2.0. Matijevic et al. (1977). Courtesy Academic Press.

rate determining step in the particle formation. The final particles were found to consist of rutile.

Figure 3-32 shows TEM-micrographs of particles formed by hydrolysis of Ti^{4+} in solutions containing HCl and Na$_2$SO$_4$ at 98°C after (a) 37 days and (b) 16 days.

There is relatively little information in the literature on the formation of small particles of titania of uniform size and shape by the hydrolysis of titanium salts. There is much more information on the formation of such particles by hydrolysis of titanium alkoxides described in Section D below.

4. Hydrolysis of Zr^{4+}

Like Ti^{4+}, Zr^{4+} and Hf^{4+} undergo hydrolysis in strongly acidic solution. In solutions containing 0.5-2 M HCl, trimers and tetramers are formed and Fig. 3-33 shows the structure of the tetramer with Cl$^-$ and Br$^-$ as counterions, in which each zirconium ion has two hydroxyl ligands, Baes and Mesmer (1976).

Table 3-7 summarizes the hydrolysis products of Zr^{4+} (and Hf^{4+}) that have been identified and their formation constants.

Figure 3-34 shows the distribution of hydrolysis products in solutions of (a) 0.1 M Zr^{4+} and (b) 0.01 M Zr^{4+} at 25°C as a function of pH. In the pH range from 2-5 the tetramer is the dominant species and is designated by the number

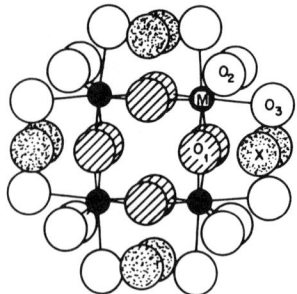

Fig. 3-33. Square planar structure of $M_4(OH)_8(H_2O)_{16}X_z^{(8-z)+}$ in solution, showing all eight positions for anion binding. M, Zr^{+4} or Hf^{+4}; X, Cl^-, or Br^-; O_1, OH oxygen, O_2 and O_3, H_2O oxygens. Muha and Vaughan (1960). Courtesy American Institute of Physics.

pair (4,8) in the figure. In Fig. 3-34c are given species complexes present in solutions saturated with zirconia, ZrO_2, as a function of pH. The solubility of ZrO_2 increases strongly at pH below 2 and also, but less strongly, at pH above 6, where the zirconate ion, $Zr(OH)_5^-$, is formed.

Table 3-7 Hydrolysis products of Zr^{4+} and Hf^{4+}.
$\log Q_{xy} = \log K_{xy} + aI^{0.5}/(1 + I^{0.5}) = bm_x.$ Baes and Mesmer (1976)

Species or phase	Log K_{xy}		a	b			$\sigma(\log Q_{xy})$	
	Zr^{4+}	Hf^{4+}		$\bar{m}_z = 0.1$	$m_z = 1.0$	$m_z = 3.0$	Zr^{4+}	$Hf4+$
MOH^{3+}	0.3	0.25	-3.066	1.4	0.61	0.34	0.05	0.3
$M(OH)_2^{2+}$	(1.7)	(2.4)	-5.110	5.0	1.96	0.95	?	?
$M(OH)_3^+$	(5.1)	(6.0)	-6.132	5.0	1.81	0.73	?	?
$M(OH)_{4(aq)}$	9.7	-10.7	-6.132	5.0	1.78	0.68	~1	~1
$M(OH)_5^-$	16.0	-17.2	5.11	4.6	1.45	0.39	?	?
$M_3(OH)_4^{8+a}$	-0.6^a		10.22		(0)		0.1	
$M_3(OH)_5^{7+a}$	3.70^a		3.066		(0)		0.1	
$M_4(OH)_6^{8+}$	6.0		4.088		(0)		0.1	
$MO_2(c)$ (log Q_{s10})	-1.9	-1.2	6.132	-5.0	-1.78	0.68	~1	~1

·z = ClO_4. [a] One or both of these trimeric species seem to form. The constants are calculated from Zielen and Connick (1956) assuming that only one is formed.
Reprinted by permission of John Wiley & sons, Inc.

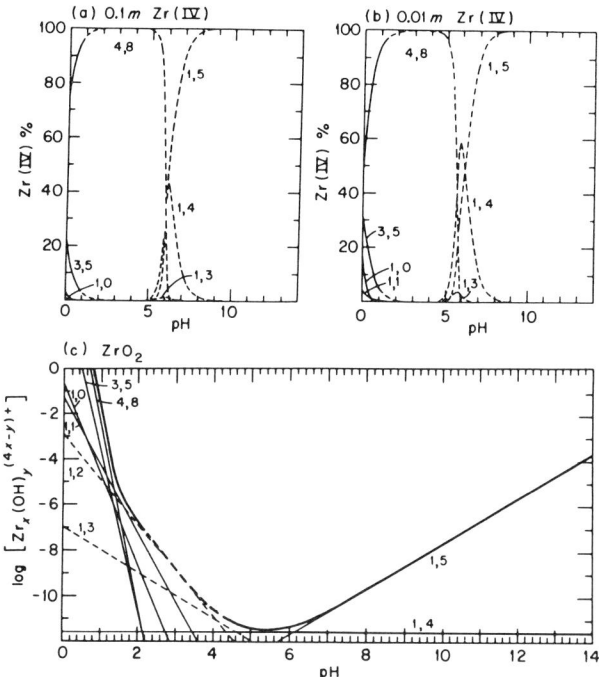

Fig. 3-34 Distribution of hydrolysis products (x,y) at I = 1 M and 25°C in (a) 0.1 M Zr(IV), (b) 0.01 M Zr(IV), and (c) solutions saturated with ZrO_2. The dashed curves in (a) and (b) denote regions supersaturated with respect to ZrO_2; the heavy curve in (c) is the total concentration of Zr(IV). Baes and Mesmer (1976). Reprinted by permission of John Wiley & Sons, Inc.

4.1. Polymerization and Formation of Colloidal Particles

High-tech ceramics of zirconia are being used in many applications where the special optical, thermal, mechanical, or electrical properties of zirconium dioxide are required. Since these properties are quite sensitive to the exact composition of the zirconia, it is important to have access to methods for preparing particles of zirconia of uniform size and well defined composition. Such particles, having the composition $Zr_2(OH)_6SO_4$, can be prepared by hydrolysis of aqueous solutions containing a mixture of $Zr(SO_4)_2$, polyvinyl pyrrolidone, dilute HNO_3 and urea at elevated temperatures, where urea decomposes and releases hydroxyl ions which promote hydrolysis of the Zr^{4+} ions, Aiken et al. (1990). Figure 3-35 shows the domain of urea and $Zr(SO_4)_2$ concentrations in which monodisperse, spherical particles of this composition will be formed.

Figure 3-36 shows TEM micrographs of particles of the composition $Zr_2(OH)_6SO_4$ made by hydrolysis of the aqueous mixture at 50°C for five hours. These particles can be converted to particles of pure ZrO_2 by calcination at 700°C. Direct preparation of colloidal, monodisperse particles of ZrO_2 by forced

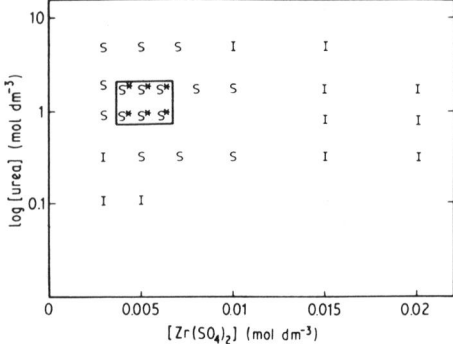

Fig. 3-35 Precipitation domain for systems containing different concentrations of $Zr(SO_4)_2$ and urea at constant concentrations of PVP (3% by weight) and HNO_3 (5×10^{-2} mol dm^{-3}). Solutions were aged for 5 h at 50°C in sealed test tubes. Enclosed area represents conditions for the formation of uniform spherical particles. Symbols: S (spheres), S* (uniform spheres) and I (coagulated spheres). Aiken et al. (1990). Courtesy Chapman & Hall.

hydrolysis has not yet been reported. However, small particles of monoclinic ZrO_2 of irregular size and shape were prepared by forced hydrolysis of aqueous solutions of $ZrO(NO_3)_2$ at 98°C for 70 h—see Fig. 3-37.

Fig. 3-36. Transmission electron micrograph of particles of $Zr_2(OH)_6SO_4$ obtained by aging for 5 h at 50°C a solution of 5×10^{-3} mol dm^{-3} $Zr(SO_4)_2$, 5×10^{-3} HNO_3 mol dm^{-3}, 1.8 mol dm^{-3} urea and 3% PVP and then calcined at 600°C for 3 h. Aiken et al. (1990). Courtesy Chapman & Hall.

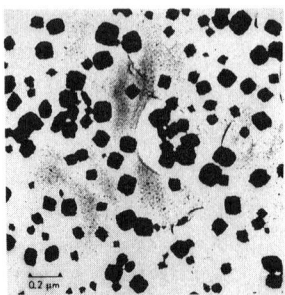

Fig. 3-37. Typical m-ZrO_2 powder produced at 98°C using $ZrO(NO_3)_2$. Bleier and Cannon (1986). Courtesy MRS & Material Research Society.

5. Hydrolysis of Cr^{3+}

Small particles of chromia are used in such applications as pigments and oxidation catalysts and basic Cr(III) sulfate is a widely used tanning agent. Cr^{3+} hydrolyzes in aqueous solutions and rapidly forms two mononuclear complexes, $Cr(H_2O)_5OH^{2+}$ and $Cr(H_2O)_4(OH)_2^+$. The formation of the polynuclear complexes $Cr_2(OH)_2^+$ and $Cr_3(OH)_4^{5+}$ is much slower, which, according to ligand field theory, depends on the fact that a large energy barrier must be overcome when Cr^{3+} changes its state of coordination. The formation of polynuclear complexes requires that the stable octahedral configuration in the mononuclear complexes changes to quadratic bipyramidal or pentagonal-bipyramidal configuration in the polynuclear complexes. The activation energy for this transformation is greatest for the transition metal ions of the d^3 type.

The distribution of species and solubility of hydrolysis products of Cr^{3+} are shown in Fig. 3-38 as a function of pH, Baes and Mesmer (1976).

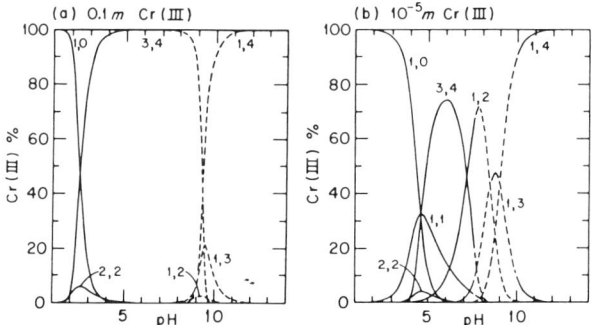

Fig. 3-38 Distribution of hydrolysis products (x,y) of ionic strength =1 M at 25°C in (a) 0.1 M Cr(III) and (b) 10^{-5} M Cr(III) as a function of pH. The dashed curves denote regions supersaturated with respect to $Cr(OH)_3$. Baes and Mesmer (1976). Reprinted by permission of John Wiley & Sons, Inc.

Polynuclear complexes like the (3,4) complex $Cr_3(OH)_4^{5+}$ can form at as low dilution as 10^{-5} M Cr(III). Figure 3-39 shows the concentration of the different Cr(III) complexes as a function of pH.

5.1. Hydrolysis and Formation of Small Particles

Matijevic et al. showed that small particles of $Cr(OH)_3$ are formed when a solution of Cr^{3+} ions is heated and the anion is either SO_4^{2-} or PO_4^{3-}, but not if the anion is Cl^- or ClO_4^-, Demchak et al. (1969), Matijevic et al. (1971), Bell and Matijevic (1975). Monodisperse sols of $Cr(OH)_3$ can be prepared at 75°C within

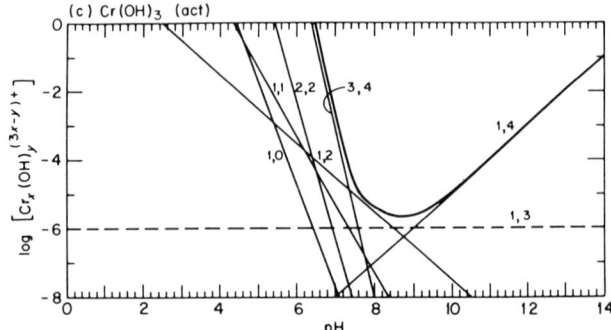

Fig. 3-39. Concentration and distribution of hydrolysis products (x,y) at ionic strength = 1 M and 25°C in solutions saturated with Cr(OH)$_3$. The heavy curve is the total concentration of Cr(III). Baes and Mesmer (1976). Reprinted by permission of John Wiley & Sons, Inc.

the domain of Cr^{3+} concentrations and molar ratios between Cr^{3+} and H$_3$PO$_4^-$ shown in Fig. 3-40. The shaded area within heavy lines denoted "HOTS" (Higher Order Tyndall Spectra) in the figure indicates conditions which produce sols having a relatively narrow particle size distribution and showing Tyndall scattering.

With SO$_4^{2-}$ as the anion, Fig. 3-41 shows the conditions of concentration of CrK(SO$_4$)$_2$ and pH which give particles of Cr(OH)$_3$ of narrow particle size distribution at different temperatures.

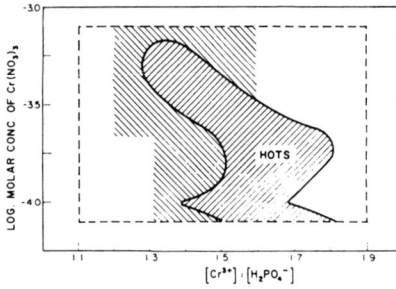

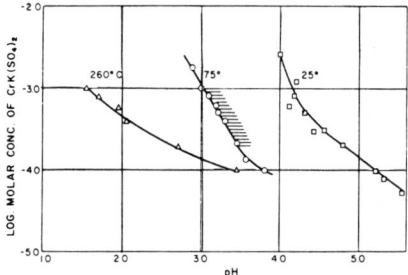

Fig. 3-40. The Cr(NO$_3$)$_3$-KH$_2$PO$_4$ domain of chromium hydroxide precipitate formation. Shaded area within the heavy lines gives conditions which produce sols exhibiting HOTS. The sols in the adjacent shaded area show some indications of spectra. Matijevic et al. (1971). Courtesy Chapman & Hall.

Fig. 3-41. Precipitation boundary of chromium hydroxide sols using chrom alum solutions at three different temperatures. The shaded area gives the conditions under which sols of narrow size distributions exhibiting HOTS are produced. Matijevic et al. (1971). Courtesy Chapman & Hall.

Fig. 3-42. Electron micrograph of chromium hydroxide sol obtained by aging of an 8 x 10^{-4} M solution of $CrK(SO_4)_2$ at 75°C for 21 h. The modal diameter of the particles, D_m = 313 nm. Matijevic et al. (1971). Courtesy Chapman & Hall.

Figure 3-42 shows a micrograph of about 300 nm $Cr(OH)_3$ particles made by heating an aqueous solution of $CrK(SO_4)_2$, $8x10^{-4}$ M, at 75°C for 21 h.

6. Hydrolysis of Y^{3+} and Lanthanide Ions, Ln^{3+}

Yttria is used in electrical technology, e.g., in capacitors and in lamp filaments, where it provides high emissivity. Ln^{3+} and Ce^{3+}, and hydrolysis complexes of these ions, are used as counter ions in the zeolites in zeolitic cracking catalysts (see Ch. 7) to impart acidity and hydrothermal stability to the catalysts.

The symbol Ln^{3+} denotes the rare earths from La^{3+} to Lu^{3+}, lutetium(III). The size of the rare earth ions of group 3B increases from Y^{3+}, 0.92Å, to Ac^{3+}, 1.18Å; but the ions in the 6th series decrease from La^{3+} to Lu^{3+}, 1.14 to 0.84Å, due to increasing nuclear charge which produces the "lanthanide contraction". The lanthanide contraction makes the behavior of the elements near the middle of the series closely similar to that of Y^{3+}. The outer, empty orbitals may, however, participate in bonding and be responsible for irregularities in the stability of hydrolysis complexes and in the stability of compounds of Ln^{3+} ions.

Ln^{3+} and Y^{3+} undergo only slight hydrolysis, producing small amounts of mononuclear and polynuclear species, before they precipitate as hydroxide above pH 6. Kumok and Serebrennikov (1975) found that the following linear correlation predicts quite well the increasing stability of the mononuclear species MOH^{2+} formed in the hydrolysis reaction $Ln^{3+} + OH^- \longrightarrow LnOH^{2+}$

$$\log Q\ LnOH \longrightarrow a \log Q\ LaOH + b \qquad (3\text{-}21)$$

where Q_{LnOH} and Q_{LaOH} are the formation quotients for $LnOH^{2+}$ and $LaOH^{2+}$ respectively, and a and b are constants. For $LaOH^{+2}$, a=1 and b=0. $M_2(OH)_2^{+4}$ (found for Nd^{3+}, Er^{3+}, and Y^{3+}), and $M_3(OH)_5^{4+}$ (found for Ce^{3+} and Y^{+3}), are considered to be the primary nuclear species, Baes and Mesmer (1976). The slight solubility of the lanthanide tri-hydroxides indicates the formation of $M(OH)_4^-$ species.

6.1. Polymerization and Formation of Colloidal Particles

Matijevic and Hsu (1987) prepared colloidal hydrous oxides of the trivalent ions of Gd, Eu, Tb, Sm, and Ce with narrow size distribution by aging solutions of lanthanide ions in the presence of urea. The morphology of the particles depends strongly mainly on several parameters: concentrations of lanthanide ions and of urea, pH, temperature, aging time, and type of anions. Table 3-8 shows the typical optimum conditions for preparing monodisperse particles at 85°C.

In a similar study, Sordelet and Akinc (1988) prepared spherical, monosized particles by homogeneous precipitation in aqueous solutions of Y^{3+} ions by reaction with the thermal decomposition products of urea. Monosized particles of yttria were obtained with yttrium ion concentrations up to 0.05 M, above which value deviation from spherical shape could be observed. Particle yield could be increased to near theoretical values by adding excess urea without deterioration of the spherical shape.

Table 3-8. Typical Optimum Conditions for the Preparation of Monodispersed Systems for Trivalent Lanthanides at 85°C. Matijevic and Hsu (1987)

Lanthanides (10^3 mole/dm^3)	Urea (mole/dm^3)	H_2SO_4 (10^3 mole/dm^3)	pH	Aging time (h)	Modal particle diam. (μm)
GdCl$_3$					
6	0.05	-	5.2	2	0.52
6	0.2	0.1	4.6	2	0.18
Gd(NO$_3$)$_3$					
5	0.5	-	-	2	0.27
TbCl$_3$					
14	1.7	-	5.8	1	0.18
6		-	5.5	1	0.28
Sm(NO$_3$)$_3$					
5	2.0	-	-	1	0.05
5	0.5	-	-	1	0.25
EuCl$_3$					
6	1.5	0.2	5.1	1	0.15
($+7 \times 10^{-4}$ mole/dm^3 HNO$_3$)					
6	1.7	0.3	5.4	1	0.15
6	0.4	0.3	5.3	1	0.25
Ce(NO$_3$)$_3$					
8	1.3	-	6.0	2	Ellipsoids

Courtesy Academic Press

7. Hydrolysis of Ce^{4+}

Ceria is used in small amounts to reduce discoloration caused by small amounts of iron (III) oxide in glass screens for color television sets. Cerium(IV) oxide is also used as an opacifier in white porcelain enamels and as an optical polishing agent. Ceria is an important catalyst in three-way catalysts for reducing emissions from motor vehicles, where it functions as a promoter for the noble metal catalysts; see Ch. 7.

The lanthanide ions are most stable in their trivalent state in water, but Ce^{4+} is also a stable ion due to the extra stability of its $4f^\circ$ configuration and is a powerful oxidizing agent. It hydrolyzes extensively in aqueous solutions to form a number of hydrolysis products, the composition of which depends on pH and concentration of Ce^{4+} ions; see Fig. 3-43.

At low concentrations of Ce^{4+} ions $CeOH^{3+}$, (1,1), is the dominant species at low pH whereas the concentration of $Ce(OH)_2^{2+}$, (1,2), increases with pH. The dimeric species $Ce_2(OH)_2^{6+}$, (2,2), and $Ce_2(OH)_3^{5+}$, (2,3), begin to form at higher concentrations of Ce^{4+} ions at low acidity. At higher pH, $Ce_2(OH)_4^{4+}$, (2,4), and $Ce_6(OH)_{12}^{12+}$ appear.

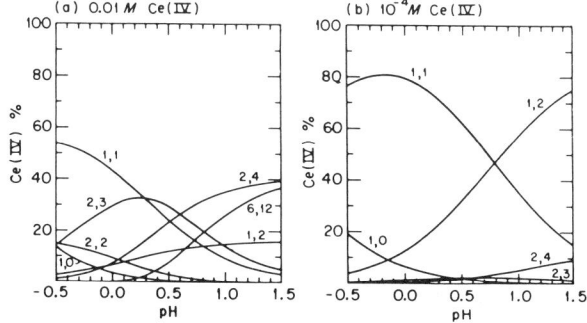

Fig. 3-43. Distribution of hydrolysis products (x,y) at I = 3 M $(ClO_4)^-$ at 25°C in (a) 0.01 M Ce(IV) and (b) 10-4 Ce(IV). Baes and Mesmer (1976). Reprinted by permission of John Wiley & Sons, Inc.

7.1. Polymerization and Formation of Colloidal Particles

Hsu et al. (1988) studied the preparation of colloidal dispersions of spherical particles of hydrous cerium(IV) oxide of narrow size distribution by heating solutions of ceric sulfates for different periods of time under carefully controlled conditions of salt concentration, ionic strength and temperature. The effect of salt concentration on the morphology of the particles formed in solutions aged for 12 h at 90°C is shown in Fig. 3-44. Within the dashed lines the solid phase is formed by homogeneous precipitation due to hydrolysis. Above the upper dashed line solids precipitate on mixing the reactants at room temperature. At lower concentrations of ceric sulfate and H_2SO_4, monosized spherical particles were formed (see Figures 3-5 and 3-6), whereas at higher concentrations of the reactants, rod-like particles were formed, designated by rectangles in the figure.

Table 3-9 shows that the particles became larger with increasing cerium and acid concentrations. X-ray diffraction analysis showed that the spherical particles made by homogeneous precipitation were crystalline. Using transmission electron microscopy, Hsu et al.(1988), were able to show that the final particles were composite in nature, formed by aggregating small spherical sub-units. Towe et al. (1967) reported a similar observation for spherical particles of hydrous ferric oxide. Matijevic and Steiner (1978), using low-angle x-ray analysis, showed that spherical hematite particles obtained by forced hydrolysis also were composed of tiny beads 30 Å in diameter.

Briois et al. (1993) also studied hydrolysis of cerium(IV) sulfate solutions and employed the techniques of extended x-ray absorption fine structure spectroscopy and small-angle x-ray scattering to follow the different steps in the hydrolysis leading to uniform particles of ceria of controlled morphology. They concluded that formation of particles proceeded in two stages. In the first stage, dimeric precursors of composition $Ce_2(OH)_2(O)_{12}$ and structure shown in Fig. 45 are formed. The twelve oxygen atoms belong to sulfate ions and/or water.

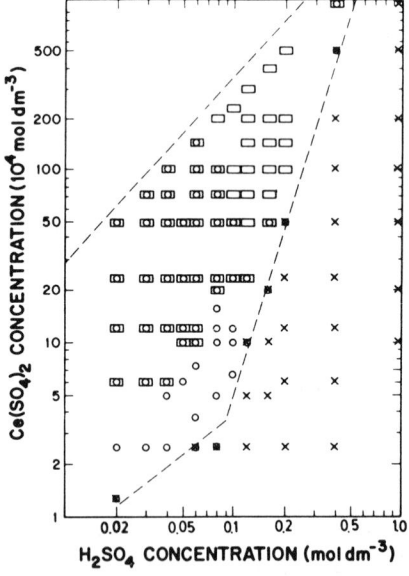

Fig. 3-44. Precipitation domain for solutions containing $Ce(SO_4)_2$ and H_2SO_4 aged at 90°C for 12 h. Symbols designating different kinds of particles: O, spheres; □, rods; ▥, rods mixed with spheres; ⊗ a very small amount of spheres; x, no particle formation. Hsu et al. (1988). Courtesy American Chemical Society.

[Ce(SO$_4$)$_2$]/10^3 mol dm^{-3}	modal particle diameter (nm) at [H$_2$SO$_4$]/mol dm^{-3}			
	0.04	0.06	0.08	0.10
0.5	30 (40)[a]	70 (110)	100 (130)	110 (140)
0.8			130 (150)	
1.0	40 (50)	70 (120)	170 (220)	140 (200)
1.2	50 (50)	80 (140)	220 (230)	190 (210)
1.6	60 (60)	80 (140)	170 (240)	150 (210)

Table 3-9. Effect of the concentrations of reactants and of aging times at 90°C on the size of particles obtained from Ce(SO$_4$)$_2$-H$_2$SO$_4$ solutions. Hsu et al. (1988)

[a] First number refers to aging for 12 h, while the number in parentheses is for 48 h of aging.
Courtesy American Chemical Society

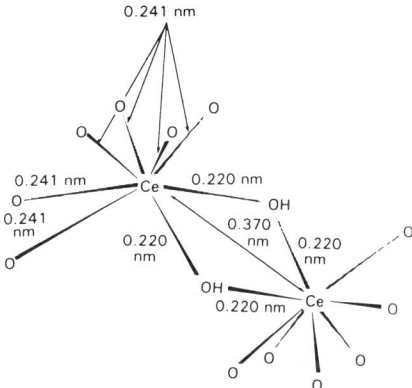

Fig. 3-45. Proposed dimeric structure for the precursors of ceria. Briois et al. (1993). Courtesy Chapman & Hall.

After the initial stage, the dimeric precursors begin to grow to colloidal particles when the temperature is raised above 60°C, and the system enters the second stage when the temperature reaches 90°C. In this stage 85% of the material is in the form of relatively monodisperse spherical colloids, 3 nm in diameter, and the rest is in the form of still smaller particles. The colloids are formed by a polymerization reaction involving deprotonization and condensation of the dimers formed at room temperature leading to the network structure shown in Fig. 3-46. The EXAFS results suggest a chain structure of composition $(Ce(OH)_2)_n^{2n+}$ where the cerium atoms are 0.366 nm apart and linked by hydroxo bridges 0.222 nm in length, which are formed by an olation mechanism; see Sect. A this chapter.

8. Hydrolysis of Zn^{2+}

Zinc oxide is used in rubber production where it lowers the vulcanization temperature and shortens the vulcanization time.   It is also used as a pigment in

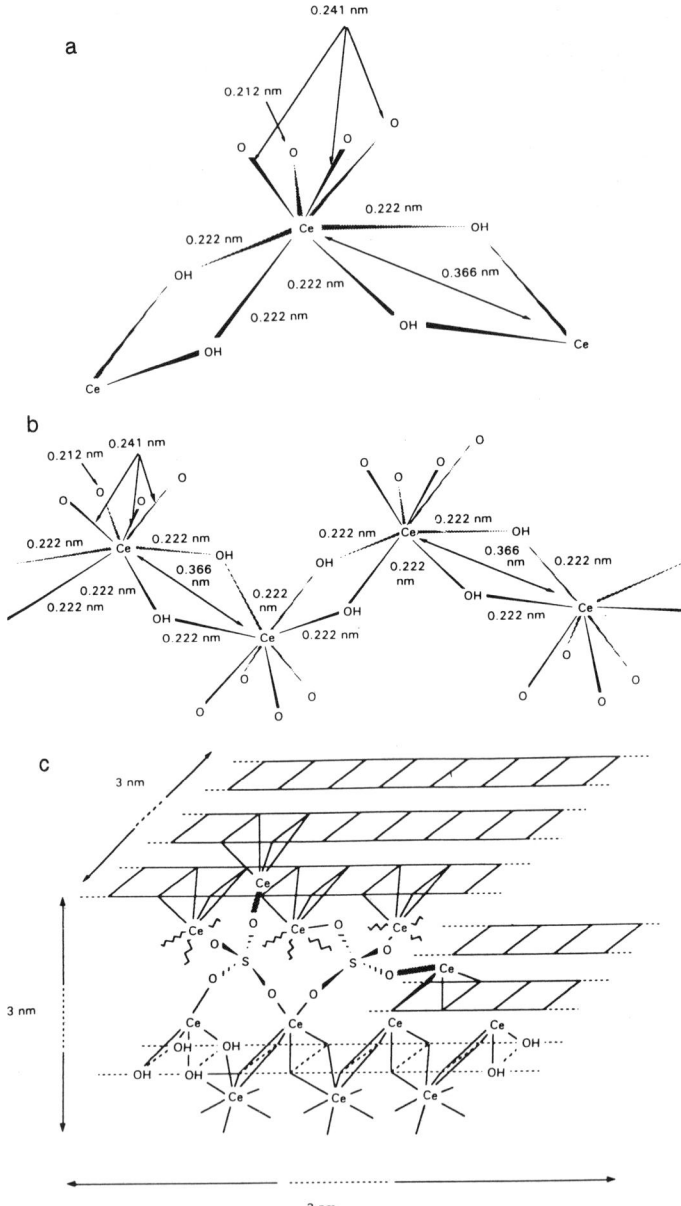

Fig. 3-46. (a) Elementary block characterized by EXAFS; (b) 3.0 nm length of chains deduced from SAXS results; (c) Possible isotropic structure of colloids in solution. Briois et al. (1993). Courtesy Chapman & Hall.

paints, "zinc white", with the twin advantages over basic lead carbonate ("white lead") that it is non-toxic and does not become discolored by sulfur compounds since zinc sulfide is also white. Varistors of zinc oxide are high-resistance, semi-conducting devices which, because of their highly nonlinear current-voltage characteristics, are used to mitigate power surges, e.g. lightning strikes, in electronic circuits and electric power transmission lines.

Zinc has the electronic configuration $(n-1)d^{10}ns^2$ in addition to the filled inner shells and there is no evidence for the existence of oxidation states higher than +2. The stabilities of the hydrolysis species of Zn^{2+} are summarized in Table 3-10, and they are well established except for $Zn(OH)_2$.

Table 3-10. Summary of Zn^{2+} hydrolysis at 25°C. Log Q_{xy} = log K_{xy} + $aI^{0.5}/(1 + I^{0.5})$ = bm_x. Baes and Mesmer (1976)

Species or phase	Log K_{xy}	a	b^*			$\sigma(\log Q_{xy})$
			$m_x = 0.1$	$m_x = 1.0$	$m_x = 3.0$	
$ZnOH^+$	-8.96	-1.022	0.6	0.31	0.21	0.05
$Zn(OH)_2$	-16.9	-1.022	0.5	0.19	0.10	?
$Zn(OH)_3^-$	-28.4	0	0.1	-0.07	-0.13	±0.2
$Zn(OH)_4^{2-}$	-41.2	2.044	-0.20			±0.1
$Zn_2(OH)^{3+}$	-9.0	1.022	-1.4	-0.4	-0.1	±0.1
$Zn_2(OH)_6^{2-}$	-57.8	1.022	0.2	-0.03	-0.1	±0.1
$ZnO(c)$ (log Q_{s10})	11.14	1.022	-0.5	-0.19	-0.10	±0.03

* Estimated for perchlorate media.
Reprinted by permission of John Wiley & sons, Inc.

The distribution of hydrolysis products at two Zn^{2+} ion concentrations, 0.1 M Zn(II) and 10^{-5} M Zn(II), as a function of pH is shown in Fig. 3-47.

The concentrations of the hydrolysis species as a function of pH in solutions saturated with ZnO are shown in Fig. 3-48.

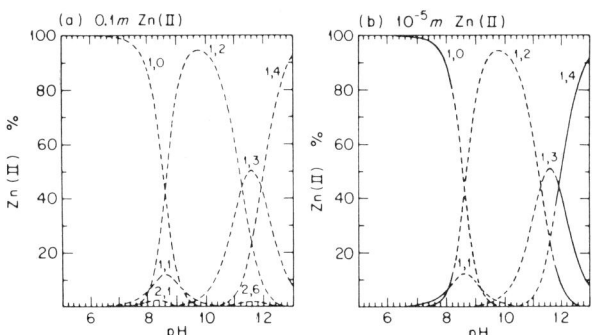

Fig. 3-47. Distribution of hydrolysis products (x,y) at ionic strength I = 1 M at 25°C in: (a) 0.1 M Zn(II) and (b) 10^{-5} M Zn(II). Baes and Mesmer (1976). Reprinted by permission of John Wiley & Sons, Inc.

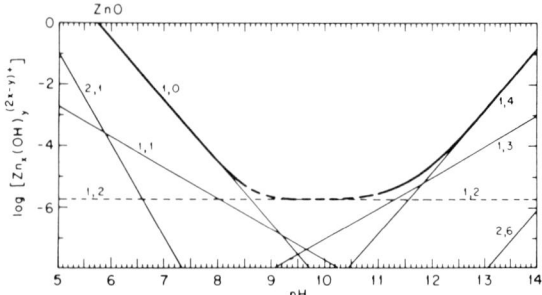

Fig. 3-48. Concentration of hydrolysis products (x,y) of ionic strength I = 1 M at 25°C in solutions saturated with ZnO. Baes and Mesmer (1976). Reprinted by permission of John Wiley & Sons, Inc.

8.1. Polymerization and Formation of Colloidal Particles

Chittofratti and Matijevic (1990) prepared particles of zinc oxide by aging solutions of zinc nitrate at elevated temperatures in the range from 90°C to 250°C and studied the effect on particle morphology of the bases NaOH, KOH, NH$_4$OH, triethanolamine (TEA), and ethylenediamine (En). Figure 3-49 shows the composition domains in which well-defined particles uniform in shape were formed in the presence of ammonia, TEA, and En.

When mixing zinc nitrate solutions with NaOH, KOH, and NH$_4$OH, precipitation occurred whereas no precipitation was observed when TEA or En was used.

Particles almost spherical in shape were obtained only with TEA as the base. The other bases produced particles with uniform, well defined, but non-spherical, geometries such as rods, prisms and ellipsoids.

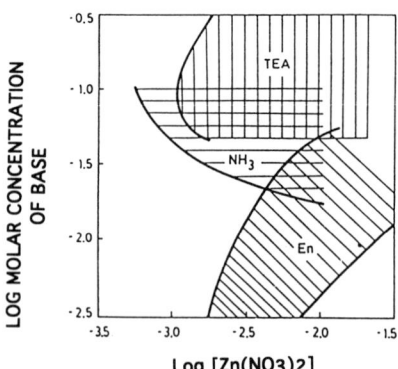

Fig. 3-49. Composition domains of well defined particles formed in solutions containing zinc nitrate and different bases by aging for 1 h at 90°C. TEA: triethanolamine; En: ethylenediamine. Chittofratti and Matijevic (1990). Reprinted from Colloids and Surfaces 48, 65 (1990) with kind permission from Elsevier Science - NL, Sara Burgerhartstraat 25, 1055 KV Amsterdam, The Netherlands.

9. Hydrolysis of Ga^{3+}

Gallium oxide is used in "bubble memories" in computers. Only trivalent gallium is stable in aqueous solutions and Table 3-11 summarizes the hydrolysis products of Ga^{+3} and their formation quotients.

The distribution of hydrolysis products at Ga^{3+} ion concentrations 0.1 to 10^{-5} M Ga(III), respectively, as a function of pH is shown in Fig. 3-50.

The concentrations of the hydrolysis species as a function of pH in solutions saturated with respect to GaO(OH) are shown in Fig. 3-51.

9.1. Polymerization and Formation of Colloidal Particles

Hamada et al. (1986) investigated the mechanism of formation of monodisperse spherical particles of hydrous gallium (III) oxide by forced hydrolysis in the presence of sulfate and nitrate ions.

Uniformly-sized particles of amorphous hydrous gallium oxide were formed by hydrolysis at 98°C for 18 h; see Fig. 3-52.

Table 3-11 Summary of Ga^{3+} hydrolysis at 25°C.
Log Q$_{xy}$ = log K$_{xy}$ + aI$^{0.5}$/(1 + I$^{0.5}$) = bm$_x$.
Baes and Mesmer (1976)

Species or phase	Log K$_{xy}$	a	b	σ(log Q$_{xy}$)
GaOH^{2+}	-2.6	-2.044	0.4	?
Ga(OH)$_2^+$	-5.9	-3.066	0.4	?
Ga(OH)$_3$	-10.3	-3.066	0.2	?
Ga(OH)$_4^-$	-16.6	-2.044	0.1	±0.3
"Ga$_{26}$(OH)$_{65}^{13+}$"a	-139.1	0	(0)	?
GaO(OH)(c)(log Q$_{s10}$)	2.9	3.066	-0.2	±0.3

a Approximate formula of one or more species formed.
Reprinted by permission of John Wiley & sons, Inc.

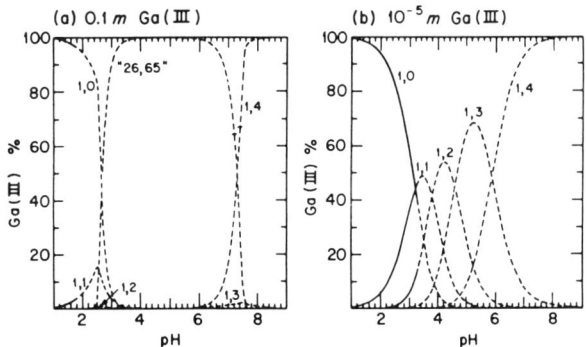

Fig. 3-50. Distribution of hydrolysis products (x,y) at ionic strength I = 1 M at 25°C in: (a) 0.1 M Ga(III); and (b) 10^{-5} M Ga(III). Baes and Mesmer (1976). Reprinted by permission of John Wiley & Sons, Inc.

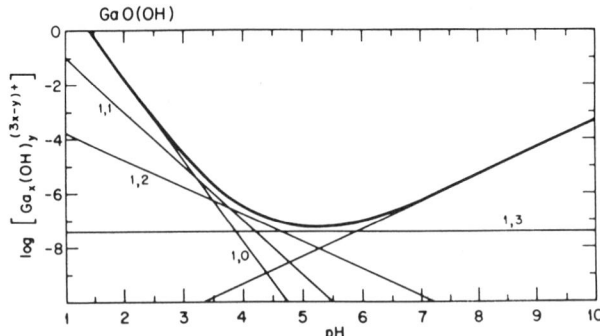

Fig. 3-51. Concentration of hydrolysis products (x,y) at ionic strength I = 1 M at 25°C in solutions saturated with GaO(OH).Baes and Mesmer (1976). Reprinted by permission of John Wiley & Sons, Inc.

Fig. 3-52. Electron micrograph of monodispersed spherical hydrous gallium(III) oxide particles obtained at 98 °C for 18 h. $[Ga^{3+}] = 9.8 \times 10^{-4}$ mol/dm^3, $2[H_2SO_4] + [HNO_3] = 1.0 \times 10^{-4}$ mol/dm^3, and $[SO_4^{2+}]_1 / [Ga^{3+}]_1 = 1.9$ respectively. Hamada et al. (1986). Courtesy Chemical Society of Japan.

However, the amorphous particles were not stable and Fig. 3-53 shows that they dissolved partially during the reaction time and rearranged to a crystalline mixture of rod-like and spherical particles.

By measuring how the fractions of mono- and polynuclear hydroxo-gallium complexes of amorphous particles varied with time, Hamada et al. (1986) suggested the following reaction path for the formation of monodisperse particles in the sulfate system.

$$\text{monomer} \longrightarrow \text{polymer} \longrightarrow \text{amorphous particles}$$

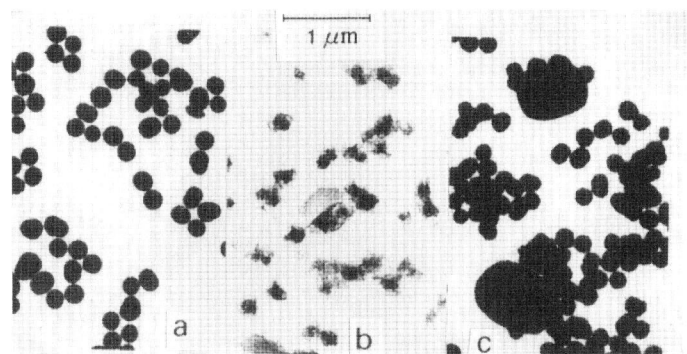

Fig. 3-53. Dissolution and recrystallization of hydrous gallium (III) oxide particles. Aging time: (a) 3.5 h; (b) 4.5 h; (c) 18 h. Hamada et al. (1986). Courtesy Chemical Society of Japan.

The first step may occur with a buildup of the polymer:

$$\text{polymer (a-b) + b monomer} \longrightarrow \text{polymer (a)}$$

where polymer (a-b) and polymer (a) refer to species of lower and higher degrees of polymerization, respectively.

They also considered the role of the sulfate ions in the formation of particles. Highly charged hydroxo-gallium(III) polymers more readily associate with sulfate ions than with univalent ions like nitrate or perchlorate ions so as to reduce the cationic charge of the polymer-free hydroxo complexes and promote further polymerization.

C. SMALL PARTICLES BY HYDROLYSIS OF METAL ALKOXIDES

In Chapter 2 methods were given for making small particles of silica of very uniform size by hydrolysis of silicon tetraalkoxides or tetraalkoxysilane. Similarly, small particles of alumina and hydrous transition metal oxides can be made by hydrolysis of metal alkoxides, $M(OR)_z$, where R is a saturated or unsaturated group. The cost of some of these alkoxides, e. g., alkoxides of Si, Al, Ti, Zr, and Mg, is now such that the production of aqueous sols of the corresponding hydrous metal oxides by this method is approaching commercial feasibility, Guglielmi and Carturan (1988); see Table 3-12.

Other hydrous metal oxides, e.g., of Co, Cu, Fe, Mn, Nb, Ni, Ta, V, and Y, can also be prepared by this method, but the production cost would probably be much too high except, maybe for very special applications.

Colloidal-size particles of boehmite of high purity for use in, e.g., catalyst applications (see Ch. 7) are produced commercially by hydrolyzing fatty

Table 3-12. Approximate costs of commercial alkoxides for large and laboratory-scale usage ($/kg). Chiu and Meehan (1974)		
	Large scale	Laboratory scale
Silicon methoxide	3-5	130[a]-630[b]
Silicon ethoxide	3-5	15[b]-670[b]
Silicon n-propoxide	3-5	200
Aluminum isopropoxide	4	15[a]-1700[c]
Titanium ethoxide	5-12	60-350
Titanium butoxide	5	40
Zirconium n- propoxide	13	30
Zirconium butoxide	13	75
Magnesium ethoxide	13	35

a 98%   b 99%   c 99.999%
Courtesy Academic Press

alkoxides of aluminum, but the resulting particles have irregular shapes and a wide particle size distribution.

The hydrolysis of alkoxides of transition metals occurs more readily and proceeds faster than that of alkoxides of silicon. This is because these metals are more electropositive than silicon, and can expand their coordination number in the process of particle formation, which silicon cannot do. One consequence of the high reactivity of the transition metal alkoxides is that they must be protected from moisture so as not to decompose.

The particles of hydrous metal oxides are formed in two main steps. In the first step, metal hydroxides are formed by a hydrolysis reaction which takes place in three stages between the metal alkoxide and water together with an alcohol.

$$\underset{\overset{|}{H}}{H\text{-}O} + M\text{-}OR \;\longrightarrow\; \underset{\overset{|}{H}}{\overset{H}{\underset{}{\diagdown}}}O{:}\longrightarrow M\text{-}OR \;\longrightarrow\; HO\text{-}M \longleftarrow O\overset{\diagup R}{\underset{\diagdown H}{}} \longrightarrow M\text{-}OH + ROH$$

$$\qquad\qquad (a)\qquad\qquad\qquad (b)\qquad\qquad (c)\qquad\qquad\qquad (d)$$

In the second main step, a dimer is formed by a condensation reaction, which may take place by one of the following competing mechanisms:

a) *Alkoxolation* is a reaction similar to the hydrolysis reaction in the first main step, but metal hydroxide takes the place of water and reacts with the metal alkoxide to form an oxo bridge between two metal atoms.

$$\underset{\overset{|}{H}}{M\text{-}O} + M\text{-}OR \;\longrightarrow\; \underset{\overset{|}{H}}{M\text{-}O}{:}\longrightarrow M\text{-}OR \;\longrightarrow\; M\text{-}O\text{-}M \longleftarrow O\overset{\diagup R}{\underset{\diagdown H}{}} \longrightarrow M\text{-}O\text{-}M + ROH$$

$$\qquad\quad (a)\qquad\qquad\qquad (b)\qquad\qquad (c)\qquad\qquad\qquad (d)$$

b) *Oxo bridge formation.* Two metal hydroxide molecules react by the same mechanism as in alkoxolation and form a dimer with an oxo bridge between two metal atoms:

$$M\text{-}O + M\text{-}OH \longrightarrow M\text{-}O\!: \longrightarrow M\text{-}OH \longrightarrow M\text{-}O\text{-}M \longleftarrow O \overset{H}{\underset{H}{\diagup}} \longrightarrow M\text{-}O\text{-}M + H_2O$$
$$\underset{H}{|}\quad (a)\qquad\qquad \underset{H}{|}\ (b)\qquad\qquad (c)\qquad\qquad\qquad (d)$$

c) *Hydroxo bridge formation.* A metal hydroxide molecule reacts with a metal alkoxide molecule or another metal hydroxide molecule to form a dimer with a hydroxo bridge between the metal ions.

$$M\text{-}OH + M \longleftarrow O \overset{H}{\underset{R}{\diagup}} \longrightarrow M\text{-}\overset{H}{\underset{}{O}}\text{-}M + ROH$$

$$M\text{-}OH + M \longleftarrow O \overset{H}{\underset{H}{\diagup}} \longrightarrow M\text{-}\overset{H}{\underset{}{O}}\text{-}M + H_2O$$

As the hydrolysis and condensation reactions proceed, particles, gels or precipitates will, depending on conditions, eventually form. Table 3-13 provides a rough guide as to how the resulting products depend on the rates of hydrolysis and condensation, Livage et al. (1988).

Table 3-13. Products and rates of hydrolysis and condensation. Livage et al. (1988)		
Hydrolysis Rate	Condensation Rate	Result
Slow	Slow	Colloids/Sols
Fast	Slow	Polymeric Gels
Fast	Fast	Colloidal Gel or Gelatinous Precipitate
Slow	Fast	Controlled Precipitation

Reprinted from Prog. Solid State Chem. **18**, 259 (1988) with kind permission of Elsevier Science Ltd., The Boulevard, Langford Lane, Kidlington, OX5 1GB, UK.

1. Factors affecting particle formation

a) Hydrolysis quotient, h, defined as: $h = [H_2O]/[M(OR)_z]$ (3-22)

If 100% of the water, in, e.g., Eqn. 3-22, is reacted, h will be equal to the hydroxyl ligand number, n; see Eqn. 3-8. Particles will form only if h is greater than 1. Gels will form if there is an excess of water.

b) Catalyst. Hydrolysis and particle formation can be controlled and modified by the use of acid or basic catalysts. Acids, such as HCl or HNO_3, will hydrolyze a metal alkoxide according to the reaction:

$$M\text{-}OR + H_3O^+ \longrightarrow M \longleftarrow\!: O \overset{H}{\underset{R}{\diagup}} + H_2O$$

and accelerate the reaction by preventing the prototropic transfer and departure of the leaving group (alcohol) as the rate-limiting steps, Livage et al. (1958)..
Provided there is enough water, all alkoxyl groups can be rapidly hydrolyzed and then undergo condensation reactions so as to form M-O-M or M-OH-M linkages in a growing structure.

Bases, such as NH_3 or NaOH, will react with a metal alkoxide according to the reaction:

$$M\text{-}OH + :B \longrightarrow M\text{-}O^- + BH^+ \quad (B\text{-}OH^-, NH_3)$$

and form metal oxo complexes, which undergo condensation reactions and form larger structures, often consisting of particulate gels, Livage et al. (1958).

c) Other factors. Temperature, type of metal, and concentration of metal alkoxide obviously affect the size and composition of the final product. Other factors, perhaps not equally obvious, are, for instance, the type and molecular weight of the organic group in the metal alkoxide compound.

2. Hydrolysis of titanium alkoxides.

Titanium alkoxides are the most thoroughly studied metal alkoxides and the most commonly used alkoxide in these investigations is tetraethylorthotitanate, titanium tetraethoxide, TEOT, or $Ti(OEt)_4$, followed by titanium tetrabutoxide and titanium tetrapropoxide. TEOT undergoes hydrolysis according to the overall reaction.

$$Ti(OEt)_4 + 4H_2O \longrightarrow Ti(OH)_4 + 4EtOH$$

which can be broken down into four steps with the following equilibrium constants

$$k_1 = \frac{[Ti(OEt)_3(OH)][EtOH]}{[Ti(OEt)_4][H_2O]} \tag{3-23}$$

$$k_2 = \frac{[Ti(OEt)_2(OH)_2][EtOH]}{[Ti(OEt)_3(OH)][H_2O]} \tag{3-24}$$

$$k_3 = \frac{[Ti(OEt)(OH)_3][EtOH]}{[Ti(OEt)_2(OH)_2][H_2O]} \tag{3-25}$$

$$k_4 = \frac{[Ti(OH)_4][EtOH]}{[Ti(OEt)(OH)_3][H_2O]} \tag{3-26}$$

The equilibrium constant for the overall reaction is:

$$K_{tot} = \prod_{i=1}^{4} k_i = \frac{[Ti(OH)_4][EtOH]^4}{[Ti(OEt)_4][H_2O]^4} \tag{3-27}$$

The first steps in the hydrolysis of TEOT that will eventually result in particles of titania with uniform sizes are oxo bridge formation.

Ti-OH + HO-Ti ——> Ti-O-Ti + H_2O oxo bridge formation

Ti-OEt + HO-Ti ——> Ti-O-Ti + EtOH   alkoxolation

The hydrolysis must proceed by these mechanisms for a certain length of time, the induction time, before nuclei are formed and particle growth can begin.

Different methods have been employed to study and follow the course of hydrolysis and particle growth in systems containing TEOT. During the early part of hydrolysis the solutions are perfectly clear, but as soon as particles, i.e. hydrolysis products of a certain minimum size begin to form, turbidity appears. Barringer et al. (1985) measured the time it took for turbidity to develop in ethanol-water solutions of TEOT undergoing hydrolysis, which they called the delay time and which is an approximation of the induction time. Figure 3-55 shows that the logarithm of the delay time depended linearly on the logarithm of the initial water concentration, water being a reactant in the hydrolysis reaction, with a slope of about 2.96. Moreover, the delay time depended linearly on the other reactant with a slope of about 1.23 (Fig. 3-56). The rate law for the hydrolysis of TEOT can therefore be written as:

$$r = k[Ti(OEt)_4][H_2O]^3 \qquad\qquad (3\text{-}28)$$

Harris and Byers (1988) studied the hydrolysis and particle growth in the same system, but at lower concentrations of TEOT by UV spectroscopy and dynamic light scattering, DLS, but obtained a completely different rate law than Eqn. 3-28. They used UV spectroscopy to measure the change in water concentration with time—perhaps a more adequate way to monitor the hydrolysis reaction than by correlating the delay time with the initial water concentrations in solutions with fairly high concentrations of TEOT; 0.10-0.20 M (Barringer and Bowen (1985)) compared to 0.0078-0.031 M (Harris and Byers (1988)).

Figure 3-56 shows that the rate of growth, determined by DLS, increased with TEOT concentration, which is as expected since the particles grow by condensation reactions involving hydrolysis products of Ti^{4+}.

Look and Zukoski (1992) reported that discrete particles of titania of uniform size can be prepared over wider ranges of TEOT and H_2O by adding small amounts of HCl to the mixture of reactants TEOT and H_2O in ethanol, see Fig. 3-57.

However, electrolyte in the form of HCl in the reaction mixture affects the morphology of the particle surface. With no HCl the particle surface is quite smooth, whereas with HCl present in the reaction mixture, the surface of the particles became rough.

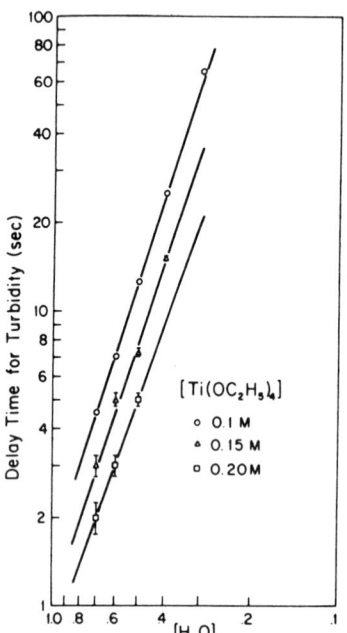

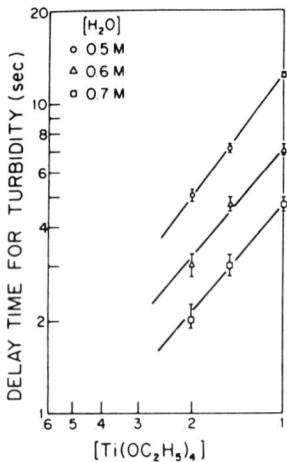

Fig. 3-54. Delay time for the initial observation of turbidity as a function of initial water concentration for the initial ethoxide concentrations of 0.10, 0.15, and 0.20 M. The average slope is 2.96 ± 0.1. Barringer and Bowen (1985). Courtesy American Chemical Society.

Fig. 3-55. Delay time for th : initial observation of turbidity as a function of initial ethoxide concentration for the initial water concentrations of 0.5, 0.6, and 0.7 M. The average slope is 1.23 ± 0.1. Barringer and Bowen (1985). Courtesy American Chemical Society.

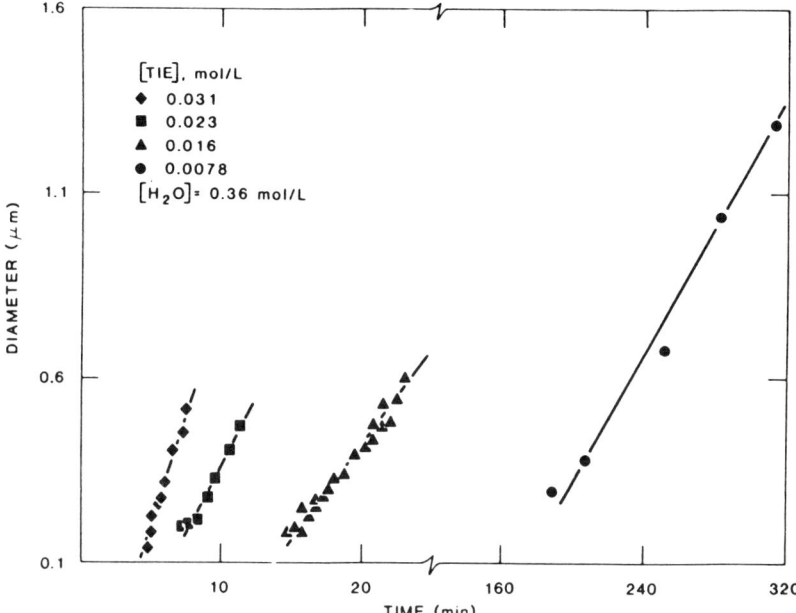

Fig. 3-56. "Titania" growth kinetics. Particle size vs. time at parametric values of TEOT concentration for hydrolysis of TEOT in water. Harris and Byers (1988). Reprinted from J. Non-cryst. Solids **103**, 49 (1988), with kind permission of Elsevier Science-NL, Sara Burgerhartstraat 25, 1055 KV Amsterdam, The Netherlands.

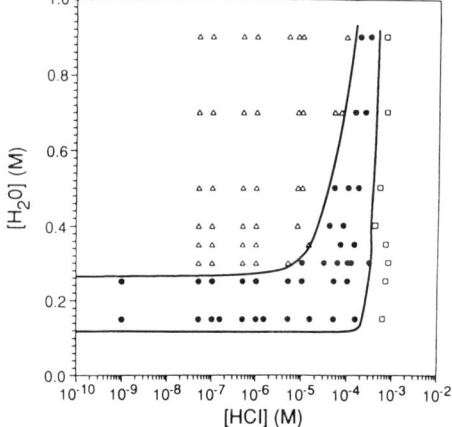

Fig. 3-57. Stability diagram showing regions of precipitates of different morphologies. Reactions carried out in ethanol for [TEOT] = 0.05M. (•) stable discrete particle precipitates; (Δ) agglomerated precipitates; (□) no observable solids after 72 h. Look and Zukowski (1992). Courtesy American Ceramic Society.

D. SMALL PARTICLES OF METAL SULFIDES

Oxygen and sulfur belong to the same group in the periodic table and form similar compounds with many metals, the calcogens. Sulfur is less electronegative than oxygen, but has a wider range of oxidation states from -II to +VI, of which -II, as in S^{2-}, and +VI, as in SO_3, H_2SO_4, or SO_4^{2-}, are the most important. Hydrogen sulfide, is readily soluble in water to give acidic solutions, thus demonstrating that it is a stronger acid than water. It forms sulfides with most metals, especially in the presence of water.

At this time there is no detailed knowledge of the mechanisms involved in the formation of small particles of metal sulfides. Furthermore, it appears that relatively little work has been published in this area in the last ten years. Nevertheless, monodisperse, discrete particles of metal sulfides are of potential interest in such applications as solar cells, pigments, and water purification.

The same factors, pH, temperature, anion type, and reactant concentration, that affect the formation of hydrous metal oxide particles also affect the formation of particles of metal sulfides, and they must be controlled in order to obtain particles of desired shape, size, and morphology. Chiu (1974, 1977, 1981) prepared monodisperse particles of ZnS, CuS, and PbS by bubbling H_2S through solutions of the metal sulfates containing ethylenediamine tetraacetic acid, EDTA, and a base such as NaOH or NH_3. The resulting particles were reasonably monodisperse, but the particle size was difficult to control, probably because addition by bubbling of one of the reactants, H_2S, was not done in a sufficiently controlled manner.

Thioacetamide, CH_3CSNH_2, TA, decomposes by releasing sulfide ions at a rate that increases with temperature. By using TA as the sulfur source, Matijevic and Wilhelmy (1982, 1984) were able to carefully control the supply of sulfide ion to the reaction mixture and make monodisperse particles of ZnS and CdS. If particle growth is not so fast that the sulfide ions are consumed as fast as they are released by the decomposing TA molecules, secondary particles will form. Fig. 3-58 shows that aging an aqueous solution containing 0.02 M $Zn(NO_3)_2$, 0.062 M HNO_3, and 0.11 M TA at 60°C for 220 min. produced monodisperse particles of ZnS, whereas secondary particles were formed if the aging time was increased by 30 min. (see Fig. 3-59)

Wilhelmy and Matijevic (1984) used the same method to make monodisperse particles of CdS; see Fig. 3-60.

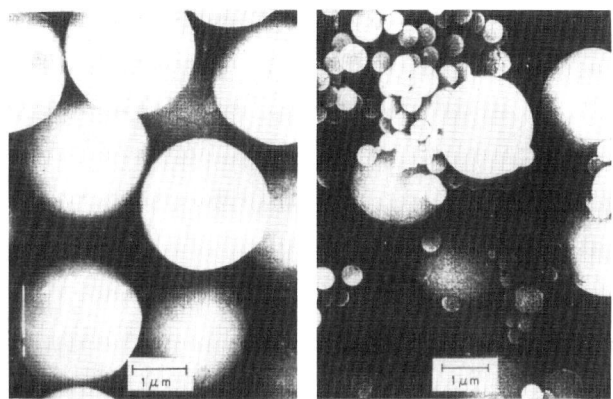

Fig. 3-58. and Fig. 3-59 (right). Particles of ZnS produced from 0.024 M $Zn(NO_3)_2$, in 0.062 M HNO_3 and 0.11 M thioacetamide at 60°C. *Left*: 220 min. aging. *Right*: 250 min. aging. Wilhelmy and Matijevic (1984). Courtesy The Royal Society of Chemistry.

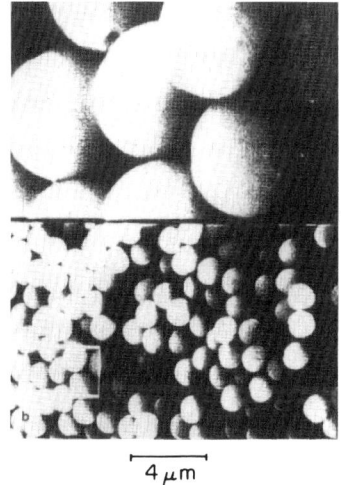

Fig. 3-60. Cadmium sulfide particles obtained by adding 6.25 ml of 0.05 M thioacetamide to 500 ml of a "seed" sol, made by aging a solution 0.0012 M in $Cd(NO_3)_2$, 0.24 M in HNO_3, and 0.005 M in thioacetamide at 26°C for 14.5 h, and further aging at 26°C for 20 min. Wilhelmy and Matijevic (1984). Courtesy The Royal Society of Chemistry.

References

Aiken, B., Hsu,W. P., and Matijevic, E., J. Mater. Sci. **25**, 1886 (1990).

Arthur, P., Canadian Patent 599,385 (1960).

Baes, C. F., and Mesmer, R. E., *The Hydrolysis of Cations*. John Wiley & Sons, New York, (1976).

Barringer, E. A., and Bowen, H. K., Langmuir, **1**, 414 (1985).

Bassett, H., and Goodwin, T. H., J. Chem. Soc., 2239 (1949).

Bell, A., and Matijevic, E., J. Inorg. Nucl. Chem. **37**, 907 (1975).

Bjerrum, N., Z. Phys. Chem., **59**, 336 (1907).

Bohnstedt-Kupletskaya, E. M., and Vlodavetz, N. I., Compt. rend. Acad. Sci. USSR **49**, 587 (1945).

Briois, V., Williams, C. E., Dexpert, H, Villain, F., Cabane, B., Deneuve F., and Magnier, C., J. Mater. Sci., **28**, 5019 (1993).

Brumberg, A. W., Lukianovits, W. M., Nemcova, W. W., Raduskevits, L. W., and Smutov, K. V., Doklady Akad. Nauk SSSR, **79**, 827 (1951).

Bleier, A., and Cannon, R. M., Mat. Res. Soc. Symp. Proc., **23**, 71 (1986).

Bottero, J.Y., and Fiessinger F., Nordic Pulp and Paper Res. J., No. 2, 81 (1989).

Bugosh, J., U.S. Patent 2,915,475 (1959) .

Bugosh, J., unpublished work (1961).

Chittofratti, A., and Matijevic, E., Colloids and Surfaces, **48**, 65 (1990).

Chiu, G., and Meehan, E., J. Colloid Interfac. Sci., **49**, 160 (1974).

Chiu, G., J. Colloid Interfac. Sci., **62**, 193 (1977).

Chiu, G., J. Colloid Interface Sci., **83**, 309 (1981).

Demchak, R., and Matijevic, E., J. Colloid Interfac. Sci., **31**, 257 (1969).

Denk, G., and Alt, J., Z. Anorg. Allg. Chem., **269**, 244 (1952).

Ervin, G., and Osborn E.F., J. Geol., **59**, 381 (1951) .

Flynn, C.M., Chem. Rev., **84**, 31 (1984).

Guglielmi, M., and Carturan, J., J. Non-cryst. Solids, **100**, 16 (1988).

Hamada, S., Bando, K., and Kudo, Y., Bull. Chem. Soc. Japan., **59**, 2063 (1986).

Hamada, S., and Matijevic, E. J., Colloid Interfac. Sci., **84**, 274 (1981) .

Harris, M. T., and Byers, C. H., J. Non-cryst. Solids **103**, 49 (1988).

Huehn, H., and Haufe, W., U.S. Patent 2,196,016 (1940).

Hsu, W. P., Rönnquist L., and Matijevic, E., Langmuir, **4**, 31 (1988).

Johansson, G., Acta Chem. Scand., **14**, 771 (1960).

Kepert , D. L., *The Early Transition Metals.*, Academic Press, London (1972).

Kumok, V.N., and Serebrennikov, V. V., Russ. J. Inorg. Chem., **10**, 1095 (1965).

Livage, J., Henry, M., and Sanchez, C., Prog. Solid State Chem., **18**, 259 (1988).

Look, J.L., and Zukoski, C.F., J. Amer. Ceramic Soc., **75**, 1587 (1992).

Matijevic, E., Lindsey A.D., and Kratohvil S., J. Colloid Interfac. Sci., **36**, 273 (1971).

Matijevic, E., Sapieszko, S. and Melville, J.B., J. Colloid Interfac. Sci., **50**, 567 (1975).

Matijevic E., Budnik, M., and Meites, L., J. Colloid Interfac. Sci., **61**, 302 (1977).

Matijevic, E., and Scheiner, P., J. Colloid Interfac. Sci., **63**, 509 (1978).

Matijevic, E., and Wilhelmy, D.M., J. Colloid Interfac. Sci., **86**, 476 (1982).

Matijevic, E. and Hsu, W.P., J. Colloid Interfac. Sci., **118** No. 2, 506 (1987).

Muha, G. M., and Vaughan, P. A., J. Chem. Phys., **33** 194 (1960).

Murphy, P.J., Posner, A.M., and Quirk, J.P., J. Colloid Interfac. Sci., **56**, 312 (1976).

Ozaki, M., Kratohvil, S. and Matijevic, E., J. Colloid Interfac. Sci., **102**, 146 (1984).

Papee, D., Tertian, R., and Biais, R., Bull. Soc. Chim. France 1301 (1958).

Pfeiffer, P., Ber., **40**, 4036 (1907).

Reichertz, P. P., and Yost, W.J., J. Chem. Phys. **14**, 495 (1946).

Sapieszko, R. S., Patel, R. C., and Matijevic, E., J. Phys. Chem., **81**, 1061(1972).

Sasvari,K. and Zalai A., Acta Geol. Acad. Sci. Hung., **4**, 430 (1957).

Sillén, L.C. and Martell A.E. (1964) *Stability Constants of Metal-Ion Complexes*, Special Publication No. 17, The Chemical Society, London. and Supplement No. 1, Special Publication No. 25 (1971).

Smith, R.M., and Martell, A.E., *Critical Stability Constants*, Vol. 4, 7, Plenum Publishing Co., New York, (1976).

Sordelet, D., and Akinc, M., J. Colloid Interfac. Sci., **122**, No. 1, 47 (1988).

Towe, K. M., and Bradley, W. F., J. Colloid Interfac. Sci., **24**, 384 (1967).

Treadwell, W. D., and Zürcher, M., Helv. Chim. Acta, **15**,980 (1932).

Weiser, H. B., and Middleton, E. B., J. Phys. Chem. **24**, 630 (1920).

Werner, A., Ber., **40**, 272 (1907).

Wilhelmy, D. M., and Matijevic, E., J. Chem. Soc. Faraday Trans. I., **80**, 563 (1984).

CHAPTER 4. CLAYS AND COLLOIDAL SILICAS

Introduction

1. Structure - Classification
   1.1. Structure of Kaolin and Halloysite
   1.2. Structure of Montmorillonite
   1.3. Structure of Attapulgite

2. Isomorphous Substitution - Ion Exchange Capacity

3. Particle Size, Shape, and Surface Area

4. Surface Charge

5. Particle Association: Flocculation and Aggregation

6. Peptization: Deflocculation - Dispersion

7. Thermal Reactions: Dehydration and Transformation

8. Synthetic Hectorites

9. Pillared Interlayered Clays - Crosslinked Smectites

10. Uses of Clays

Introduction

Clays can lay claim to being the starting point for fine particle technology in the history of mankind. Their use in constructing shelters and in pottery manufacture far predates recorded history. Their particular crystalline structure which confers the plastic-like rheological behavior necessary for forming bricks, plugging holes, and making pottery, together with their permanence and ready availability all led inevitably to their extreme importance in mankind's struggle to survive and prosper.

In addition, clays provide the mechanical and chemical environment for almost all plant growth. Their ion exchange capabilities play a major role in plant growth, being a reservoir of calcium, potassium, and nitrogen, as well as providing the proper physical environment of porosity, water retention, and aeration. Their industrial uses number in the hundreds, with the bulk being in building products, pottery, specialty cements, catalysts, paints, adsorbents, paper fillers, insulating materials, and personal products such as medicines, cosmetics, and detergents.

In a technical sense clays are fine particles (with a substantial fraction less than less than 2 microns) of three main mineral types characterized by a sheet-like or lath-like structure as noted in the following.

In this chapter we will describe the structure, properties, and uses of the technically important clays kaolin, halloysite, montmorillonite, and attapulgite.

1. Structure - Classification

Clays are hydrous silicates with layered structures belonging to the larger group of phyllosilicates of which micas are probably the most widely known. Many of their properties derive from their fine-particle character. Their morphology, usually thin platelike crystals, depends on their underlying atomic arrangement. Bailey (1980) defines clay minerals as:

"Clay minerals belong to the family of phyllosilicates and contain continuous two-dimensional tetrahedral sheets of composition T_2O_5 (T= Si, Al, Fe^{3+}, ...) with tetrahedra linked by sharing three corners of each with a further corner pointing in any direction. The tetrahedral sheets are linked in the unit structure to octahedral sheets and to groups of coordinated cations or individual cations."

Figure 4-1a shows that the characteristic features of phyllosilicate structures are the continuous two-dimensional tetrahedral T_2O_5 sheets, the basal oxygens of which form a layered pattern. The contiguous octahedral sheet, in which octahedra are linked by sharing edges—see Fig. 4-1b—contains the apical oxygens at the fourth corner. The plane joining the tetrahedral and octahedral sheets contain the shared apical oxygens of the tetrahedra and unshared OH groups that lie in the projection at the center of each hexagonal ring of tetrahedra. The octahedral sheet usually contains coordinating ions such as Al^{3+}, Mg^{2+}, Fe^{2+}, and Fe^{3+}, but other ions can also occur.

A so-called 1:1 layer is formed by linking one octahedral sheet to a tetrahedral sheet. The exposed surface of the octahedral sheet consists of OH groups—see Fig. 4-1c. The figure also shows that the octahedral sheet can accommodate a second tetrahedral sheet to form a 2:1 layer, both surfaces of which consist of the hexagonal pattern of basal oxygens. Due to isomorphic substitution—see below—the two types of layers may not be electrically neutral, in which case charge balance is maintained by interlayer material which may be distinctive cations as in the mica group, hydrated cations as in vermiculites and smectites, or octahedrally coordinated hydroxide groups as in the chlorite minerals. A structure unit consists of the composite unit layer plus interlayer.

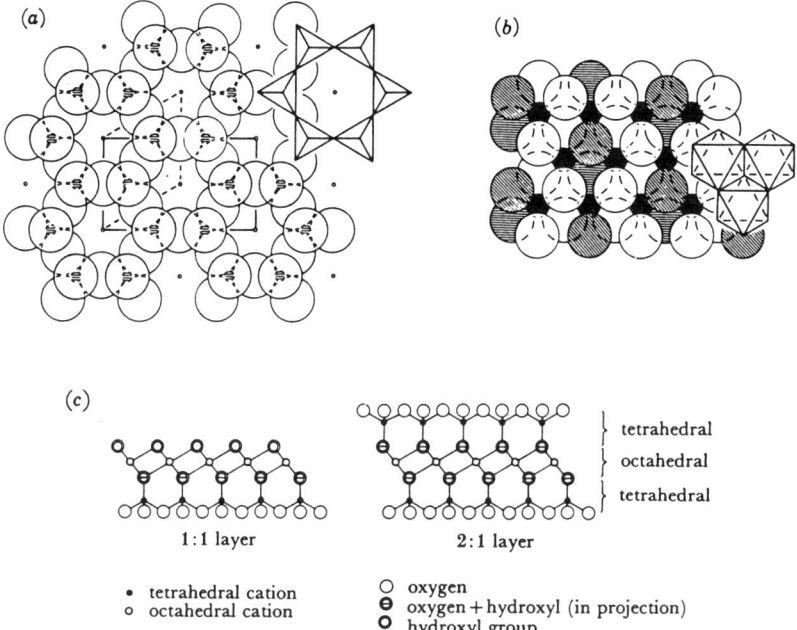

Fig. 4-1 (a) Plan of ideally hexagonal tetrahedral sheet; alternative hexagonal P- (dashed) and orthogonal C- cells (full lines) are shown. (b) Octahedral sheet with inner hydroxyls of 2:1 layer shaded. (c) 1:1 and 2:1 layers (from Bailey 1980b). Courtesy of the Mineralogical Society.

The terms *plane, sheet, layer,* and *structure unit* suggest increasingly thicker parts of the layered arrangement.

Table 4-1 shows the classification of clay minerals and charge of the layers, which indicates that interlayer material, often cations, is present in the structure.

Table 4-1. Classification of Clay Minerals and Related Phyllosilicates Brown (1984)		
Structure type	Group	Charge*
1:1 layer	serpentine-kaolin	ca. 0
2:1 layer	talc-pyrophyllite	ca. 0
	smectite	ca 0.2-0.6
	vermiculite	ca. 0.6-0.9
	mica	ca 1.0
	brittle mica	ca. 2.0
	chlorite	variable
2:1 inverted ribbons	sepoilite-palygorskite	variable

* Negative charge per formula unit layer, $O_5(OH)_4$ for 1:1 layer minerals and $O_{10}(OH)_2$ for mica, brittle mica, vermiculite and smectite.
With permission of The Royal Society of London.

1.1. Structure of Kaolin and Halloysite

The structure of kaolin, $(OH)_8Al_4Si_4O_{10}$, consists of layers made up of a hexagonal siloxane sheet and an octahedral sheet containing Al^{3+}, as shown in Fig. 4-2. It is a 1:1 layer structure and can be compared to a stack of open-faced sandwiches, each consisting of a layer of bread (siloxane layer) and a layer of butter (alumina layer).

The layers making up a platelet of kaolinite are about 0.72 nm thick and are continuous in the a- and b-directions and are stacked in a given order along the c-axis. The number of layers can vary from 25 layers for a French kaolinite to 75 layers for a Georgia kaolinite.

The variation in mineralogical composition with particle size of primary clays, e.g., two clays from Cornwall corresponding roughly to a commercial filler grade and a commercial coating grade, is shown in Table 4-2. The finer fraction has the greater kaolinite content and the smaller mica content with both fractions having a very low mica content.

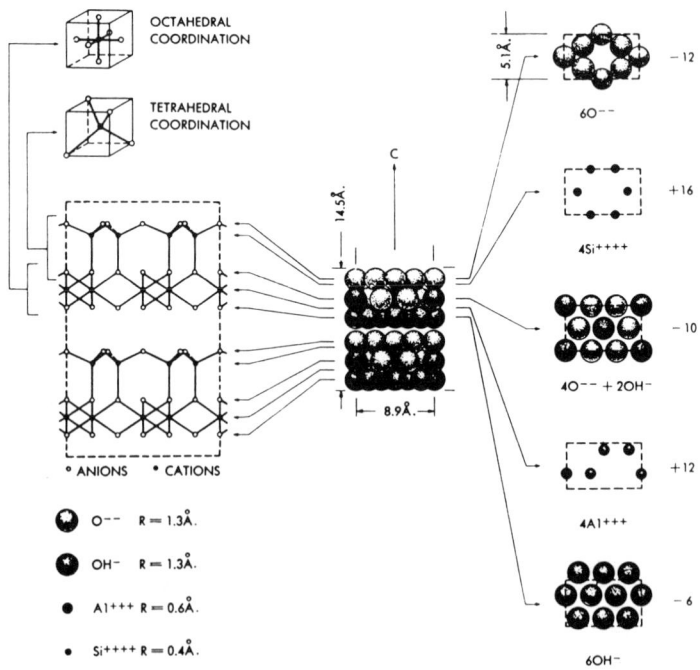

Fig. 4-2 Structure of Kaolin. Norton (1952). Courtesy Addison-Wesley Publishing Co.

Table 4-2. Analyses of kaolins prepared from Cornish matrix Jepson (1984).

	composition (% mass)	
	under 20 μm	under 10 μm
kaolinite	83	90
mica	14	8
feldspar	2	1
Fe_2O_3	0.9	0.6
TiO_2	0.05	0.06
K_2O	1.9	1.1

With permission of The Royal Society of London.

Table 4-3. Analyses of two kaolins (under 10 μm) prepared from Georgia matrixes, Jepson (1984).

	composition (% mass)	
	central Georgia	east Georgia
kaolinite	>95	>95
mica	≈1	≈2
Fe_2O_3	0.25-0.60	0.8-1.5
TiO_2	1.0-2.0	1.5-3.0
K_2O	≈0.1	0.2

With permission of The Royal Society of London.

The mineralogical composition of secondary clays, shown in Table 4-3 for two Georgia clays, varies much less with particle size.

Halloysite is a member of the kaolinite group and it occurs in two forms: $(OH)_8Si_4Al_4O_{10} \cdot 4H_2O$ and $(OH)_8Si_4Al_4O_{10}$. The hydrated form is usually a dense porcelain-like hard clay that dehydrates at temperatures slightly above 100°C to a white or light-colored porous, friable or almost fibrous-like material. The dehydrated form is similar to kaolinite in composition and mineral structure, while the hydrated form has a c-axis spacing greater than that of kaolinite.

1.2.   Structure of Montmorillonite

The 2:1 layer structure of montmorillonite is shown in Fig. 4-3. Magnesium has been partially substituted for Al^{3+} in the octahedral layer, giving rise to an overall negative charge of the structure, which is balanced by exchange cations, e.g., as Na^+ and Ca^{2+} in the figure. The unique feature of the smectite structure is that water and other polar molecules, such as certain organic molecules, can enter between the unit layers, causing the lattice to expand in the c-direction. The c-axis dimension of montmorillonite is therefore not fixed, but varies from about 9.6 Å, when no polar molecules are between the unit layers, to essentially complete separation of the individual layers in some cases. In Fig. 4-3 the c-axis dimension is 14 Å since water is present between the layers. It is more difficult to separate the unit layers with Ca^{2+} than with Na^+ as exchange ions.

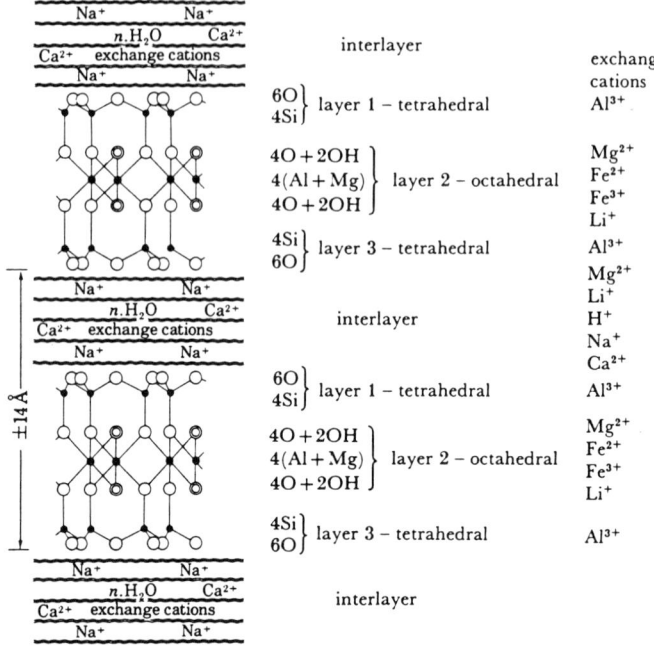

Fig. 4-3 Structure of smectite clay minerals. The chemical composition shown is that of montmorillonite. Various ions that may be present in other smectite clay species are shown in column on right. Odom (1984). Courtesy of The Royal Society.

It should be noted here that clay mineralogy nomenclature is sometimes ambiguous. In 1932 Kerr showed that certain clay materials which had been described as smectite are actually montmorillonite. He concluded that "it seems in the best interests of science to continue the use of montmorillonite and drop that of smectite." According to Grim (1968): "The name montmorillonite is used currently both as a group name for all clay minerals with an expanding lattice, except vermiculite, and also as a specific mineral name."

The chemical composition $M_y^{n+} nH_2O(Al_{2y}Mg_y)Si_4O_{10}(OH)_2$ is a montmorillonite clay, in which every third Al^{3+} in the octahedral layer has been replaced by a Mg^{2+} and yM^{n+} balances the excess negative charge of the structure. Between the unit layers in the structure there are nH_2O molecules.

1.3. Structure of Attapulgite

Attapulgite, like sepiolite and palygorskite, are clays consisting of particles having a lath-like structure. The structure consists of a double chain of tetrahedrons of silicon and oxygen (Si_4O_{11}) running parallel to the long axis. Upper and lower parts of each double chain are linked by a layer of magnesium atoms in sixfold coordination with the chains forming a network of strips which

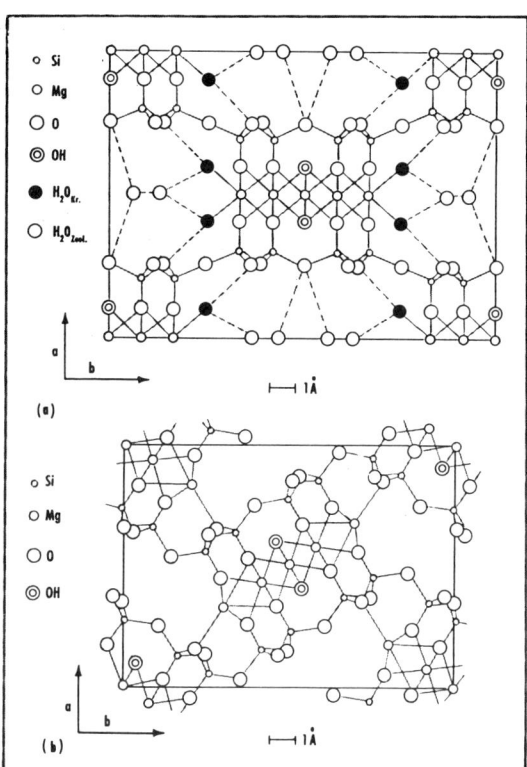

Fig. 4-4. Structure of attapulgite projected on the 001 plane (a) hydrous form (b) anhydride. Haden and Schwint (1967). Courtesy American Chemical Society.

are joined together only along the edges. One can imagine the over-all structure as a channeled wall in which every second brick is missing.

The structure of the unit cell of attapulgite is shown in Fig. 4-4, a and b, Haden and Schwint (1967). The composition of the ideal cell is $(OH_2)_4(OH)_2Mg_5Si_8O_{20} \cdot 4H_2O$. The crystal structure of the water-free clay is essentially the same as that of the hydrated clay, except for the tilted lattice and change in the a and b dimensions of cells. Each crystal unit normally contains eight water molecules. Four molecules are coordinated to the extreme magnesium ion as crystal water and the other four, zeolitic water, are held in loosely bound wandering fashion in open channels. When the zeolitic water is removed, each channel has an estimated cross-section of about 3.7 by 6.0 Å.

Since the structure of attapulgite consists of three-dimensional chains, it cannot swell in water. Also, cleavage parallel to the 110 plane along the Si-O-Si bonds holding the strips together gives an unusual needlelike shape to the attapulgite particles. Electron micrographs show attapulgite to occur in elongated

laths and bundles of laths. The individual laths can be several microns in length
with widths of 50 to 100 Å. Sometimes individual laths are bent so that an
interwoven structure occurs, Grim (1968).

2. Isomorphous Substitution - Ion Exchange Capacity

In many types of clays other metal ions of about the same size can be
substituted for silicon in the tetrahedral layer and for aluminum in the octahedral
layer; see Fig. 4-3. Silicon or aluminum ions are often replaced by metal ions of
lower charge, causing the crystalline framework to assume a negative charge,
which is neutralized by cations on or near the particle surface. Thus Al^{3+} can be
substituted for Si^{4+} in the tetrahedral layer and Mg^{2+}, Li^+, Fe^{3+}, Cr^{3+}, and Zn^{2+}
for Al^{+3} in the octahedral layer of montmorillonite; see Fig. 4-3. The excess
negative charge can be compensated by cations located close to the outside
surfaces of the clay particles, or between the interlayer surfaces in swollen clays,
since they are too large to fit into the crystal lattice.

Ion exchange and the exchange reaction are of great importance in all the
fields in which clay minerals are used. As an example, the retention and
availability of potash in fertilizers depends on cation exchange between the clay
mineral and the potash salt.

The total amount of such exchangeable cations can be determined
analytically and will then be a measure of the cation exchange capacity, CEC, of
the clay, which is usually expressed as milliequivalents of exchangeable cations
per dry gram, or 100 grams of clay. Isomorphous substitution affects the
composition and the extent of it can be found from the chemical formula for the
clay. Thus, montmorillonite, a specific mineral of the smectite group has the
following typical composition:

$$(OH)_4Si_8(Al_{3.34}Mg_{0.66})O_{20}$$
$$\downarrow$$
$$Na_{0.66}$$

in which the charge on the aluminosiloxane layer arises from a substitution of
Mg^{+2}, 0.66 equivalents, for some of the Al^{3+} in the octahedral layer, and this is
balanced by the exchangeable sodium ions, 0.66 equivalents, indicated by the
arrow. The cation exchange capacity of montmorillonite depends on the degree of
isomorphous substitution and ranges from 0.7 to 1.3 meq/ g. A strong acid cation
exchange resin and zeolite A (see Ch. 5) each have CEC's of about 5 meq/ g.
About 80% of the ion exchange capacity is due to charges resulting from
structural substitution, and about 20% is due to charges from broken bonds at the
edge of crystals. Exchangeable ions play an important role in the commercial use
of smectite clays. Montmorillonite with Na^+ as the exchangeable ion has a high

swelling capacity. The hydration associated with Na^+, which promotes the development of many oriented water layers on interlamellar surfaces, may produce swelling to the extent of complete separation of the structure into unit layers. In other words, the clay will be completely dispersed or peptized—see § 4.7—and the dispersion will be a true colloidal solution, attaining high viscosity at relatively low solids content. Conversely, montmorillonite containing Ca^{2+} or Mg^{2+} as exchangeable ions will show only a small degree of swelling even when fully hydrated, cf. § 1.2 and Fig. 4-3.

The degree of isomorphous substitution and, consequently, the cation exchange capacity of kaolin is much lower than for montmorillonite. Careful measurements show that the cation exchange capacity of kaolin is typically about 2 meq per 100 g clay, corresponding to about 1 Na^+ per nm^2 of surface, Jepson (1984).

The cation exchange capacity of natural attapulgite is typically between 20 to 30 meq per 100 g clay, which makes it higher than that of kaolin, but lower than that of montmorillonite.

3. Particle Size and Shape

The shape and size distribution of clay particles are of great importance in most commercial uses. Particle size is expressed as the equivalent spherical diameter even though the clay particles may be quite different from spherical in shape. Electron microscopy has greatly aided the study of the various clay minerals. Modern techniques and instruments provide resolutions of better than 5Å and magnifications of about 250,000.

Transmission electron microscopy of kaolin clays shows that they are largely composed of pseudo-hexagonal plates, schematically shown in Fig. 4-5.

Particle size measurements of Cornish kaolin showed that the plate width varied from about 8µm down to at least 0.1µm. The aspect ratio, which is the ratio of plate width to thickness, varied, through the size distribution, from about 10:1 at the coarse end to about 50:1 at the fine end, Jepson (1984).

The particle size distributions in Fig. 4-6 show that Cornish primary kaolins are much coarser than sedimentary kaolins from Georgia. The particles of the latter clays are much closer to hexagonal in shape. Their aspect ratios lie between 6:1 and 10:1 and vary little through the size range.

Electron microscopic examination of a kaolin fraction exceeding 5 µm ("equivalent spherical diameter") revealed the existence of many exfoliated booklets, which were up to 50 µm thick and 20 µm across, Jepson (1984). Figure 4-7 shows schematically that such large particles can be reduced in size to small "booklets" and even "pages" by vigorous shearing of the clay slurry.

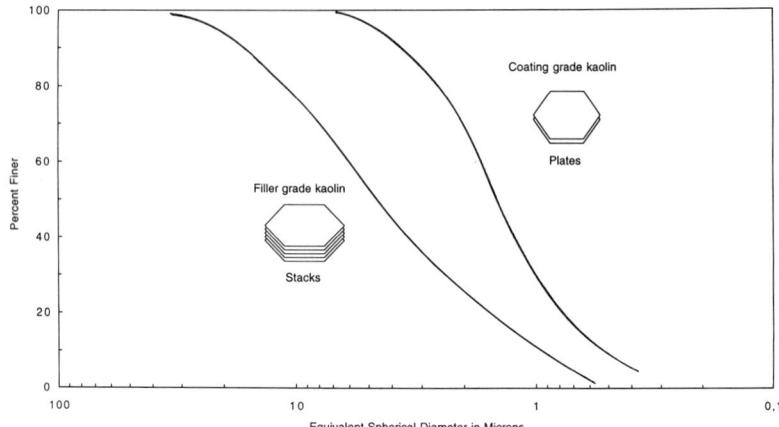

Fig. 4-5. Typical particle size distributions of filler and coating grades of kaolin.

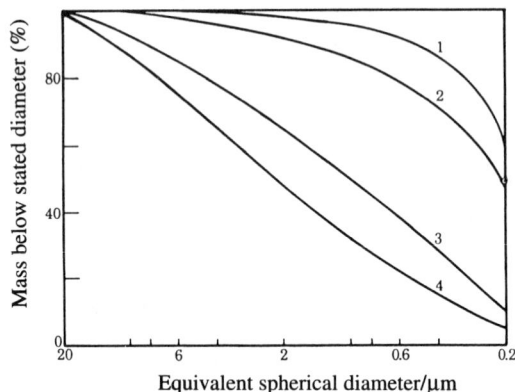

Fig. 4-6. Cumulative particle size distribution curves of kaolins. Curves (1) and (2), sedimentary kaolins from East Georgia; curve (3), sedimentary kaolin from Central Georgia; curve (4), primary kaolin from Cornwall. Jepson (1984). Courtesy The Royal Society.

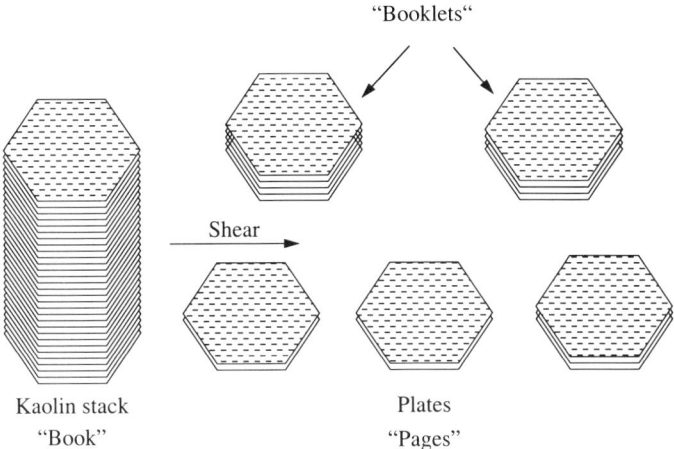

Fig. 4-7. Diagrammatic illustration of the splitting of kaolinite booklets or stacks into thin plates by the delamination process.

The shape of montmorillonite (smectite) particles may range from rhombic to hexagonal lamellar to lath-like, and also to fiber shapes. The particle size may be as large as 2μm and as small as 0.2 μm, with an average size of about 0.5μm.

Due to aggregation, the effective particle size of smectite clays may be considerably larger than the ideal particle size, and the surface area is thus less than ideal. Important properties in various industrial applications of smectite clays such as ion exchange, viscosity, and decreased fluidity depend on the effective size of the clay particles. Smectites containing sodium as the major exchangeable ion yield the smallest effective particle size and the largest surface area in clay-water systems because swelling pressure tends to disrupt and separate interlocked layers.

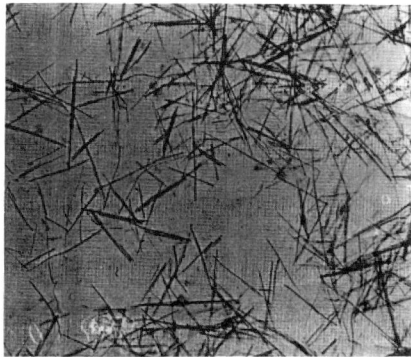

Fig. 4-8. Electron micrograph of attapulgite clay showing the characteristic bundles of needles. Haden and Schwint (1967). Courtesy American Chemical Society.

Table 4-4. Properties of industrial clays, Haden & Schwint (1967)			
	Attapulgus clay	Bentonite	Kaolin
Principal mineral	attapulgite	montmorillonite	kaolinite
Crystal structure	chain	3-layer sheet[a]	2-layer sheet
Particle shape	needle	flake	plate
Surface area	high	medium	low
Sorptivity	high	medium	low
Decolorizing power	high	medium	low
Binding power	medium	high	low
Color	gray to pink	gray to white	white
Brightness	low	variable	high
Thickening power	high	high	low
Effect of electrolytes	little or none	flocculates	flocculates

a Swelling bentonites are saturated with sodium ions and have a low ratio of magnesium to aluminum. They expand on sorption of water. Nonswelling bentonites are calcium saturated and have a high ratio of magnesium to aluminum. Courtesy American Chemical Society.

The typical attapulgite needle has a length ranging from about 0.5 to 1 μm and a width from about 0.02 to 0.06 μm. The needles may separate from each other in aqueous dispersions, but, as shown in Fig. 4-8, they usually remain in bundles, which are the principal structural feature of attapulgite products.

Table 4-4 shows important properties of the principal industrial clays attapulgite, montmorillonite, and kaolin.

4. Surface Charge

In platelike clay particles at least one of the flat layer surfaces is often negatively charged due to isomorphous substitution, while the edge surfaces could be positively charged.

At the edges of the plates the tetrahedral silica sheets and the octahedral alumina sheets are disrupted, and primary bonds are broken. This situation is analogous to that on the surface of silica and alumina particles in silica and alumina sols. The part of the edge surface where an octahedral aluminum layer has been broken may be compared with the surface of an alumina particle which is positively charged at pH below about 7. The part of the edge surface where a tetrahedral layer has been disrupted is similar to the surface of a silica particle which is negatively charged at pH above 2. Silica, however, has a great affinity for Al^{3+} ions, which are always present due to the slight solubility of the clay. The Al^{3+} ions adsorb on the silica sites making the edge surface positive over a wide pH range.

However, since the area of the edges is usually small relative to that of the basal surfaces—about 1% for smectites (Dyal and Hendricks, 1950)—the net charge on the structure is largely determined by the extent of isomorphous substitution occurring in the tetrahedral and octahedral sheets.

5. Particle Association: Flocculation and Aggregation

In general, a particle can obtain a surface charge by: (1) adsorption of an ion for which the solid acts as a reversible electrode, e.g., H^+ or OH^- for metal oxides; (2) preferential adsorption, e.g., by chemisorption, of an anion or cation from an electrolyte solution wherein the charge of the surface varies with the electrolyte concentration as given by the adsorption isotherm; and (3) isomorphous substitution, as in the case of clay particles.

The surface charge forms one layer of the electrical double layer. The opposite sign compensating charge is accumulated in the solution near the surface and is rather diffuse due to the competing electrostatic and diffusion forces—see Ch. 6.

In a flocculating suspension of platelike clay particles, three types of particle association occur: face-to-face (FF), edge-to-face (EF), and edge-to-edge (EE), Van Olphen (1977). Electrical interaction energies for the three modes of association are governed by three different combinations of the two double layers. Also, the van der Waals interaction energy is different for the three types of association because a different geometry pertains to the summation of the attraction between all the atom pairs of the apposite plates, Vold (1954). The net potential curves of interaction for the three types are therefore different. As a result, the three types of association do not necessarily occur simultaneously or to the same extent when a clay suspension flocculates. The three types of association lead to different structures. Face-to-face association results in thicker and possibly larger flakes, while EF and EE association yields three-dimensional voluminous card-house structures. The effect of the various associations on the properties of the flocculated suspensions, which are of practical interest, are quite different.

The various modes of particle association are shown in schematic form in Fig. 4-9. Here it is noted that even though the three types of association are really three modes of flocculation in the "colloid chemistry" sense, only the EF and EE association types yield agglomerates which can be called flocs. The thicker particles resulting from FF association cannot be properly termed "flocs"; therefore it has been suggested that a different term be used for this kind of association. The terms "parallel aggregation", "oriented aggregation", or simply "aggregation" have been suggested, of which the latter is the most common, though non-specific.

It has also been proposed that only the dissociation of EF and EE joined particles be termed "deflocculation," and that the splitting of FF associated "aggregates" into thinner flakes be described as *dispersion* of the clay.

Flocculation and "aggregation" do not necessarily proceed simultaneously or to the same extent under certain conditions. A system may be well "dispersed" but not deflocculated, or it may be deflocculated but not very well "dispersed".

Several multilayer particles may be described as "aggregated," but a single particle with layer stacking can also be so described since no differentiation can be applied, for example, to two entities one of which consists of a particle with ten stacked layers and the second two particles each of which has five layers. Thus a calcium montmorillonite particle, which usually contains more layers per particle than a sodium montmorillonite particle, is called a less well "dispersed" clay or a more "aggregated" clay, Van Olphen (1977).

6. Peptization: Deflocculation–Dispersion

Clays can be dispersed in water to dispersions with a solids content that depends strongly on the type of clay; for kaolin it is about 30-40 wt %, for attapulgite almost 10 wt %, and for montmorillonite only 3-5 wt %. In such dispersions three-dimensional "card house" structures can be formed, which appear at lower solids content for montmorillonite and attapulgite than for kaolin. Attapulgite and montmorillonite are therefore also used as thickening agents in various applications—see Ch. 10.2. In clay technology, however, it is often of greater interest to be able to produce as concentrated clay slurries as possible. To that end dispersing agents are used such as Na-phosphates, Na-silicates, Na-polyacrylates, and basic Al-salts, which at low concentrations—often less than 1 wt % based on the amount of clay—deflocculate and disperse the clay particles.

For instance, when a kaolin clay is added under suitable conditions to an aqueous solution of tetrasodium pyrophosphate, TSPP, that contains a small amount of sodium hydroxide, a solids content of about 70 wt. %, or 47 % by volume, can be attained at a TSPP addition of 0.3 % and pH 8. The deflocculation of kaolin with TSPP, Na-silicate and Na-polyacrylate, is due to Ca^{2+} and Mg^{+2} exchangeable cations being displaced from the clay surface by Na^+ ions to form complexes with the polyanions in solution. Moreover, the polyanions are strongly adsorbed on the kaolin surface, probably at exposed Al atoms on the particle edges and at steps on the flat particle surfaces. The negative charges built up together with the mechanical shear applied during mixing reduce or eliminate face-to-face, edge-to-face, and edge-to-edge associations and cause the particles to separate (compare Fig.4-9).

With suitable dispersing agents, suspensions of attapulgite and montmorillonite containing 25% and 12% solids, respectively, can be made.

Mattsson and Otterstedt (1996) studied the effectiveness of different dispersing agents with regard to reducing the viscosity of slurries of kaolin (Supreme from ECC, UK), Na montmorillonite (Hydrocol® HS from Allied Colloid, UK), and attapulgite (Attagel® 50 from Engelhard, USA). Thick slurries, i.e., having viscosities in the range 100,000 to 200,000 cp, of the three

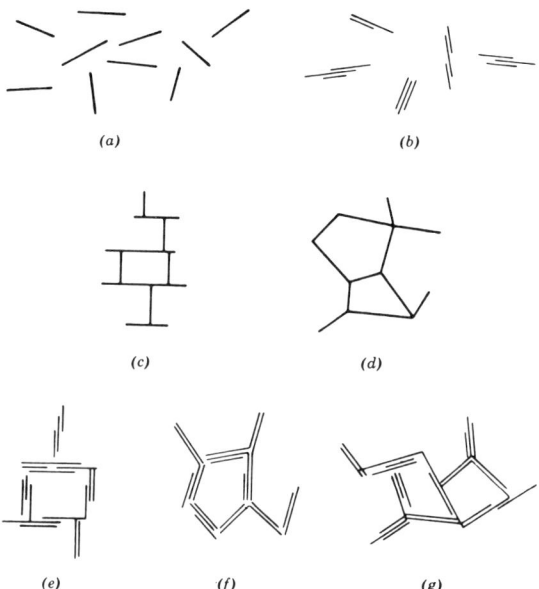

Fig. 4-9. Modes of particle association in clay suspensions, and terminology. (a) "dispersed" and "deflocculated." (b) "aggregated" but "deflocculated" (face-to-face association, or parallel or oriented aggregation). (c) Edge-to-face flocculated but "dispersed." (d) Edge-to-face flocculated and "aggregated." (f) Edge-to-edge flocculated and "aggregated." (g) Edge-to-face and edge-to-edge flocculated and "aggregated." Van Olphen (1977). Reprinted by permission of John Wiley & Sons, Inc.

clays were prepared, which corresponded to a solid content of about 47 wt. % for kaolin, about 11 wt % for Na-montmorillonite, and about 18 wt % for attapulgite. Small amounts of the dispersing agents 3.3 ratio Na-silicate, sodium polyacrylate (Dispex® N40 from Allied Colloid, UK), and basic Al-chloride (Locron® L from Hoechst, Germany) were successively added to the slurries and the viscosity was recorded after each addition.

Figures 4-10a,-b and-c show that the viscosity falls rapidly to a plateau, which for some of the dispersing agents can be fairly wide, with increasing amounts of dispersant. Figure 4-10a shows that 3.3 ratio Na-silicate and Dispex® N40 are about equally effective dispersing agents for kaolin, but that it requires about five times more Locron® L on a weight basis to reach the plateau of low-viscosity. A Locron® L solution consists of polycations of aluminum, primarily $Al_{13}O_4(OH)_{24}(H_2O)_{12}^{7+}$ (see Ch. 3), which will give the kaolin particles a positive charge by adsorbing on the flat negatively charged particle surfaces.

For montmorillonite, Figs. 4-10b, 3.3 ratio Na-silicate is a somewhat more

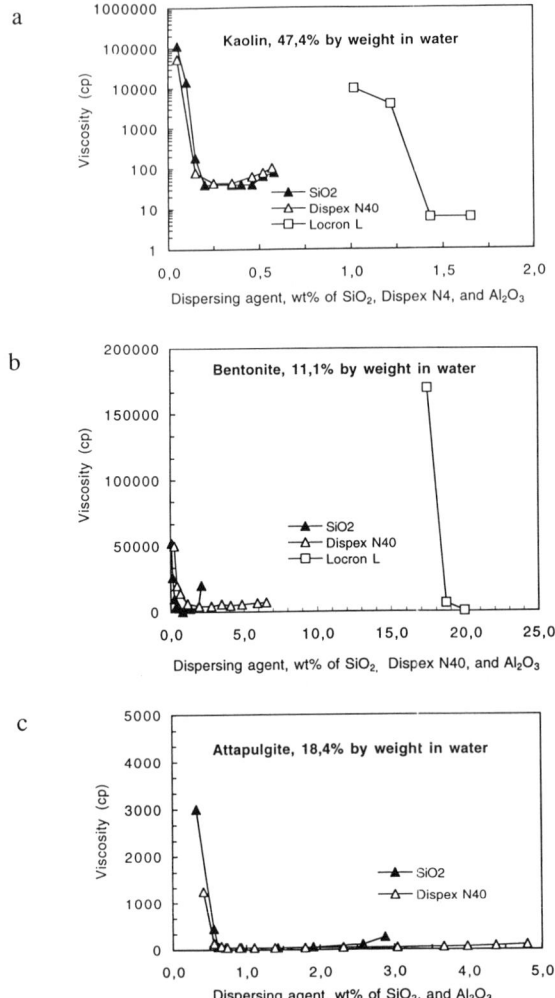

Fig. 4-10. Effect of dispersing agent on viscosity of clay slurries.
Mattsson and Otterstedt (1996).

effective dispersing agent than Dispex® N40 because less of it is required to reduce the viscosity to the plateau of low viscosity, of which a portion is at a lower viscosity than that for Dispex® N40. Compared with kaolin, montmorillonite requires even more Locron® L, both on an absolute basis and relative to 3.3 ratio Na-silicate, to reduce the viscosity to the minimum value, indicating several differences between the two types of clay. The total surface

area, which is about 10 m^2/g for Supreme® and about 450 m^2/g for Hydrocol®
HS (actually it would be about 800 m^2/g if the Hydrocol® HS were dispersed
down to the elementary 2:1 layers), and the fraction of it that can be attributed to
the flat particle surfaces is much greater for Hydrocol® HS than for Supreme®.

Furthermore, the degree of isomorphous substitution is much higher in
montmorillonite than in kaolin, requiring more aluminum polycations to reverse
the negative charge on the flat bentonite particle surfaces.

Figure 4-10c shows that Dispex® N40 is a somewhat better dispersing
agent for attapulgite than 3.3 ratio Na-silicate.

When much more dispersing agent is added than is needed to reach the
plateau of minimum viscosity, the counter ions, often sodium ions, in the aqueous
phase will act as bridges or links between the oppositely and uniformly charged
clay particles, and the viscosity will increase.

7. Thermal Reactions: Dehydration and Transformation

Differential thermal analysis of kaolin minerals shows a prominent
endothermic effect in the region 550-700°C, due to the dehydration of the
mineral, followed by two exothermic reactions in the temperature ranges 950°C
to 1000°C and 1150°C to 1250°C, respectively, which corresponds to the
formation of new material.

The thermal reactions of kaolin can be expressed in the following way,
Richardson (1961):

At 550°C: $Al_2O_3 \cdot 2SiO_2 \cdot 2H_2O \longrightarrow Al_2O_3 \cdot 2SiO_2 + 2H_2O$
                    kaolin mineral   metakaolin

At 850°C to1050°C:
        $Al_2O_3 \cdot 2SiO_2 \longrightarrow \gamma\text{-}Al_2O_3 + SiO_2$ (with Al_2O_3 in solid solution)

At 900°C and up: $\gamma\text{-}Al_2O_3 + SiO_2 - Al_2O_3 \longrightarrow 3Al_2O_3 \cdot 2SiO_2$
                             solid solution   mullite

Smectite clays contain two forms of water, adsorbed and crystalline. The
clay loses its adsorbed water, most of which is interlayer water—see Fig. 4-3—in
the temperature range 100°C to 200°C. The amount of adsorbed water loss
depends on the type of exchange ion, the pretreatment of the sample, and also, but
to a lesser degree, on the structure of the smectite. Figure 4-11 shows that when
Na^+ is the predominant exchange cation, adsorbed water is usually lost in one
stage, whereas when Ca^{2+} and Mg^{2+} are predominant, the adsorbed water may be
lost in two stages, Odom (1984).

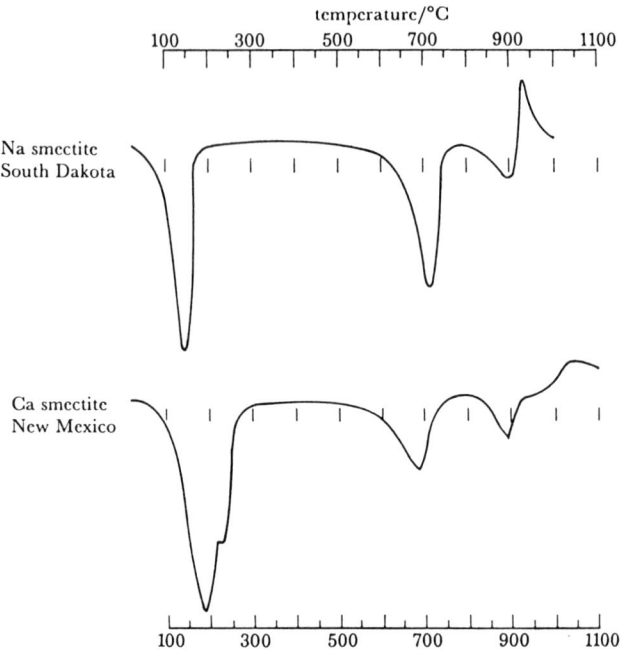

Fig. 4-11. Differential thermal analysis patterns illustrating the single-stage loss of adsorbed water that is typical of Na smectites and the two-stage loss of adsorbed water that is shown by Ca smectites. Odom (1984). Courtesy The Royal Society.

Since loss of crystalline water, in the form of OH, involves a breakdown of the silicate structure, the temperature range in which this loss occurs depends primarily on crystalline structure, strength of cation-OH bonds, and chemical composition. For montmorillonite with low substitution of Fe and Mg for Al and having Na^+ as the predominant exchange cation, the loss starts gradually at 550°C, reaches a peak at about 680-700°C, and ends at about 750°C; see Fig. 4-11. Large substitution of Fe or Mg for Al causes a reduction in the temperature of loss of OH.

If the adsorbed water is completely removed from smectite clays, rehydration will occur to an extent which depends on the conditions of rehydration and the properties of the clay, primarily the nature of the exchange cation. Montmorillonite with Ca^{2+} and Mg^{2+} as exchange cations will rehydrate more readily than the clay with Na^+ as exchange cation. This effect reflects differences in hydration properties of the exchange cations.

Dehydration of attapulgite also proceeds in stages; see Fig. 4-12. Water adsorbed on the external surface on the attapulgite particles is lost up to 100°C. The three peaks between 100°C and 600°C are characteristic for all types of attapulgite and the first one represents loss of zeolitic water, i.e., water adsorbed in pores and channels of the crystalline structure. The anhydride, which is

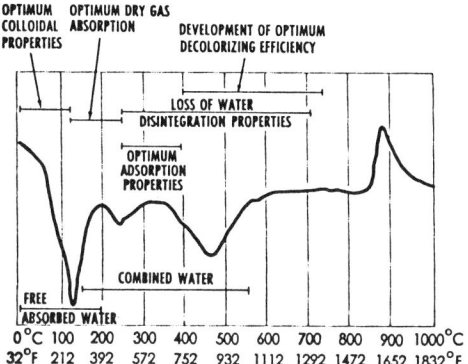

Fig. 4-12. Differential thermal analysis of attapulgite. Haden and Schwint (1967). Courtesy American Chemical Society.

crystalline though its structure does not contain pores and channels (as does regular attapulgite), forms between 400°C to 500°C. At 600°C the clay is almost fully dehydrated and above 700°C further shrinkage takes place. The attapulgite structure is destroyed and new phases, mainly enstatite, are formed, Haden and Schwint (1967).

8. Synthetic Hectorites

A range of synthetic hectorites is manufactured by Laporte under the trade name Laponite®, a sodium magnesium lithium silicate. These resemble the natural mineral hectorite in both structure and composition. This material has a number of distinct advantages compared with natural clays. It is free from crystalline silica impurities present in natural clays and is chemically pure and consistent in its properties due to its synthesis under controlled conditions.

Laponite® used as a rheology control agent is many times more efficient than natural products, and it can also be dispersed rapidly to give colorless, transparent, highly thixotropic gels without the need for chemical dispersants, elevated temperatures or high-shear mixing. It can be used with a wide range of thickeners to give various rheological effects, Doyle and Barlas (1995).

9. Pillared Interlayered Clays—Crosslinked Smectites

Montmorillonites pillared with $N^+(CH_3)_4$ and $N^+(C_2H_5)_4$ cations were first reported by Barrer and MacLeod (1955). The molecular sieving properties of these materials were well known, but their potential as catalysts was neglected, perhaps because of their limited thermal stability. During the oil crises in the late

perhaps because of their limited thermal stability. During the oil crises in the late seventies, refineries were looking for cracking catalysts with larger pore structures than those of zeolite Y, able to accommodate the large molecules present in heavy oils, and became interested in heat-stable, two-dimensional sieve-like materials like those prepared by Barrer and co-workers—see Chapter 7.

Montmorillonites were interlayered, or crosslinked with polycations of Al and Zr, colloidal alumina, or colloidal titania—see Chapters 3 and 7. These colloidal-size pillars propped up the clay layers to a structure easily accessible to molecules as large as 1,3,5-trimethylbenzene (d=7.6 Å) and also having acceptable thermal stability.

Catalytic cracking of heavy feedstocks requires cracking catalysts with a pore structure more accessible to "resid" molecules than e.g. zeolite Y—see Chapter 7. The desire to crack such feedstocks has provided the stimulus to develop new types of pillared interlayered clays which have:
• larger spacings between the clay sheets
• improved thermal and hydrothermal stability
• improved selectivity and novel catalytic activity

Pillar Compositions

The original pillars in crosslinked smectites were based primarily on Al_{13} and Zr_4 polycations—see Chapter 3—that give rise to several different structures. New pillar compositions (Vaughan, 1988) include:
• new groups of polymers of $(TiO_2)_x$
• modifying some of the original polycations by substituting some of the Al with secondary elements
• small clusters or colloid compositions of transition metal oxides or metal hydroxides.

Table 4-5 compares the properties of clays pillared with different "props" and Fig. 4-13 shows models of various inorganic polymers that may be used to pillar clays—see also Chapters 3 and 7.

Table 4-5. Comparison of the properties of Al, Zr, Si, and Ti-pillared clays, Vaughan (1988) Courtesy of the American Chemical Society

Polymer Species	(001) Clay Spacing	Surface Area m^2/g
Al_{13}	17.8Å- 18.5Å	400
$(Si_8)_n$	19Å	300
Zr_4	18.0Å	260
$(Zr_4)_x$	21.5	290
$(Zr_4)_y$	21.5 + 25.6	250
$(TiO_2)_a$	25Å	300
$(TiO_2)_b$	29Å	330

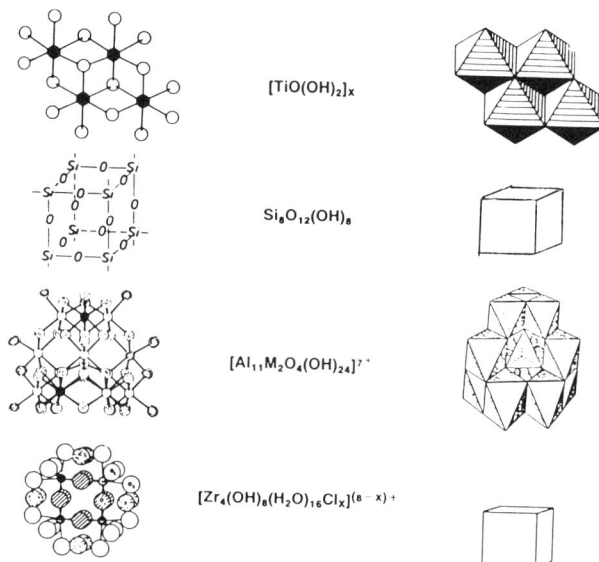

Fig. 4-13. Models of various oxy-hydroxy inorganic polymers possibly used to pillar several layer structures, including complexes of Ti, Si, Al; Al, M; and Zr. Vaughan (1988) Courtesy American Chemical Society.

10. Uses of Clays

The major uses of clays in processes and products are given in Tables 4-6 and 4-7. The world market for clays is projected to reach 180.5 million metric tones by the year 2000 with an estimated average growth rate of about 5% per year. The United States is the largest and fastest growing clay market in the world (Global Industry Analysts, Inc., 1996).

Table 4-6. Major Uses of Clays in Processes		
Process/Application	Clay	Function - Property
Drilling muds	smectites	Viscosity control, sealing drill holes by swelling
	attapulgite	Viscosity control under salt-rich conditions
Filtering, decolorizing and clarifying	smectites attapulgite	Adsorption, flocculation, coagulation
Civil Engineering	smectites	Grout cracks in rocks, impede water or chemical waste movement through sand, suuport of excavation by providing viscosity, thixotropy, impermeability and plasticity

Table 4-7. Major Uses of Clays in Products		
Product	Clay	Function - Properties
Paper	kaolin, halloysite	Filler: opacity, body, cost-economy, printability Coating: gloss, opacity, printability
Rubber	kaolin, halloysite	Filler-extender: strength, abrasion resistance, rigidity, cost
Paint	kaolin, halloysite smectites attapulgite	Extender: opacity, rheology, flattening, cost Suspending and thickening agent Thickening, antisag, and levelling agent
Plastics	kaolin	Smoother surface, more attractive finishes, dimensional stability, resistance to chemical attack
Ceramics	kaolin, halloysite	High green and dry strength, color, low conductivity, high dielectric constant, low power loss
Foundry moulding sand	smectites	Binder: green compression strength, hot compression strength, flowability and durability
Pellets -iron ore -animal feed	smectites	Binder: pellet development, green and dry strength, water adsorption Increase nutritional benefits animals get from feed
Adhesives, sealants, putties	attapulgite	Thickening, thermal stability, prevents bleaching
Pharmaceuticals	attapulgite	Adsorption of toxin, bacteria and alkaloids, control of neutralization rate
Pet litter boxes	smectites	Adsorbent

Kaolin. Kaolin has many important industrial uses—principally in paper, rubber, ceramics, paints, and plastics. It is important because of its relatively low cost and unique blend of properties which include chemical inertness over a wide pH range, whiteness, good covering or hiding power when used as a pigment or extender in coated films, softness, non-abrasiveness, low heat and electrical conductivity.

Paint, paper, and rubber products account for the biggest volume of kaolin usage even though its best-known and historically first application was likely in ceramics, especially where a white color is desired. In porcelain it is used for good suspension characteristics and whiteness. In making electrical insulators it is used for its low conductivity, low power loss, good plasticity, high dielectric constant, and high fired strength.

Kaolin is used as a filler and coating pigment in paper production—see Ch. 9. Worldwide, the paper industry consumes 75% of the total kaolin production. Major paper uses are in supercalendered paper such as those in magazines where the filler can be up to 25 % of the total weight, and in light-weight coated papers used in mail order catalogues with as much as 33% kaolin.

Kaolin is used in the rubber industry as a filler and extender in both synthetic and natural rubber. Incorporated in the latex mix, kaolin improves properties such as strength, abrasion resistance, and stiffness while lowering the cost.

Kaolin is an effective extender in paints due to its chemical inertness, high covering power, favorable rheological behavior, whiteness, and low cost. Furthermore, it is available in a wide range of particle sizes which can be used in a variety of paints, and has excellent suspension properties. The larger-particle kaolins are used in paints where a dull or "flat" finish is desired, whereas fine-particle kaolins are used for high-gloss paints. Water-beneficiated kaolins disperse easily in water and are thus especially suitable in latex paints. By treating kaolin chemically, it can be made organophilic and thereby suitable for use in exterior oil-base paints. Calcined kaolin is an excellent extender for TiO_2, especially in latex paints, because it reduces costs and simplifies formulation of the paint.

As with many natural products, significant differences in performance in some uses are noted in grades from different sources since the chemical composition may differ and different impurities occur in material from different locations. Work by Helaly et al. (1994) showed that styrene butadiene rubber formulations utilizing Egyptian kaolin as a filler, compared to a commercial imported kaolin sample, showed improvements in physico-mechanical properties of SBR along with improved stability to oxidative aging at a reduced cost.

Purification of kaolins from different mines can be carried out, but it is difficult and expensive because kaolinite and the major gangue minerals in the ore are usually very fine. A study by Wenqi and Jizu (1994) established that separation by selective dispersion and flocculation successfully purified the kaolin and that improvements in kaolin quality were obtained.

Kaolin is used extensively as a filler in reinforced plastics where its platy nature and viscosity-increasing effect prevent fiber "blooming" or the appearance of fibers at the molded surface. Grades used with fiberglass preforms and mats should have low fractions of particles greater than 44 μm because larger particles filter onto the mats to cause non uniform filler distribution, crazing and poor physical properties. Air-classified grades are commonly used in pre-mix compounds, but coarser, water-classified grades are used in sheet molding compounds, Ferrigno and Florea (1987).

Whereas hydrous kaolins are extremely soft and non abrasive because of their platy character and the absence of massive particles with well-defined sharp edges, calcined kaolin contains very hard particles with rough edges and fewer platy particles to provide lubricity.

Figure 4-14 shows a typical particle size distribution for a calcined kaolin clay (Snow*Tex® 45 from U. S. Silica) where the average particle size is 1.5 μm with a packed bulk density of 29.4 lb/ft^3.

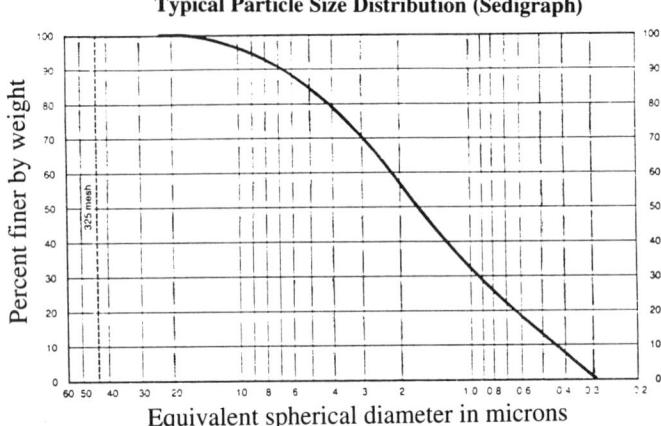

Typical Particle Size Distribution (Sedigraph)

Fig. 4-14. Typical particle size distrbution for a calcined kaolin clay (Snow*Tex® 45 from U. S. Silica Co.), packed bulk density 470.9 kg/m³, refractive index 1.62,specific gravity 2.58. Courtesy U. S. Silica Co.

Inasmuch as kaolin is at least as resistant to corrosive compounds as most resistant polymers, it usually improves chemical resistance to a higher extent than the polymer alone provides. However, the naturally acid pH of the mineral causes catalysis and decomposition with epoxies and vinyls unless the acid sites are neutralized or reacted. In addition, the hydrophilic nature of kaolin surfaces and the large amounts of air contained by the powder make dispersion difficult with non-polar polymers. Thus kaolin has not become a principal filler in melt-processed polymers. Surface treatment of the kaolin can overcome these problems to produce rapid and uniform dispersion.

Incorporation of kaolin into polyvinyl chloride (PVC) as a reinforcing agent makes the plastic more durable. Calcined and partially calcined kaolins are used as a filler in PVC wire insulation to improve electrical resistivity.

There are also many other applications of kaolin which are not close to the high-tonnage uses, but which are important due to this clay's special properties, ready availability or cost effectiveness. These uses include the following:

- Adhesives
- Bleaching
- Cosmetics
- Crayons
- Filter aids
- Catalyst preparations
- Detergents
- Porcelain enamels
- Roofing granules
- Linoleum

- Textiles
- Detergents
- Fertilizers
- Plaster
- Insecticides

Attapulgite. The unusually high specific surface area of attapulgite is a key factor in its high adsorption capacity. The surface area together with its needle structure, usually arranged in bundles, largely determines its uses. The specific surface area ranges from as high as 210 m^2/g for colloidal grades to as low as 125 m^2/g for thermally activated adsorbent grades. It is the principal member of a group of sorptive clays known collectively as fuller's earth.

The cation exchange capacity of natural attapulgite is rather low, 20-30 meq/100 g, which is somewhat higher than the values for kaolin, but about one half the values for montmorillonite.

The slurries of attapulgite are viscous and thixotropic, but unlike bentonite (a mixture of clays, principally montmorillonites, with some biotite, orthoclase, and other minerals of the original volcanic rock), attapulgite is not flocculated by electrolytes.

The principal applications of colloidal grades of attapulgite generally involve modification of rheological properties such as thickening, gelling, and stabilizing. For example, attapulgite is widely used in latex and oleo-resinous paints as a thickening, antisag, and leveling agent. The attapulgite is generally incorporated at 5 to 10 lb/100 gal paint, but in gelled paints the loading can be as much as 30 lb/100 gal. Such paints have excellent film build on sharp edges, and heavy films can be obtained if desired.

Decolorization of organic liquids by adsorption of the impurities is one of the oldest uses of attapulgite. In the direct contact process, from 0.5 to about 30 % attapulgite is mixed with the liquid at temperatures up to 315°C and is later removed by filtration. In the percolation process the liquid is passed through a bed of attapulgite at a somewhat elevated temperature. A wide range of liquids from oils, fats, waxes, resins, vitamins, beers, water, to industrial wastes and sewage can be purified by attapulgite treatment.

The high adsorptive power of attapulgite together with its nontoxicity lead to its use in pharmaceutical products, particularly intestinal preparations where it is superior to other clays in adsorbing diptheria toxin, bacteria, and alkaloids.

It is also used in antacid preparations where it helps to control the neutralization rate and in veterinary cases for the treatment of dysentery in large animals.

Attapulgite is a useful thickener for starch-based corrugating adhesives because without a suitable thickener, the loss of viscosity during the application of starch adhesives often requires frequent adjustment of glue-roll clearances.

Other applications include microcrystalline wax and joint sealing compounds based on alkyd resins, chlorinated rubber, and oils at attapulgite loadings of 8 to 20%; and in putties and glazing compounds where it prevents bleeding at loadings of about 5 wt. %.

Attapulgite is an efficient thickener for aqueous and organic liquids, especially for high-temperature uses because of its insolubility. It is well suited in thickening lubricating oils to grease with 6 to 15 % attapulgite. Many other liquids such as ethanol, isopropyl alcohol, ketones, ethers, esters, chlorinated aliphatic hydrocarbons, linseed oil, soybean oil, wax compositions, and liquid polyesters are also effectively thickened with attapulgite. Stable gels are made from aqueous solutions containing 5 to 30% ammonia by addition of 2 to 10% attapulgite. These gels can be used as vehicles in paint, ink, and cleansing compositions.

Attapulgite is extensively employed in drilling mud compositions where salt formation are drilled because these compositions with attapulgite retain their viscosity in such cases whereas those using bentonite, hectorite, and other montmorillonite clays do not.

Montmorillonites (Smectites). The many industrial and chemical uses of smectite clays depend on one or more of the unique properties of these materials. The broad range of uses stem from their physiochemical properties of which many are not exhibited by other known natural products. These properties are the result of their extremely small crystal size, variations in internal chemical composition, large cation exchange capacity, large surface area that is chemically active, interactions with inorganic and organic liquids, and variations in types of exchangeable ions and surface charge. Smectite clays interlayer cations are exchangeable and interlayer surfaces and cations are hydratable.

Drilling muds account for a large amount of smectite clay use. Smectite clays for this use must meet the American Petroleum Institute (A.P.I.) or the Oil Companies Materials Association (O.C.M.A.) standards for grit, viscosity and other rheological properties, and fluid loss. Certain natural Na or Na-exchanged Ca smectites have the potential for meeting the A.P.I. or O.C.M.A. specifications.

Smectite clays also find extensive usage as a binding agent for the pelletization of animal feeds, iron ore and other fine-grained solids. When added at the 2% level as a binder in animal feeds, the clay reduces die friction during the pelletization process and increases the nutritional benefits the animals obtain from the feed. Both Na and Ca smectites are used for feed pelletization.

During the 1970's the largest single use of smectite clays was in pelletizing the iron ore derived from taconite. In that process the beneficiation is achieved by grinding the ore to a fine particle size so that impurities can be removed in a fluid medium. The higher grade ore is then pelletized using smectite clays as binders. Both natural Na and Na-exchanged smectite clays find use in this process. The Na smectites give the pellets the required green and dry strength and a high strength after calcination.

Natural Na smectites and some Na-exchanged Ca smectites have the necessary properties of viscosity, thixotropy, plasticity, impermeability, high swelling capacity, and good dispersibility for civil engineering applications such as grouting cracks in rocks, impeding water and chemical waste movement through, sand, gravel, and permeable soils, and providing nonmechanical support of excavations, Odom (1984).

A new variation for sodium bentonite clay use in containing chemical waste is in high performance environmental liners consisting of a layer of sodium bentonite between a woven and non-woven geotextile which are needlepunched together (CETCO, 1996). Upon hydration the bentonite swells and changes into a dense, monolithic mass with a very low hydraulic conductivity (about 1×10^{-9} cm/s—abour one hundred times lower than a typical compacted clay layer). These geosynthetic clay liners (GCLs) are manufactured in large rolls that require just a simple overlapped seam. Similar liners can be used for secondary containment of above-grade fuel storage tanks, for surface impoundments in decorative lakes and aeration lagoons, and in landfill covers.

Clay products containing smectite clays and other clays with adsorbent properties are extensively used for pet litter, adsorbing oil and grease, as a carrier for some types of agricultural insecticides, and as packaging desiccants.

Another major use of smectite clays is in the treatment of animal and mineral oils and greases, decolorizing vegetable oils, and in wine clarification. Usually acid treated Ca smectites are used for filtering and decolorizing oils because they have an enhanced decolorizing ability.

Colloidal organic impurities in wine carry a positive charge, and these particles are coagulated by mixing a small amount of a negatively charged smectite clay into the wine. For this use Na smectites with a light color and high dispersability are preferred. Clarification of beer, vinegar, and fruit juices is also carried out using smectites, Odom (1984).

Among the newer possibilities for smectites are uses in new types of catalysts. Pillaring of smectites by complex inorganic cations yields porous solids that possess some of the properties of zeolites—strong acidity and regular porosity—since the size of the micropores is determined by the size of the pillars, Vaughan et al. (1979, 1988). Well-defined pore structure in the range 1-2 nm can be obtained. A variety of catalysts have thus been prepared that are active for cracking, Vaughan et al. (1979, 1988), alkylation, Ocelli et al. (1985), dehydrogenation, Pinnavera et al., (1983), and epoxidation, Choudary et al. (1990), depending on the nature of the pillars, Fetter et al. (1994).

The key problem in understanding the physico-chemical properties of pillared clays as sorbents and catalysts is the determination of the structure of the hydroxymetal cationic pillars and the way in which they are distributed within smectite interlayer spaces.

Furthermore, the work of Lou and Huang (1993), shows that silicious Al-interlayers in smectite can be formed through the intercalation of hydroxy-aluminosilicate cations or "proto-imogolite", as well as that of imogolite tubes. These interlayer complexes also have potential application in industrial catalysis. Kikuchi and Matsuda (1988) investigated the synthesis of different pillared interlayered clays and their catalytic properties in a variety of chemical reactions.

In addition to the above major uses, specialty uses of smectite clays include the following, Odom (1984);

- Adhesives
- Atomic waste disposal—adsorption of radioactive isotopes of strontium and cesium with subsequent fixation against ground water leaching
- Emulsifying, suspending, and stabilizing agents
- Greases - stabilizing the gel properties of lubricating greases
- Inks - to control consistency, penetration, and misting during printing
- Medicines, pharmaceuticals, cosmetics
- Paint - suspending and thickening agents in both oil and water-based paints
- Catalysts
- Soaps - detergent action and water softening; also cleaning and polishing compounds
- Ceramics - improvement of vitrification and color properties; improve suspension characteristics in kaolin slips; increase plasticity in structural clay products during extrusion
- Anti-stick and drying agents - prevention of sticking of various grains and fertilizers
- Seed coating - improvement of germination
- Water clarification - removal of industrial oils and organic contaminants
- Mortar mixes - improvement of water retention, plasticity, and general workability.

References

Bailey, S. W., Am. Miner., **65**, 1 (1980).

Bailey, S. W., *Crystal Structures of Clay Minerals and their X-ray Identification*, Ed. G. W. Brindley and G. Brown, Ch. 1. Mineralogical Society, London (1980b).

Barrer, R. M. and MacLeod, D. M., Trans. Faraday Soc., **51,** 1290 (1955).

Brown, G., Phil. Trans. R. Soc. Lond., **A311**, 221 (1984).

CETCO Colloid Environmental Technologies Company brochure Geosynthetic Clay Liners, GLC-1, Arlington Heights, IL (1996)

Choudary, B. M., Valli, V. L. K., and Durga Prasid, A., J. Chem. Soc. Chem. Commun., 1186 (1990).

Doyle, J., and Barlas, J., Polymers Paint Colour Journal, **185**, July, 15 (1995).

Dyal, R. S. and Hendricks, S. B., Soil Sci. **69**, 421 (1950).

Ferrigno, T. H., and Florea, T. G., *Handbook of Fillers for Plastics*, Eds. H. S. Katz and J. V. Milewski. Van Nostrand Reinhold, New York (1987).

Fetter, G., Ticht, D., Massiani, P., Dutartre, R., and Figueras, F., Clay and Clay Minerals, **42**, 161 (1994).

Global Industry Analysts Inc., Global Business Report (1996).

Grim, R. E., *Clay Mineralogy*, 2nd ed., McGraw-Hill, New York (1968).

Haden, W. L. and Schwint, I. R., Industrial and Engineering Chem., **59**, No. 9, 59 (1967).

Helaly, F. M., El-Sawy, S. M., and Abd El-Ghaffar, M. A., J. Elastomers and Plastics, **26**, 335 (1994).

Jepson, W. B., Phil. Trans. Royal Soc. London. A311, 411 (1984).

Kikuchi, E., and Matsuda, T., Catalysis Today **2**, 297 (1988).

Lou, G. and Huang, P. M., Clays and Clay Minerals, **41**, 38 (1994).

Mattsson, M., and Otterstedt, J-E. A., unpublished results (1996).

Richardson, H. M., *The X-ray Identification and Crystal Fractures of Clay Minerals*. Ed. G. Brown, Mineralogical Society,136, London (1951).

Norton, F. H., *Elements of Ceramics*, Addison-Wesley, Cambridge (1952).

Ocelli, M. L., Innes, R. A., Hwu, F. S. S., and Hightower, J. W., Applied Catalysis **14**, 69 (1985).

Odom, I. E., Phil. Trans. Royal Soc. London. **A311**, 391 (1984).

Patterson, S.H. and Murray, H. H., American Institute of Mining, Metallurgical, and Petroleum Engineers, Inc., *Industrial Minerals and Rocks*, Ed. S. I. Leopold, New York, 519 (1975).

Pinnavera, T. J., Mortland, M., and Endo, T., U. S. Patent 4,367,163 (1983).

Van Olphen, H., *An Introduction to Clay Colloid Chemistry*, John Wiley, New York, 95 (1977).

Vaughan, D. E. W., Amer. Chem. Soc. Symp. Ser. #368, Ed. W. H. Flank and T. E. Whyte, 308 (1988).

Vaughan, D. E. W., Lussier, R. J., and Magee, J. S., U. S. Patent 4,176,090 (1979).

Vold, M. J., J. Colloid Sci., **9**, 451 (1954).

Wenqi, G., and Jizu, Y., J. Wuhan University of Technology, Materials Science Ed., **9**, 39 (1994).

CHAPTER 5. ZEOLITES

Introduction

A. CONVENTIONAL ZEOLITES

B. COLLOIDAL ZEOLITES

Introduction

Zeolites are materials which are characterized by having a structure consisting of channels, cages, and cavities, the diameters of which are large enough to admit small molecules and ions, but sufficiently small to prevent the entry of larger ones.

The term zeolite was coined by the Swedish mineralogist Axel Cronstedt in 1756 as a name for a remarkable crystalline aluminosilicate material that appeared to boil when heated; hence the name zeolite from *zeolen*, to boil, and *lithos*, stone, in Greek. Since Cronstedt's time at least 40 naturally occurring zeolites have been identified. The extent of zeolite structural chemistry has, however, been broadened tremendously by the success of laboratory syntheses which have contributed more than 150 types of zeolites that have no natural counterparts. The diversity of interest in zeolite chemistry over the last 35 years has been fueled by the economic rewards of industrial applications of zeolites, by ongoing developments in synthetic procedures, and by application of new techniques to zeolite characterization.

The open framework structure of zeolites has pores and cages of molecular dimensions, 3-13Å, that select only those molecules physically small enough to pass through—hence their common name of molecular sieves. They also orient those molecules as they gain access to the internal voids of the crystallite. These materials are the ultimate result of the quest for ever smaller reaction vessels. Chemistry carried out in them is affected by the confines in which it is performed. One negative charge per aluminum is present on the framework and is compensated by loosely attached, and therefore ion-exchangeable, cations. These cation-exchange sites within the internal void space of the crystallite allow the straightforward introduction of active metal ions for catalysis.

The term zeolite means different things to different people. To research scientists in many different fields they are extraordinarily interesting materials to study. The refining and petrochemical industries could not cost effectively utilize oil to make oil-based products without using zeolites as catalysts. Many separation processes rely on the molecular sieving and ion-exchanging properties of zeolites.

In this chapter we will discuss conventional zeolites and then colloidal zeolites, an entirely new type of materials which, in addition to the properties of conventional zeolites, also have properties characteristic of small particles.

For a more complete discussion of all aspects of zeolite technology the reader is directed to the seminal works of Barrer (1978) and Breck (1984). The engineering side of their use in adsorption technology is covered well in the books by Yang (1987) and Ruthven (1984). Meier and Olson (1992) give complete crystallographic details in their book on zeolite structure types.

A. CONVENTIONAL ZEOLITES

1. Structure and Properties

The properties of zeolites can be derived from their structure. In order to understand the properties it is necessary to know the structure of zeolites and the building blocks making up their structure.

Zeolites are tectosilicates, i.e., the basic building blocks forming the three-dimensional framework structures are electrostatically charged tetrahedra of silica, SiO_4^{4-}, and alumina, AlO_4^{5-}, that share vertices. All vertices of a given tetrahedron are shared with adjacent tetrahedra. The only limitation on how tetrahedra are connected to form the three-dimensional frameworks characteristic of zeolites is that two AlO_4^{5-}-tetrahedra cannot share vertices—Loewenstein's rule. When zeolites are formed from aqueous solutions, as is usually the case, they also contain water, which can be removed by heat treatment without collapsing the structure.

The individual tetrahedra are always close to regular, but since the angle between tetrahedral species, T, (silicon or aluminum) and oxygen, i.e., the T-O-T angle, can vary between 130° to 180°, they can be combined into a variety of framework structures. Figure 5-1 shows how tetrahedra can be combined to various secondary and tertiary building blocks, and how these in turn can combine to form different distinctive framework structures. In this way shape selectivity can be built into the structures—such as A, ZK4, sodalite and faujasite as typical examples— in a controlled fashion so as to accomodate the reactants and the products.

The counter ions, e.g., Na^+, also called non-framework ions, are located at certain principal sites in the pores and cavities (Fig. 5-2), from which they can be more or less readily exchanged for other cations. Cations that prefer higher coordination numbers usually occupy the S(I) sites. Adjacent S(I) and S(I') sites are not simultaneously occupied by cations, whereas almost all the S(II) sites usually are occupied since they are situated on the wall of the supercage.

Figure 5-3 shows that zeolite framework structures may be divided into one-, two-, and three-dimensional structures. More than fifty distinct structures have been identified in the zeolite world. The first zeolites studied by scientists were natural zeolites, of which about 40 different structures are known. The most important zeolites, however, are man-made (synthetic). Table 5-1 shows some commercially available natural and synthetic zeolites.

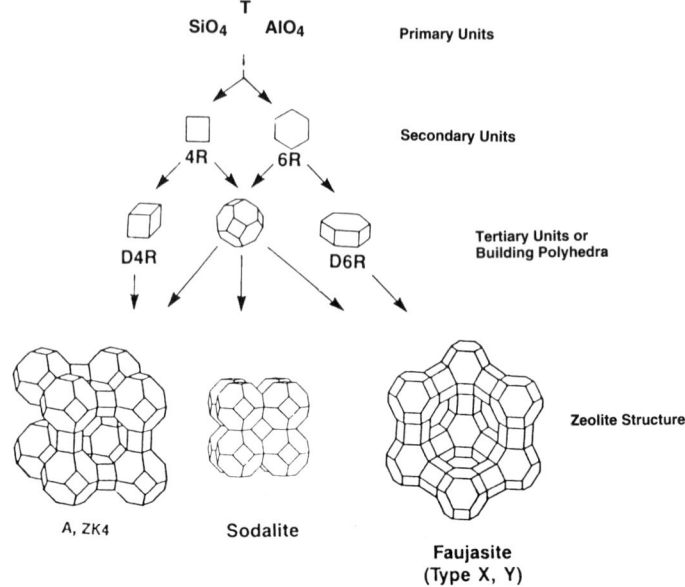

Fig. 5-1. Zeolites X and Y, zeolites A and ZK$_4$, and sodalite may be viewed as being assembled from primary (i.e. TO$_4$), secondary, tertiary (cubes or double-four (O4) rings, hexagonal prisms or double-six (O6) rings etc.) building units. Vaughan (1988). Courtesy American Institute of Chemical Engineers

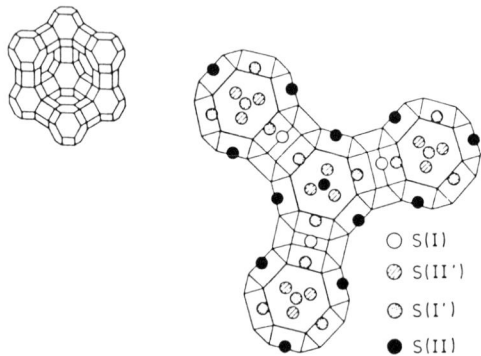

Fig. 5-2. The principal cation sites in faujasite (zeolites X and Y). Site S(I) is at the center of the hexagonal biprisms (D6R) which connect the β-cages (forming a diamond lattice of β-cages). The S(II) sites are in the supercages, but sites S(I') and S(II') are within the β-cages. Thomas (1984). Courtesy Springer Verlag.

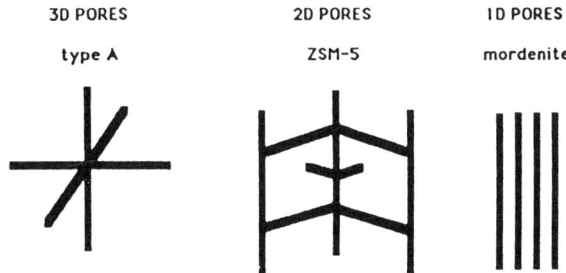

Fig. 5-3. Three commercial zeolites with different pore dimensionalities. Type A has three intersecting channels running through the structure; ZSM-5 has two intersecting channels—one straight and the other sinusoidal; and mordenite has a single channel system, resembling a pack of soda straws. Vaughan (1988). Courtesy American Institute of Chemical Engineers.

Table 5-1. Commercially available zeolites, Vaughan (1988). Courtesy American Institute of Chemical Engineers.

	Pore Size Å	Composition		Sorption Capacity (wt. %)			Vendors
		Si/Al	Cation	H_2O	nC_6H_{14}	C_6H_{12}	
Zeolite							
Faujasite							
X	7.4	1-1.5	Na	28	14.5	16.6	GLPTU
Y	7.4	1.5-3	Na	26	18.1	19.5	GLPTU
US-Y	7.4	>3	H	11	15.8	18.3	GLPTU
A	3	1.0	K, Na	22	0	0	GLPTU
A	4	1.0	Na	23	0	0	GLPTU
A	4.5	1.0	Ca, Na	23	12.5	0	GLPTU
Chabazite	4	4	*N*	15	6.7	1	GU
Clinoptilolite	4x5	5.5	*N*	10	1.8	0	A
Erionite	3.8	4	*N*	9	2.4	0	AM
Ferrierite	5.5x4.8	5-10	H	10	2.1	1.3	T
L type	6	3-3.5	K	12	8	7.4	LTU
Mazzite	5.8	3.4	Na, H	11	4.3	4.1	U
Mordenite	6x7	5.5	*N*	6	2.1	2.1	AU
Mordenite	6x7	5-6	Na	14	4.0	4.5	PTU
Mordenite	6.7	5-10	H	12	4.2	7.5	PTU
Offretite	5.8	4	K, H	13	5.7	2.0	U
Phillipsite	3	2	*N*	15	1.3	0	A
Silicalite	5.5	∞	H	1	10.1	0	U
ZSM-5	5.5	10-500	H	4	12.4	5.9	M

A=Anaconda Minerals, Letcher Minerals, Double Eagle Minerals, a.o.
G=W. R. Grace & Co.; L=Laporte PLC; M=Mobil Oil Co.; P=PQ Corp.; T=Toyo Soda; U=Union Carbide Corp.

2. Zeolites and Molecular Sieves

Zeolites, because of their porous structure consisting of interconnected cavities of molecular dimensions, are synonomously called "molecular sieves".

Up until as recently as 1982 "molecular sieves" would be restricted to aluminosilicates in a chemical sense. It is known, however, that zeolites are not the only microporous solids containing cavities and channels of molecular dimensions. Partial or complete isomorphous substitution of silicon or aluminum, or both, is possible, giving rise to many other tetrahedrally bonded T-atoms (see preceding section) besides Si and Al. Such substitution thus gives rise to families of, for example, aluminophosphates (Wilson et al., 1982), gallosilicates (Barrer et al., 1959), silicoaluminophosphates (Lok et al., 1984), and ferrisilicates (Ratnasamy et al., 1985)—see Fig. 5.4.

Today, several elements in addition to the previously known Si and Al, have been found which yield compositions having molecular sieve properties.

Szostak (1989) states that a molecular sieve framework is based on an exclusive three-dimensional network of oxygen ions containing almost always tetrahedral, T, sites. Other cations than Si and Al, which compositionally define the zeolite molecular sieves, can occupy theses sites. They need not be isoelectronic with Si^{4+} or Al^{3+}, but must be able to form T-sites. In Table 5-2 are listed cations presently known to occupy such sites.

The framework charge of zeolite molecular sieves is negative, but it is not necessary that a molecular sieve framework exhibits any charge; cf. Si, Ge, Mn, and Ti in Table 5-2. Molecular sieves containing only Si^{4+} in the tetrahedral sites, e.g. silicalite, will have a neutral framework and display a very hydrophobic surface and no ion exchange capacity. The net charge on the $AlPO_4$ molecular sieve is also zero since the charges of the framework AlO_2^- and PO_2^+ units balance out to give an uncharged molecular sieve framework with no ion exchange capacity. The common name *zeolite* will be taken to be synonymous with *zeolite molecular sieve* in this book.

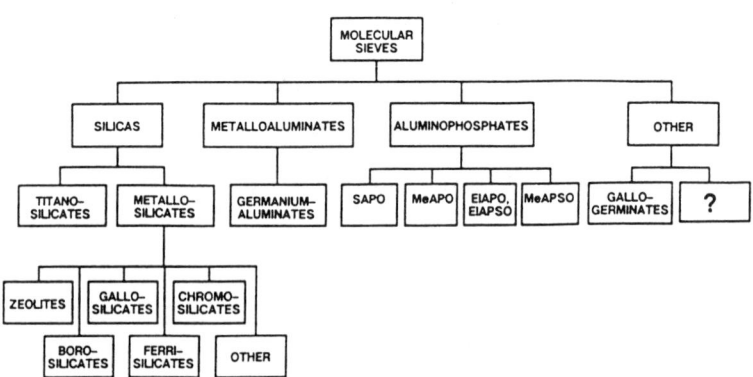

Fig. 5-4. Classification of molecular sieve materials indicating the extensive variation in composition. The zeolites occupy a subcategory of the metallosilicates. Szostak (1989). Courtesy Chapman and Hall.

Table 5-2. Cations that may form molecular sieve framework structures and the metal oxide charge possible. Szostak (1989). Courtesy Chapman & Hall.

	M
$(M^{2+}O_2)^{2-}$	Be, Mg, Zn, Co, Fe
$(M^{3+}O_2)^{-}$	Al, B, Ga, Fe, Cr
$(M^{4+}O_2)^{0}$	Si, Ge, Mn, Ti
$(M^{5+}O_2)^{+}$	P

3. Properties and Applications of Zeolites

The structure and composition of zeolites are responsible for their special properties and the utilization of these properties in various applications. Commercial applications of zeolites exploit different aspects of zeolite chemistry. As ion exchangers, zeolite frameworks facilitate the mobilities of the non-framework cations. As sorbents and molecular sieves, zeolites provide hydrophilic or, as e.g., in silicalite, organophilic, micropores of controlled dimensions and accessibilities. In catalysis, zeolites make highly active sites available as well as the possibility of controlling reaction outcomes by shape-selective constraints.

3.1. Ion Exchange

The framework structure in zeolites has a negative charge which is compensated by cations held in channels and cavities of the zeolite. These ions are coordinated to the oxygen ions making up the walls of the cavities. Part of the so-called zeolitic water may also be coordinated to these metal ions. The open structure and the relatively weak forces acting between framework and the counter ions makes it easy to exchange other cations for them.

Table 5-3. Characteristic properties of zeolites and organic cation exchangers.

Property	Organic cation exchangers	Zeolites
Ion exchange capacity	~ 5 meq/g	1-6 meq/g
Selectivity	$Me^+ < Me^{++} < Me^{+++}$	$Me^{+++} < M^+, Me^{++}$
Exchange rate	high	generally high
pH	the whole pH range	pH > 3 (most zeolites dissolve in acid solutions)

The most noticeable difference between zeolitic ion exchangers and strongly acidic organic cation exchangers of, for instance, sulfonic acid type is the high cation selectivity of the former. Generally, an organic ion exchanger has greater affinity for highly charged ions than for ions of low electrical charge, whereas this is not the case for zeolites. Compared with zeolites, organic ion exchangers are relatively poor at separating metal ions according to size. Table 5-3 compares some properties of zeolites and organic cation exchangers.

The property of size-selective ion exchange in zeolites is apparently a consequence of the well-defined size of the channels metal ions and other cations must pass through for ion exchange to occur. The actual space at the coordination site in the zeolite is also of importance—those ions, the size of which fits the given geometry of the site, will be held more strongly than those ions which do not fit as well. The great variety of zeolite structures makes it possible to tailor zeolites for a special ion exchange separation process.

The major use of zeolite ion exchangers is in low-phosphate detergents, Flank (1980), in which zeolite A is used in partial replacement for sodium tripolyphosphate builders, and in water softeners, Olson (1984). Zeolites are also used in agriculture, particularly in Japan, and in certain waste-water treatments, Flank (1980) and Olson (1984).

The ion exchange properties of zeolites have also been utilized in nuclear power technology to selectively remove the radioactive isotopes Sr^{90} and Cs^{137} from solutions containing radioactive waste.

The ion exchange capacity of zeolites can be used to introduce transition metal ions into the structure for subsequent catalytic applications (see Ch. 7).

3.2. Adsorption

When two immiscible phases (solid-gas or solid-liquid) are in contact, the concentration of the gas or liquid at the interface is usually greater than its value in the bulk of the fluid phase. The deposition of molecules from a gas or a liquid onto a solid surface is known as adsorption. This attachment process partially neutralizes the attractive force of the solid surface and as a result causes the attractive force of the solid surface to be a minimum at equilibrium, thereby reducing the surface tension.

There are at least two types of adsorption which may be distinguished from each other according to the basic nature of the interaction between the atoms of the adsorbate and the solid. These are concisely described below for gas-phase adsorption.

Physical Adsorption. The interaction between the weak forces of polarization between adsorbate and surface atoms of the van der Waal's type is characterized by low heats of adsorption in the range of something less than about 40 kJ/mol, about the same as that for the liquefaction of gases. This type of adsorption takes place at temperatures near or below the boiling point of the gas. Since physical adsorption does not require any activation energy, it is a fast process and does not require immediate contact of the adsorbate atoms with the solid surface, thus allowing multiple layers of adsorbate beyond the monolayer formed immediately adjacent to the surface. Hobson (1967) concluded from experimental data that physical adsorption is qualitatively independent of pressure and is determined largely by the nature of the surface.

Chemisorption. Chemisorption involves the sharing of electrons of surface and adsorbate atoms, thereby satisfying the valency requirements of the surface. In this case a weak form of chemical bonding takes place with resulting larger heats of adsorption from about 40-400 kJ/mol. Chemisorption takes place over a wide temperature range, may require activation energy, and usually leads to a pronounced change in the nature of the surface.

However, inasmuch as the two types of adsorption can take place simultaneously, the interpretation of data is often complex and some cases are not cleanly resolved.

Adsorption has many practical applications for zeolites. The use of adsorption as a purification process antedates by hundreds of years its understanding in physico-chemical terms, but its large-scale commercial use is a relatively recent development which only became feasible when improved adsorbents of good selectivity, especially zeolites, became available.

3.2.1. Adsorbent Properties. Economic factors underlie the identification of the key adsorbent properties for commercial adsorption processes: 1) selectivity, i.e. the preferential adsorption of one component or class of similar components, from a feed mixture, which can be based on differences in adsorption equilibrium, adsorption rates, or size exclusion (sieving); 2) capacity—which must be high enough that the equipment size falls into the practical range; 3) high specific surface area—usually a porous material with micropores —which allows for a tolerable pressure drop in bulk; 4) preferably physical adsorption rather than chemisorption for reasons of higher capacity and easier reversibility.

3.2.2. Zeolites as Adsorbents. The microporous crystalline structure of zeolites ideally suits them for use as adsorbents. Of the roughly 50 different zeolite framework structures known, only a few have so far found use in commercial adsorption processes. The three main types given in Table 5-4 account for most of the adsorbents market.

Essentially, these three types are distinguished by the number of oxygens in the rings: 8-member zeolite A (free aperture 4.3Å in unobstructed Ca^{++} (5A) form, 3.8Å in the Na^+ (4A), and 3.0Å in the K^+ (3A)); 10-member (6Å) ZSM-5, and 12-member (8.1Å) faujasite). Whereas the 3A, 4A, and 5A types are useful for size-selective adsorption of relatively small molecules, the X and Y types are useful for relatively large molecules, while the 10-member types silicalite and ZSM-5 handle the intermediate sizes. Size alone, however, is not the only factor: in the X and Y types the free apertures are the same, but the Si/Al ratio, which controls the cation density, changes the adsorptive characteristics.

Table 5-4. Structural details of important zeolite adsorbents. Ruthven (1988). Courtesy American Institute of Chemical Engineers

	Zeolite A*			Faujasite*		Pentasil**
	3A	4A	5A	Zeolite X	Zeolite Y	Silicalite/ZSM-5
Unit cell contents*	$K_{12}[AlO_2 \cdot SiO_2]_{12}$	$Na_{12}[AlO_2 \cdot SiO_2]_{12}$	$Ca_6[AlO_2 \cdot SiO_2]_{12}$	$Na_{86}[(AlO_2)_{86}(SiO_2)_{106}]$ $Na_{56}[(AlO_2)_{56}(SiO_2)_{136}]$	$(SiO_2)Na_n(SiO_2)_{96}(AlO_2)_{(96-n)}$	
Unit cell dimensions*		cubic: 12.3Å	–	cubic: 12.5 Å	cubic:12.35Å	20.1Å* orthorhombic 19.9Å** 13.4Å***
Si/Al	–	0.9-1.0	–	1.0-1.5 (higher in dealum. forms)	1.5-3.0	∞-10
Framework density (g/cm³)	–	1.27	–	1.31	1.25-1.29	1.76
Crystal density (g/cm³)	1.69	1.52	1.48	1.54	~1.42	1.76
Brief framework description	cubic array of sodalite cages linked by 4 rings			Tetrahedral array of sodalite cages linked by 6 rings		stacking of pentasil units to give 10-ring channel system
Sp. micro-pore vol.	–	0.47	–	0.51	0.48	0.33
Micropore system Pore geometry	3-dimensional large cages connected through 8 rings			3-dimensional large cages (12.5Å) connected through 12 rings		3-dimensional sinusoidal channels in one plane-cylindrical in perpendicular direction 5.4x5.6 and 5.1x5.5
Pore diameter	3.0	3.8	4.3	8.1[†]	8.1	
Largest molecules	H_2, H_2O	C_2H_6, Xe	CF_4, n-paraffins	$(C_4H_9)_3N$, dimethyl naphthalenes		CCl_4, m-xylene

* dehydrated   [†]reduced somewhat in Ca^{++} form   **pseudo cell   ***ideal value including sodalite cages

3.2.3.   Selectivity of Zeolite Adsorbents. In determining selectivity of zeolite adsorbents the key point is the sharp sorption cut-off, which occurs due to the uniformity of the molecular-size micropores. This creates the possibility of size-selective molecular sieve separations, as shown in Fig. 5-5 for some commercial zeolite adsorbents. The main pore-size determinant is the framework structure, but fine tuning is afforded by ion exchange in determining the actual free aperture, especially for the smaller 8-ring sieves. Figure 5-6 shows sorption cut-off versus Na^+-Ca^{++} exchange in zeolite A while Fig. 5-7 shows the silicalite cut-off for increasing molecular size.

Early research efforts were motivated by the molecular sieve effect and resulted in several industrial-scale separations such as the separation of linear from branched and cyclic hydrocarbons with 5A sieve adsorbent, but more important in the development of highly selective adsorbents are the polar properties of the zeolites together with the unique affinities promoted by the accommodation afforded to a particular shape of molecule by a pore of the right geometry.

Aluminum-rich zeolites high affinity for water combined with a saturation capacity of about one-quarter of their weight make their use as desiccants attractive, especially for drying gases where very low dew points are needed; but on the other hand the water is so strongly adsorbed that regeneration requires higher temperatures. For that reason applications requiring a moderate dew point often can utilize silica gel or alumina more economically. When a zeolite desiccant is used, it is usually 4A because it is the cheapest commonly available

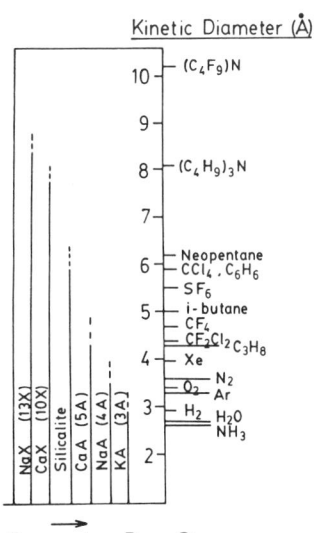

Fig. 5-5. Correlation between effective pore size of zeolite in adsorption at 77 to 420 °K as determined from the Lennard-Jones force constants. Adapted from Breck (1974). Reprinted by permission of John Wiley & Sons, Inc.

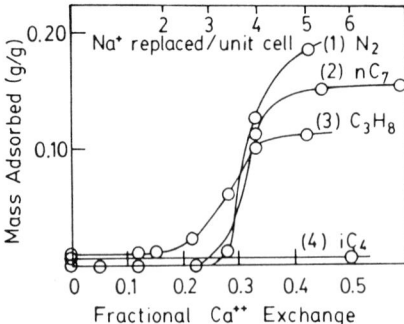

Fig. 5-6. Sorption cut-off in zeolite A as a function of Na^+-Ca^{++} exchange. (1) N_2 at 15 torr, -196°C; (2) n-heptane, 45 torr, 25°C; (3) propane, 250 torr, 25°C; (4) isobutane, 400 torr, 25° C. D.W. Breck et al. (1974). Reprinted by permission of John Wiley & Sons, Inc.

synthetic zeolite. Another factor which must be taken into account in desiccant selection is that most desiccants are catalytically active, so that drying reactive gas streams with these materials might result in undesirable surface reactions. The use of the molecular sieve 3A which has K^+ ions replacing Na^+ ions in the 4A sieve avoids that problem because the opening of the 3A sieve is obstructed by the larger K^+ ion, so that hydrocarbons are prevented from penetrating at an appreciable rate while water penetration is hardly affected. The use of 3A sieve in drying reactive gas streams is now well established.

In addition, there is another option in drying reactive gas streams wherein a *pore-modified* 4A sieve is used. Subtle structural changes in the 4A lattice, possibly caused by rearrangement of the cations, are brought about by exposure to steam at elevated temperatures. This *pore-closed* 4A sieve is similar to the 3A variety in being able to exclude most larger molecules while retaining a high capacity for water.

It is worthwhile to distinguish between *kinetic selectivity* and *equilibrium selectivity*. Kinetic selectivity takes advantage of the fact that in penetrating the pores of an adsorbent, larger molecules diffuse more slowly than smaller species, so that in an appropriately designed process, the faster-diffusing species is the one that is preferentially adsorbed, even though at equilibrium the reverse may be true in a two component system. A case in point is that of using 4A sieve to separate nitrogen and oxygen. The equilibrium selectivity favors nitrogen, but the smaller size of oxygen gives it the higher kinetic selectivity.

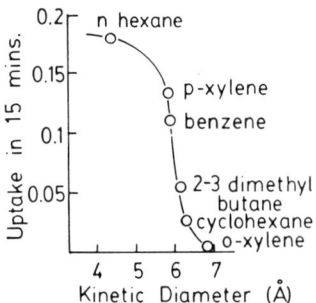

Fig. 5-7. Sorption cut-off in silicalite showing decrease in sorption rate with increasing kinetic diameter. Courtesy I. D. Harrison, H. F. Leach, and D. A. Whan (1984).

Equilibrium selectivity depends on the nature of the molecular interactions between adsorbate and surface. Van der Waals interactions which depend on induction forces (polarization-type forces), orientation forces (between permanent dipoles), and dispersion forces (ever-present long-range forces which may be attractive or repulsive which are caused by the interactions between the finite dipole moments arising from the instantaneous positions of the electrons about the nuclear protons) are operative for any sorbate on any surface whereas electrostatic forces depend on permanent charges and dipoles. Van der Waals forces depend on the surface geometry, but are only slightly affected by the chemical nature of the adsorbent. The van der Waals forces increase as the polarizability of the sorbate molecule increases, which in turn is approximately proportional to the molecular weight. Electrostatic forces depend on both an electric field at the surface and a dipolar or quadrupolar adsorbate. For more detailed discussion of these forces and their quantification for specific cases, see Israelachvili (1994).

On a nonpolar homogeneous surface such as activated carbon or some high-silica zeolites, van der Waals forces are dominant, with the result that if sieving effects are excluded, molecules are adsorbed with a force roughly in proportion to the adsorbate molecular weight. Most zeolites are polar adsorbents, due to their exchangeable cations and thus molecules such as water or ammonia (high dipoles), carbon dioxide, nitrogen (quadrupolar), or aromatic hydrocarbons (interactive π-layer) are adsorbed more strongly than nonpolar species of comparable molecular weight. Increasing the affinity for polar or quadrupolar molecules can generally by achieved by increasing the cation charge and decreasing the cation radius in the zeolite, e.g., by replacing a monovalent cation with a small divalent cation with increase in the local field gradient. That effect may be lessened appreciably by residual moisture, which is itself adsorbed more strongly on highly charged cations with consequent reduction in the affinity for less polar molecules.

Most zeolites are strongly hydrophilic due to the strong interaction between the highly polar water molecule and the cation and the framework aluminum ion, whereas the high-silica zeolites are hydrophobic by comparison because water is adsorbed on them less strongly than most organics. Water, with its low molecular weight and relatively low polarizability, is adsorbed less strongly than most polarizable organics on a nonpolar surface since the van der Waals forces largely dominate the adsorptive interaction when there is no electrostatic contribution.

The effect of pore size and shape can also be important in determining selectivity because the van der Waals interaction energy will be maximized for a molecule which fits into the channel so tightly that it is surrounded on all sides by the pore wall at a distance approximating the potential minimum. An example is the comparison of the adsorption of n-hexane and benzene on silicalite where heats of sorption of 16.8 kcal/mol (n-hexane) and 13.2 kcal/mol (benzene) were measured, Ruthven (1988).

Recently, an approach to predict selectivity was described by Richards et al. (1995) using molecular modeling software to predict adsorption isotherms for cases of interest on zeolites to cost-effectively screen potential molecular sieve structures to separate nitrogen and oxygen. Specifically, an anhydrous lithium-only type of zeolite-X selected from four faujasite Li-X structures found in a literature search was modeled using the *Crystal Builder* program module. This involved inputting the space group and unit cell parameters along with atom type and x-y-z coordinates for the main lattice framework followed by introducing the cations. After a satisfactory structure was built, a force field for the lattice framework and adsorbate consisting of a Lennard-Jones van der Waals interaction with a Coulombic term was introduced. Simulations were run for a temperature of 300° K using simulations based on a statistical mechanics sorption model over a simulated pressure range of 0.5 to 4 bars for both nitrogen and oxygen. The results indicated that the adsorption of nitrogen in the zeolite lattice was always greater than that for oxygen at the chosen conditions. Using advanced visualization features of the program yielded further useful information on the reason for the higher selectivity for nitrogen.

3.2.4. Adsorption Processes. Adsorption processes depend on preferentially adsorbing one or more components from a mixed feed stream followed by recovery of either or both the desorbate or the purified raffinate. A classification of adsorption processes may be made on the basis of:

1. gas or liquid operation
2. method for adsorbent regeneration
3. scheme for contacting the feed with the adsorbent bed.

The critical step in an adsorption separation process is commonly the regeneration of adsorbent. Having chosen operating conditions favorable for adsorption of the desired component(s), the objective is suddenly changed to thermodynamically favor desorption. This shift in equilibrium is usually done by one of the following methods:

1. thermal swing with a preheated gas
2. pressure swing by reducing the total pressure in the adsorber at constant temperature
3. inert purge stripping without change in temperature or pressure
4. displacement gas purge with a gas or vapor that adsorbs about as strongly as the adsorbate.

Table 5-5 summarizes information on factors governing choice of regeneration methods while Tables 5-6, and 5-7 present industrial examples of separation processes using zeolites. Further details can be found in Yang (1987).

As is typical of separation processes, economic design entails both rate behavior (usually controlled by diffusional processes in commercial adsorbers) and equilibrium considerations, which influence the efficiency in reaching the ultimate capacity of the adsorbent.

Table 5-5. Factors governing choice of regeneration method. Ruthven (1988). Courtesy American Institute of Chemical Engineers.

Method	Advantages	Disadvantages
Thermal swing	Good for strongly adsorbed species; small change in temperature gives large change in q (amount sorbate per unit mass sorbent) Desorbate may be recovered at high concentration Gases and liquids	Thermal aging of desorbent Heat loss means inefficiency in energy usage Unsuitable for rapid cycling so adsorbent cannot be used with maximum efficiency In liquid systems high latent heat of interstitial liquid must be added
Pressure swing	Good where weakly adsorbed species is required in high purity. Rapid cycling- efficient use of adsorbent	Very low pressure may be required. Mechanical energy more expensive than heat Desorbate recovered at low purity
Purge stripping	Operation at constant temperature and total pressure	Large purge volume required.
Displacement	Good for strongly held species. Avoids risk of cracking reactions during regeneration. Avoids thermal aging of adsorbent.	Product separation and recovery needed (choice of adsorbent is crucial).

Table 5-6. Purification processes with zeolite adsorbents. Ruthven (1988). Courtesy American Institute of Chemical Engineers.

Process	Feed	Gas or Liquid	Adsorbent	Process Details
Drying	natural gas	G	4A	T-swing
Drying	air	G	4A	T-swing (also P-swing)
Drying	refrigerants (chlorocarbons)	G/L	modified 4A	non-regenerative or T-swing
Drying	cracked gas	G	3A	T-swing
Drying	organic solvents	L	3A	T-swing
Drying	acid gas	G	chabazite	T-swing
CO_2 removal	air in submarines/space-craft	G	4A	P-swing (vacuum)
H_2S removal	sour gas	G	CaA or Ca chabazite	T-swing
SO_x, NO_x removal (Purasiv)	air	G	silicalite	T-swing
Kr^{85} removal	air	G	silicalite or de-alum. H-mordenite	chromat-ographic
I^{129} removal	air	G	AgX, Ag mordenite	T-swing
Concentration of alcohols	dilute aqueous alcohol	L	3A, 4A	T-swing

Table 5-7. Bulk separation using zeolite adsorbents. Ruthven (1988) Courtesy American Institute of Chemical Engineers.

	Process	Feed	Gas or Liquid	Adsorbent	Process Details
Linear Paraffin Separation	Isosiv	C_6-C_{10} Distillate	G	5A	PSA with vacuum desorption
"	BP		G	5A	PSA with vacuum desorption
"	TSF	C_{10}-C_{16} Kerosene	G	5A	Displacement-light naphtha
"	Ensorb	C_{10}-C_{16} Kerosene	G	5A	Displacement-NH_3
"	Elf-N-Iself	Light Naphtha	G	5A	Chromato-graphic
"	Molex	Kerosene	L	5A	Sorbex
Aromatics Separation	Parex	C_8 aromatics	L	Sr-BaX	Sorbex-PDEB desorbent
"	Ebex	C_8 distillate	L	Sr-LX	Sorbex-Toluene desorbent
Olefins from Paraffins	Olex	C_4 distillate	L	- -	Sorbex
Air Separation	O_2 production	Air	G	5A/13X	PSA (equilibrium separation}
"	N_2 production	Air	G	4A	PSA (kinetic separation)
Mono-saccharide Separation	Sarex	Corn syrup, product of starch hydrolysis	L	CaY	Sorbex

3.3. Catalysis

In the introduction to this chapter we stated that zeolites are the ultimate extrapolation in the quest for ever smaller reaction vessels. Here we shall give examples of the uses of zeolites as catalysts in commercially important processes.

The two-century span between Cronstedt's discovery of natural zeolites and their use in 1959 as cracking catalysts was mainly due to the relative scarcity of natural zeolites, such as faujasites, which have pores large enough to treat the broad range of organic molecules in crude oil. Linde X and Linde Y zeolites (synthetic faujasites) were the first used, and Linde Y still is, the most widely used zeolite catalyst.

The importance of faujasite-type zeolites compared with conventional amorphous silica-alumina cracking catalysts may be gauged from the estimate that in the period 1959-1976 the use of these zeolites in catalytic cracking saved the petroleum industry more than $3 billion in the United States alone, Chen and Degnan (1988). Since the introduction of those zeolites as catalysts many new synthetic zeolites have been developed for commercial processes, largely in the petrochemical industry.

The major areas of interest for using these unique zeolites are:

1. Petroleum refining
2. Synfuels production
3. Petrochemical manufacture
4. NO_x abatement

Chen and Degnan (1988) noted that the six most important properties of zeolites for use as heterogeneous catalysts are:

1. Well-defined crystalline structure
2. High internal surface areas (> 600 m^2/g)
3. Uniform pores which are different in different zeolites
4. Good thermal stability
5. Ability to adsorb and concentrate hydrocarbons
6. Highly acidic sites when ion exchanged with protons

Table 5-8 shows commercially important zeolites used in catalytic applications.

The acidity of zeolites originates from the protons that are required to maintain electrical neutrality in the structure. The size of zeolite pores is

Table 5-8. Commercially important zeolites used in catalytic applications*		
Zeolite	Channel system**	Cavity+
Large pore Faujasite (Linde type X and Linde type Y)	(12) 7.4, 3-Dimensional	6.6, 11.4
Mordenite	(8) 2.9 x 5.7, 1D (12) 6.7 x 7.0, 1D	Interconnected Channels
L	(12) 7.1, 1D	Unidimensional
Medium pore ZSM-5	(10) 5.4 x 5.6, 1D (10) 5.1 x 5.6, 1D	Interconnected Channels
Synthetic Ferrierite	(8) 3.4 x 4.8, 2D (10) 4.3 x 5.5, 1D	Interconnected Channels
Small pore Erionite	(8) 3.6 x 5.2, 2D	6.3 x 13
* Modified from Derouane, E.G. Catalysis on the Energy Scene, p. 1, Elsevier, Amsterdam (1968). Courtesy American Institute of Chemical Engineers. ** Number of oxygen atoms constituting the smallest ring determining pore size (in parentheses), pore diameter(s) in Å, and number of directions in which the channel runs. + Cavity free dimension(s) in Å		

determined by the number of tetrahedral units or, alternatively, oxygen atoms, required to form the pore; and the nature of the cations that are present in or at the mouth of the pore.

It is common to group zeolites into three size classes as shown in Table 5-9. Zeolites with more than one pore system are classified according to their largest accessible pore.

The number and strength of the acid sites depend in a complex way on the nature and concentration of tetrahedral trivalent X groups, their location, and the nature and concentration of exchangeable cations present. Zeolites are often classified by their silica to alumina ratios. Usually, the higher the silica to alumina ratio, the more thermally stable the zeolite is. Zeolites with silica to alumina ratios greater than 10:1 are classified as high-silica materials, of which none are known in nature. Amorphous aluminosilicate was formerly widely used as a cracking catalyst. Studies of the acidity of solids show that the Brönsted acid strength distribution for amorphous aluminosilicates and hydrogen-Y, HY, zeolites are closely similar. On the other hand, the number of Brönsted acid sites is greater for HY, but this difference does not seem to be able to explain the much greater activity of HY.

Study of adsorbed xenon with NMR shows that for a given gas pressure the molecule adsorbed in a zeolite is subjected to an "apparent pressure" about 100 times greater than the pressure on a molecule adsorbed on a planar surface, Fraissard (1980). This effect, due to the unique structure of zeolites, may explain why, for active sites of the same chemical nature and concentration, zeolites have much greater catalytic activity than the surface of conventional catalysts.

Also, the surface within zeolite crystals is different from ordinary crystal surfaces. Although it is a real surface in a physical sense because adsorbed molecules are in direct contact with it, it is also part of a crystal fully surrounded by the crystal lattice, with the result that both the "surface" atoms and the molecules adsorbed within the zeolite crystal are subject to the zeolite crystal field. Indeed, the zeolite surface has properties of both the ordinary crystal surface and the crystal lattice. The strong polar environment of the zeolitic surface caused by both zeolite cations and the crystal field of the surrounding anionic crystal lattice results in a strong polarizing interaction between the zeolite crystal and adsorbed molecules.

Recently Davis, Jones, and Tsuji (1998) described a technique to functionalize the interior of zeolites with organic groups, thus allowing their use

Table 5-9. Classification of zeolites based on pore size		
Pore size	Number of tetrahedra	Maximum free diameter
small	6, 8	4.3 Å
medium	10	6.3 Å
large	12	7.5 Å

as catalytic molecular sieves whose crystalline pores are selective for reactants or products of a particular size and shape.

3.3.1. Zeolite Catalysts in the Refining and Petrochemical Industries

In the broad sense it was realized, even before the Arab oil embargo in 1973, that liquid fuels will eventually have to be obtained from sources other than petroleum, that is, from shale, tar sand bitumen, and coal. Generally, the fuel quality can be related to its atomic hydrogen to carbon ratio. While methane has a ratio of 4, crude petroleum has a ratio of about 1.7 and coal about 0.8. Simplistically, upgrading these fuels may be seen as a hydrogen addition or carbon subtraction to yield a higher H/C ratio. In addition, since concentrations of compounds with elements like sulfur, nitrogen, and oxygen, with deleterious effects such as catalyst poisoning and toxicity, are present in the fuel to be upgraded, processing must also be able to greatly reduce concentrations of those compounds. Zeolites play an important role in the refining of petroleum and petrochemical processing.

In a fully integrated refinery zeolite catalysts are used in fluidized catalytic cracking, hydrocracking, reforming, C_5/C_6 isomerization, selective cracking and dewaxing processes and in the synfuels production process. In the petrochemicals industry they are used in the processes for xylene isomerization, toluene disproportionation, ethylbenzene synthesis, toluene alkylation, para-ethyltoluene and para-methylystyrene synthesis and methanol to olefins.

Fluidized Catalytic Cracking. Called by some the "heart" of the modern refinery whose goal is gasoline production, the fluidized catalytic cracking unit converts a wide range of heavier petroleum to lighter boiling gasoline and light fuel oils. Major byproducts are liquefied petroleum gas, fuel gas, heavy fuel oil, and petroleum coke. The objective of most FCC operations is to maximize the throughput and octane of the gasoline fraction while minimizing the byproducts, Chen and Degnan (1988).

The catalytic cracking process grew out of a discovery by Houdry in 1927 that certain naturally occurring silica-alumina clays catalyzed the cracking of high molecular weight hydrocarbons to give a good yield of gasoline. Large-scale construction of catalytic crackers began in the 1940s with this process quickly superseding thermal cracking for gasoline production because of its higher yields and improved product quality.

Using rare earth cation-exchanged zeolite X was an important improvement in this application because, in addition to its increased catalytic activity, it increased gasoline production at the expense of both light gas and coke largely because of its enhanced hydrogen transfer activity. This changed the nature of the gasoline produced over this catalyst from the primary gasoline product of olefins + naphthenes to a product (after hydrogen transfer) of aromatics plus paraffins, Chen and Degnan (1988).

In addition, the gasoline produced with the zeolite catalyst is more stable than the primary gasoline product, so that overcracking of the primary product to light gases and coke can be avoided. Thus, in the cracking scheme below, substituting rare earth cation-exchanged zeolite X for silica-alumina as the cracking catalyst, enhanced k_0 and suppressed k_2. Later zeolite X was replaced by zeolite Y in this application because of the better hydrothermal stability exhibited by the higher silica/alumina ratio of zeolite Y. Currently, typical FCC catalyst consists of about 10-40 wt. % zeolite Y dispersed in an amorphous matrix of clay or silica-alumina. Other types used here are ZSM-5 and Ultrastable Y (USY), all of which exhibit good selectivity and high conversion rate. Further details of zeolite cracking catalysts can be found in Ch. 7.

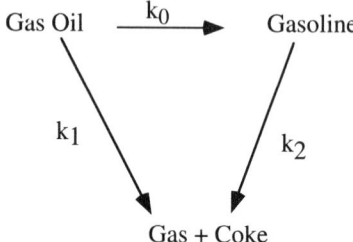

Hydrocracking. Hydrocracking, the second largest application for zeolite catalysts, involves cracking at high partial pressures of hydrogen in the presence of supported metal and acid functions. Catalytic cracking proper increases the hydrogen content of the product by taking out carbon, while hydrocracking adds hydrogen. The large-pore zeolites, X, Y, and mordenite, most widely used in hydrocracking, showed improvement over other acid catalytsts for this use because they are more resistant to sulfur and nitrogen poisoning. Mordenite is satisfactory for hydrocracking naphtha, while for heavier feeds the faujasite-type zeolites in rare-earth-exchanged or high-silica forms are preferred.

Reforming. Reforming is a process for converting low-octane paraffins and naphthenes to substantially higher-octane isoparaffins and aromatics over a bifunctional (i.e., acid and noble metal) catalyst. Historically, zeolites have not been used in reforming applications because they are so acidic that they tend to crack the naphtha to lower molecular-weight fragments. Although conventional supported bimetallic reforming catalysts such as $PtRe/Al_2O_3$ do an excellent job on "full-range" naphtha, they do a poor job in aromatizing lighter, C_6 and C_7 paraffins. Chevron developed a process using a PtBa/zeolite L catalyst to convert normal hexane and heptane to benzene and toluene at high yields. A full evaluation of this zeolite in this application is not yet available.

C_5/C_6 Isomerization. With the octane shortfall caused by the phase-out of tetraethyl lead in gasoline, many refiners added pentane and hexane isomerization capacity which use both noble metal containing amorphous and zeolite catalysts at moderate pressures in the presence of hydrogen. The zeolite catalysts used, for example in Shell's Hysomer process, Benesi (1965), is Pt on mordenite in an alumina binder, which exhibits good resistance to poisoning by impurities such as sulfur and water.

Dewaxing. Low concentrations of long-chain linear and slightly branched paraffins, which have relatively high melting points, adversely affect the pour point of the oil and thus must be removed or converted in distillate fuels whose quality is related to paraffin content. Commonly used are ZSM-5, ZSM-23, and mordenite.

Mobil's distillate dewaxing process uses ZSM-5 to selectively crack the long-chain linear paraffins to gasoline and liquefied petroleum gas (LPG), Chen et al. (1977). This process, applied commercially since 1974, extends the usable upper boiling point of distillate fuels and can make very high-quality jet fuels when followed by a hydrotreating step, Chen and Degnan (1988).

Selective Cracking. In the mid-1960s a process known as *Selectoforming* was first commercialized to give refiners the option to selectively crack low-molecular-weight paraffins to LPG (mainly propane) instead of isomerizing or aromatizing that fraction. The catalyst, erionite—an eight-membered ring, small-pore zeolite— excludes aromatics and branched or cyclic paraffins thereby, concentrating the higher octane components.

The octane improvement attained by Selectoforming is limited by the concentration of normal paraffins in the reformate. Additional octane increases require the conversion of singly branched paraffins as well. Thus the less constrained catalyst ZSM-5 was substituted for erionite and a new process, *M-forming*, was developed by Mobil in the mid-70s, Chen et al. (1987).

Synfuels Production. In 1976 Mobil introduced a zeolite-catalyzed process (MTG) for converting methanol to gasoline, which received a great deal of attention because it was the first alternative in 40 years for converting coal and natural gas to liquid fuels. This process was successfully producing 14,500 bbl/d of premium gasoline from natural gas in a plant at Motonui, New Zealand in the early 1980s. The chemistry of the process was reviewed by Chang (1983).

Medium-pore zeolites such as ZSM-5 are essential in this process since small-pore zeolites require severe operating conditions and produce no aromatics while larger-pore zeolites produce heavy aromatics ($>C_{10}$) and deactivate rapidly.

Toluene Disproportionation. Over the past fifteen years the considerably lower price of toluene relative to benzene and xylenes, especially p-xylene, offered a strong incentive to develop a process to disproportionate toluene. Zeolites turned out to be far more effective catalysts for this reaction than conventional solid acid catalysts such as amorphous silica-alumina. Initially rare-earth-exchanged zeolite X was used, but that gave way to the inherently more stable mordenite and ZSM-5, which also produced less polymethylbenzene and o- and m-xylenes.

Ethylbenzene Synthesis. Around 90% of all ethylbenzene, the precursor for styrene, is produced via acid-catalyzed alkylation of benzene with ethylene, with the rest obtained directly from petroleum. Before introduction of the zeolite ZSM-5 Mobil-Badger ethylbenzene process in 1976, the two commercial routes to ethylbenzene were a liquid-phase alkylation with pure ethylene using $AlCl_3$ and a vapor-phase alkylation using diluted ethylene with alumina-supported BF_3. The ZSM-5 process has major advantages: (1) elimination of the corrosion and waste disposal problems of AlCl3 and BF3, (2) improved selectivity for ethylbenzene (>99%) and greater than 95% heat recovery.

Para-ethyltoluene and Para-methylstyrene Synthesis. Another ZSM-5 catalyzed process is the alkylation of toluene with ethylene in the para position at a high yield of better than 97% p-isomer with about 3% m- and o-isomers with a high toluene conversion. Catalytic dehydrogenation converts the mixture to the corresponding methylstyrenes and hydrogen.

Toluene Alkylation with Methanol. Alkylation of toluene can be carried out readily over acidic Friedel-Crafts catalysts such as AlCl3 or over amorphous silica-alumina to produce equilibrium mixtures of xylenes and ethylbenzene with the reaction rate being directly related to catalyst acidity. However, using zeolites as the catalyst, e.g. ion-exchanged faujasites, it is possible to produce up to 50% p-xylene as compared to the 24% in the equilibrium mixture at almost total toluene conversion. Modified large-crystal ZSM-5 catalyst has also been described for this reaction, Kaeding et al. (1981), with a high selectivity for the p-isomer of greater than 90% of all xylenes.

Methanol to Olefins. In the methanol to gasoline (MTG) process, light olefins are valuable intermediates. Olefin products are favored by higher reaction temperatures, shorter contact times, and lower pressures. Various products can be made in high yields with different catalysts. Propylene is the most abundant olefin product when ZSM-5 is the catalyst, while smaller-pore zeolites such as erionite can be used to make ethylene in high yields, and the SAPO (silicoaluminophosphate) zeolites (Union Carbide) can be used to produce olefins at higher than 90% yields with no aromatics.

3.3.2. Zeolites as NO$_x$ Abatement Catalysts. An important use of zeolites such as mordenite or zeolite Y is in the Selective Catalytic Reduction (SCR) of NO and NO2 in one of several competing processes for reducing NO$_x$ emissions. In SCR processes NH3 is used as a reducing agent to convert NO$_x$ to N2 and H2O. Reaction temperatures are in the 200 to 450°C range. Zeolites as NOx abatement catalysts for diesel-powered trucks and cars with lean-burn engines are discussed in Ch. 7.

4. The Synthesis of Conventional Zeolites

Following the development of Schoeman (1994) we summarize this topic in relation to the key points of physico-chemical principles.

Zeolites are synthesized from active hydrated aluminosilicate gels in an alkaline environment. The preparation of amorphous aluminosilicate mixtures is achieved by mixing a silica- and an alkaline aluminate solution, which normally results in the precipitation of a gel, followed by hydrothermal treatment at an elevated temperature (60 - 300°C), Barrer (1982), Ciric (1968), Bodart et al. (1986), Freund (1976), and Szostak (1989). In general, zeolites may be represented by the empirical formula $M_{2/n}O \cdot Al_2O_3 \cdot xSiO_2 \cdot yH_2O$ where x is usually equal or greater than 2. The synthesis mixtures are characterized by an excess of alkali which is apparent in the examples quoted below. With excess alkali it is meant that the molar ratio $R_{2/n}O \cdot Al_2O_3 > 1.0$ where R is the alkali- or alkaline earth metal.

Suitable silica sources are, for example, hydrated silicates, precipitated silica powders and colloidal silica sols. The alumina source may be aluminum alkoxides, aluminum salts, aluminum oxides, aluminum hydroxide or metal aluminates. The necessary alkalinity is supplied by additions of alkali hydroxides, alkaline earth hydroxides or organic bases or combinations thereof. The zeolite literature abounds with reports on the synthesis of zeolites and the reader is referred to Barrer (1982), Szostak (1989), and Breck (1974) for a review of these. The discussion below highlights the relation between crystal size and synthesis conditions.

4.1. Synthesis from Sodium Aluminosilicate Hydrogels

The preparation of silica - alumina synthesis gels for the synthesis of zeolite Na-Y wherein the alkali is added as sodium hydroxide is described by Breck (1964). The composition field for suitable aluminosilicate gels from which zeolite Na-Y crystallizes is shown in Fig. 5-8. A method, which one may term representative for the synthesis of zeolite Na-Y, entails aging of a gel, (Breck (1964), patent example 13) with a Na_2O/Al_2O_3 molar ratio between 2.2 and 3. Following hydrothermal treatment, zeolite Na-Y crystallizes with a particle size such that the zeolite may be separated from the mother liquor by conventional filtration methods. Generally, the primary particle size of conventionally synthesized zeolite is between 1 and 7 μ, although the particle size is a function of

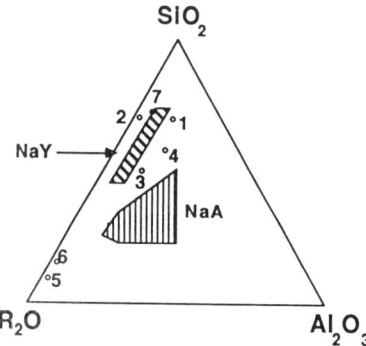

Fig. 5-8. The gel compositional fields in mol % with which the zeolite type Y and type A may be synthesized. R=Na or TMA. Water content in all mixtures is 88-99 mol %. The shaded fields show the compositional fields for zeolite NaY and NaA, Breck (1974), and point 1 indicates the composition in patent example 13, Breck (1964). Point 2, composition of a clear homogeneous solution yielding NaY, Ueda et al. (1984). Points 3 and 4, composition of TMA-Na solutions yielding Na-Y, Schoeman et al. (1993). Points 5 and 6, clear homogeneous solutions yielding NaA, Wenqin et al. (1986), and point 7, the composition yielding zeolite HS, Schoeman (1994).

numerous parameters such as temperature, alkalinity and the nature of the raw materials. These aspects are discussed in §4.5 to §4.7

According to Breck (1974), when solutions of aluminate and polysilicate anions are mixed, the composition and the structure of the amorphous gel thus produced is controlled by the size and the structure of the polymerizing species. It has been shown that the mixing of a sodium aluminate solution with an amorphous silica sol (an example of a highly polymeric silica source) results in a solid amorphous silica-rich gel due to the destabilization of the colloidal silica sol by the high electrolyte content (Na^+) in the mixture, Ginter et al. (1992 and 1992a). Subsequent dissolution of the silica particles in an aging step releases silica anions into solution which form an insoluble complex with the alumina in solution thus forming an aluminum-rich aluminosilicate solid phase whereby a complex heterogeneous system is created. The composition of the liquid phase must as a result be a function of the dissolution of the solid amorphous phases present in the mixture. Since it is generally accepted that nucleation occurs in the solution phase, Zhdanov (1971), Kerr (1976), and Lowe (1988), it follows that gelation controls the nucleation of the zeolite crystallites.

Hydrothermal treatment of a freshly prepared aluminosilicate gel, i.e., without a preceding aging step, can result in a low yield of a poor-quality zeolite presumably because the molar composition of the aqueous solution-phase actively participating in the crystallization is very different from that of the bulk composition. Aging of the amorphous aluminosilicate gel aids in increasing the content of ionic species in solution and produces a liquid phase with a composition more like the bulk composition, thus allowing for increased yields of zeolite, Breck (1964). Aging of aluminosilicate gels has also been shown to result in an increased number of crystals and therefore a smaller ultimate particle size, Bronic et al. (1988). This might be due to the higher content of ionic species formed in the solution phase as a result of the depolymerization of the hydrogel during the aging step.

4.2. The Seeding Technique

Elliott and McDaniel (1972) and Elliott (1979) describe a method whereby zeolite Na-Y is synthesized without aging of the synthesis mixture. A seed mixture, which is obtained as a gel with a relatively high alkalinity, 4.7 to 6.6 M NaOH corresponding to a Na_2O/Al_2O_3 molar ratio of 15 to 17, is aged and thereafter added to a synthesis mixture which, upon hydrothermal treatment, yields zeolite Na-Y. The product, in the form of a sediment, is separated from the mother liquor and dried to a powder after a washing sequence.

Highly active crystallization initiator slurries introduced by Albers et al. (1973) contain an alkali metal oxide, for example, sodium hydroxide, silica, alumina, water and small amounts of a compound selected from the group consisting of boron, vanadium, phosphorus, molybdenum, tungsten, germanium and gallium. The SiO_2/Al_2O_3 molar ratio in the slurry is 15 and the alkalinity is 5.6 M NaOH corresponding to a Na_2O/Al_2O_3 molar ratio of 16. The aged initiator slurry is added in small amounts to alkaline aluminosilicate mixtures and reacted typically at 100°C to form zeolite within a short period of time. An example of this technique was shown to yield zeolite Na-Y with a particle size of 200 nm after hydrothermal treatment of the gel at 100°C in 5 hours.

This example may be used to emphasize an important point. The product obtained from this synthesis exhibits rapid sedimentation which indicates clearly that although the primary particle size is 200 nm, the primary particles are aggregated to form secondary particles with a size far in excess of 200 nm. In other words, the product cannot be termed colloidal zeolite, at least not in the context of this book. In this respect, there are several reports in the literature where it is apparent that zeolite particles are secondary particles composed of smaller primary particles, Rajagopalan et al. (1986), Chutoransky and Dwyer (1973), Otake (1994), and Adnadjevic et al. (1990).

It is also notable that the above mentioned seed mixtures or initiators possess no detectable crystallinity when examined by standard X-ray diffraction analysis techniques. The seeding technique may be employed to control the ultimate crystal size as shown by Lechert (1984). This was shown by using a synthesis gel which upon hydrothermal treatment does not yield zeolite Na-Y unless seeds of zeolite Na-X are added. Nucleation of zeolite Na-Y does not take place during the hydrothermal treatment of the gel, but rather growth upon the seed material proceeds such that the nutrient is distributed over the available seed material. The amount of seed material thus determines the ultimate particle size in the product. The seed material had a particle size of 230 nm and the ultimate crystal sizes were 600 nm, 1200 nm and 2600 nm with seed amounts of 4.4, 0.44 and 0.044 wt.%, respectively.

It is apparent from these examples from the literature that the alkali is present in excess, i.e. there is more alkali in the gel than is necessary to neutralize the net negative charge that arises upon incorporation of aluminum in the zeolite structure. It is also apparent that the crystallization products are characterized by relatively large crystal sizes and relatively broad particle size distributions. These facts would indicate that although the aluminosilicate gels contain appreciable ionic species content, the amount of these species is only such that sub-micron crystals are synthesized. Increasing the alkali content in order to increase the ionic content in the solution phase does not necessarily result in a smaller crystal size, Zhdanov (1971). Instead, as discussed by Lechert (1984), the alkali content in the sodium aluminosilicate gel has a much stronger influence on the type of zeolite that is crystallized than has the silica to alumina ratio of the synthesis mixture. In other words, increasing the alkali content may result in the crystallization of an unwanted zeolite phase instead of a reduced particle size of the desired zeolite phase. It is, however, possible to increase the alkali content in certain aluminosilicate mixtures so as to fall within a well defined compositional field and still crystallize zeolites such as zeolite Na-A and Na-Y. This is the topic of the next section.

4.3. Clear Homogeneous Sodium Aluminosilicate Solutions

It is possible to prepare highly alkaline aluminosilicate solutions in the form of clear solutions as shown by Ueda and Koizumi (1979). The aqueous solutions are characterized by a high SiO_2/Al_2O_3 molar ratio, for example 70 to 100, and with an alkalinity in the range 3 to 7 M NaOH. These compositions differ markedly from those employed to synthesize conventional zeolite from the aluminosilicate gels discussed in §2.2.1 and as shown in Fig. 5-8. From these clear homogeneous solutions, analcime crystallizes with an average particle size of between 15 and 25 μ. A further example of a zeolite that crystallizes in clear homogeneous synthesis solutions is Na-Y as described by Kashara et al. (1986).

The clear aqueous solution with a SiO_2/Al_2O_3 molar ratio of 10 and an alkalinity of 9.3 M NaOH produces a fine material that grows, agglomerates and precipitates as an XRD amorphous solid with ultra-fine crystallites as shown by ^{29}Si and ^{27}Al MAS NMR analyses. As a result of the particle size, the growing crystals sediment, thus allowing separation of the crystalline phase by means of conventional filtration methods.

Prolonged aging of these solutions leads to phase transformations, for example aging 6h at 80°C leads to the phase transformation of zeolite A to hydroxysodalite and aging 24h at 40°C results in the transformation of zeolite X to hydroxysodalite. Dilution of the clear solution that yields zeolite Na-Y, Ueda et al. (1984), results in the precipitation of an amorphous aluminosilicate gel due to the fact that the mixture composition is no longer within the composition field that allows the preparation of a clear aluminosilicate solution.

These and other examples of crystallization from clear, highly alkaline aluminosilicate solutions show that, although it is possible to synthesize zeolite Na-Y and Na-A from clear solutions in the presence of high sodium contents, increased alkalinity does not necessarily allow for the synthesis of small zeolite crystals as a result of the increase in the content of ionic species in solution, Ueda et al. (1980, 1984, 1985), Wenqin et al. (1986).

4.4. Zeolite Synthesis in the Presence of Organic Cations

The use of the tetramethylammonium cation in a zeolite synthesis, the first organic cation employed as the source of alkali in zeolite synthesis, was reported by Barrer and Denny (1961). In their work, the tetramethylammonium hydroxide, TMAOH, was stirred with freshly prepared amorphous alumina gel into Syton® silica sols. The resulting aqueous gels with the molar composition 3 NMe$_4$ • OH Al$_2$O$_3$ n SiO$_2$, (NMe$_4$ • OH = TMAOH, n = SiO_2/Al_2O_3 molar ratio and varied from n = 1 to 9) were heated in stainless steel autoclaves in the temperature range 100°C, 9 days to 300°C, 3 days. The solids content in these gels is expressed as 0.75g dry weight to 2.0g dry weight per 10 ml of water. The products obtained were boehmite, sodalite, montmorillonite, harmotone and the zeolites, Linde type A, and an analog of faujasite. The type A and faujasite were obtained in the temperature range 100 to 150°C with the SiO_2/Al_2O_3 molar ratio in the range 2 to 6. No particle sizes were disclosed but the authors did reveal that fine, white crystals formed that could be centrifuged off to obtain low yields of pure crystalline phases. The montmorillonite tended to persist as a sol which could not be centrifuged off after a centrifuge time of 3 or 4 hours.

From the work of Barrer and Denny, it is clear that sodium was not deliberately added to the synthesis gel and the sodium content is not specified. It is also generally accepted that the zeolites A and faujasite require the presence of sodium in order to crystallize, Lok et al. (1983). One can conclude that the sodium in the work of Barrer and Denny is derived from the silica source—a colloidal silica sol with the trade name, Syton®, Barrer and Denny (1961).

Barrer et al. (1967) describe a method for the synthesis of nitrogenous zeolites A, X, and Y using organic bases such as tetramethylammonium hydroxide, TMAOH, as the source of alkalinity. In the patent description, the presence of sodium as a necessary component in the synthesis mixture is recognized, but the sodium content is not specified. The sodium is either derived from the soft glass reactor vessel or from a colloidal silica sol having ca. 30 wt. % SiO_2 and which contained up to several tenths of a wt. % Na_2O. No other specifications are made available. It is, however, well known that colloidal silica sols are manufactured with differing sodium contents—see Ch. 2 § D—and therefore the essential information concerning the sodium content in the silica source is not available.

An effect of organic cations on the synthesis product is a reduction in the average particle size, Bodart et al. (1986), Vaughan and Strohmaier (1986), Vaughan (1988). This was shown by Bodart et al. (1986) where the average particle size of zeolite Y synthesized in a sodium aluminosilicate gel is 0.8 μ whereas the same mixture supplemented with TMACl yields an average particle size of 0.5 μ.

Another effect of the presence of organic cations in zeolite synthesis mixtures is to produce more siliceous materials than has been shown possible in the presence of only NaOH. As a result, much effort has been focused on this branch in zeolite synthesis. Besides the tetramethylammonium ion, other tetraalkylammonium ions such as tetraethyl-, Wadlinger et al. (1967), tetrabutyl- and tetrapropylammonium-, Argauer and Landolt (1972), cations have been employed to synthesize a variety of zeolites and zeolite-type materials. Other organic materials include organic amines, which have resulted in the synthesis of the aluminophosphates, Wilson, Flanigan et al. (1982), Wilson, Messina et al. (1982), alcohols, ketones, morpholine and glycerol, Lok et al. (1982). The role of organic molecules in zeolite synthesis is discussed in numerous literature reports as for example by Jacobs and Martens (1987) and in a review by Lok et al. (1983).

A material of interest is the nominally aluminum-free form of ZSM-5, Argauer and Landolt (1972), TPA-silicalite-1, first synthesized by Grose and Flanigen (1977) from a tetrapropylammonium silicate mixture with a pH between 10 and 14. TPA-silicalite-1 can be crystallized in clear homogeneous solutions as described by Cundy et al. (1990). The crystals obtained in their investigation were separated from the mother liquor by sedimentation and washed free from excess alkali by filtering through a 0.2-μ filter membrane. Separation according to this technique was possible due to the fact that the product consisted of particles with an average particle size of greater than 1 μ. The particle size distribution could be determined by means of optical microscopy. A particularly interesting feature concerning this material is the fact that it is synthesized in the absence of aluminum and renders itself suitable for the study of zeolite nucleation and growth mechanisms. These aspects, as well as other interesting aspects concerning this material, are described in detail in the literature, Lowe (1988), Cundy et al. (1990), Fegan and Lowe (1986), Den Ouden and Thompson (1992).

4.5. The Effect of Temperature on Crystal Size

It is well known that a reduction in the crystallization temperature favors the formation of a larger number of crystals and a smaller particle size in the ultimate product, Breck (1964), Cundy et al. (1993). The rate of zeolite crystallization is, however, dependent on temperature since the rate decreases with decreasing temperature. Furthermore, although the relative reduction in particle size may be significant, we have not found literature citations describing the synthesis of colloidal zeolite via a crystallization temperature reduction.

4.6. The Effect of the Silica Source on Crystal Size

The effect of the choice of the silica source in the crystallization of zeolite NaX is most clearly illustrated by the results of Hamilton et al. (1993). They have shown that the crystallization time and the final average crystal size are dependent on the properties of the silica source. These properties included metal impurities and the degree of dispersion (solubility) of the silica in an alkaline NaOH solution. The variation in nucleation as a result of the nature of the silica source was responsible for the difference in crystallization time and final particle size. In particular, the final crystal size which is inversely related to the number of nuclei, could be correlated with the impurity level in the silica source. Particle sizes in their work varied from 10 to 100 μ. Freund (1976) shows that an active silica source results in a smaller average particle size. An active silica source means a silica source whose activity is related to the impurities present. The effect of an active silica source is to promote the rapid formation of nuclei. Active silicas have also been investigated by Lowe and MacGlip (1980) who show that the activity of such silicas is related to the presence of traces of aluminum compounds.

In the work of Meise and Schwochow (1973), the influence of the silica source in terms of the silicas specific surface area was such that the average particle size decreases with increased surface area. The specific surface area was determined by Carman's permeability method, Carman and Malherbe (1950), Carman (1954). The BET specific surface area did not show any correlation with crystal size. The reason for the crystal size dependence on the silica source was ascribed to differences in nucleation behavior.

4.7. The Effect of Additives on Crystal Size

Numerous additives besides those necessary for the crystallization of zeolite may be included in the synthesis mixture to influence particle size. Examples of such additives that may be employed to reduce crystal size are dyestuffs, Shiralkar et al. (1991), Buckley (1951), alcohols such as methanol, Maher and Scherzer (1970) and polyvalent metal cations such as molybdenum and gallium, Albers et al. (1973), iron, Verduijn (1992), and cobalt and nickel, Verduijn (1992a). On the other hand, the synthesis of large single crystals may also be favored by supplementing the synthesis mixture with additives. The addition of triethanolamine enables the synthesis of zeolite Na-A with a particle size of up to 60 μ, Charnell (1971), Scott et al. (1989).

4.8. Summary

In conclusion it is evident that the crystallization of zeolite occurs in an alkaline medium in the presence of a solid amorphous aluminosilicate gel and, given the correct conditions, from a clear homogeneous solution. In either case, the alkali content in the synthesis mixture is in excess and neither gel systems nor clear solution systems are a sufficient criteria that need be fulfilled if small crystals of zeolite are to be synthesized. The number of parameters that can be varied in an attempt to produce small crystals are numerous and the effects due to parameter changes are poorly understood. Small crystals may be synthesized by varying the parameters discussed above, but there are no reports in the literature on the synthesis of discrete colloidal crystals in the colloidal size range as is defined in this thesis work.

As mentioned in §3.3, there are numerous examples in which it has been shown that the crystal size of the zeolite can influence selectivity and activity in various catalytic reactions. This was elegantly shown for the hydroxylation of phenol over a zeolite TS-1 catalyst, van der Pol et al. (1992). The low activity of this catalyst observed for larger crystal sizes (> 1 μ) is ascribed to intraporous diffusion limitations. In the same work, van der Pol et al. (1992), showed that the product distribution is also influenced by crystallite size and hence, they recommend the use of small crystals, < 0.5 μ, to increase catalyst efficiency. The unique properties of zeolites briefly noted in §1, may also be exploited in a diverse field of applications such as sensor materials, zeolite-polymer composites, thin films and membranes. It is within such potential areas of application that colloidal zeolite may be a suitable choice of material. Well-defined colloidal zeolite systems should be particularly suited to investigate the mechanisms of zeolite nucleation and growth. The approach in such an investigation, as reported by Schoeman (1994), would differ from those reported in the literature, Ginter et al. (1992, 1992a), Cundy et al. (1990, 1993), Den Ouden and Thompson (1992).

B. COLLOIDAL ZEOLITES

Colloidal zeolites are new and unique materials which have been developed and studied at the Department of Engineering Chemistry, Chalmers University of Technology. In his doctoral dissertation Schoeman (1994) gave an excellent account of the methods he developed to prepare colloidal zeolites which we use as the basis for our summary here.

Zeolite synthesized according to the method of hydrothermal treatment of an alkaline aluminosilicate gel yields crystals in the size range ca. one to several microns. Such crystals are termed *conventional zeolites* in this work. These particle sizes are generally considered small, but if compared to the generally accepted colloidal size range, it is evident that these crystals fall within the range of the upper colloidal limit and commonly exceed colloidal dimensions.

There are several reports in the zeolite literature dealing with adsorption and diffusion in zeolites of varying crystal size, van der Pol (1992), Voogd and van Bekkum (1989, 1990), Camblor et al. (1989). In the above-mentioned application areas of catalysis and detergency, as well as several others, it becomes apparent upon closer inspection that the ability to control the particle size would be desirable. It is also apparent from the zeolite literature as well as the general colloidal science literature that zeolite has not been considered as a colloidal material. Here we consider the synthesis of zeolite in the form of discrete colloidal zeolite crystals in suspension.

1.   Synthesis of Colloidal Zeolites

The preparation of zeolites N-Y and N-A will be briefly described. In all of the syntheses the synthesis solutions were hydrothermally treated directly following preparation except for the TPA-silicate and ZSM-5 solutions containing tetraethoxysilane. Those solutions were allowed to undergo hydrolysis with the aqueous TPAOH solution, which is complete within 12 hours, before hydrothermal treatment. For the preparation of colloidal hydroxysodalite and ZSM-2 we refer to the original work by Schoeman et al. (1994b, 1995a).

1.1   Zeolite N-Y and Zeolite N-A

The preparation of colloidal zeolites N-Y and N-A was described in detail by Schoeman et al. (1994a). Here we summarize their procedures with an example which yields both N-Y and N-A.

The synthesis mixtures were prepared by adding a TMA-aluminate solution to the silica sol with strong mixing to ensure that a clear homogeneous solution free of gel particles was obtained.

The TMA-aluminate solution was prepared by precipitation of an $Al(OH)_3$ gel from an aqueous $Al_2(SO_4)_3$ solution with ammonia. The gel was filtered by suction, redispersed in water, and filtered again. Three such rinse steps sufficed to remove essentially all remaining ammonia and sulfate ions. The wet $Al(OH)_3$ filter cake was added to a pre-weighed 8.4 M TMAOH solution and was stirred until a clear homogeneous TMA-aluminate solution was obtained. This solution was weighed once more to determine the weight of the $Al(OH)_3$ filter cake. It was thereby possible to calculate the water content in the filter cake assuming that it consisted of $Al_2O_3 + H_2O$. The remaining water was added to the TMA-aluminate solution, which, after including the water content of the silica sol, gave the correct water content in the final reaction mixture as shown in Tables 5-10 and 5-11.

In those cases where additional sodium was required over and above the sodium present in the silica sol, a NaOH solution was added to the TMAOH solution before the addition of the $Al(OH)_3$ filter cake. In certain experiments, the sodium content in the silica sol was reduced by means of ion exchange using a cation-exchange resin, Dowex® HCR-S (H+). The sodium content reduction was monitored by the decrease in pH, but in all cases, the final sodium content was determined by elemental analysis. The ion-exchanged silica sol remained as a stable colloidal suspension at least for a period of three weeks, i.e., the silica sol,

Table 5-10. The product distribution as determined by XRD as a function of sodium content in a reaction mixture with the molar composition (2.5-x) $(TMA)_2O:xNa_2O:Al_2O_3:3.4 SiO_2:370 H_2O$. Schoeman et al. (1994a).

Run	Na_2O/Al_2O_3 molar ratio	Product	
		% Y	%A
A1	0.072	60	40
A2	0.10	40	60
A3	0.20	0	100
A4	0.40	0	100
A5	0.45	0	100
A6	0.50	0	100
A7	0.04	100	0

Table 5.11. The product distribution as determined by XRD as a function of sodium content in a reaction mixture with the molar composition (1.62-x) $(TMA)_2O:xNa_2O:Al_2O_3:3.62 SiO_2:246 H_2O$. Schoeman et al. (1994a).

Run	Na_2O/Al_2O_3 molar ratio	Product	
		% Y	%A
B1	0.420	0	100
B2	0.077	75	25
B3	0.044	100	0
B4	0.0.077/0.42*	0	100

* Na_2O/Al_2O_3 molar ratio at the start of crystalization was 0.077. Sodium was added as five equal additions with 2 h intervals starting after a crystalization time of 48 h to obtain a final Na_2O/Al_2O_3 molar ratio of 0.42

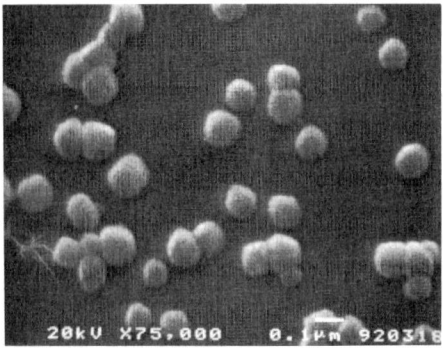

Fig. 5-9. SEM photograph of the product, zeolite Y, in run A7. Magnification 75,000X.
Schoeman et al. (1994a)

as in the case of the TMA-aluminate solution, was a clear solution.

The TMA-aluminate solution was added to the silica sol with vigorous stirring to obtain a clear homogeneous solution. These solutions were heated quiescently at 100°C in polypropylene bottles with reflux in a polyethylene glycol bath.

The SEM photograph in Fig. 5-9 confirms the average particle size of 95 nm measured by light scattering. The specific surface area was measured to be 523 m^2/g.

1.2. Zeolite TPA-Silicalite-1

The synthesis mixtures were made by adding the alkali, tetrapropylammonium hydroxide, to the silica source, TEOS, with strong mixing, Persson et al. (1994) and Schoeman et al. (1994d). The solutions were shaken for 12 hours on a gyratory shaker to allow hydrolysis of the TEOS. The synthesis mixture of the composition: 9 TPAOH : 0.1 Na_2O : 25 SiO_2 : 480 H_2O : 100 EtOH, yielded, after hydrothermal treatment at 98°C for about 24 hours, a colloidal suspension of discrete TPA-silicalite-1 particles (SiO_2/Al_2O_3 > 3000). The average particle size was 95 nm as determined by dynamic light scattering. The BET surface area of a purified, freeze-dried and calcined (600°C) sample of the preparation was measured as 445 m^2/g. This agrees with data obtained from the literature and indicates a well-crystallized sample.

1.3. Zeolite (Na, TPA)-ZSM-5

Persson et al. (1995) reported the synthesis of colloidal ZSM-5 zeolite at 98° C with $SiO_2:Al_2O_3$ ratios > 80. The ultimate average particle size of the zeolite microcrystals was determined to be 166 nm by dynamic light scattering. By increasing the reaction temperature of the synthesis mixture up to 160° C, Eriksson and Otterstedt (in press) were able to extend the $SiO_2:Al_2O_3$ ratios down to 35. The steps in their procedure are outlined below.

1. Hydrolysis of TEOS
   16.47 g TPAOH was mixed with 17.3 g TEOS and the mixture was placed
   in a gyrating shaker at room temperature for 13 hours.
2. Sodium aluminate solution
   11.66g $NaAlO_2$ was dissolved in 24.82 g twice-distilled water at about 50°
   C. The warm solution was filtered.
3. Reaction mixture
   As much filtered aluminate solution as was needed to give Zeolite ZSM-5
   of desired ratio was added to 13.18 g TPAOH. Water was added so that the
   amount of water was the same and constant in all the runs. The solution
   containing the aluminum source was added to the solution of hydrolyzed
   TEOS, the silica source, under vigorous stirring. The reaction mixture had
   the following molar composition:

   $(9TPAOH:(0.377-0.871)Na_2O:25\ SiO_2:(0.312-0.714)Al_2O_3:100EtOH:480H_2O)/306.9$

   The lower and higher molar ratios of Al_2O_3, 0.312/306.9 and 0.714/306.9,
   gave $SiO_2:Al_2O_3$ ratios of 80 and 35 respectively, in the final zeolite
   crystals.
4. Nucleation
   The reaction mixture was heated in a water bath at 97°C for 2 hours to get
   nucleation started.
5. Crystallization
   The reaction mixture was transferred to an autoclave, which was heated in
   an oven for about 20 hours at a temperature sufficient to give the desired
   $SiO_2:Al_2O_3$ ratio.
6. Isolation of zeolite crystals
   After crystallization was complete, the reaction mixture was centrifuged at
   about 7000 rpm. The supernatant solution was discarded and the zeolite
   crystals were washed with twice-distilled water. This procedure was
   repeated four times.
   The washed zeolite was freeze dried for 24 hours and then was calcined at
   550°C for 2 hours.

Figure 5-10 shows the $SiO_2:Al_2O_3$ ratio of colloidal ZSM-5 zeolites as a
function of reaction temperature.

Figure 5-11 shows that the size of crystallites of ZSM-5 of ratio 40, 60,
and 90 fall in the range from 250 to about 500 nm. The particle size appears to
increase with decreasing ratio. The effect of the size and ratio of colloidal ZSM-
5, ion exchanged with copper ions, on its ability to eliminate NO_x from vehicle
exhausts under oxygen-rich conditions, will be described in Ch. 7.

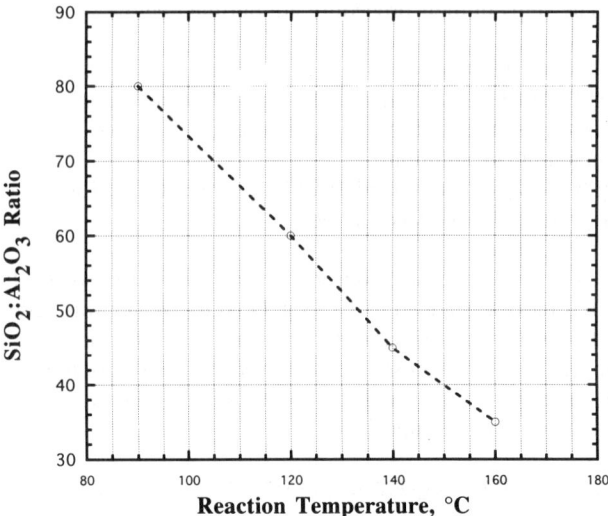

Fig. 5-10. $SiO_2:Al_2O_3$ ratio of colloidal ZSM-5 zeolites as a function of reaction temperature.Eriksson et al. (in press). Courtesy Butterworth Scientific.

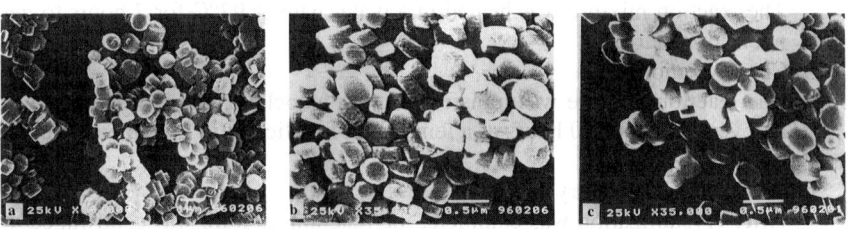

Fig. 5-11. SEM micrographs of colloidal ZSM-5 crystallites of $SiO_2:Al_2O_3$ ratios a) 40; b) 60; and c) 80. Eriksson et al. (in press). Courtesy Butterworth Scientific.

2.   Properties of Colloidal Zeolites

The discussion below deals with the properties of colloidal zeolite as well as the nature of the crystallizing system that yields these colloidal particles. The first topic treated is a comparison of this system to conventional zeolite systems as well as to the systems yielding the colloidal metal oxides α-alumina, zinc oxide, and chromium oxide, Matejevic and Schreiner (1978), Chittofrati and Matejevic (1990), Demchak and Matejevic (1969) . This is followed by a discussion on the system itself and thereafter, a discussion dealing with the colloidal properties of the zeolite crystallites and the possibilities that arise as a result of these properties.

2.1. Colloidal Zeolites *vis a vis* other Colloidal Metal Oxides and Conventional Zeolites

Matijevic (1991) pointed out that the synthesis of monodisperse materials via the technique of *forced hydrolysis* is sensitive to factors such as salt concentration, pH, nature of the anion and temperature; see Ch. 3 §B. These factors are also known to influence the synthesis of conventional zeolite and hence colloidal zeolite as well. Of interest is the sensitivity of the preparation methods described here to these parameters. The following points serve to illustrate the robust nature of these preparations. As we shall show, the alumina source, a freshly precipitated aluminum sulfate, is rinsed to remove essentially all of the sulfate. Removal of the sulfate can be shown by the absence of a $BaSO_4$ precipitate in the filtrate following rinsing. Partially rinsed $Al(OH)_3$ gels prepared from aluminum sulfate and aluminum nitrate (without quantifying the amount of sulfate or nitrate remaining in the gel) have resulted in similar results compared to those runs using well rinsed gels, Schoeman (1993). The effect of the synthesis temperature on the particle size of silicalite is given by Schoeman (1994d). The trends observed are to be expected. Although the effects of pH and salt concentration were not specifically studied, it does not appear that highly sensitive techniques are required in the preparation of the synthesis mixtures.

A major difference between the zeolite sols described here and the metal oxide sols described in the literature, Matijevic (1981, 1984, 1986), Sugimoto (1987), is that the latter are synthesized in very dilute solutions. As an example, aluminum hydroxide particles are synthesized by keeping a 0.002 M or 0.07 wt. % solution of $Al_2(SO_4)_3$ at 97°C for 48 hours, Matijevic (1981). The particle size obtained is ca. 1 μ with a very narrow particle size distribution. Ferric phosphate particles are prepared by aging a very dilute solution of ferric perchlorate (0.00080 M) and phosphoric acid (0.030 M), Matijevic (1981). It is therefore of interest to take a closer look at the solids content in the synthesis mixtures as well as in the as-synthesized zeolite sols. The aluminosilicate concentrations calculated as $SiO_2 + Al_2O_3$ in the synthesis mixtures for the synthesis of colloidal zeolite sols are 4.4 to 6.7 wt. %, Schoeman et al. (1994a), 14.7 wt. %, Schoeman et al. (1994b), and 14.8 wt. %, Persson et al. (1994). In comparison, the synthesis of zeolite Na-Y according to Breck (1964) is achieved from gels with solid contents in the range 17.5 (patent example 1) to 25 wt. % $SiO_2 + Al_2O_3$ (patent example 13).

The colloidal zeolite content is 6 wt. %, Schoeman et al. (1994b) and 10 wt. %, Persson et al. (1994). The zeolite content from conventional zeolite syntheses such as those used for the synthesis of zeolite Na-Y, Breck (1964) is in the range 3.6 (patent example 1) to 20 wt. % (patent example 13) assuming all alumina is consumed during the crystallization. Schoeman and co-workers (1994a and 1994b) purified the colloidal zeolite sols by ultracentrifugation in order to remove the soluble unreacted aluminosilicate This post-synthesis treatment may also be used to concentrate the zeolite sol. A colloidal suspension of ZSM-2, particle size 46 nm, was concentrated to a solids content of 25 wt. % based on the calcined weight, at 600°C, Schoeman (1993). If it is recalled that this material has a porosity of ca. 50 vol. %, the solids content would be ca. 40 wt. % (including the water in the pore structure). In this respect, it can be noted that commercial silica sols, Ludox® SM and Ludox®TM, have solid contents of ca. 30 and 50 wt. % respectively; see Ch. 2 §D. In other words, the zeolite sols described here are comparable to conventional zeolite synthesis mixtures and commercial silica sols both in terms of the solid content in the synthesis mixture (reactants) and in the ultimate product.

It was mentioned in §4.2 and §4.3 that a zeolite synthesis gel characterized by a relatively high alkali content yields relatively smaller crystals. The alkalinity of the synthesis mixtures that yield colloidal zeolite are not markedly different from the alkalinities used for the synthesis of conventional zeolite with possibly the exception of TPA-silicalite-1. In actual fact, the alkalinity expressed as the concentration NaOH in the sols, Schoeman et al. (1994a), for the synthesis of zeolite N-Y and N-A is less than those in the patent examples given by Breck (1964). The alkalinities in his patent examples 1 and 13 are 3.3 M and 2.7 M NaOH, respectively, while the alkalinity in the sols prepared by Schoeman et al. (1994) is 0.75 M TMAOH + NaOH.

2.2. What is Meant by a Clear Homogeneous Solution

The clear homogeneous solutions employed by Ueda and co-workers and discussed in §4.3 are clear in the sense that no visible solid aluminosilicate phase is present. According to Ueda and Koizumi (1979), the Tyndall phenomenon of colloidal dispersions was not observed before crystallization of analcime began, thus indicating that no colloidal material was present. The absence of Tyndall light scattering is, however, no measure of homogeneity as shown in the work of Hamilton et al. (1993). Silicate solutions were filtered through a 0.2-μ polysulfone filter membrane and the turbidity of the solution was determined by means of light scattering. The results showed that particulate material was present although the solutions appeared to be clear. In this respect, it may be noted that a diluted commercial amorphous silica sol in which the particle size is 7 - 8 nm, Ludox® SM, ca. 10 wt. % SiO_2, appears to be a clear solution. In other words, seemingly clear solutions may be heterogeneous. Light scattering can, however, be used to investigate the nature of these solutions.

The colloidal zeolite crystals synthesized according to the methods described in this chapter have been prepared in what we have termed clear homogeneous solutions, which is apparent due to the fact that Tyndall light scattering is not evident during the period prior to the onset of crystallization. This would indicate that no solid amorphous material is present during this period. Recognizing that these observations are not sufficient criteria that must be fulfilled in order to term a solution a homogeneous solution, light scattering has been employed to determine whether colloidal material is present or not, as discussed by Schoeman et al. (1994a, 1994b) and Persson et al. (1994).

It is, however, not known at which particle size-concentration value the method of light scattering fails to yield information due to instrumental limitations. There is a lower limit in terms of both particle size and particle concentration that may be detected. As shown in the work of Schoeman et al. (1994a), the dissolution of silica particles during hydrothermal treatment could be monitored up to a certain stage. No scattered light could be detected at the stage corresponding to a silica particle size of ca. 10 nm and a corresponding estimated particle concentration of 9×10^{16} per cm3. These results show that one should place strict criteria on what a clear homogeneous solution is, even if no colloidal material is visible to the eye.

2.3. Linear Growth Rates

Low growth rates were reported by Schoeman et al. (1994a), and were found to be most noticeable in the crystallization of TPA-silicalite-1. Possible reasons for this were discussed by those authors. A topic not discussed there is the question of the solubility of the colloidal silicalite crystals. This is an important aspect that has a profound impact on the growth of small particles. It is a fact that small particles grow at a slower rate due to their high solubility and hence one would expect that the initial growth stage would be characterized by a lower rate of growth than the intermediate stage of growth. As a result, the curves showing the increase in particle size versus synthesis time would be S-shaped and not entirely linear as has been shown in the work of Persson et al. (1994) and Schoeman et al. (1994d). The discussion below, restricted to TPA-silicalite-1 and amorphous silica, serves to illustrate the complexity of this topic. The Ostwald-Freundlich equation applied to size-dependent solubility (known as the Thomson-Gibbs effect) may be expressed as

$$\ln\left(S_d \Big/ S_i\right) = \frac{4\gamma\nu}{RTd}$$

as discussed in Chapter 2.

Values of the interfacial surface energy γ are not known with any accuracy and hence it is difficult to predict the size-dependent solubility with this equation. An attempt has been made to obtain a rough estimate of this parameter by reviewing the literature.

It is reported that the value of γ decreases with decreasing particle size, Tolman (1949), Heicklen (1976); and furthermore, according to Den Ouden and Thompson (1991), a lower value of γ would be expected at higher temperatures. In their analysis of the crystallization of TPA-silicalite-1 with the population balance model, Randolph and Larson (1988) obtained a value of 31×10^7 J / cm2. It is interesting to note that the chemical composition of their crystallization mixture is very similar to that used by Schoeman et al. (1994d), although the ultimate crystal sizes differ markedly. In view of the interfacial surface energy dependence on temperature and particle size, one would expect a low value of the interfacial surface energy and, as seen in Fig. 5-12, the effects predicted by the Ostwald-Freundlich equation for a value of the interfacial surface energy of 30 x

10^7 J/cm^2, curve 3, would not be of significance for particle sizes larger than ca. 35 nm.

The crystalline forms of silica exhibit lower solubilities than amorphous silica, Barrer (1989). This trend is reinforced by the results of a study of the solubility of dehydrated silica gels in anhydrous alcohols at 500°C which shows that the solubility drops markedly with an increase in the molecular weight of the alcohol, Kitahara (1969). (When TPA-silicalite-1 was synthesized in the presence of ethanol, the latter was present as a result of the hydrolysis of tetraethoxy silane).

These facts indicate that the colloidal silicalite crystals would be less soluble than amorphous silica particles of the same size at a given pH (pH < 10.5). On the other hand, one would expect the solubility of a porous silica to be higher than that of a dense (non-porous) silica due to the high local solubility of the silica surface as a result of the small radius of curvature, Charles (1964). Since it is not advisable to draw any conclusions from these conflicting trends, it is of interest to consider the solubility of amorphous silica.

According to Iler (1979), the interfacial surface energy of amorphous silica is probably 46×10^{-7} J/cm^2. The corresponding solubility curve for amorphous silica based on this value at a temperature of 298°K is shown in Fig. 5-12 as curve 2. One would expect the solubility curve for the amorphous silica system to be shifted towards curve 1 for higher temperatures since solubility increases with increasing temperature. In the event that the value of the interfacial surface energy in the TPA-silicate-1 system is comparable to that of

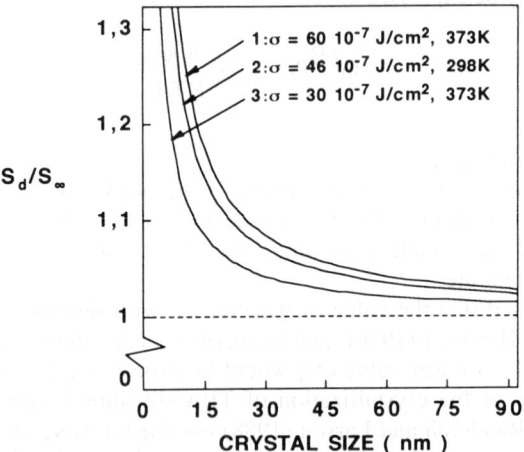

Fig. 5-12. The solubility curves, 1 and 3, of colloidal silicalite crystals, expressed as a ratio of the solubility of a particle with diameter d to that of a particle of infinite radius, as a function of particle size at 373°K. Curve 2 is the corresponding solubility curve for amorphous silica based on a value of σ of 46×10^{-7} J/cm^2 at 298°K. Schoeman (1994). [Note: The interfacial tension on the graph is designated σ, whereas we have used γ throughout the text.]

the amorphous silica system, the latter corrected for temperature, the effects of the Ostwald-Freundlich equation would be significant for crystal sizes as large as ca. 75 nm as shown by curve 1 in Fig. 5-12. Note that the ultimate particle size in the standard run is only 95 nm. As a result, crystal growth would exhibit a non-linear growth behavior throughout almost the whole period of growth. However, one cannot draw any conclusion as to a reasonable value of the interfacial surface energy for crystalline TPA-silicalite-1 and any attempt to predict or estimate the value of the interfacial surface energy would be subject to great error. As seen in Fig. 5-12, relatively small changes in the value of the interfacial surface energy result in quite different solubilities.

A closely related topic to linearity in growth behavior is that of colloidal zeolite stability, which will be discussed below.

2.4. The Stability of Colloidal Zeolites

The stability of colloidal dispersions is usually thought of in terms of the tendency of the particles to form secondary particles by aggregation. This phenomenon is of great importance in many areas of zeolite technology and is treated briefly here There is however another aspect concerning stability that is a well known phenomenon within zeolite science, i.e. Ostwald's rule of successive transformations.

Many of the zeolites that are of interest are metastable phases which may recrystallize to thermodynamically more stable phases under certain conditions with prolonged hydrothermal treatment. For example, zeolite Na-A crystallizes rapidly under highly alkaline conditions, but under these conditions, the more stable zeolite hydroxysodalite replaces zeolite Na-A via the dissolution of the latter and a solution phase crystallization of the former, Barrer (1989). It is therefore surprising that in Schoeman's studies only one case of this phenomena was observed, Schoeman et al. (1995a). From the results obtained in the investigation of the crystallization of ZSM-2 in the presence of lithium and sodium cations, it can be concluded that the ZSM-2 shows a limited stability over a period of time, ca. 12 h, whereafter the structure dissolves while the nitrogenous form of the edingtonite zeolite, (Li, Na)-E appears. The phase transformation to zeolite (Li, Na)-E was not observed in those runs where sodium was absent in the synthesis mixture.

It has been reported that ZSM-4 is the thermodynamically favored phase that may crystallize in the system employed to crystallize zeolite N-Y with TMA and Na as the source of the cations, Szostak (1989). This has not been observed in the work of Schoeman et al. (1995a). Similarly, in the synthesis of high-silica zeolites such as TPA-silicalite-1, α–quartz generally appears as the more stable phase, Szostak (1989). This has not been observed in spite of prolonged hydrothermal treatment of the silicalite sols. Although no definite explanation can be given to why these colloidal crystals of metastable phases are stable, it can be noted that in all those runs producing "stable" colloidal crystals, the alkali cation content is at a minimum except in the case of those runs where ZSM-2 crystallizes in the presence of both Na and Li. This point will be discussed in § 2.5.

According to Lowe (1983), a large pH rise in the synthesis mixture following completion of crystallization is usually associated with a pure stable product. The change in pH during the synthesis of TPA-silicalite-1 was monitored by Schoeman et al. (1994d), and recorded as 12.5 before hydrothermal treatment and 13.1 following the completion of crystallization. This large increase in pH would indicate that the product crystals possess a reasonable degree of stability. The incorporation of the template, TPA+, in the zeolite crystal lattice has a stabilizing effect on the structure. The relative stabilities of various cationic forms of ZSM-5 has been discussed by Araya and Lowe (1986) who show that the TPA+ cation is a truly structure-directing template for the synthesis of ZSM-5. According to Zones and Santilli (1992), the specific structure/fit relationship between the template and the zeolite structure stabilizes the lattice at high solution pH. The zeolite achieves a lower energy state rendering it less susceptible to Ostwald ripening phenomena.

Although no studies have been performed in this area on sub-colloidal TPA-silicalite-1 crystals (< 50 nm), it is tempting to put forward the argument that the above discussion may be extended to such colloidal sized silicate crystals. If so, the linearity of the growth curves presented by Persson et al. (1994) would be a fair representation of the course of events during the initial stages of TPA-silicalite-1 crystallization. In recent work by Twomey et al. (1994), TPA-silicalite-1 was crystallized from clear TPA-silicate solutions and the crystallization was monitored by *in situ* light scattering. Although the recorded growth rates differ markedly from those reported by Persson et al. (1994)—presumably due to the different compositions and silica source, the linearity of the growth curves extends into the same colloidal size range as in that work.

In conclusion, the linearity of the crystal growth curves appears to be a reasonable result indicating that the colloidal crystals of TPA-silicalite-1 exhibit a relatively low solubility at the particle sizes recorded, although this is probably not the case for particle sizes less than ca. 20 nm.

The colloidal zeolite crystals exhibit stability against aggregation in the as-synthesized form. In this form, the electrolyte content is high as seen by the high pH, typically higher than 12.5. It is well known that high electrolyte contents in colloidal suspension destabilize the sol resulting in flocculation. This is not observed in these colloidal suspensions. The probable reason is that the flocculation behavior is a function of the type of electrolyte, and the tetraalkylammonium ions are adsorbed on the crystal surface providing steric stabilization, Persson et al. (1994, 1996). Steric stabilization is not afforded by alkali cations but since the content of these is low, they do not have a destabilizing effect on the colloidal crystals.

Purification of TPA-silicalite-1 sols by ultracentrifugation has revealed that flocculation occurs if the pH of the concentrated (> 1 wt. %) suspension drops to ca. 8. The rate of flocculation increases dramatically if such purified sols are diluted such that the pH in the diluted sols approaches 7. The reason for this behavior can be explained by the surface charge of TPA-silicalite-1 crystals as seen in Fig. 5-13. The isoelectric point is at pH 5 according to Fig. 5-13, but this value has been found to be as high as 7 in certain measurements. As a result, purified TPA-silicalite-1 sols are susceptible to flocculation in the pH interval 5 to 8. The aluminum-containing zeolites do not exhibit an isolectric point above pH 4. Below this pH, the zeolite structure dissolves almost instantaneously and measurements in this pH range are not possible.

With reference to Fig. 5-13, it is important to note that the electrophoretic mobilities have been converted to a zeta potential with the Smoluchowski equation, Hunter (1987). This is not entirely correct since the crystal sizes of the zeolites investigated are less than 200 nm—see Ch. 6 § A3. For such particle sizes, the equation of Hückel should be employed. The purpose of Fig. 5-13 is, however, primarily to show that the colloidal zeolite may be characterized with respect to surface charge, as are other colloidal materials. Since it is possible to determine the zeta potential of these colloidal zeolite crystals, it is not unreasonable to take this parameter (or even surface potentials), and variations thereof as a result of different synthesis environments, into account during the study of zeolite growth. This was not done in the work just cited, but it is an interesting aspect.

A direct consequence of the lack of stability of colloidal zeolite with respect to the electrolyte content is two-fold. First, susceptibility to increased electrolyte contents is a characteristic of colloidal systems, thus emphasizing the colloidal nature of these materials. Second, it would appear that the destabilizing effect of cations restricts the variation in the cationic form of the zeolite. As is seen in Fig. 5-13, the surface charge of the aluminum zeolites is at a minimum in the pH range 4 - 5. It is, however, in this pH region ion-exchange of, for example, zeolite Na-Y to RE-Y is performed. The fact that high cation contents

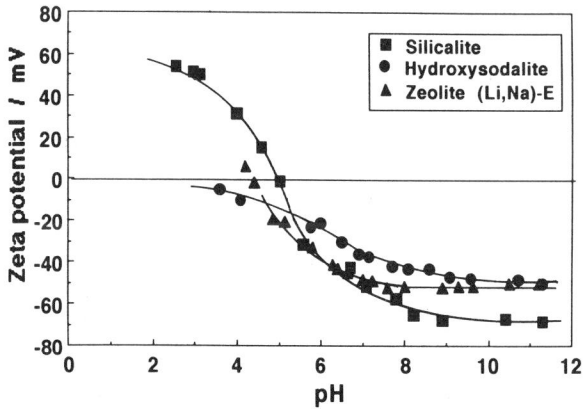

Fig. 5-13. The surface charge of colloidal zeolite expressed as the zeta potential as a function of pH. The ionic strength in all suspensions is 10mM NaCl. Measurements were performed at 25°C. Schoeman (1993).

are present during conventional ion-exchange operations at elevated temperatures, as well as the possible presence of di- and tri-valent cations which are more destabilizing than monovalent cations (Schulze-Hardy rule) requires a more appropriate method for the ion-exchange of colloidal zeolite than for conventional zeolite. The colloidal zeolite may be ion-exchanged to other cationic forms by means of the addition of a strong cationic resin in a suitable form to the colloidal suspension with mixing. This method allows for the ion-exchange of the colloidal particles without a change in the colloidal properties, i.e. particle size, Schoeman (1993). This is possible in those cases where the organic cation may be ion-exchanged as is the case with ZSM-2, Schoeman et al. (1995a), but not in the case of TPA-silicalite-1.

2.5. The Relation between Nucleation, Number of Crystals and Crystal Size

It is interesting to inquire as to the likely reasons that discrete colloidal zeolite crystals are formed in these solutions. Numerous possible contributing factors may be excluded from the present discussion since these have been shown not to affect the particle size to a large extent, e.g., the silica and the alumina source, low temperatures, alkalinity and additives such as polyvalent cations and dyestuffs. The fact that the synthesis solutions are clear homogeneous solutions does not appear to be a significant factor either since, as shown in §4.3, such solutions may yield large zeolite crystals.

A feature that is most notable is the identification of a growth-limiting nutrient in each system. In the synthesis of faujasite and zeolite N-A, the growth-limiting nutrient is sodium while the growth-limiting nutrient in the synthesis of sodalite and TPA-silicalite-1 is alumina and silica respectively. The growth-limiting nutrient in the synthesis of ZSM-2 has not been identified as yet due to the complex nature of the crystallizing system. As in the synthesis of conventional zeolite, the growth limiting nutrient places a limit on the ultimate particle size of the colloidal zeolite. The existence of a growth limiting nutrient is, however, not the sole criteria that needs to be fulfilled in order to synthesize colloidal zeolite since, as discussed in §4.1, the nucleation event controls the number of crystals. A relatively large number of nuclei allow for a smaller ultimate particle size and *vice versa*. It is reasonable therefore to state that nucleation is responsible for the large number of crystals produced in the synthesis of colloidal zeolites.

What is the reason for the nucleation behavior? A key feature of the synthesis solutions employed in the synthesis of colloidal zeolite is the low alkali contents. The investigation performed by Hayhurst et al. (1988) is interesting in this respect. In their work on the effect of hydroxide on the growth rate in silicalite synthesis they concluded that nucleation and crystallization are inhibited by the presence of sodium in the reaction mixture. They proposed that there is a preferential association of the sodium cation to the silica gel compared to the TPA cation, thus forming stable sodium silicate species. This association must be broken before the species responsible for crystallization can be formed. In the absence of sodium cations and using TPAOH as the sole source of alkali, a more facile transformation of the gel to silicalite is permitted. It should be noted that the differences in crystallization and nucleation rates for comparable synthesis gels recorded in their work could be due to differences in, for example, pH.

One must, however, consider the observations made by Fegan and Lowe (1986) concerning the crystallization of TPA-silicalite-1. They show that more nuclei are formed at higher alkalinities where the alkalinity is supplied as NaOH. In contrast, at relatively high alkalinities they state that this may not be the case. The fact that higher alkalinities favor smaller crystals has been mentioned in §4.2, and, as shown by Persson et al. (1996), it is probably the OH⁻ concentration that is responsible for this effect. If the alkalinity is supplied solely as NaOH, the negative effects of the sodium cation (negative with respect to enhanced nucleation) might become predominant at high alkalinities, thus outweighing the positive effects of the increased OH⁻ content. It has also been shown that sodium retards the crystallization of ZSM-5 at constant alkalinity, Chao et al. (1981).

Although it may be somewhat speculative to describe the role of sodium as nucleation inhibiting—the situation is probably far more complicated, it is nevertheless tempting to suggest that a significant contribution towards the nucleation kinetics (large number of crystal produced presumably in a short time) found in the synthesis of colloidal zeolite is due to the low alkali metal contents in the mixtures, Schoeman et al. (1993, 1994a, 1994b, 1994d) and Persson et al. (1994).

Another contribution to the success in the synthesis of discrete colloidal zeolite crystals is the lack of aggregation of the primary particles. The probable reason for this is, as discussed in § 2.4, the stabilization imparted to the crystals by steric stabilization afforded by tetraalkylammonium cations. The destabilizing effect of alkali cations is absent due to the fact that the crystallizing systems contain low alkali contents.

2.6. Crystal Size Tailoring of Colloidal Zeolites

It is difficult to control the particle size of colloidal particles within a few nanometers due to the nature of the synthesis procedure—a process entailing nucleation followed by growth. Examples of such processes are those of the type, *forced hydrolysis*, Matijevic (1981), and the Stöber synthesis of colloidal silica, Stöber et al. (1968). If we restrict ourselves to silica, there is a simple method by which the size of the particles may be tailored. If the size of the "seeds" and their number is known, the amount of silica required to build up the particles to a predetermined size can be calculated and build-up of the seeds is made possible under controlled conditions—see Ch. 2 § D. There are numerous examples of this technique in the literature, Coenen and DeKruif (1988), Bechtold and Snyder (1951), Mindick and Vossos (1970), Albrecht (1972), Irani (1974), Bergna (1981)—see also Ch. 2 §D.

The ability to tailor the size of the zeolite crystals in a controlled fashion was demonstrated by Schoeman et al. (1993, 1994a, 1994b). Besides the possible impact that these results may have within the area of applications, this principle may be used to investigate the growth mechanism of zeolite. One advantage would be, as pointed out by Kacirek and Lechert (1976), that it is possible to separate the nucleation event and the growth stage and still be able to investigate the mechanisms in the colloidal size range.

2.7. Colloidal Zeolite Organosols

The hydrophilic nature of the colloidal suspensions is clearly evident. A property of hydrophilic colloids is that they can be introduced to certain non-aqueous but still polar media such as alcohols. In completely nonpolar media such as benzene or paraffins, it is hardly possible to obtain a stable colloidal suspension without flocculation of the primary particles. This is true for the colloidal zeolite suspensions as well. This poses a problem if these microporous crystals are to be used, for instance, in the fabrication of thin films or membranes, Jia et al. (1992). Adhering to the general methods employed in colloidal science for the preparation of organosols with an aqueous sol as the starting material, Iler (1957), it is possible to prepare a colloidal TPA-silicalite-1 organosol essentially free from water (0.2 wt. % H_2O by Karl-Fischer titration) by azeotropic distillation with n-propyl alcohol, Schoeman (1993). The particle size in the TPA-silicalite-1 organosol remains unchanged compared to that in the starting aqueous sol *viz* 185 nm. The significance of this result is that the silicalite crystals exist as discrete colloidal particles in an organic medium with a shelf life of at least six months. Not only does this example show that the colloidal particles withstand the conditions employed during the preparation of the organosol, but that there exists a flexibility in the post-synthesis stage (cf. crystal size tailoring, Schoeman et al. (1993)) that increases the potential of these colloidal materials.

References

Adnadjevic, J., Vukicevic, J., Filipovic-Rojka, Z., and Markovic, Zeolites, **10**, 699 (1990).

Albers, E. W., Edwards, G. C., and Vaughan, D. E. W., U.S. Patent 3,755,538 (1973).

Albrecht, W. L., U.S. Patent 3,673,104 (1972).

Araya, A., and Lowe, B. M., Zeolites, **6**, 111 (1986).

Argauer, R. J., and Landolt, G. R., U. S. Patent 3,702,886 (1972).

Barrer, R. M., Brynman, J. W., Bultitude, E. W., and Meier, W. M., J. Chem. Soc., Pt. 1, 195 (1959).

Barrer, R. M., and Denny, P. J., J. Chem. Soc., 971 (1961).

Barrer, R. M., Denny, P. J., and Flanigen, E. M., U.S. Patent 3,306,922 (1967).

Barrer, R. M., *Zeolites and Clay Minerals as Sorbents and Molecular Sieves*, Academic Press, NY (1978).

Barrer, R. M., *Hydrothermal Chemistry of Zeolites*, Academic Press, London (1982).

Barrer, R. M., *Zeolite Synthesis*, (Eds. M. L. Occeli and H. E. Robson), ACS Symp. Ser. **398**, 11 (1989).

Bechtold, M. F., and Snyder, O. E., U.S. Patent 2,574,902 (1951).

Benesi, H. A., U.S. Patent 3,190,939 (1965).

Bergna, H. E., U.S. Patent 4,272,409 (1981).

Bodart, P., Nagy, J. B., Gabelica, Z., and Derouane, E. G., J. Chim. Phys., **83**, 777 (1986).

Breck, D. W., U.S. Patent 3,130,007 (1964).

Breck, D. W., *Zeolite Molecular Sieves*, John Wiley & Sons, NY (1974).

Breck, D. W., *Zeolite Molecular Sieves*, Krieger, Malabar, FL (1984).

Bronic, J., Subotic, B., Smit, B., and Despotovic, L., *Innovation in Zeolite Material Science*, (Eds. P. J. Grobet, W. J. Mortier, E. F. Vansant, and G. Schulz-Ekloff), Studies in Surface Science and Catalysis, vol. 37, Elsevier, Amsterdam, 107 (1988).

Camblor, M. A., Corma, A., Martinez, A., Mocholi, F. A. and Perez Pariente, J., Applied Catalysis, **55**, 65 (1989).

Buckley, H. E., *Crystal Growth*, John Wiley & Sons, NY (1951).

Carman, P. C., and Malherbe, P. le R., J. Soc. Chem. Ind., **69**, 134 (1950).

Carman, P. C., J. Oil Chem. Ass., **37**, 165 (1954).

Chang, C. D., Cat. Rev. Sci. Eng. **25** (1), 1 (1983).

Chao, K-J., Tasi, T. C., and Chen, M. S., J. Chem. Soc., Faraday Trans. 1, **77**, 547 (1981).

Charles, R. J., J. Am. Chem. Soc., **47**, 154 (1964).

Charnell, J. F., J. Crystal Growth, **3**, 291 (1971).

Chen, N. Y., and Degnan, T. F., Chem. Eng. Progress, **84**(2), 32 (1988).

Chen, N. Y., Garwood, W. E., and Heck, R. H., Ind. Eng. Chem. Res., **26**, 706 (1987).

Chen, N. Y., Gorring, R. L., Ireland, H. R., and Stein, T. R., Oil and Gas J., **75**, (23), 165 (1977).

Chittofrati, A., and Matijevic, E., Colloids and Surfaces, **48**, 65 (1990).

Chutoransky, P., and Dwyer, F. G., Adv. Chem. Ser., **101**, 549 (1973).

Ciric, J., J. Colloid Interface Sci., **28**, 315 (1968).

Coenen, S., and DeKriuf, C. G., J. Colloid Interface Sci., **124**, 104 (1988).

Cundy, C. S., Lowe, B. M., and Sinclair, D. M., J. Crystal Growth., **100**, 189 (1990).

Cundy, C. S., Lowe, B. M., and Sinclair, D. M., Faraday Discuss., **95**, 235 (1993).

Davis, M. E., Jones, C. W., and Tsuji, K., Nature, **393**, 52 (1998).

Demchak, R. and Matijevic, E., J. Colloid Interface Sci., **31**, 257 (1969).

Den Ouden, C. J. J., and Thompson, R. W., J. Colloid Interface Sci., **143**, 77 (1991).

Den Ouden, C. J. J., and Thompson, R. W., Ind. Eng. Chem. Res., **31**, 369 (1992).

Elliott, C. H., and McDaniel, C. V., U.S. Patent 3,539,099 (1972).

Elliott, C. H., U.S. Patent 4,164,551 (1979).

Eriksson, L., Löwendahl, L., Gevert, B., Törncrona, A., and Otterstedt, J-E. A., Applied Catalysis, (in press).

Fegan, S. G., and Lowe, B. M., J. Chem. Soc., Chem. Comm., 437 (1984).

Fegan, S. G., and Lowe, B. M., J. Chem. Soc., Faraday Trans. 1, **82**, 785 (1986).

Fegan, S. G., and Lowe, B. M., J. Chem. Soc., Faraday Trans. 1, **82**, 801 (1986).

Feoktisova, N. N., Zhdanov, S. P., Lutz, W., and Bulow, M., Zeolites, **9**, 136 (1989).

Flank, W. H., Ed. ACS Symposium Ser. 135 (1980).

Fraissard, J. *Catalysis by Zeolites*, 343, (Ed. B. Imelik), Elsevier, Amsterdam (1980).

Freund, E. F., J. Crystal Growth, **34**, 11 (1976).

Ginter, D. M., Bell, A. T., and Radke, C. J., Zeolites, **12**, 742 (1992).

Ginter, D., M., Went, G. T., Bell, A. T., and Radke, C. J., Zeolites **12**, 733 (1992a).

Grose, R. W., and Flanign, E. M., U.S. Patent 4,061,724 (1977).

Hamilton, K. E., Coker, E. N., Sacco, A., Dixon, A. G., and Thompson, R. W., Zeolites, **13**, 645 (1993).

Harrison, I. D., Leach, H. F., and Whan, D. A., Proc. Sixth Int. Conf. on Zeolites, (Eds. K. Olson and A. Bisio), Butterworths, Guilford, 479 (1984).

Hayhurst, D. T., Nastro, A., Aiello, R., Crea, F., and Giordano, C., Zeolites, **8**, 416 (1988).

Heicklen, J., *Colloid Formation and Growth*, Academic Press, NY, 50, (1976).

Hobson, J. P. *The Solid-Gas Interface I*, Marcel Dekker, NY, 447 (1967).

Hoppe, R., Schulz-Ekloff, G., Rathousky, J., Strek, J., and Zukal, A., Zeolites, **14**, 126, (1994).

Hunter, R. J., *Foundations of Colloid Science*, Clarendon Press, Oxford, vol. 1, 557 (1987).

Iler, R. K., U.S. Patent 2,801,185 (1957).

Iler, R. K., *The Chemistry of Silica*, John Wiley & Sons, New York (1979).

Irani, F. A., U.S. Patent 3,789,009 (1974).

Israelachvili, J. N., *Intermolecular and Surface Forces*, 2nd ed, Academic Press, London, (1994).

Jacobs, P, A., and Martens, J. A., *Synthesis of High-silica Aluminosilicate Zeolites*, Studies in Surface Science and Catalysis, vol. 33, Elsevier, Amsterdam (1987).

Jia, M-D., Peinemann, K-V., and Behling, R-D., J. Membr. Sci., **73**, 119 (1992).

Kacirek, H., and Lechert, H., J. Phys. Chem., **80**, 1291 (1976).

Kaeding, W. W., Chu, C., Young, L. B., Weinstein, B., and Butter, S. A., J. Cat., **67**, 159 (1981).

Kasahara, S., Itabashi, K., and Igawa, K., *New Developments in Zeolite Science and Technology*, Proc. 7th Int. Conf. on Zeolites, (Eds. Y. Murakami, A. Lijima, and J. W. Ward), Kodansha/Elsevier, Tokyo, 185 (1986).

Kerr, G. T., J. Phys. Chem., **70**, 1047 (1966).

Kitahara, S., Nippon Kagaku Zasshi, **90**, 237 (1969).

Lechert, H., *Structure and Reactivity of Modified Zeolites*, (Eds. P. A. Jacobs, N. I. Jäger, P. Jiru, V. B. Kazansky, and G. Schulz-Ekloff), Studies in Surface Science and Catalysis, vol. 18, Elsevier, Amsterdam, 107 (1984).

Kageyama, N., Ueda, S., and Koizumi, M., J. Chem. Soc. Jpn. Chem. Ind. Chem., **9**, 1510 (1984).

Lin, D. H., Ducarme, V., Coudurier, G., and Verdine, J. C., *Zeolites as Catalysts, Sorbents and Detergent Builders - Applications and Innovations*, (Eds. H. G. Karge and J. Weitkamp), Studies in Surface Science and Catalysis, vol. 46, Elsevier, Amsterdam, 615 (1989).

Lok, B. M., Messina, C. A., Patton, R. L., Gajek, R. T., Cannan, T. R., and Flannigan, E. M., J. Am. Chem. Soc. **106**, 6092 (1984).

Lowe, B. M., and MacGlip, N. A., Proc. of the 5th Int. Conf. on Zeolites, (Ed. L. V. C. Rees), Heyden, London, 85 (1980).

Lowe, B. M., Zeolites, **3**, 300 (1983).

Lowe, B. M., *Innovation in Zeolite Material Science*, (Eds. P. J. Grobet, W. J. Mortier, E. F. Vansant, and G. Schulz-Ekloff), Studies in Surface Science and Catalysis, vol. 37, Elsevier, Amsterdam, 1 (1988).

Maher, P. K., and Scherzer, J., U.S. Patent 3,516,786 (1970).

Matijevic, E., and Schreiner, P., J. Colloid Interface Sci., **63**, 509 (1978).

Matijevic, E., Acc. Chem. Res., **14**, 22 (1981).

Matijevic, E., *Ultrastructure Processing of Ceramics, Glasses and Composites*, (Eds. L. L. Hench and D. R. Ulrich), John Wiley & Sons., New York, 334, (1984).

Matijevic, E., *Science of Ceramic Chemical Processing*, (Eds. L. L. Hench and D. R. Ulrich), John Wiley & Sons, NY, 463 (1986).

Meier W.M. and Olson, D.H., *Atlas of Zeolite Structure Types*, 3rd edition, Butterworth-Heineman, London (1992).

Meise, W. and Schwochow, F. E., *Molecular Sieves*, (Eds. W. N. Meier and J. B. Uytterhoeven), Adv. Chem. Ser., **121**, 169 (1973).

Mindick, M., and Vossos, P., U.S. Patent 3,538,015 (1970).

Oil and Gas J., **84** 31, Dec. 22/29 (1986).

Olson, D. H., and Bisio, A., Eds. *Proceedings of the Sixth International Zeolite Conference*, Butterworths, Surrey (1984).

Otake, M., Zeolites, **14**, 42 (1994).

Persson, A. E., Schoeman, B. J., Sterte, J., and Otterstedt, J-E. A., Zeolites, **14**, 557 (1994).

Persson, A. E., Schoeman, B. J., Sterte, J., and Otterstedt, J-E. A., Zeolites, **15**, 611 (1995).

Persson, A. E., Schoeman, B. J., Sterte, J., and Otterstedt, J-E. A., *Synthesis of Porous Materials*, (Eds. M. Ocelli and H. Kessler), Marcel Dekker, NY, 159 (1996).

Rajagopalan, K., Peters, A. W., and Edwards, G. C., Applied Catalysis, **23**, 69 (1986).

Randolph, A. D. and Larson, M. A., *Theory of Particulate Processes*, 2nd ed., Academic Press, San Diego (1968).

Ratnasamy, P., Borade, R. B., Sivasanker, S., Shiralkar, V. P., and Hedge, S. G., Acta Phys. Chem. **31**, 137 (1985).

Richards, A. J., Maginn, S. J., Leusen F. J. J., and Bick, J., Cerius^2® Molecular Modeling Software for Industrial Applications, *Data Visualization in Molecular Science*, J. E. Bowie, Addison-Wesley, Reading (1995).

Ruthven, D. M. *Principles of Adsorption and Adsorption Processes*, Wiley, NY, (1984).

Ruthven, D. M., Chem. Eng. Progress, **84**(2), 32 (1988).

Schoeman, B. J., *Synthesis and Properties of Colloidal Zeolite*, Ph. D. Dissertation, Chalmers University of Technology, Göteborg (1994).

Schoeman, B. J., unpublished results (1993).

Schoeman, B. J., Sterte, J., and Otterstedt, J-E. A., J. Chem. Soc., Chem. Comm.,**12**, 994 (1993).

Schoeman, B. J., Sterte, J., and Otterstedt, J-E. A., Zeolites, **14**, 110 (1994a).

Schoeman, B. J., Sterte, J., and Otterstedt, J-E. A., Zeolites, **14**, 208 (1994b).

Schoeman, B. J., Sterte, J., and Otterstedt, J-E. A., Zeolites, **14**, 568 (1994d).

Schoeman, B. J., Sterte, J., and Otterstedt, J-E. A. , *Zeolites and Microporous Crystals*, (Eds. T. Hattori and Y. Yashima), Studies in Surface Science and Catalysis, 49, Kodansha/Elsevier, Tokyo (1994c).

Schoeman, B. J., Sterte, J., and Otterstedt, J-E., J. Colloid and Interface Sci.,**170**, 449 (1995a).

Schoeman, B. J., Sterte, J., and Otterstedt, J-E., J. Porous Mater. 1, 185 (1995b).

Scott, G., Dixon, A. G. Sacco, A., and Thompson, R. W., *Zeolites: Facts, Figures, Future*, (Eds. P. A. Jacobs and R. A. van Santen), Studies in Surface Science and Catalysis, vol. 49, Part A, Elsevier, Amsterdam, 363, (1989).

Shiralkar, V. P., Joshi, P. N., Eapen, M. J., and Rao, B. S., Zeolites, **11**, 511 (1991).

Stöber, W., Fink, A., and Bohn, E., J. Colloid Interface Sci., **26**, 62 (1968).

Sugimoto, T., Adv. Colloid Interface Sci., **28**, 65 (1987).

Szostak, R., *Molecular Sieves - Principles of Synthesis and Identification*, van Nostrand Reinhold, New York, 2 (1989).

Thomas, J.M., Proceedings of the Eighth International Congress on Catalysis, Verlag Chemie, Berlin, **1**, 31 (1984).

Thomas, J. M., *Chemistry and Physics of Solid Surfaces VI*, (Eds. R.Vanselow and R. Howe), Springer Verlag, Berlin, 107 (1986).

Tolman, R. C., J. Chem. Phys., **17**, 333 (1949).

Twomey, T. A. M., Mackay, M., Kuipers, H. P. C. E., and Thompson, R. W., Zeolites, **14**, 162 (1994).

Udea, S. and Koizumi, M., Am. Miner., **64**, 172 (1979).

Udea, S., Murata, H., Koizumi, M., and Nishimura, H., Am. Miner., **65**, 1012 (1980).

Ueda, S., Kageyama, N., and Koizumi, M., *Proc. of the 6th Int. Conf. on Zeolites*, (eds. A. Bio and D. H. Olson), Butterworths, Surrey (1984).

Ueda, S., Kageyama, N., and Koizumi, M., Kobayashi, S., Fujiwara, Y., and Kyogoku, Y., J. Phys. Chem., **88**, 2128 (1984).

van der Pol, A. J. H. P., Verdujn, A. J., and van Hoof, J. H. C., Proc. from the 9th Int. Zeolite Conf., (Eds. R. van Ballmoos, J. B. Higgins, and M. M. J. Treacy), Butterworth-Heinemann, Montreal, 607 (1992).

Vaughan, D. E. W., and Strohmaier, K. G., *New Developments in Zeolite Science and Technology*, Proc. of the 7th Int. Conf. on Zeolites, (Eds. Y. Murakami, A. Lijima, and J. W. Ward), Kodansha/Elsevier, Tokyo, 207 (1986).

Vaughan, D. E. W., Chem. Eng. Progress, **84**, 25 (1988).

Verduijn, J. P., PCT Int. Appl. 92/13 799 (1992).

Verduijn, J. P., PCT Int. Appl. 92/14 680 (1992a).

Voogd, P., and van Bekkum, H., *Zeolites as Catalysts, Sorbents and Detergent Builders - Applications and Innovations*, (Eds. H. G. Karge and J. Weitkamp), Studies in Surface Science and Catalysis, vol. 46, Elsevier, Amsterdam, 519 (1989).

Voogd, P. and van Bekkum, H., Applied Catalysis, **59**, 311 (1990).

Wadlinger, R. L., Kerr, G. T., and Rosinski, E. J., U.S. Patent 3,308,069 (1967).

Wenqin, P., Ueda, S., and Koizumi, M., *New Developments in Zeolite Science and Technology*, Proc. of the 7th Int. Conf. on Zeolites, (Eds. Y. Murakami, A. Lijima, and J. W. Ward), Kodansha/Elsevier, Tokyo, 177 (1986).

Wilson, S. T., Lok, B. M., and Flanigan, E. M., U.S. Patent 4,310,440 (1982).

Wilson, S. T., Lok, B. M., Messina, C. A., Cannan, T. R., and Flanigan, E. M., J. Am. Chem. Soc. **104**, 1146 (1982).

Yang, R. T., *Gas Separation by Adsorption Processes*, Butterworths, Boston, (1987).

Zhdanov, S. P., *Molecular Sieve Zeolites*, Adv. Chem. Ser., **101**, Am. Chem. Soc., Washington, D. C., 20 (1971).

Zhdanov, S. P., and Samuelevich, N. N., Proc. of the 5th Int. Conf. on Zeolites, (Ed. L. V. C. Rees), Heyden, London, 75 (1980).

Zones, S. I., and Santilli, D. S., Proc. from the 9th Int. Zeolite Conf., (Eds. R. van Ballmoos, J. B. Higgins, and M. M. J. Treacy), Butterworth-Heinemann, Montreal, 171 (1992).

CHAPTER 6. SURFACES OF SMALL PARTICLES

A. Stability of Colloidal Dispersions
1. Colloidal Stability
2. Steric Stabilization - Steric Repulsion
3. Electrostatic Stabilization - Electrostatic Forces
    3.1. Electrostatic Interactions in Colloidal Systems
    3.2. Stern Potential and Zeta Potential
    3.3. Debye Length and Ionic Strength
    3.4. Zeta Potential and Surface Charge Density from
    Electrostatic Measurements
        3.4.1. Zeta Potential from Electrophoresis Experiments

B. The Surface of Small Particles in Aqueous Solutions
1. Coordination Chemistry at the Metal Oxide-Water Interface
2. Acid-Base Characteristics of the Hydrous Metal Oxide Surface
3. Adsorption of Metal Ions on the Hydrous Metal Oxide Surface
4. Adsorption of Anions on the Hydrous Metal Oxide Surface
5. Isoelectric Point, i. e. p., and point of Zero-charge, p. z. c.
    5.1. Factors Affecting i. e. p. and p. z. c.
        5.1.1. Effect of Dehydroxylation
        5.1.2. Effect of Contaminants
    5.2. Electrophoretric Mobility as a Function of pH
6. Surface Properties of Metal Sulfides in Aqueous Solutions

C. The Surface of Silica
1. Properties and Characteristics of the Silica Surface
    1.1. Types of Surface Hydroxyl Groups
    1.2. Concentration of Silanol Groups
    1.3. Dehydration of the Silica Surface
    1.4. Rehydroxylation of the Silica Surface
    1.5. Stability of Colloidal Silica
2. Modification of the Silica Surface
    2.1. Introduction of Aluminosilicate Sites
    2.2. Charge Reversal
    2.3. Nonionic Reactions of the Silica Surface
        2.3.1. Esterification
        2.3.2. Diazotization
        2.3.3. Halogenation
        2.3.4. Amination
        2.3.5. Grignard Reactions
        2.3.6. Reactions with Organic Silicon Compounds
    2.4. Organic Molecules Hydrogen Bonded to the Silica Surface

CHAPTER 6. SURFACES OF SMALL PARTICLES

Modern surface physics, using techniques such as x-ray photoelectron spectroscopy, XPS, low-energy ion scattering, LEIS, Auger Electron spectroscopy, AES, secondary ion mass spectroscopy, SIMS, electron microscopy of different types, thermal desorption spectroscopy, and several other techniques, makes possible a detailed characterization of the surface of small particles in particular. Common to all these techniques, however, is that the material must be studied as a dry solid and often under very low pressure. Many of the small particles discussed in this book, on the other hand, are born in water and in most applications they are used, at least initially, in the form of aqueous dispersions. The surface of the small particles will, of course, be affected by the water and indeed other components present in the dispersion. Moreover, the surface properties of small particles in aqueous dispersions can often be drastically changed by deliberate modification of the particle surface. The most important methods for studying the properties of small particles in aqueous dispersions are acid-base titration, electrophoresis, and adsorption of cations, usually metal cations and anions, onto the particles.

A. Stability of Colloidal Dispersions

When dealing with dispersions of small particles, two closely related questions arise, which are both of fundamental and practical importance and which depend on the nature of the particle surface. The first question is concerned with the conditions under which small particles will remain in the dispersed state. We must know the answer before we can develop methods for making stable dispersions. Secondly, we must know the conditions under which a dispersion of small particles will flocculate or coagulate so as to avoid or eliminate colloids—an important consideration, e.g., in the filtration of precipitates and in water purification where it is often necessary to induce aggregation.

1. Colloidal Stability

A dispersion of small particles represents a state of higher free energy than that corresponding to the material in bulk, since the surface free energy, due to the high specific surface area, is so much higher for the colloidal system.

The total free energy, ΔG, between particles at infinite separation and at a separation H, is a sum of several contributions, Everett (1988):

$$\Delta G = \Delta G^{att}(vanderWaals) + \Delta G^{rep}(shortrange) + \Delta G^{rep}(electrostatic)$$
$$+\Delta G^{rep}(steric \tag{6-1}$$

Lennard-Jones calculated the total potential energy of interaction between a pair of atoms or molecules as

$$\Delta G = \Delta G^{rep} + \Delta G^{att} = (B'/d^{12}) - (A'/d^6) \tag{6-2}$$

A' in the ΔG^{att} -term $= (3/4) h\upsilon\alpha^2$ where h is Planck's constant, α is the polarizability of the atom or molecule, and υ is a characteristic frequency in the ultraviolet corresponding to the first ionization potential.

The potential energy of interaction between two particles can also be calculated by assuming that every molecule in one particle interacts with each molecule in the other according to a Lennard-Jones potential and that the total free energy of interaction is obtained by summing the contributions from all possible particle pairs of molecules. The repulsive contribution can be neglected except for molecules on opposing surfaces. Using this method, Hamaker calculated the free energy of attraction for unit area of surface of two infinite plates a distance H apart:

$$\Delta G^{att} = \frac{-A_H}{12\pi H^2} \tag{6-3}$$

where A_H is called the Hamaker constant. A_H is closely related to A' of Eqn. 6-2 and is given by:

$$A_H = \frac{3}{4}h\upsilon\alpha^2\pi^2q^2 = A'\pi^2q^2 \tag{6-4}$$

where q is the number of molecules per unit volume of particles. The value of A_H generally varies between 10^{-20} J and 10^{-19} J.

When the close-range repulsion term is taken into account, the total free energy curves for particles interacting through London-van der Waals forces can have the various forms shown in Fig. 6-1, Everett (1988).

It was pointed out above that a colloidal dispersion is a system of high free energy and therefore has a tendency to spontaneously reach a state of lower energy, for instance by flocculation, unless there is a sizable energy barrier to prevent collapse of the colloidal state. If such a barrier exists, the system is unstable but can exist for a long time. Some of the gold sols made by Michael Faraday over 150 years ago can still be seen at the Royal Institute in London.

Curve iii in Fig. 6-1 represents the total interaction free energy in a colloidal system resulting from a combination of attractive (i) and repulsive (ii) interactions. In (a) states of separation and contact are maintained by a high energy barrier, the primary maximum P, created by strong repulsive interaction. If weaker repulsive interactions cause the barrier to decrease in height to a value not much larger than a few kT, the system can climb the primary maximum P and pass over into the primary minimum M_1. There is no energy barrier in (c) and the system can pass directly over into the primary minimum.

The question of preparing and stabilizing colloidal systems is therefore closely related to ways of creating energy barriers of such height that they can prevent collapse of the colloidal state. On the other hand, there are situations where we want to destroy a colloidal system, as for instance, in water purification by filtration or in the preparation of gels, and then it must be known how to change the conditions so as to reduce or eliminate the energy barrier.

The energy required to bring a colloidal system over the energy barrier into the primary free energy minimum arises from the particles' Brownian, or thermal, motion. The mean translational energy, i.e. the thermal energy, of colloidal particles is of the order 3/2kT, where k is Boltzmann's constant,

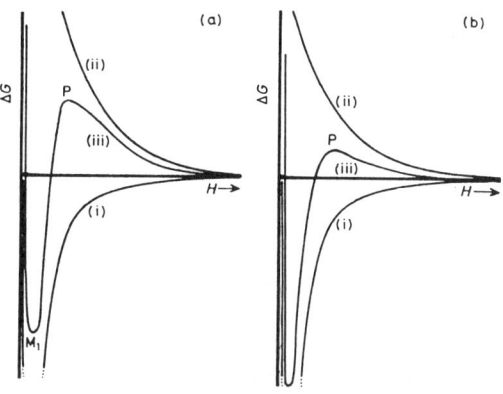

Fig. 6-1. Some possible forms of the total-interaction free energy (iii) resulting from a combination of attractive (i) and repulsive (ii) contributions. In these diagrams account is also taken of the short-range repulsive forces when the surfaces are almost in contact (Born repulsion). Everett (1988). Courtesy of the Royal Society of Chemistry.

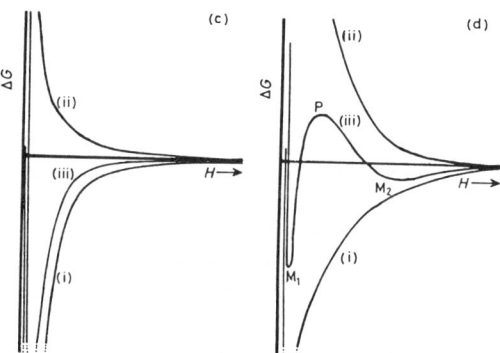

1.38×10^{-23} J /(molecule °K), and T is the absolute temperature. At 300°K two particles collide with an energy of the order 10^{-20}J, but there is a finite probability that a particle at any given moment may have higher or lower energy. The likelihood, however, that the total energy of collision is many times, say ten times larger than kT, is very small. If, therefore, the energy barrier is sufficiently high compared with kT, the colloidal dispersion will remain in a metastable state for an indefinite time; the colloid is said to be stable.

Instability will result if the ratio between the barrier height and kT is reduced to about 1-2, and this can be accomplished in different ways; for instance by raising the temperature. In actual practice it has been found that the barrier height is a sensitive function of several factors such as the composition of the medium, temperature, and pressure. Reduction of the barrier height leading to instability and flocculation is therefore more often brought about by changing one or several of these factors.

Figure 1d points to a situation that is unique to colloidal systems. Aggregation is prevented by a high energy barrier, but preceding this is a comparatively shallow secondary minimum. Small, relatively weakly bound aggregates, flocs, may form if this secondary minimum is of the order of a few kT. Since their lifetime is relatively short, a kinetic equilibrium will be established between single particles and flocs, which is often described as weak flocculation or secondary minimum flocculation. An important characteristic of this phenomenon is that although the flocs cannot be broken by the Brownian motion, they can be dispersed by applying external hydrodynamic forces such as vigorous stirring. Thixotropy, which is an important property of, e. g., paints and lacquers, arises from secondary minimum flocculation.

2. Steric Stabilization - Steric Repulsion

The stability of colloidal systems can also be enhanced by contribution from the term ΔG^{rep}(steric) in Eqn. 6-1. Through practical experience it has been known for a very long time that some natural polymers such as gelatin can stabilize colloidal dispersions. The earliest example is probably the use of natural gums in preparation of ink in China and Egypt 2000-3000 years ago. According to Faraday and later scientists, these natural polymers act by forming a protective layer around each colloidal particle thus preventing them from flocculating.

In recent years systematic studies of the stabilizing action of many synthetic polymers have led to a good understanding of the mechanism of this type of stabilization, which has been named steric stabilization. Adsorbed layers of stabilizing polymers can act either by reducing the attractive van der Waals forces or increasing the repulsive forces between the particles.

3. Electrostatic Stabilization - Electrostatic Forces

Electrostatic stabilization involves methods of modifying the surface of small particles so as to increase the term ΔG^{rep}(electrostatic) in Eqn. 6-1. An electrically charged particle surface plays an important role in stabilizing many colloidal systems and Everett (1988) has described the following mechanisms involved in charging the surface of colloidal particles, which are schematically shown in Fig. 6-2. These are:

i) *Ionization of surface groups.* Surfaces can contain acidic or basic groups whose dissociation brings about negatively charged or positively charged surfaces, respectively—see Fig. 6-2 a and b. The magnitude of the surface charge depends on the acidic or basic strengths of the surface groups and on the solution pH. A point of zero charge, p. z. c., can be attained, i.e., a surface charge of zero, by suppressing the surface ionization by decreasing the pH in the acidic case, or by increasing the pH in the basic case. Varying the pH can cause many metal oxides to show amphoteric behavior so as to create both positively and negatively charged surfaces.

ii) *Differential solution* of ions from the surface of a sparingly soluble crystal. Silver iodide has a solubility product of: $[Ag^+][I^-] = K_S = 10^{-16}$ (mol/dm^3)^2. Thus, if equal amounts of silver and iodide ions dissolved, $[Ag^+] = [I^-] = 10^{-8}$ mol /dm^3, and the surface charge would be zero. However, it is found that Ag$^+$ ions dissolve more than the I$^-$ ions do, thereby leaving a negatively charged surface. On the addition of Ag$^+$ ions as, e.g., from AgNO$_3$, the preferential solution of silver ions is suppressed and the charge becomes zero when $[Ag^+] = 10^{-5.5}$. Further addition of silver ions results in a positively charged surface because it is now iodide ions that are preferentially dissolved. In other words, the surface charge depends on the relative concentrations of Ag$^+$ and I$^-$ ions. Conversely, an equivalent description of the process can be given in terms of preferential adsorptions at the surface.

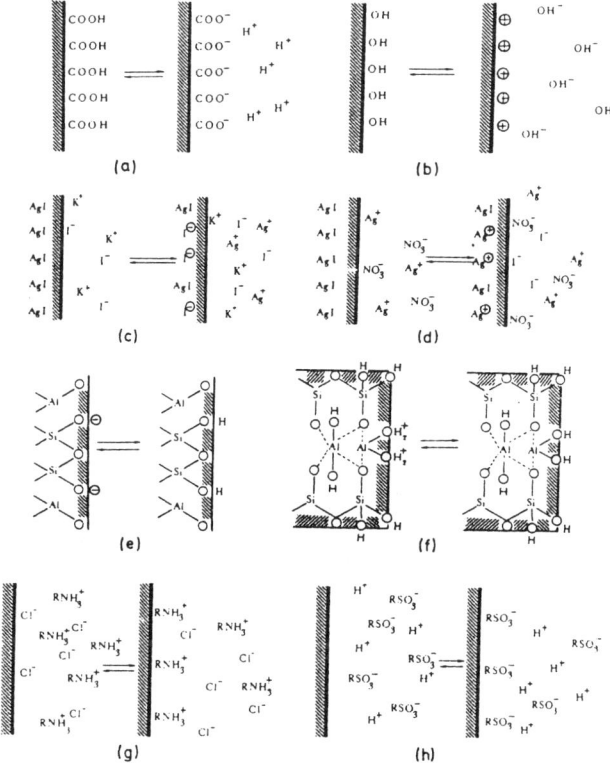

Fig. 6-2. Origin of surface charges by: (a) ionization of acid groups to give a negatively charged surface; (b) ionization of basic groups to give a positively charged surface; (c) differential solution of silver ions from an AgI surface; (d) differential solution of iodide ions from an AgI surface; (e) isomorphous substitutions in a clay surface to give a negatively charged surface; (f) breaking a clay crystal to give a positively charged edge; (g) specific adsorption of a cationic surfactant; (h) specific adsorption of an anionic surfactant. Everett (1988).Courtesy of The Royal Society of Chemistry.

iii) *Isomorphous substitution*. As shown in Chapter 4, a clay may exchange an adsorbed, intercalated, or structural ion with one of lower valency to produce a negatively charged surface. Thus Al could replace Si in the surface, tetrahedral, layer of a clay to produce a negative surface charge as given in Fig. 6-2e. Here the p. z. c. can be attained by reducing the pH because the added H^+ ions combine with the negative charges on the surface to form OH groups.

iv) *Charged crystal surfaces*. When a crystal is broken, it can happen that surfaces with different properties are exposed. For example, if a kaolinite platelet is broken, the exposed edges contain AlOH groups which combine with H^+ ions to yield a positively charged edge—Fig. 6-2f. Here this charged edge may coexist with negatively charged basal surfaces, leading to interesting special properties. In that case there will be no single p. z. c. and each surface will have its own characteristic value.

v) *Specific ion adsorption*. Ions of surface active agents may be specifically adsorbed, leading to positively charged surfaces (cationic surfactants) or negatively charged surfaces (anionic surfactants).—Fig. 6-2h.

3.1. Electrostatic Interaction in Colloidal Systems

Electrostatic interactions between charged particles and between such charged particles and molecules and ions is a central theme of colloidal science. The surface charge of particles, possible charging mechanisms being ionization, ion adsorption or ion dissolution, as described above, influences the distribution of nearby ions in the polar, usually aqueous, medium. Ions of opposite charge (counter-ions) are attracted towards the surface and ions of the same charge (co-ions) are repelled away from the surface. Together with the mixing tendency of thermal motion, this leads to the formation of an electric double layer consisting of the charged surface and a neutralizing excess of counter-ions over co-ions distributed in a diffuse way in the aqueous medium, as shown schematically in Fig. 6-3, Shaw (1970).

The Stern layer is a fairly immobile layer of ions that adhere strongly to the surface of the colloidal particle, which may include water molecules. The second layer, the diffuse layer, or the Gouy layer, extends beyond the Stern layer a distance represented by $1/\kappa$, which is called the Debye length; see below and also Fig. 6-3. The radius of the sphere that envelops the Stern layer is called the radius of shear, and is the major factor determining the mobility of the particles. The electric potential at the surface of shear relative to its value in the distant, bulk medium is called the ζ-potential or the electrokinetic potential; see below.

The primary role of the electric double layer is to confer kinetic stability to colloidal particles. Colliding particles break through the double layer and coalesce only if the collision is sufficiently energetic to disrupt the layers of ions and solvating molecules, usually water molecules, or if thermal motion has swept away the surface accumulation of charge. This may happen at high temperatures, which is why sols may precipitate when they are heated; see above.

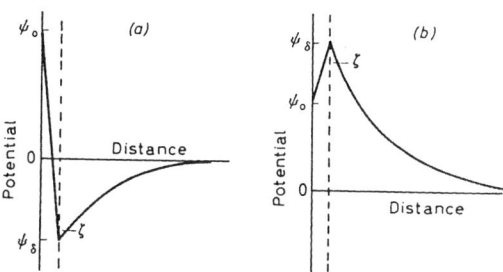

Fig. 6-3.(a) Reversal of charge due to the adsorption of surface-active or polyvalent counter-ions. (b) Adsorption of surface-active co-ions. Shaw (1970). Courtesy Butterworths.

Quantitative treatment of the electric double layer is a very difficult problem. The simplest quantitative treatment of the diffuse part of the double layer of a flat surface is that due to Gouy and Chapman, based on the following assumptions:

a) The surface is flat, of infinite extent, and uniformly charged.

b) The ions in the diffuse part are point charges distributed according to the Boltzmann distribution.

c) The solvent influences the double layer only through its dielectric constant, which is assumed to have the same value throughout the diffuse part.

d) There is only a single electrolyte, which is symmetrical and has charge number z.

The solution to the differential equation obtained by combining the Poisson and Boltzmann equations with the boundary condition $\psi = \psi_0$ when x=0; and $\psi=0$, $d\psi/dt=0$ when x=∞ are taken into account, can be written in the form

$$\psi = \frac{2kT}{ze} \ln \frac{1 + \gamma \exp[-\kappa x]}{1 - \gamma \exp[-\kappa x]} \tag{6-5}$$

Here the Debye screening length $1/\kappa$ emerges naturally and is given by

$$\frac{1}{\kappa} = (\frac{\varepsilon_r \varepsilon_0 kT}{\Sigma e^2 N_A c_i z_i^2})^{1/2} \tag{6-6}$$

where

e = electronic charge

N_A = Avogadro's number

c_i = concentration of ion i in the bulk [mol/m^3]

z_i = valency of ion i

ε_r = relative dielectric constant

ε_0 = permittivity of vacuum [F/m]
k = Boltzmann's constant [J/K]
T = absolute temperature [K]

The coefficient γ relates to the surface potential ψ_0 through

$$\gamma = \frac{\exp[ze\psi_0 / 2kT] - 1}{\exp[ze\psi_0 / 2kT] + 1} = \tanh\left(\frac{ze\psi_0}{4kT}\right) \qquad (6\text{-}7)$$

The Gouy-Chapman theory describes the properties of a charged planar interface by relating the surface charge density σ and the surface potential ψ_0

$$\sigma = (8kTc_0\varepsilon_0\varepsilon_r)^{1/2} \sinh(ze\psi_0 / 2kT) \qquad (6\text{-}8)$$

By linearizing the Poisson-Boltzmann equation, Eqn. 6-5 is reduced to

$$\psi = \psi_0 \exp(-\kappa x) \qquad (6\text{-}9)$$

The Gouy-Chapman theory thus shows that in the diffuse layer the potential decays exponentially with distance with a decay constant given by the Debye length $1/\kappa$; see also Fig. 6-3b.

In the Stern layer, where there are ions specifically adsorbed on the particle surface, the Gouy-Chapman theory is not valid. Specifically adsorbed ions are those which are held so strongly to the surface by electrostatic and/or van der Waals forces that they cannot be thermally agitated. The potential changes faster than by exponential decay from ψ_0, i.e. the surface or wall potential, to ψ_δ, the Stern potential, in the Stern layer, and then to zero in the double layer. The concept of the Stern layer is not uncontroversial among workers in the field, but we use it to bring out the fact that the Gouy-Chapman theory is not valid for that region of the diffuse layer which contains specifically adsorbed ions.

3.2. Stern Potential and Zeta Potential

The spheres that envelop the immobile layers of ions that stick tightly to the surface of the colloidal particles define the surface of shear, which is a region of rapidly changing viscosity and is the major factor determining the mobility of the particles. The precise location of the shear plane, though not known, is reasonably assumed to be situated somewhat further out from the surface than the Stern plane; see Fig. 6-3. The zeta potential, or the electrokinetic potential, ζ, is therefore somewhat lower in absolute value than the Stern potential ψ_σ, but it is common to assume that the two potentials are identical. Compression of the diffuse layer, which happens at high electrolyte concentrations, causes the potential to drop very rapidly behind the shear plane and accentuates differences between ζ and ψ_σ. Also, the adsorption of non-ionic surfactants may result in the plane of shear being located at a relatively large distance from the Stern plane and a zeta potential significantly lower than ψ_σ.

3.3. Debye Length and Ionic Strength

The Debye length, $1/\kappa$, according to Eqn. 6-9 is the distance from the surface at which the potential drops to ψ_0/e and it decreases as the ionic strength of the medium increases according to the following equation, which is Eqn. 6-6 for a 1-1 electrolyte at 25° C with appropriate values of the constants inserted.

$$1/\kappa = 3.05 \times 10^{-10} (cz^2)^{-1/2} \qquad (6\text{-}10)$$

The Debye length is about 1 nm and 10 μm for colloidal solutions containing 10^{-1} and 10^{-3} mol/dm^3 of a 1-1 electrolyte, respectively. For unsymmetrical electrolytes the Debye length can be calculated by taking z to be the counter-ion charge number.

At high ionic strengths the potential falls to its bulk value within a short distance. In this case there is little electrostatic repulsion to hinder the close approach of two colloidal particles and coagulation, or flocculation, readily occurs by the action of van der Waals forces. The term cz^2 in Eqn. 6-10 denotes the ionic strength of the medium and it is increased by the addition of ions, especially those of high charge which act as flocculating agents. This is the basis of the empirical Schulze-Hardy rule, which states that hydrophobic colloids are flocculated most effectively by ions of opposite charge and high charge number.

3.4. Zeta Potential and Surface Charge Density from Electrokinetic Measurements

Since it is difficult to obtain a reliable estimate of a colloidal particle's surface potential and charge density, electrokinetic measurements are instead used to characterize the double layer. However, since electrokinetic phenomena are directly related to the nature of the mobile part of the electric double layer, they can only give information about the zeta potential and the charge density at the surface of shear, but not about ψ_σ and ψ_0 or the charge density at the surface of the particle. As already discussed and also shown in Fig. 6-3, the value of ζ, however, often does not differ very much from that of ψ_σ.

Electrokinetic is the general term applied to four phenomena which result when attempts are made to shear off the mobile part of the electric double layer from a charged surface and which play a role in colloid science. Electrophoresis and sedimentation potential involve the motion of charged particles in a stationary liquid. Electro-osmosis and streaming potential involve the flow of liquid along a stationary charged surface. We will deal only with electrophoresis here.

3.4.1. Zeta Potential from Electrophoresis Experiments In an

electrophoresis experiment an electrical field, E, across the system of small particles causes them to move with a velocity, v, which can be expressed by means of the mobility, u = v/E.

In order to calculate zeta potentials from measurements of particle mobility one needs the relation between u and ζ, which depends on the ratio $R/(1/\kappa) = \kappa R$ between the particle radius, R, and the Debye length. One can derive such a relation for the following three cases.

1. $\kappa R \ll 1$. In this case the particles are treated as point charges in an unperturbed electric field and the following relation, the Hückel equation, can be derived

$$\zeta = \frac{2}{3}\eta u/(\varepsilon_r \varepsilon_0) \qquad (6\text{-}11)$$

It is, however, unlikely that the Hückel equation would be very useful for calculating the zeta potential of small particles in aqueous systems. Particles of radius 10 nm dispersed in water would, for instance, require an electrolyte concentration as low as 10^{-5} mol/ dm^3 to give $\kappa R = 0.01$.

2. $\kappa R \gg 1$. The particles are in this case large and in the derivation of the following relation, the Helmholtz-Smoluchowski equation, are treated as flat surfaces.

$$\zeta = \frac{\eta u}{\varepsilon_r \varepsilon_0} \qquad (6\text{-}12)$$

3. The Hückel equation (Eqn. 6-11) and the Helmholtz-Smoluchowski equation (Eqn. 6-12) differ by a factor of 1.5. They are valid in the two limits $\kappa R \ll 1$ and $\kappa R \gg 1$, respectively, which, however, seldom are strictly applicable to colloidal systems. Instead, in the range $1 < \kappa R < 100$, zeta potentials of colloidal particles can be more accurately calculated by using the Henry equation

$$\zeta = \frac{3}{2}\frac{\eta u}{\varepsilon_r \varepsilon_0} f(\kappa R) \qquad (6\text{-}13)$$

$$\text{where } f(\kappa R) = 1.5 - \frac{9}{2\kappa R} + \frac{75}{2\kappa^2 R^2} - \frac{330}{\kappa^3 R^3}$$

Törncrona and Otterstedt (1990) used the Henry and Helmholtz-Smoluchowski equations and a computer program called MacMobility developed in the Mathematics Department of the University of Melbourne, to calculate the zeta potential of aqueous silica sols from mobility measurements.

Figure 6-4 shows how the zeta potentials, calculated by the three different methods, of a 200-nm silica sol at pH 5 (a) and a 500-nm silica sol at pH 4 (b)

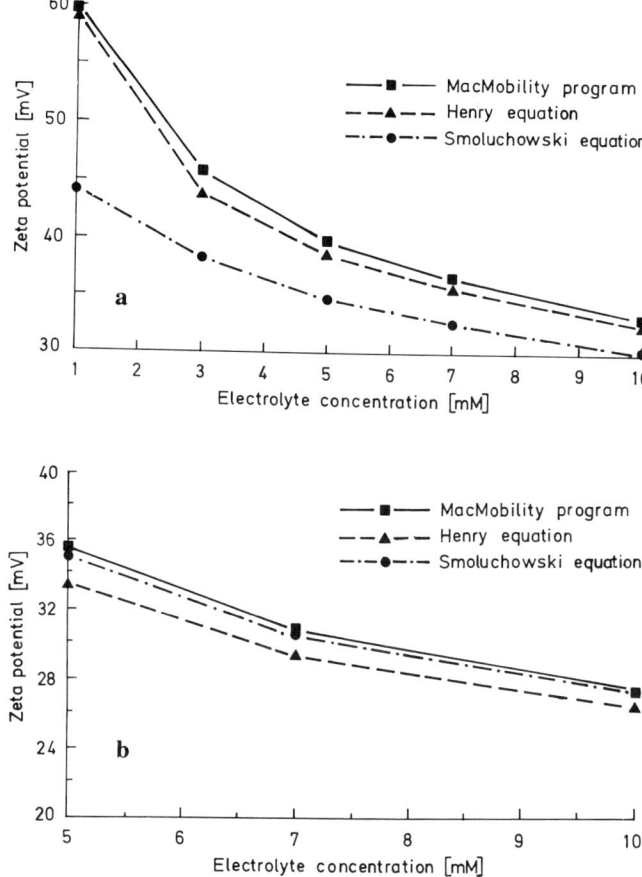

Fig. 6-4. Comparison of ζ-potential vs. electrolyte concentration calculated by three different methods for (a) 200-nm silica sol at pH 5; and (b) 500-nm silica sol at pH 4. Törncrona and Otterstedt (1990).

vary with electrolyte concentrations.

Compared with the MacMobility program, both the Henry and Helmholtz-Smoluchowski equations underestimate the zeta potential, but the latter equation much more so than the Henry equation. The Helmholtz-Smoluchowski equation works much better for the 500-nm sol than for the 200-nm sol, suggesting that the larger particles come closer to meeting the limit of this equation, i.e., being flat surfaces.

3.4.2. Surface Charge Density from Electrophoresis Experiments.

The surface charge density can be calculated from the surface potential ψ_0, or the Stern potential, ψ_σ; see Fig. 6-3. Approximating these potentials by the ζ potential, and assuming that the electric double layer is small relative to the particle diameter, i.e., that the particles are treated as flat surfaces, the surface charge can be calculated from the equation

$$\sigma_0 = (8n_0\varepsilon_r\varepsilon_0 kT)^{1/2}\sinh(\frac{ze\psi_0}{2kT})\ [c/m^2] \qquad (6\text{-}14)$$

Taking into account the curvature of the particles yields the more exact equation:

$$\sigma_0 = -\varepsilon_r\varepsilon_0 kT\kappa 2\sinh(\frac{ze\psi_0}{2kT}) + \frac{4zek\tanh(\frac{ze\psi_0}{2kT})}{R}\ [c/m^2] \qquad (6\text{-}15)$$

where R is the particle radius.

Using the MacMobility program to obtain the zeta potential, Törncrona and Otterstedt (1990) calculated the surface charge density of 200-nm and 500-nm silica sols at different pH's and at electrolyte concentrations of 3 and 10 mM from Eqns. 6-14 and 6-15. Table 1 shows that the surface charge densities calculated from both equations, as expected, increase with pH and electrolyte concentration, and that the increase with electrolyte concentration is more pronounced at higher pH.

The agreement between Eqn. 6-14 (upper values) and Eqn. 6-15 (lower values) is good for both sols, but somewhat better for the larger particles, again suggesting that these particles are quite closely approximated by a flat surface.

Table 6-1. Surface charge densities of (a) 200-nm and (b) 500-nm silica sol particles calculated from Eqn. 6-14 (upper values) and Eqn. 6-15 (lower values) as a function of pH and electrolyte concentration. Törncrona and Otterstedt (1990)

	(a) 200-nm SiO_2 sol		(b) 500-nm SiO_2 sol	
	Electrolyte concentration, mM		Electrolyte concentration, mM	
pH	3	10	5	10
4	4.7000	5.052	6.215	6.585
4	4.934	5.199	6.310	6.660
5	6.521	8.108	10.367	8.236
5	6.820	8.330	10.504	8.326
6	10.015	12.704	16.026	17.833
6	10.406	13.016	16.200	17.987
7	13.817	17.237	21.984	20.275
7	14.273	17.615	22.182	20.440
8	17.057	21.042	23.603	21.425
8	17.551	21.462	23.805	21.594
9	23.308	40.861	43.874	40.285
9	23.852	41.398	44.111	40.499
10	38.151	77.660	71.579	50.585
10	38.755	78.274	71.834	50.812

B. The Surface of Small Particles in Aqueous Solutions

The nature of the surface of hydrous metal oxides depends on the interaction of the surface with various ionic species such as H^+, OH^-, metal ions, ligand anions, and metal-ligand complexes. Adsorption of protons may bring about dissolution of metal oxides species adsorbed on the surface of particles of hydrous metal oxides and even cause the particles to dissolve. Adsorption of charged species will change the surface charge and zeta potential of the particles, and may even cause charge reversal if they are present in sufficient amounts. Interaction of the particle surface with various charges will almost always affect the stability of the particles. Knowledge of the thermodynamics and reaction kinetics of the surface by hydrous metal oxides will allow and facilitate the preparation of stable sols of such materials. To study the properties of the surface of small particles of hydrous metal oxides, the most important methods are acid-base titration, electrophoresis, and the adsorption of cations and anions onto the particle surface.

1. Coordination Chemistry at the Hydrous Metal Oxide-Water Interface

In aqueous solutions the surface of hydrous metal oxides is normally covered with surface hydroxyl groups, S-OH, which are ampholytes. Adsorption of H^+ or OH^- ions thus depends on whether these groups act as Brönsted acids, Eqn. 6-16, or Brönsted bases, Eqn. 6-17:

$$S\text{-}OH + (OH^-) = S\text{-}O^- + H^+ (+ H_2O) \qquad (6\text{-}16)$$

$$S\text{-}OH + H^+ = S\text{-}OH_2^+ \qquad (6\text{-}17)$$

Deprotonated surface hydroxyl groups may act as Lewis acids, and adsorption of metal ions may therefore involve one or two groups per metal ion:

$$S\text{-}OH + M^{z+} = S\text{-}OM^{(z-1)+} + H^+ \qquad (6\text{-}18)$$

$$2S\text{-}OH + M^{z+} = (S\text{-}O)_2 M^{(z-2)+} + 2H^+ \qquad (6\text{-}19)$$

Since adsorbed metal ions can coordinate several ligands (octahedral coordination is the most common configuration for transition metal ions in aqueous solution), such ligands can be bound to the surface.

$$S\text{-}OH + M^{z+} + 2L = S\text{-}OML_2^{(z-1)+} + H^+ \qquad (6\text{-}20)$$

Such metal complexes are called ternary surface complexes of type A.

Anions can be adsorbed on the hydrous metal oxide surface by so-called ligand exchange:

$$S\text{-}OH + L = S\text{-}L^+ + OH^- \tag{6-21}$$

$$2S\text{-}OH + L = S_2L^{2+} + 2OH^- \tag{6-22}$$

Ligands can also adsorb to form a surface complex of type B:

$$S\text{-}OH + L + M^{z+} = S\text{-}LM^{(z+1)+} + OH^- \tag{6-23}$$

Figure 6-5 shows five different ways that H_2O and OH^- can coordinate to one or two surface metal ions of a hydrous metal oxide surface in aqueous solution.

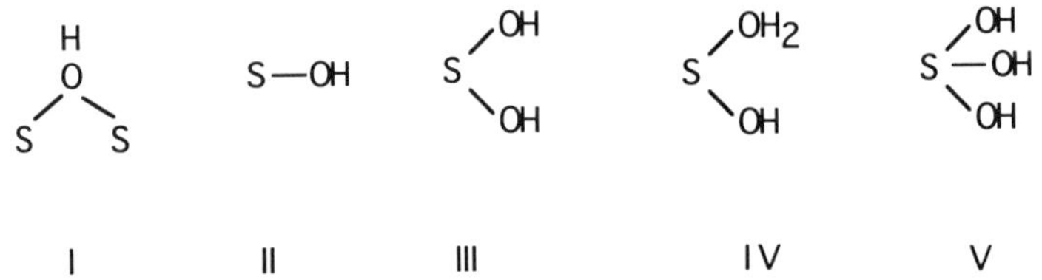

Fig. 6-5. Various ways of coordinating OH^- and H_2O to surface metal ions. Schindler and Stumm (1987). Reprinted by permission of John Wiley & Sons, Inc.

The type of coordination depends primarily on the coordination number of the surface metal ion and Table 6-2 shows that coordination type V has not been observed for OH^- on surface metal ions with coordination numbers 4 or 6. The acid strength of OH^- adsorbed on the surface depends on the type of coordination

Table 6-2. Some Coordinative Environments of Metal Ions in Hydrated Surfaces of (Hydr)oxides, Schindler and Stumm (1987). Reprinted by permission of John Wiley & Sons, Inc.

Coordination Unit [a]	Stoichiometry	Pertinent Surface Group
(a) S^{3+}, Coordination number 6		
$SO_{4/4}(OH)_{2/2}$	$S_2O_3 \cdot H_2O$	I
$SO_{2/4}(OH)_{4/2}$	$S_2O_3 \cdot 2H_2O$	I
$S(OH)_{6/2}$	$S_2O_3 \cdot 3H_2O$	I
$SO_{2/4}(OH)_{2/2}(OH)(OH_2)$	$S_2O_3 \cdot 4H_2O$	IV
$S(OH)_{4/2}(OH)(OH_2)$	$S(OH)_3 \cdot H_2O$	IV
(b) S^{4+}, Coordination Number 4		
$SO_{3/2}OH$	$SO_2 \cdot \frac{1}{2}H_2O$	II
$SO_{2/2}(OH)_2$	$SO_2 \cdot H_2O$	III
$SO_{1/2}(OH)_3$	$SO_2 \cdot \frac{3}{2}H_2O$	IV

a The suffixes indicate structurally the nearest neighboring atoms, i.e. in $S(OH)_{6/2}$ every metal ion S^{3+} has 6 neighboring OH groups and every OH group has 2 neighboring S^{3+} metal ions and quantum chemical calculations show that hydroxyl groups in bridge coordination, type 1 in Fig. 6-5, are more acidic than those in coordination represented by type II, Schindler and Stumm (1987).

It is likely that a given surface contains hydroxyl groups coordinated in several different configurations. The way H_2O and OH^- are coordinated to a surface may well depend on the method of preparation of small particles of metal oxides and their surface properties may therefore depend on how they were made.

The ampholytic behavior of the surface of hydrous metal oxides can be described by the following two acid constants:

$$K_{a1}^S = \{SOH_2^+\}/\{SOH\}[H^+] \quad [dm^3/mol] \tag{6-24}$$

$$K_{a2}^S = \{SO^-\}[H^+]/\{SOH\} \quad [mol/dm^3] \tag{6-25}$$

where { } denotes the concentration of surface species (-OH, -O⁻, and $-OH_2^+$) in moles per kg adsorbent.

For adsorption of metal ions on the surface the following equilibrium constants can be derived:
(from Eqn. 6-18)

$$*K_1^S = \frac{\{SOM^{(z-1)+}\}[H^+]}{\{S-OH\}[M^{z+}]} \tag{6-26}$$

(from Eqn. 6-19)

$$*\beta_2^S = \frac{\{(SO)_2M^{(z-2)+}\}[H^+]^2}{\{SOH\}^2[M^{z+}]} \quad [kg/dm^3] \tag{6-27}$$

Expressed as mol/dm^3, the concentration of surface species is:

$$[SOH] = \frac{A}{V}\{SOH\} \tag{6-28}$$

where A = kg adsorbent and V = volume of water phase (dm^3).

The overall equilibrium constant for adsorption of metal ions on the surface of a hydrous metal oxide, K_{ads}^S, can be written as a product of two factors

$$K_{ads}^S = K_{int}^S \cdot K_{coul}^S \tag{6-29}$$

where K_{int} refers to the chemical reaction between the metal cation and the reacting group on the surface, and K_{coul} accounts for the electrostatic interaction between the metal cation and the surface. The free energy of adsorption can be obtained from the equilibrium constant by the relation:

$$\Delta G_{ads} = -RT \cdot \ln K^S_{ads} \tag{6-30}$$

Equations 6-29 and 6-30 show that ΔG_{ads} can be written as a sum of two terms:

$$\Delta G_{ads} = \Delta G_{int} + \Delta G_{coul} \tag{6-31}$$

$$\text{where } \Delta G_{int} = -RT \cdot \ln K^S_{int} \tag{6-32}$$

$$\text{and } \Delta G_{coul} = zF\psi_0 = -RT \ln K^S_{coul} \tag{6-33}$$

ΔG_{coul} represents the energy required to bring a metal ion of valency z from the bulk of the solution to the particle surface whose potential is ψ_0.

Equations 6-30 to 6-33 show that the overall equilibrium constant is:

$$\ln K^S_{ads} = \ln K^S_{int} - \frac{zF}{RT}\psi_0 \tag{6-34}$$

Since ψ_0 cannot be experimentally determined, models have to be used to describe how the potential varies with the distance from the surface so as to be able to calculate the surface potential from experimentally available data. The three models, I-III, in Fig. 6-6 show the potential as different functions of distance from the surface. The terms ΔG_{int} and ΔG_{coul} in Eqn. 6-31 are different

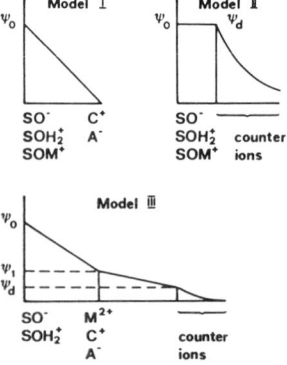

Fig. 6-6.Schematic presentation of the surface-solution interface for the case of an oxide suspended in a solution containing the divalent metal ion M^{2+} and the supporting electrolyte C^+A^-. For each model the potential is shown as a function of distance from the surface; the assumed positions of the adsorbed ions are indicated. Schindler and Stumm (1987). Reprinted by permission of John Wiley & Sons, Inc.

Method/Model	log K_{a1}'	log K_{a2}'	Reference
Extrapolation b	7.2	-9.5	Hohl and Stumm (1976)
Model I[b]	7.40	-9.24	Westall and Hohl (1980)
Model II	7.66	-8.98	Westall and Hohl (1980)
Model III	7.33	-9.31	Westall and Hohl (1980)

Table 6-3. Acidity Constants of Surface Al—OH Groups Computed with Different Double-Layer Models[a], Schindler and Stumm (1987). Reprinted by permission of John Wiley & Sons, Inc.

[a] Experimental data from Hohl and Stumm (1976).

[b]The differences originate from the fact that α values observed in linear regressions to obtain log K_{a1}' and log K_{a2}' are usually not identical, whereas fittings on the basis of model I assume the same value of κ (and thus α) for the entire pH range.

2. Acid-Base Characteristics of the Hydrous Metal Oxide Surface

The acid constants K_{a1}^{s} and K_{a2}^{s} in Equations 24 and 25 describe the ampholytic nature of the surface of hydrous metal oxides and refer to the following two equations, respectively:

$$S\text{-}OH_2^+ = S\text{-}OH + H^+ \tag{6-35}$$

for which the equilibrium constant is $(K_{a1}^{s})^{-1}$ and

$$S\text{-}OH = S\text{-}O^- + H^+ \tag{6-36}$$

for which the equilibrium constant is K_{a2}^{s}.

Attempts have been made to compare the acid constant of a certain metal surface with that of a surface hydroxyl group of the corresponding hydrous metal oxide, but, not unexpectedly, the constants turned out to be different. Thus, monosilicic acid, $Si(OH)_4$, has a pK_a of about 10, whereas the silanol groups on the surface of amorphous silica particles have a pK_a of about 6.7; see Ch. 2. The reasons for the difference are that there are more than one type of hydroxyl group on the surface of hydrous metal oxides—see Fig. 6-5—and that cations and anions can adsorb or form ion pairs on the surface, which change its acid-base characteristics. James and Parks (1987) used a model, which takes into account the actual proton transfer reaction and separates it from other phenomena such as ion pair formation, to calculate acid constants of hydrous metal oxides that did not differ from those of the corresponding metal hydroxides by more than an order of magnitude.

Table 6-4 shows acid constants of surface hydroxyl groups on various metal oxides with K_{a1}^s and K_{a2}^s denoted as $K_{a1(int)}^s$ and $K_{a2(int)}^s$, respectively, in the table. Note that the variation of the acid constant of TiO_2 (rutile) with LiCl concentration, which most likely is due to ion pair formation according to the following reactions:

$$TiOH_2^+ + Cl^- = TiOH_2^+Cl^- \tag{6-37}$$

$$TiO^- + Li^+ = TiO^-Li^+ \tag{6-38}$$

Table 6-4. Acidity Constants of Surface Hydroxyl Groups (298.2 °K) Adapted from James and Parks (1987). Courtesy Plenum Press.

Group	Solid	Ionic Medium	log $K_{a1(int)}^s$	log $K_{a2(int)}^s$
Al—OH	γ-Al_2O_3	0.1 M $NaClO_4$	7.2	-9.5
	d-Al_2O_3[a]	0.1 M $NaClO_4$	7.4	-10.0
Al(OH)(OH$_2$)	γ-Al(OH)$_3$	0.1 M $NaClO_4$	5.24[a]	-8.08[b]
Si—OH	SiO_2(am)	0.1 M $NaClO_4$		-6.8
		0.2 M KNO_3		-6.53
		1.0 M $LiClO_4$		-6.57
		1.0 M $NaClO_4$		-6.71
		1.0 M CsCl		-5.71
Ti—OH	TiO_2			
	Anatase	3.0 M $NaClO_4$	4.98	-7.80
	Rutile	1.0 M $NaClO_4$	4.13	-7.39
		10^{-3} M LiCl	2.75[a]	-9.1[b]
		10^{-2} M LiCl	3.25	-8.9
		0.1 M LiCl	3.6	-8.4
Zr—OH	ZrO_2	1.0 M KNO_3	5.67[a]	-7.91
Th—OH	ThO_2	1.0 M $NaClO_4$	5.15[a]	-7.90[b]
Fe—OH	Fe(OH)$_3$(am)	I = 0.1	6.6	-9.1
	α-FeOOH	0.1 M $NaClO_4$	6.4[a]	-9.25[b]

[a] Material similar to that classified as δ-Al_2O_3

[b] See original reference

3. Adsorption of Metal Ions on the Hydrous Metal Oxide Surface

Adsorption of metal ions according to equations 6-18 and 6-19, with equilibrium constants $^*K_1^s$ and $^*\beta_2^s$, respectively, according to Equations 6-26 and 6-27, depends strongly on pH. The adsorption of various metal ions on TiO_2 (rutile), Fig. 6-7, and on SiO_2, Fig. 6-8, increases from less than 10% to 100% within 1-2 pH units.

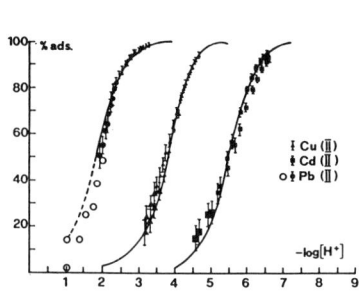

Fig. 6-7. Adsorption of some divalent metal ions on TiO_2 (rutile), Fürst, 1976. For each metal ion there is an interval of 1-2 pH units where the ex-tent of adsorption rises from zero to almost 100%. Schindler et al. (1976). Courtesy Academic Press.

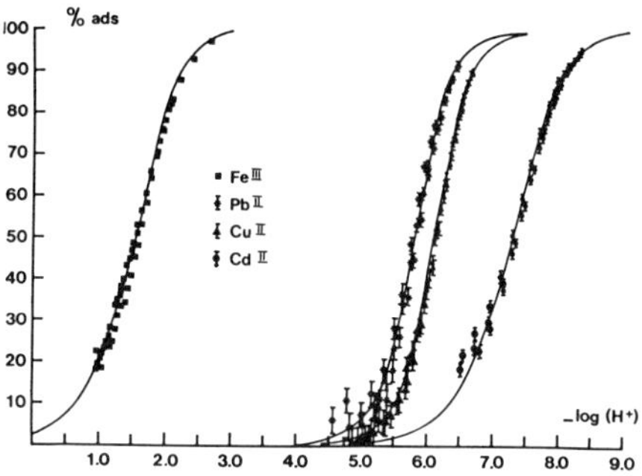

Fig. 6-8. Adsorption of Fe(III), Cu(II), Cd(II), and Pb(II) on hydroxylated silica as a function of log [H^+]. Schindler et al. (1976). Courtesy Academic Press.

Stability constants of different metal complexes on the surface of γ-Al_2O_3, TiO_2, δ-MnO_2 and Fe_2O_3 are shown in Table 6-5.

Group	Solid	M^{z+}	Ionic Medium	$\log {}^*K^S_{1(int)}$	$\log {}^*\beta^S_{2(int)}$
Al—OH	γ-Al_2O_3	Ca^{2+}	0.1 M $NaNO_3$	-6.1	—
		Mg^{2+}	0.1 M $NaNO_3$	-5.4	—
		Ba^{2+}	0.1 M $NaNO_3$	-6.6	—
		Pb^{2+}	0.1 M $NaClO_4$	-2.2	-8.1
		Cu^{2+}	0.1 M $NaClO_4$	-2.1	-7.0
Si—OH	SiO_2(am)	Mg^{2+}	1 M $NaClO_4$	-7.7	-17.5
		Fe^{3+}	3 M $NaClO_4$	-1.77	-4.22
		Cu^{2+}	1 M $NaClO_4$	-5.52	-11.19
		Cd^{2+}	1 M $NaClO_4$	-6.09	-14.20
		Pb^{2+}	1 M $NaClO_4$	-5.09	-10.68
Ti—OH	TiO_2 (rutile)	Mg^{2+}	1 M $NaClO_4$	-5.90	-13.13
		Co^{2+}	1 M $NaClO_4$	-4.30	-10.60
		Cu^{2+}	1 M $NaClO_4$	-1.43	-5.04
		Cd^{2+}	1 M $NaClO_4$	-3.32	-9.00
		Pb^{2+}	1 M $NaClO_4$	0.44	-1.95
Mn—OH	δ-MnO_2	Ca^{2+}	0.1 M $NaClO_4$	-5.5	—
Fe—OH	Fe_3O_4	Co^{2+}	I = 0	-2.44	-6.71

Table 6-5. Stability Constants of Metal Complexes (298.2°K) Adapted from Schindler and Stumm (1987). Reprinted by permission of John Wiley & Sons, Inc.

The adsorption of metal ions on the surface of hydrous metal oxide particles can be described by two different mechanisms, According to the first mechanism, the metal ion is hydrolyzed in the solution before it is adsorbed.

$$M(H_2O)_6^{z+} + H_2O = M(H_2O)_5(OH)^{(z-1)+} + H_3O^+ \qquad (6\text{-}39)$$

$$M(H_2O)_5(OH)^{(z-1)+} + SOH = SOM(H_2O)_5^{(z-1)+} + H_2O \qquad (6\text{-}40)$$

In the second mechanism the hydrated metal ion is first adsorbed on the surface before it undergoes hydrolysis.

$$M(H_2O)_6^{z+} + SOH = SOM(H_2O)_5^{(z-1)+} + H_3O^+ \qquad (6\text{-}41)$$

In general it is not possible to disentangle the two mechanisms by using stoichiometric or thermodynamic considerations. Kinetic data on the adsorption of metal ions on γ-Al$_2$O$_3$, however, support the second mechanism, Eqn. 6-42; that is, the adsorption of metal ions is associated with protons leaving the surface. In most studies of the adsorption as a function of pH, however, the metal ion concentration was so low—in the range 0.01 to 1 μM, that only mononuclear complexes could form; see Ch. 3.

At higher concentrations a significant fraction of the metal ions would be present as polycations, which would adsorb strongly and in larger numbers than mononuclear complexes on a negatively charged surface. By very carefully raising pH, the metal ions can be made to precipitate onto the particle surface and thus increase the thickness of the adsorbed layer. This is a technique that can be used to change the nature of a particle surface by coating it with a uniform layer of a suitable adsorbate; see Section C of this chapter for more details.

4. Adsorption of Anions on the Hydrous Metal Oxide Surface

As in the case of cations, the adsorption of anions also depends strongly on pH and is favored by low pH, as is illustrated for the system of F$^-$ on γ-Al(OH)$_3$ (bayerite) in Fig. 6-9.

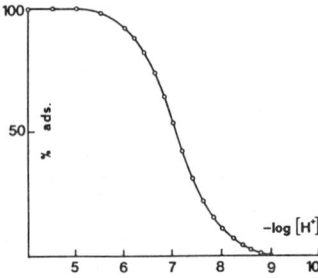

Fig. 6-9. Adsorption of F$^-$ on γ-Al(OH)$_3$(calculated from data given by Pulfer et al. (1984).Since adsorption of anions is coupled with a release of OH$^-$ ions, the extent of adsorption increases with decreasing pH. Schindler and Stumm (1987). Reprinted by permission of John Wiley & Sons, Inc.

Stability constants for the adsorption of various anions on α-FeOOH and γ-Al$_2$O$_3$ according to Eqn. 6-42 are shown in Table 6-6.

$$SOH + H_2A = SHA + H_2O \qquad (6\text{-}42)$$

Table 6-6. Intrinsic Stability Constants of Anion Complexes (295°K, ionic stregth not specified). Schindler and Stumm (1987). Reprinted by permission of John Wiley & Sons, Inc.

Equilibrium		$K^S_{(int)}$
1. α-FeOOH		
>FeOH + F$^-$	= >FeF + OH$^-$	-4.8
>FeOH + SO$_4^{2-}$	= >FeSO$_4^-$ + OH$^-$	-5.8
>FeOH + SO$_4^{2-}$	= (>Fe)$_2$SO$_4$ + 2OH$^-$	-13.5
>FeOH + HAc	= >FeAc + H$_2$O	2.9
>FeOH + H$_4$SiO$_4$	= >FeSiO$_4$H$_3$ + H$_2$O	4.1
	= >FeSiO$_4$H$_2$ + H$_3$O$^+$	-3.3
>FeOH + H$_3$PO$_4$	= >FePO$_4$H$_2$ + H$_2$O	9.5
>FeOH + H$_3$PO$_4$	= >FePO$_4$H + H$_3$O$^+$	5.1
>FeOH + H$_3$PO$_4$ + H$_2$O	= >FePO$_4^{2-}$ + 2H$_3$O$^+$	-1.5
2>FeOH + H$_3$PO$_4$	= (>Fe)$_2$PO$_4$H + 2H$_2$O	8.5
2>FeOH + H$_3$PO$_4$	= (>Fe)$_2$PO$_4$ + H$_2$O + H$_3$O$^+$	4.5
2. γ-Al$_2$O$_3$		
Benzoic acid		
	>AlOH + HA = >AlA + H$_2$O	3.7
Catechol		
	>AlOH + H$_2$A = >AlAH + H$_2$O	3.7
	>AlOH + H$_2$A = >AlA$^-$ + H$_3$O$^+$	-5
Phthalic acid		
	>AlOH + H$_2$A = >AlAH + H$_2$O	7.3
	>AlOH + H$_2$A = >AlA$^-$ + H$_3$O$^+$	2.4
Salicylic acid		
	>AlOH + H$_2$A = >AlAH + H$_2$O	6.0
	>AlOH + H$_2$A = >AlA$^-$ + H$_3$O$^+$	-0.6

5. Point of Zero Charge, p. z. c., and Isoelectric Point, i. e. p.

The surface of small particles of hydrous metal oxides, and for that matter, the surface of colloidal particles in general, can be characterized by specifying the point of zero charge or the isoelectric point of the particles. In aqueous systems, the p. z. c. is the pH where the surface charge is zero, i.e. where the surface potential, ψ_0, is zero, and the i. e. p. where the electrical mobility of the particles is zero, i.e. where the zeta potential is zero. The p. z. c. and the i. e. p. of colloidal systems, are, in general, not equal; compare Fig. 6-3. The point of zero charge is the more fundamental double layer property, but cannot be

Table 6-7. IEPS and Oxidation State, Parks (1965). Courtesy of the American Chemical Society.	
M_2O	$IEPS^* > pH\ 11.5$
MO	$8.5<IEPS<12.5$
M_2O_3	$6.5<IEPS<10.4$
MO_2	$0<IEPS<7.5$
M_2O_5, MO_3	$IEPS<0.5$

* IEPS refers to IEP of solids

determined experimentally. Instead, the isoelectric point is used to study and characterize the flocculation and aggregation behavior of colloidal systems. It can be determined by measuring the electric mobility as a function of pH when small monovalent cations are adsorbed on the particles.

Of the many factors that affect the p. z. c. and i. e. p. of a certain hydrous metal oxide, the oxidation number of the metal ion has the greatest influence and Table 6-7 shows typical values of the p. z. c. of oxides, in which the oxidation number of the metal ion varies between +1 and +6. Oxides of metals of low oxidation number, e.g. alkali metal oxides (M_2O), have an i. e. p. above 11.5. The higher the oxidation number, the lower will be the p. z. c. of the hydrous metal oxide. Metal oxides with high oxidation numbers, e.g. U_2O_5, have very low p. z. c.'s.

5.1. Factors Affecting i. e. p. and p. z. c.

5.1.1. Effect of Dehydroxylation.
Dehydroxylation of particles of hydrous metal oxides, which often is accomplished by calcination and subsequent redispersion of particles, usually leads to a decrease of the p. z. c. Table 6-8 shows that dehydroxylation of hydrous oxides of iron, aluminum, and titanium will lower the i. e. p. by 1.5 to 2.5 pH units. On a dehydroxylated metal oxide surface there are metal ions of low oxidation number which behave as Lewis acid sites. When such a surface is redispersed in water, molecules of water will immediately coordinate to these acid sites by dissociative chemisorption and hydroxyl groups will thus be reintroduced on the metal oxide surface as illustrated in Fig. 6-10. The difference in p. z. c. between original and rehydroxylated metal oxide surfaces may be due to the fact that there are different types of hydroxyl groups on the two surfaces; see Fig. 6-5.

Table 6-8. Variation of IEPS with Hydration, Parks (1965). Courtesy of the American Chemical Society.				
Compound	IEPS	Compound	IEPS	ΔIEPS
Fe_2O_3	6.7	Fe_2O_3 (hydrous)	8.6	1.9
Al_2O_3	6.7	Al_2O_3 (hydrous)	9.2	2.5
TiO_2	4.7	TiO_2	6.2	1.5

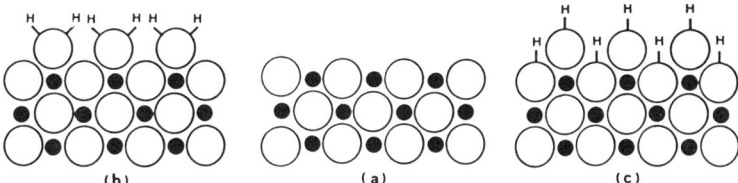

Fig. 6-10. Cross section of the surface layer of a metal oxide. (●) metal ions; (O) oxide ions. (a) Surface ions show low coordination and exhibit Lewis acidity. (b) In the presence of water, the surface metal ions may coordinate H_2O molecules. (c) Dissociative chemisorption leads to a hydroxylated surface. Stumm et al. (1980).

5.1.2. Effect of Contaminants. Ideally, determination of the p. z. c. and the i. e. p. of the surface of a hydrous metal oxide should be carried out under such conditions that protons and hydroxyl ions are the only ions present in the aqueous phase and the hydrous metal oxide itself should be free of impurities, i.e., contain only the metal, hydrogen and oxygen. In actual practice, it is difficult to meet these conditions because the i. e. p. is often obtained by adsorbing small cations and anions on the particles, accomplished by titrating the colloidal system with an electrolyte, and measuring the mobility as a function of pH.

Adsorption of anions and cations changes the i. e. p. and the p. z. c. of the surface of particles of hydrous metal oxides, and indeed of all types of colloidal particles. Anions decrease the i. e. p. because more protons, i.e., more acid, is required to neutralize the negative charge of the anions adsorbed on the surface. Di- and trivalent anions like sulfate ions and phosphate ions lower the i. e. p. much more than monovalent ions like chloride and nitrate ions because they can coordinate directly to the hydrous metal oxide surface and to some extent replace the hydroxyl ions on the surface. Compared with monovalent anions, protons are needed to neutralize these hydroxyl ions and the anions coordinated to the surface and part of its structure. It is possible to distinguish between simply adsorbed anions and anions incorporated into the surface structure by measuring the i. e. p. before and after careful washing of the particles, which can be accomplished by centrifuging the dispersion repeatedly. If the i. e. p. increases after washing, the anions were simply adsorbed, whereas if it is unaffected, they had become part of the surface structure.

Adsorbed metal cations cause the i. e. p. to shift toward the i. e. p. of the hydrous oxide of the metal making up the cation. Usually, this means that the i. e. p. will shift toward higher pH since cationic contaminants often consist of metal cations of oxidation number +1 or +2, the oxides of which have isoelectric points at high pH; see Table 6-7. Simply adsorbed, but not structurally incorporated cations, can also be removed by washing the colloidal particles.

Vordonis et al (1987) found that doping of γ-Al_2O_3 with very small amounts of Li^+ and Na^+ increased the i. e. p. of the hydrous aluminum oxide from 5.30 to 9.80, whereas doping with minute quantities of F^- decreased it to 3.40. They also found that heating a dispersion of γ-alumina particles increased their i. e. p., whereas cooling had the opposite effect.

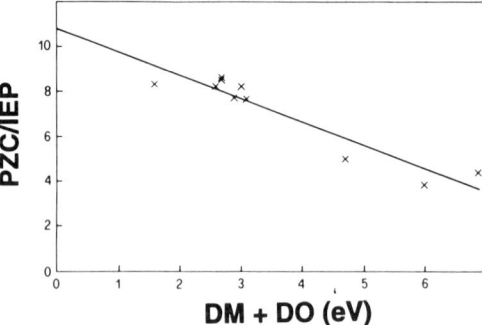

Fig. 6-11. Experimental values of P. Z. C./I. E. P. vs (DM + DO) values. Cattania et al. (1993). Reprinted from Colloids and Surfaces 48, 65 (1990) with kind permission from Elsevier Science - NL, Sara Burgerhartstraat 25, 1055 KV Amsterdam, The Netherlands.

Recently, attempts have been made to correlate x-ray photoelectron spectroscopy, XPS, chemical shifts with the p. z. c. and i. e. p. of several metal oxides, Cattania et al. (1993). The difference, DM, between the XPS binding energy, BE, of an electron ejected from a particular level in a metal cation and the XPS binding energy of the same electron in the metal can be considered a measure of the acidity of the cation. Thus, DM = (cation BE - metal BE) ev. Since the most basic oxides exhibit an XPS 1s-binding energy of about 530 ev, the basicity of the surface oxygens of an oxide can be described by the relation $DO=(O1s \cdot BE-530.0)$. Cattania et al. considered that the overall acid-base behavior of a metal oxide surface could be estimated from the sum of the metal acidity and the oxygen basicity, i.e., DO + DM.

Figure 6-11 shows that there appears to be a linear correlation between experimentally obtained i. e. p./p. z. c. values and DM + DO. If this method gives relevant information about double-layer properties—it must be kept in mind that the XPS measurements were performed on dry materials under high vacuum—it may provide a means of determining the p. z. c., an elusive property of the double-layer concept.

5.2. Electrophoretic Mobility as a Function of pH

The electrophoretic mobilities of three CeO_2-sols made by hydrolysis of $Ce(SO_4)_2$ in dilute H_2SO_4 as a function of pH are shown in Fig. 6-12.

Sol A was washed with distilled water only and sol B with dilute NaOH. Sol C was obtained by separating the particles of sol B by centrifugation, calcining them at 600°C for 2 h, and redispersing them in water. The i. e. p. of sols A, B, and C were 5.9, 6.1, and 5.2, respectively. The mobilities of sol A at low pH are much lower than for the other two sols, whereas at pH > i. e. p., the values for the three systems are in reasonable agreement. The leaching of sulfate ions from

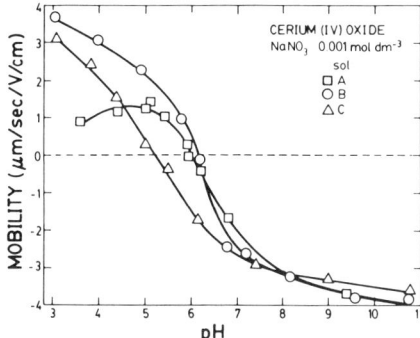

Fig. 6-12. Electrophoretic mobilities of CeO_2 sols as a function of pH in 1.0×10^{-4} mol dm^{-3} NaNO$_3$. Sols: A ($\square$), B (O), C ($\triangle$). Hsu et al. (1988). Courtesy American Chemical Society.

the particles in sol A by acid is most likely responsible for the decrease in the mobility as pH is lowered.

The shift in the i. e. p. of sol C illustrates that calcination can change the OH/O^{2-} ratio on the surface, which determines the acidity of the hydroxyl groups there.

Figure 6-13 shows the mobility as a function of pH of dispersions of rod-like particles of β-FeOOH and of particles of α-Fe$_2$O$_3$ of different shapes, made

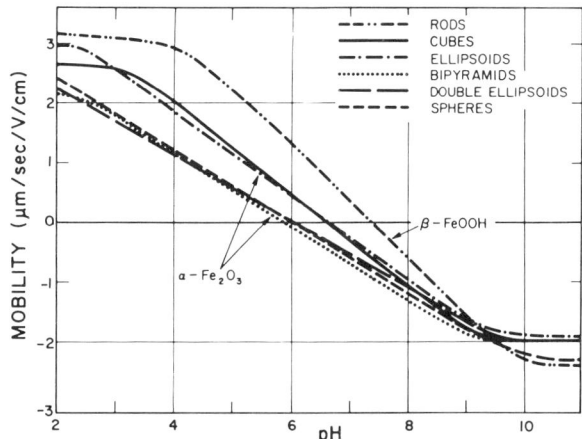

Fig. 6-13. Electrophoretic mobilities of different iron (III) hydrous oxide particles as a function of pH. β-FeOOH sols: rods, formed from solution of 0.09 M in FeCl$_3$ and 0.01 M in HCl aged 24 h at 100° C (– – –·· – –); α-Fe$_2$O$_3$ sols containing different shaped particles formed in solutions as given below: cubes, 0.09M in FeCl$_3$ and 0.01 M in HCl aged 24 h at 150°C (—); ellipsoids, 0.018 M in Fe(NO$_3$)$_3$ and 0.05 M in HNO$_3$ aged 24 h at 100°C (— · —); bipyramidal, 0.018 M in Fe(ClO$_4$)$_3$ and0.05 M in HClO$_4$, aged 3 days at 100°C (····); double ellipsoids, 0.018 M in FeCl$_3$ and 0.05 M in HCl aged 1 week at 100°C (— — —), and spheres, 0.0315 M in FeCl$_3$ and 0.005 M in HCl aged 2 weeks at 100°C (- - -). Matijevic and Scheiner (1978). Courtesy Academic Press.

by hydrolysis of ferric solutions with Cl^-, NO_3^-, and ClO_4^- as anions, Matijevic and Scheiner (1978). The i. e. p. of Fe_2O_3 varies between 6.0 and 6.6, whereas that of β-FeOOH is 7.3. The difference in the i. e. p. of the various particles is probably not due to varying particle shape, but more likely due to the fact that the particles were made from reactant mixtures of different compositions and therefore may have somewhat different surface compositions.

Garg and Matijevic (1988) covered cores of hematite with chromium hydrous oxide and Fig. 6-14 shows that the coating on the hematite particles is uniform.

Figure 6-15 shows that the coated hematite particles have the same i. e. p. as chromium hydrous oxide particles, but different from that of the hematite cores. Coating particles of one composition with a layer of a different composition is an effective and useful way of modifying the surface of particles, as will be discussed in Section C of this chapter.

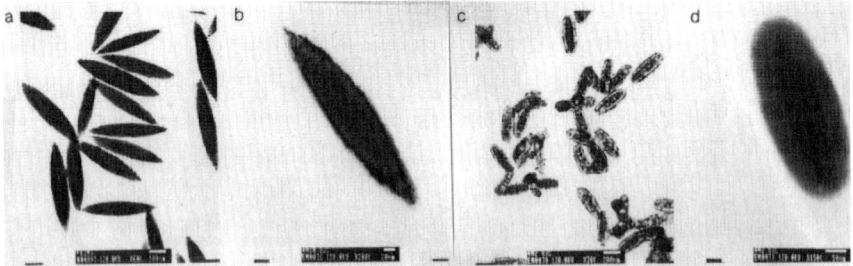

Fig. 6-14. Transmission electron micrographs (TEM) of (a) hematite cores, (b) hematite particles coated with chromium (hydrous) oxides by method I using 10 mg dm^{-3} of the cores, and (c,d) coated hematite particles under the same conditions using 2 mg dm^{-3} of the cores. Note marker in each micrograph for different magnifications. Garg and Matijevic (1988). Courtesy American Chemical Society.

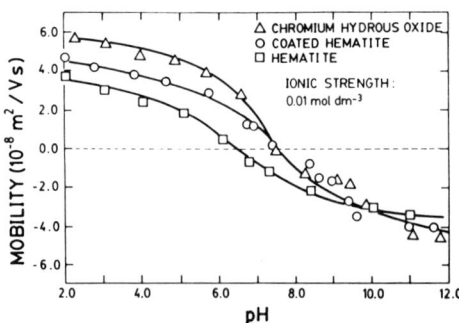

Fig. 6-15. Electrophoretic mobilities of chromium hydrous oxide particles. Garg and Matijevic (1988). Courtesy American Chemical Society.

6. Surface Properties of Metal Sulfides in Aqueous Solution

Like metal oxides, metal sulfides are minerals of great commercial importance. Upgrading and purification of metal oxide ores by techniques such as flotation have benefited greatly from the results of electrokinetic studies of such ores. Relatively little, however, is known about the surface of sulfide minerals and the knowledge is often inconsistent.

In reactions occurring at the metal oxide-water interface it is usually assumed that the surface hydroxyl groups are the active sites, although other, unspecified, groups may also be involved. H^+ and OH^- are considered to be the potential determining ions, the p.d.i.'s, on the surface of hydrous metal oxides.

These ions will also play the role of pdi's on the surface of metal sulfides. The H^+ and OH^- ions form Brönsted sites on the surface, whereas Cd^{2+} and S^{2-} form Lewis sites. Figure 6-16 shows a schematic representation of p.d.i.'s, or surface functional groups, on a CdS surface.

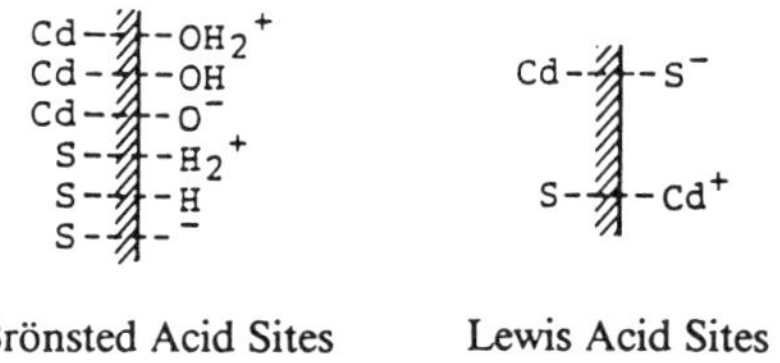

$$
\begin{array}{ll}
Cd - \!\!\!\!\! / \!\!\!\!\! - OH_2{}^+ & \\
Cd - \!\!\!\!\! / \!\!\!\!\! - OH & Cd - \!\!\!\!\! / \!\!\!\!\! - S^- \\
Cd - \!\!\!\!\! / \!\!\!\!\! - O^- & \\
S - \!\!\!\!\! / \!\!\!\!\! - H_2{}^+ & \\
S - \!\!\!\!\! / \!\!\!\!\! - H & S - \!\!\!\!\! / \!\!\!\!\! - Cd^+ \\
S - \!\!\!\!\! / \!\!\!\!\! - {}^- & \\
\end{array}
$$

Brönsted Acid Sites   Lewis Acid Sites

Fig. 6-16. Brönsted and Lewis acid site formation on the surface of CdS. Park and Huang (1987). Courtesy Academic Press.

The surface of cadmium sulfide contains Cd atoms with coordinatively unsaturated sites, which are expected to strongly chemisorb water.

$$\underline{CdS} + H_2O = \underline{CdSH_2O} \tag{6-43}$$

Proton transfer to adjacent sulfur atoms leads to the formation of hydroxyl groups on the Cd sites and thiol groups on the sulfur sites.

$$\underline{CdSH_2O} = \underline{CdOHSH} \tag{6-44}$$

The zeta potential of hydrous cadmium sulfide as a function of pH at various concentrations of $NaClO_4$ is shown in Fig. 6-17, Park and Huang (1987). It is governed by the H^+ or OH^- ions and decreases, as expected from the Gouy-Chapman theory—see Fig. 6-4—with electrolyte concentration. The isoelectric point of the three systems is the same, pH 7.5, suggesting that the Na^+ and ClO_4^- are inert ions and do not adsorb on the surface; compare with Fig. 6-13.

Figure 6-18 shows that the zeta potential, and also the i. e. p., increases when strongly adsorbing ions like Cd^{2+} are added to the aqueous dispersion of CdS (the same dispersion as in Fig. 6-17).

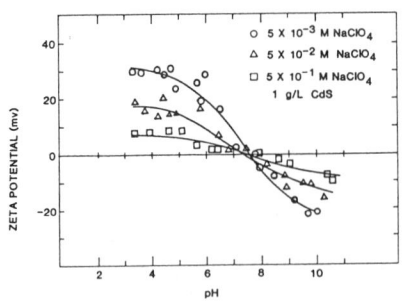

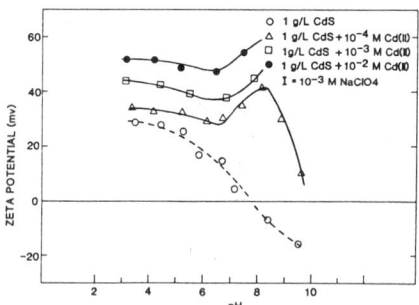

Fig. 6-17. Zeta potential measurements of CdS(s) in NaClO₄ electrolytes. Park and Huang (1987). Courtesy Academic Press.

Fig. 6-18. Effect of Cd(II) ions on zeta potential of CdS(s). It is noted that continuing increases in Cd(II) (as Cd(ClO₄)₂) increases the positive surface charge of CdS(s). Park and Huang (1987). Courtesy Academic Press.

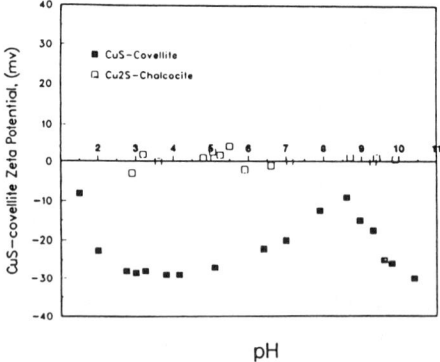

Fig. 6-19. Zeta potential of cooper sulfide as a function of pH. Experimental conditions: 1 g/dm³ solid, 0.05 M NaClO₄, 30-min hydration time. Liu and Huang (1992). Courtesy American Chemical Society.

The i. e. p.'s of a number of synthetic and natural metal sulfides were summarized by Liu and Huang (1992) and are shown in Table 6-9. The values are generally lower than the i. e. p.'s of the corresponding metal oxides, emphasizing the fact that metal sulfides interact with H^+ and OH^- ions in a different way than metal oxides. This is further brought out by Fig. 6-19, which shows that the i. e. p. of Cu_2S in the form of chalcocite under anaerobic conditions is about zero and does not vary with pH, indicating that neither H^+ nor OH^- ions are potential determining ions.

In other studies, however, it has been found that chalcocite has a negative charge between pH 3 and 10, Lekki and Lakowski (1971) and Oestreicher and McGlashan (1972). These experiments were done under aerated conditions and chalcocite could then be converted to covellite, a naturally occurring CuS:

$$Cu_2S\ (s) = CuS\ (s) + Cu^{2+} + 2e^-$$

In air, the surface of the particles of the chalcocite suspension was therefore covered with a layer of covellite, yielding the electrokinetic characteristics of covellite.

Table 6-9. Summary of Electrokinetic Studies of Metal Sulfides. Liu and Huang (1992). Courtesy American Chemical Society.

Sample	Method	IEP
ZnS(s), synthetic	zeta meter	3.0
ZnS(s), synthetic	microelectrophoresis	6.7
ZnS(s), synthetic	zetasizer	8.0
ZnS(s), sphalerite	zetasizer	3.0
ZnS(s), synthetic	microelectrophoresis	6.8
ZnS(s), synthetic (wurzite)	zeta meter	8.5
ZnS(s), sphalerite	zeta meter	<3.0
CdS(s), synthetic	zeta meter	7.5
CdS(s), synthetic	zeta meter	3.8
CdS(s), synthetic (hawleyite)	zeta meter	7.0
CdS(s), synthetic	titration	7.6
PbS(s), galena	electrophoresis	<3.0
PbS(s), synthetic (galena)	electrophoresis	5.0
PbS(s), galena	zeta meter	<3.0
PbS(s), synthetic (galena)	zeta meter	<3.0
HgS(s), cinnabar	titration and adsorption	3.5
HgS(s), synthetic (cinnabar)	zeta meter	<3.0
HgS(s), synthetic (metacinnabar)	zeta meter	5.1
HgS(s), synthetic (metacinnabar)	zeta meter	7.0
NiS(s), synthetic	microelectrophoresis	<3.0
NiS(s), synthetic	zeta meter	<3.0
CuS(s), synthetic (covellite)	zeta meter	<3.0
CuS(s), covellite	zeta meter	<3.0
$Cu_2S(s)$, chacocite	zeta meter	e
FeS(s), synthetic	zeta meter	5.7
FeS(s), marcasite	zeta meter	<3.0
FeS_2, pyrite	zeta meter	2.5
CoS(s), synthetic	zeta meter	<3.0
MnS(s), synthetic	zeta meter	<3.0
Sb_2S_3, stibnite	zeta meter	<3.0
As_2S_3, orpiment	zeta meter	<3.0
MoS_2, molybdenite	zeta meter	<3.0
S(s), synthetic	zeta meter	<3.0

C. The Silica Surface

In Section B of this chapter we discussed the nature and characteristics of the surface of hydrous metal oxides, which include silica, from a more formal point of view. Here we will give a more detailed account of the surface of silica.

Silica is a technically and commercially very important and widely used material, and compared with other metal oxides much more detailed knowledge of its surface is available. The properties of amorphous silicas of high specific surface area, and indeed of all hydrous metal oxides of high surface area, depend on the chemical nature of their surface. In describing the surface of silica and methods of modifying it, we hope that the silica surface will serve as a model that will allow us to better understand the practical importance of silica and other hydrous metal oxides in the technology of catalysts, ceramics, and adsorbents, and as fillers, pigments, binders and thickening agents in many different applications.

1. Properties and Characteristics of the Silica Surface

The chemistry and properties of the silica surface have been the subject of numerous studies. Bergna (1994) recently summarized the present state of knowledge of the surface of silica.

1.1. Type of Surface Hydroxyl Groups
In the system SiO_2-H_2O water is primarily present as surface hydroxyl groups, silanol groups, but could in some cases also exist as structural water within the structure, for instance in minute pores, of colloidal silica particles. Silanol groups can form on the surface of silica by essentially two processes. In Section B of Ch. 2 it was noted that supersaturated monomeric silicic acid, $Si(OH)_4$ condensed to polymeric forms that grew to spherical colloidal particles containing $\equiv Si-OH$ groups on the surface. Depending on the conditions, the surface may retain all or lose some of the silanol groups. Surface hydroxyl groups can also form as a result of rehydroxylating a dehydroxylated surface, when it is treated with water under various conditions—see sections 1-3 and 1-4 below.

The silanol groups will act as centers for molecular adsorption when they interact specifically with adsorbate capable of forming hydrogen bonds, e.g., water, with the surface hydroxyl groups; or more generally, by donating or accepting electrons.

Using old and new techniques such as diffuse reflectance Fourier transform (DRIFT) infrared spectroscopy, ^{29}Si cross polarization magic-angle nuclear magnetic resonance spectroscopy (CP MAS NMR), and deuterium-exchange methods investigators have identified the groups shown in Fig. 6-20 with the exception of the silanetriol group, Bergna (1994).

Isolated silanol groups are OH groups sufficiently far away from adjacent silanols to prevent hydrogen bonding. Vicinal silanols are hydroxyl groups sufficiently close so as to form hydrogen bonds with one another. In geminal silanols two OH groups are bonded to a Q^2 atom. Silanetriols, that is three hydroxyl groups bonded to a Q^1 silicon site, have not yet been identified.

The surface hydroxyl groups are responsible for the high affinity of the silica for water, making it a very effective drying agent. Water molecules can hydrogen bond to the silanols in stages and form clusters on the silica surface.

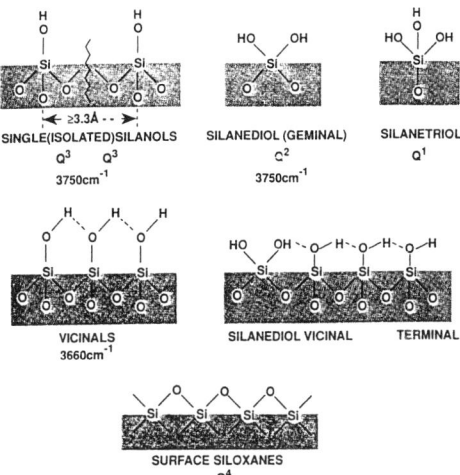

Fig. 6-20. Silanol groups and siloxane bridges on the surface of colloidal silicas. Characteristic infrared bands at 3750 and 3660 cm^{-1} are shown for single and vicinal groups. Q^n terminology is used in NMR; n indicates the number of bridging oxygens (-O-Si) bonded to the central silicon (n = 0-4). Bergna (1994). Courtesy American Chemical Society.

Kondo et al. (1992) observed that when silica in the form of hydrogel, xerogels and aerogels was aged at pH of about 11, a new surface hydroxyl IR band appeared at around 3500 cm^{-1}. By studying the behavior of this band, they concluded that the exchange of deuterium ions for protons occurred readily, substitution of cations such as Co^{2+} did not take place, and that surface modification with monochlorotrimethylsilane was not possible—see §2-3 below.

These hydroxyl groups therefore appeared to be less reactive than isolated silanol groups, and Kondo suggested that they were instead vicinal groups. Kondo and co-workers also determined the solubility of alkali-treated silica, and Fig. 6-21 shows that it has much lower solubility, about 70 ppm, than conventional microporous silica, about 170 ppm, which was made at pH 6.

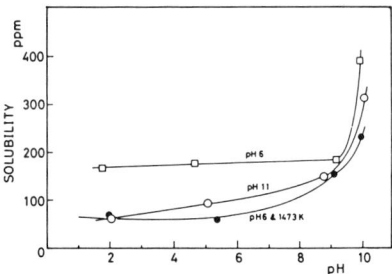

Fig. 6-21. Solubility of mesoporous silica gel, (□); alkali-treated macroporous gel at 303° K, (O); macroporous silica gel heat treated at 1473° K, (●). Kondo et al. (1992). Reprinted from Colloids and Surfaces 48, 65 (1990) with kind permission from Elsevier Science - NL, Sara Burgerhartstraat 25, 1055 KV Amsterdam, The Netherlands.

1.2. Concentration of Silanol Groups

The number of surface hydroxyl groups per unit area can be determined by various methods and until quite recently there was no consensus as to the value of the silanol number, i.e., the number of silanol groups per nm^2 of silica. Iler (1979) and De Boer and Vleeskens (1958) reasoned that the concentration of surface hydroxyl groups could be estimated from the cristobalite crystal structure and using the 1-1-1 plane of the octahedral face, calculated that the silanol number, α_{OH}, is 4.6 OH nm^{-2}—see Fig. 6-22.

The silanol number can also be determined experimentally. The silica sample is first washed with water of pH 3 to remove Na^+ ions since they often occur in pairs with hydroxyl ions. After drying the silica between 110°C and 150°C—see Table 6-10—to remove physically adsorbed, i.e., hydrogen-bonded water, the specific surface area is determined by the BET method. The silanol number can then be calculated by determining the weight loss after calcining the silica sample at about 1100°C for a few hours. Table 6-10 shows that the silanol number varied between 4 and 10 OH nm^{-2}, a variation large enough to require an explanation.

Zhuravlev (1993), in reviewing the work on surface characterization of amorphous silica in the former USSR, reported that differential thermal analysis, DTA, of silica samples showed a peak at 150°C, which was associated with physically adsorbed water. It is true that most of the physically adsorbed water will be removed by heating the samples to 150°C, but some may remain on the silica surface up to about 200°C and obscure determination of the silanol number,

Reference	Type of SiO_2	Drying Method			Specific Surface Area (m^2/g)	Method of Determination	α_{OH} OH/nm^2
		°C	Atm.	Time			
Shapiro and Weiss	Gel	155	-	-	300	Diborane	7.9
Lowen and Broge	Deionized colloid	110	Air	16 h	182	Ignition, 1100 °C	10.0
Iler	Deionized colloid	120	Air	4 days	160	Ignition, 1100 °C	8.0
Iler	Deionized colloid	120	Air	4 days	169	Ignition, 1100 °C	8.1
		150	Air	2 h	160	Ignition, 1100 °C	6.3
Erkelens and Linsen	Mallinckrodt	120		17 h	591	Ignition, 1200 °C	9.6
						Infrared	8.6
	Mallinckrodt, heated to 800°C, soaked in water, 5x	120		17 h	166	Ignition, 1200 °C	5.2
						Infrared	4.0
Davydov, Kiselev, and Zhuravlev	Rehydrated gels, aerogels, aerosils	150	-	-	39-750	Combination methods	5.0-5.7

Table 6-10. Hydroxyl Groups per Square Nanometer by Different Methods. Iler (1979). Reprinted by permission of John Wiley & Sons, Inc.

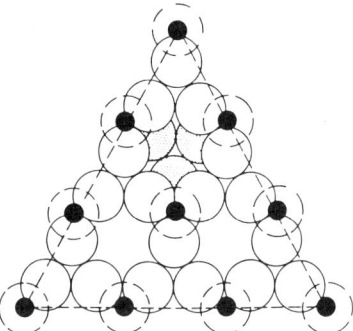

Fig. 6-22. Atomic arrangement of the (1.1.1) octahedral face of cristobalite: 4.55 SiOH nm^{-2}. Large circles, oxygen atoms; small circles, silicon atoms at surface; dashed circles, position of hydroxyl groups on the surface, attached to the underlying silicon atoms. The atomic sizes are not to scale. De Boer and Vleeskens (1958). Reprinted by permission of John Wiley & Sons, Inc.

α_{OH}. Returning to Table 6-10, it can be seen that the highest values of α_{OH}, 10.0 and 9.6, were obtained for samples that had been dried at 110°C and 120°C, at which temperatures all of the physically adsorbed water was not removed.

Moreover, the sample from Mallinckrodt had a specific surface area of 591 m^2/g, indicating that it contained very small pores, in which water was held so strongly by capillary forces that it could not be removed by merely heating the silica at 120°C.

Zhuravlev used 100 different samples of silica, differing in the conditions of their preparation and in their specific surface areas, from which physically adsorbed water had been completely removed by heating the samples in vacuo from room temperature to about 200°C, to determine the silanol number of a fully hydroxylated silica surface. Fig. 6-23 shows that the silanol number is independent of the specific surface of the silica samples and has the average value

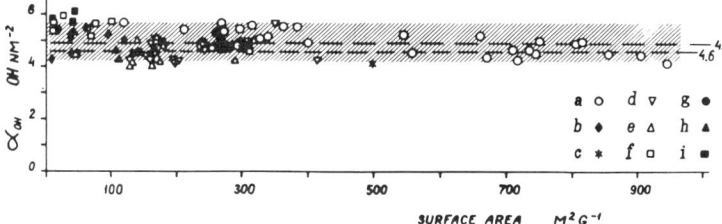

Fig. 6-23. Concentration of the surface hydroxyl groups (the silanol number) α_{OH} for silicas having different specific surface areas, S, when the surface has been hydroxylated to a maximum degree. Symbols a-i indicate different types of amorphous silica. The shaded area is the range of experimental data (100 samples of SiO$_2$ with different S values from 9.5 to 950 m^2/g. Broken lines are average values of the silanol number (231 independent determinations), α_{OH} = 4.9 OH groups nm^{-2} (arithmetic mean); $\alpha_{OH,av}$ = 4.6 OH groups nm^{-2} (least squares method). Zhuravlev (1993). Reprinted from Colloids and Surfaces **74**, 71 (1993) with kind permission from Elsevier Science - NL, Sara Burgerhartstraat 25, 1055 KV Amsterdam, The Netherlands.

of 4.9, the arithmetic mean, and 4.6 OH groups nm^{-2}, from least squares calculations, respectively.

The value of 4.6 OH groups per nm^2 is in good agreement with the silanol number calculated from the 1-1-1 plane of the octahedral face of crystobalite— see Fig. 6-22—suggesting that on the surface of fully hydroxylated silica each Si atom holds one hydroxyl group.

1.3. Dehydroxylation of the Silica Surface

Zhurvalev (1993) used the deuterium exchange method to follow the silanol number decrease with temperature when silicas were heated under vacuum. Figure 6-24 and Table 6-11 show that α_{OH} decreases monotonically with temperature, but faster in the range from 200 to about 450°C, region IIa in the figure, than in the range between 450°C and 1100°C, region IIb in Fig. 6-24.

Table 6-11 shows that α_{OH} decreases from 4.60 at 180-200°C to <0.15 at 1100°C and θ_{OH}, the surface coverage by hydroxyl groups, from 1.0 to <0.03.

The intensity of the IR absorption band of the hydroxyl groups was used to speciate these groups on the surface. Figure 6-25 shows that on the fully hydroxylated surface 60% and 40% of the silanols are vicinal and isolated groups, respectively. At 400°C the vicinal groups have disappeared whereas the isolated groups remain, although in very small amount, up to 1100°C.

Figure 6-21 shows that alkali treatment of a silica gel at pH of about 11 and heat treatment at 1100°C cause about the same decrease in the solubility of silica.

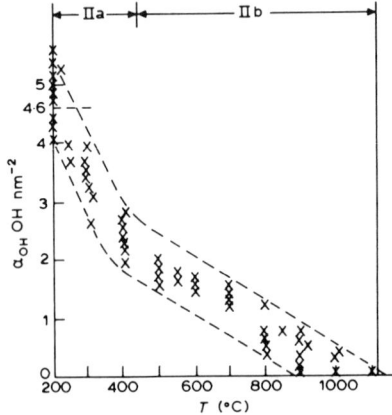

Fig. 6-24. Silanol number as a function of the temperature of pretreatment in vacuo for different samples of SiO$_2$. The broken lines delimit the range of experimental data (16 samples with different S from 11 to 905 m^2/g). The subregions of dehydroxylation are II$_a$ from 200 to about 450°C, and IIb from 450 to about 1100°C. Zhuravlev (1993). Reprinted from Colloids and Surfaces **74**, 71 (1993) with kind permission from Elsevier Science - NL, Sara Burgerhartstraat 25, 1055 KV Amsterdam, The Netherlands.

Table 6-11. Values of α_{OH} and θ_{OH} following treatment in vacuo of amorphous silica at different temperatures, with the initial state corresponding to the maximum degree of surface hydroxylation. Zhuravlev (1993). Reprinted from Colloids and Surfaces **74**, 71 (1993) with kind permission from Elsevier Science - NL, Sara Burgerhartstraat 25, 1055 KV Amsterdam, The Netherlands.

Temperature of vacuum treatment (°C)	Silanol number $\alpha_{OH, av}$ (OH groups per nm^2)	Degree of coverage with OH groups (θ_{OH})
180-200	4.60	1.00
300	3.55	0.77
400	2.35	0.50
500	1.80	0.40
600	1.50	0.33
700	1.15	0.25
800	0.70	0.15
900	0.40	0.009
1000	0.25	0.05
1100	<0.15	<0.03

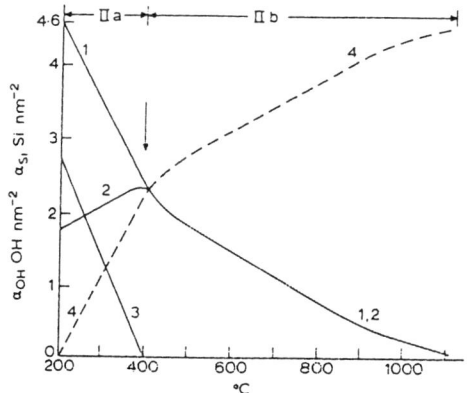

Fig. 6-25. Different types of surface groups as a function of pretreatment of silica in vacuo: curve 1, average concentration of all the OH groups; curve 2, concentration of the free OH groups; curve 3, concentration of OH groups bound through the hydrogen bond; curve 4, concentration of surface Si atoms that are part of the siloxane bridges. The arrow indicates the combined data obtained by the deuterium-exchange method and the IR spectroscopic method. Zhuravlev (1993). Reprinted from Colloids and Surfaces **74**, 71 (1993) with kind permission from Elsevier Science - NL, Sara Burgerhartstraat 25, 1055 KV Amsterdam, The Netherlands.

1.4. Rehydroxylation of the Silica Surface

A dehydroxylated silica surface can be fully rehydroxylated by reacting it with water in the liquid or vapor state, increasing the silanol number from a low value—see Fig. 6-24 and Table 6-10, to 4.6 OH groups nm^{-2}. The rate of dehydroxylation is fast for silicas that have been dehydroxylated in the temperature range 200-450°C, region IIa in Fig. 6-24, since the concentration of siloxane bridges is still low and each $\equiv Si - O - Si \equiv$ bridge is surrounded by silanol groups, which weakens the Si-O bond in the bridge. The rehydroxylation process involves physical adsorption of water molecules on the silanol groups followed by dissociative chemisorption on the adjacent silanol bridges. The rate of rehydroxylation, however, decreases rapidly with the temperature of dehydroxylation and becomes very low in liquid water at room temperature when αOH has been reduced to 0.66 OH groups, corresponding to a calcination temperature of 800-900°C according to Table 6-10. Under these conditions it took 5 years to completely rehydroxylate the silica surface, Zhuravlev (1993)! At the upper end of region IIb in Fig. 6-24 the concentration of siloxane bridges will be high and there will be only a few isolated silanols on the surface—see Fig. 6-25—but these will act as centers for physical adsorption of water molecules. Around these centers new silanol groups will gradually form as the physically adsorbed water molecules dissociatively chemisorb on neighboring siloxane bridges.

1.5. Stability of Colloidal Silica

The DLVO (Deryagin-Landau-Verwey-Overbeek) theory has been used since the 1940s to describe and explain colloidal stability. In the DLVO theory colloidal stability is determined by a balance between double-layer repulsion, which increase exponentially with decreasing distance between the particles and van der Waals attraction, Evans and Wennerström (1994). The theoretical rationalization of the Schulze-Hardy rule was an early success of the DLVO theory and the potential energy diagrams in Fig. 6-1 are schematic representations of the theory. The DLVO theory, however, failed to predict and explain the stability behavior of colloidal silica, shown in Fig. 6-26.

Colloidal silica has a local stability maximum at the point of zero charge, where the DLVO theory predicts that colloidal systems have minimum stability. The stability of a silica sol decreases with pH and reaches a minimum around pH 6, after which the sol enters a region of high stability between pH 8 and about 10.5. According to Iler (1979), colloidal silica will gel when the particles can come so close together that siloxane bridges will form between the particles with

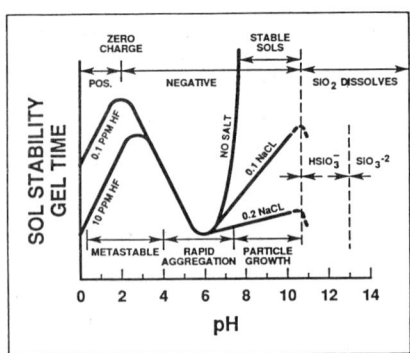

Fig. 6-26. Effects of pH in the colloidal silica-water system. Iler (1979). Reprinted by permission of John Wiley & Sons, Inc.

hydroxyl ions acting as catalysts. Since the point of zero charge, or the isoelectric point of silica, occurs at low pH, around pH 2, the concentration of hydroxyl ion is low; hence a low rate of gel formation and high stability at the isoelectric point.

The stability decreases as the concentration of hydroxyl ions increases with pH and reaches a minimum at pH about 6; however the surface charge of the silica particles increases—see Fig. 6-30, and at about pH 8 it becomes so highly negative that the particles can no longer come close enough together for inter-particle siloxane bridges to form, and colloidal silica enters a region between pH 8 and 10.5 of almost unlimited stability.

The stability-versus-pH curve in Fig. 6-26 is actually the result of two opposing mechanisms; one represented by curve 1 in Fig. 6-27, showing that the stability decreases linearly with pH, and the other symbolized by curve 2 showing that the stability increases exponentially with pH, i.e. implicitly with the negative charge of the particles.

Thus, silica sols do not, as opposed to colloidal particles of the hydrous metal oxides discussed in Ch. 3, conform to the predictions of the DLVO theory that the point of zero charge represents a minimum of stability. Silica is also different in another aspect, namely that the Si-O-Si bond, i.e., the siloxane bridge, is much more stable and inert to acidic attack than other metal oxides. In Ch. 3 it was described how small particles of hydrous metal oxides other than silica were formed by forced hydrolysis of the corresponding, often very acidic, metal salt solutions. The process of particle formation can, however, be reversed by adding acid, under appropriate conditions of time and temperature, to the sol of hydrous metal oxide particles. Silica particles, or for that matter, a gel network of silica particles, on the other hand, cannot be dissolved in acids (with the exception of HF). The reason why colloidal silica does not conform to the DLVO theory may therefore be the special role played by hydroxyl ions in catalyzing the formation of siloxane bridges between silica particles; which is especially active and pronounced in the pH range 2-6.

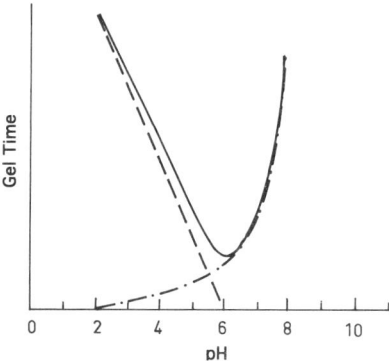

Fig. 6-27. Gel time is the superposition of two opposing mechanisms: a linear attraction term and an exponential repulsion term. Otterstedt (1995).

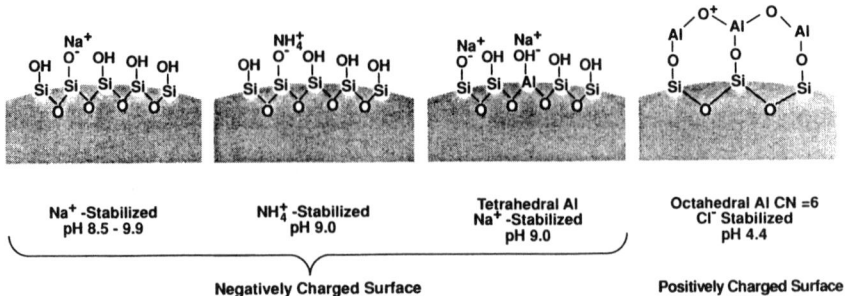

Fig. 6-28. Sodium- and ammonium-stabilized and aluminum-modified surfaces of colloidal silicas. Bergna (1994). Courtesy American Chemical Society.

As was described in Ch. 2, commercial, high-solids silica sols are stabilized by adding alkali so that their pH falls into the range 8-10. Fig. 6-28 shows sodium hydroxide and ammonia-stabilized silica sols with the cations Na^+ and NH_4^+ near the negatively charged particle surface.

The amount of alkali required to bring the sol into the pH range 8-10 is directly proportional to the specific surface area of the silica particles, but even for sols of particle size as small as 4-5 nm, the amount of alkali ions introduced as counter-ions to the hydroxyl ions will be quite small and not enough to destabilize the sol. From the density of the sol (1.1 g/cm^{-3}). Eqns. 1-15 and 1-16, and the fact that the hydroxyl ion concentration on the surface of silica at pH 9 (Na^+ only as counter ions) is 0.63 OH$^-$/nm^2, Iler (1979), it can be readily calculated that [Na^+] in a 5-nm sodium hydroxide stabilized silica sol containing 15% SiO_2 is about 0.1M, which is not enough to gel or coagulate the sol.

Alkali ions, as well as other cations, will, however, coagulate silica sols if present in sufficiently high concentrations. Milonjic (1992) and Matijevic and Allen (1969, 1970) studied the critical coagulation concentration, c.c.c., variation with pH—see Table 6-12.

The difference between the c.c.c.'s of chloride and sulfate solutions was explained to be due to differences in activity coefficients of the electrolytes.

Figure 6-29, which includes data from Milonjic, and Matijevic and Allen, shows the critical sorption curve for colloidal silica. Above the curves, the added and adsorbed electrolyte will coagulate the sol, whereas this will not happen in the electrolyte-pH domain below the curves. The c.c.c.-values for the different electrolytes fall on the same curve, although at a given pH, they may differ considerably between two electrolytes; as, e.g., between Li^+ and Cs^+ at pH of about 7.8.

Table 6-12. Critical coagulation concentrations (c.c.c) of alkali metal electrolytes for colloidal silica as a function of pH, T=298°K. Milonjic (1992). Reprinted from Colloids and Surfaces **63**, 113 (1992) with kind permission from Elsevier Science - NL, Sara Burgerhartstraat 25, 1055 KV Amsterdam, The Netherlands.

c.c.c (mol /dm^3)	LiCl	Li$_2$SO$_4$	NaCl	Na$_2$SO$_4$	KCl	CsCl	Cs$_2$SO$_4$
0.10	-	12.0	11.9	11.8	-	-	-
0.25	11.5	11.4	10.6	10.6	-	7.6	7.6
0.50	11.0	10.8	9.7	9.9	-	7.1	7.3
0.75	-	-	-	-	8.3	-	-
1.00	10.0	10.1	8.7	9.5	7.6	7.0	7.3
1.25	8.8	9.6	8.1	-	-	7.0	7.6
2.50	7.9	9.0	7.6	-	7.0	7.0	8.1
4.00	5.8	-	6.8	-	7.0	7.1	-

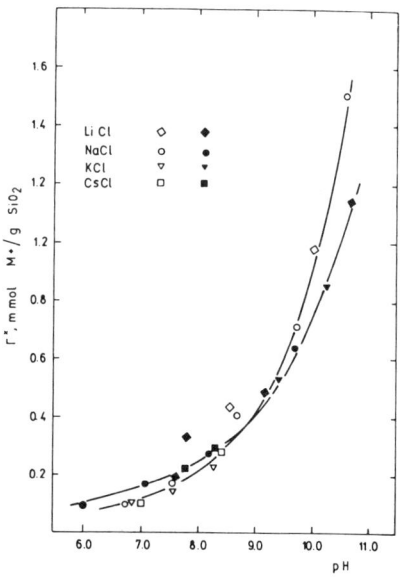

Fig. 6-29. Critical sorption curve for colloidal silica. Open symbols: Milonjic data; filled symbols: data of Allen and Matijevic (1969 and 1970). Milonjic (1992). Reprinted from Colloids and Surfaces, **63**, 113 (1992) with kind permission from Elsevier Science - NL, Sara Burgerhartstraat 25, 1055 KV Amsterdam, The Netherlands.

2. Modification of the Silica Surface

In many uses of silica sols and aggregated structures of small silica particles such as gels and powders, it is desirable to modify the silica surface so as to improve performance in a particular application. The types of modification that are described here are:

1. Introduction of surface aluminosilicate ions
2. Charge reversal
3. Nonionic reactions with the surface
4. Interaction with hydrogen-bonding organic molecules

5. Surface coating (treated separately in Section D, this chapter).

Some of these modifications are relatively modest, e.g., enhancing the negative charge on the particle surface, but others will completely change the chemical nature of the surface. In some cases the modifications can also be applied to other hydrous metal oxide particles than silica.

2.1. Introduction of Aluminosilicate Sites

Iler (1979) noted that there is a peculiar relationship between aluminum and silicon; probably because both elements can assume a coordination number of 4 toward oxygen and, as ions (Al^{3+} and Si^{4+}), have approximately the same diameter.

When the particle surface of silica sols is modified with aluminosilicate ions, the surface will have a fixed, pH-independent negative charge that will make the sols more stable towards gelling at low pH than the sols from which they were prepared. They will also be more stable at high pH because the negative aluminosilicate site will hinder the depolymerization of silica catalyzed by hydroxyl ions. Aluminosilicate ions, $SiAlO_4^-$, on the surface of silica structures are very acidic and act as very effective acid catalyst in, e.g., fluid catalytic cracking of oil.

Alexander and Iler (1961) describe a process for modifying the properties of a silica sol by treating it with sodium aluminate. In a typical preparation, a 22-nm sol, e.g., Ludox® TM (Du Pont), containing 20% SiO_2 was deionized to a pH of 8.5 with a strong-acid cation exchange resin in the hydrogen form. A quantity of 2500 grams of this sol was placed in a beaker and 9.12 g of sodium aluminate solution (prepared by diluting 2.914 g $NaAlO_2$ in 6.205 g deionized water) was added dropwise to the vortex of the vigorously stirred silica sol over a period of about 25 minutes. The resulting sol had a pH of 9.5. Assuming that all the added aluminum formed aluminosilicate sites, there would be 0.3 such sites per nm^2 of

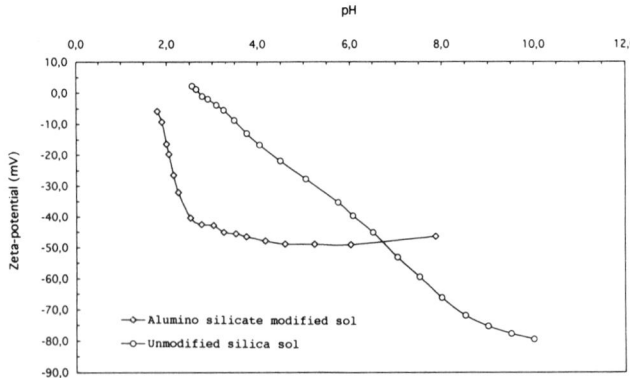

Fig. 6-30. Zeta potential of unmodified (o) and aluminosilicate modified silica sol (□) as a function of pH. Asplund and Otterstedt (1995).

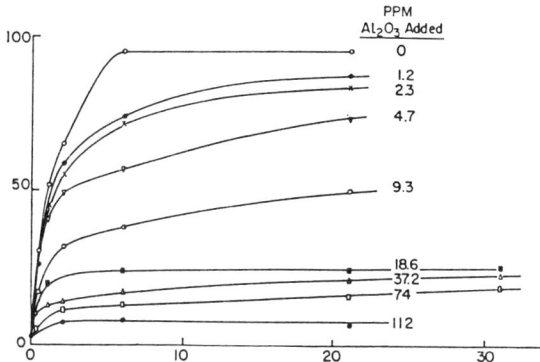

Fig. 6-31.Increasing concentration of soluble silica with time, as silica dissolves from colloidal particles into an initially unsaturated solution in water in the presence of different concentrations of alumina in system. Ordinate: dissolved SiO_2; abscissa: days at 25°C. Iler (1973). Courtesy Academic Press.

silica surface. Figure 6-30 shows that the zeta potential of the modified sol remains negative down to pH of about 3.

Iler (1973) studied the effect of aluminosilicate sites, introduced by treating the silica with aluminum citrate solutions, on the solubility of amorphous silica. Figure 6-31 shows that the rate of dissolution, and the solubility, were drastically reduced when the concentration of aluminosilicate sites on the surface increased from 0 to about 1.3 Al atoms per nm^2, corresponding to 112 ppm Al_2O_3 added to the sol; see Fig. 6-32.

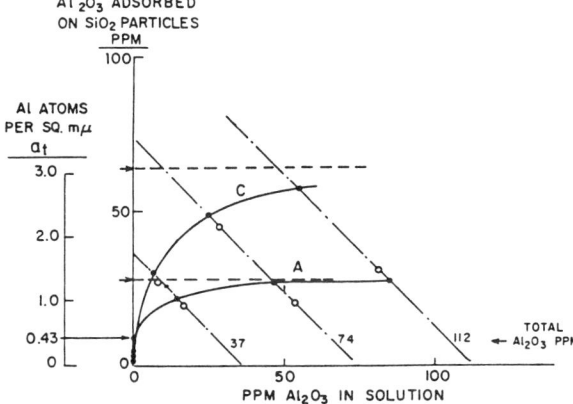

Fig. 6-32. Relation between amount of Al_2O_3 in system adsorbed on SiO_2 particles and amount of Al_2O_3 in equilibrium in solution. Ordinates: Al_2O_3 in system adsorbed on SiO_2 particles, ppm Al atoms adsorbed per square millimicron, a_t; abscissa: ppm Al_2O_3 in solution; (A) Series A, initially unsaturated solutions; (C) Series C, initially saturated solutions; (o) after 30 days at 25°C; (•) after 20 days at 25°C, 3 days at 58°C, 4 days at 25°C. Iler (1973). Courtesy Academic Press.

In Fig. 6-32 the amount of adsorbed alumina is plotted against the concentration of alumina remaining in solution for three different levels of total aluminá in the system. About 9 ppm alumina, or 0.43 Al atoms per nm^2 appears to be irreversibly adsorbed. Iler used the Langmuir adsorption isotherm to estimate that the maximum of adsorbed alumina was 1.3 Al per nm^2 in series A (the sol was initially unsaturated with respect to monomeric silica when the aluminum citrate solution was added), represented by the lower broken line in Fig. 6-32, and 3.1 Al/ nm^2 in series C (initially saturated sols), represented by the upper broken line in the figure.

The surface of silica contains about 8 Si atoms/nm^2, Iler (1979), which is also the theoretical maximum number of aluminosilicate ions in the surface. Aluminizing the surface of silica by using aluminum citrate, and also sodium aluminate, thus results in a surface coverage of Al that is much less than that corresponding to a Langmuir monolayer. The fact that the aluminosilicate sites are all negatively charged probably ensures a fairly regular distribution on the surface, Iler (1973).

Iler (1976) also found that surface aluminosilicate ions will improve the stability of silica sols toward gelling. Figure 6-33 shows that a 12-nm silica sol (Ludox® HS, 30% solids) will gel in acid solution. Electrolytes will further reduce the surface charge and the sol will gel even more rapidly. If only 4% of the surface, however, is covered with aluminosilicate ions, as in Ludox® AM, the particles will retain a negative charge at low pH and be stable even in the presence of electrolyte. At pH 8.5 and higher, both sols are negatively charged and stable except at very high electrolyte concentrations.

Figure 6-28 shows schematically a tetrahedral aluminosilicate site on the surface of silica. Note that the sodium ions are readily exchangeable.

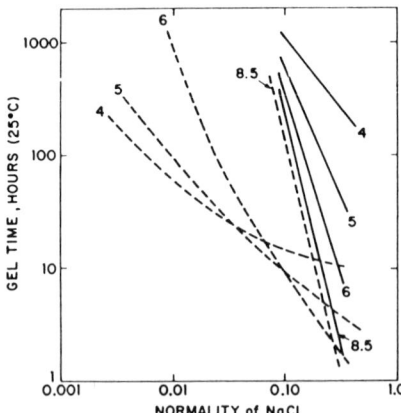

Fig. 6-33. Stability of aluminosilicate modified slica sol (Ludox® AM) versus unmodified sol. (Ludox® HS). Solid lines = Ludox® AM; dotted lines = Ludox® HS; pH indicated. Iler (1976). Courtesy Academic Press.

In their patent, Alexander and Iler (1959) also describe how the surface of silica can be modified with negatively charged sites of zincate, stannate, or plumbite ions.

In an analogous manner, silica can be adsorbed on the surface of alumina particles, forming an aluminosilicate surface that is much less soluble than either oxide alone, Hingston and Raupach (1967).

2.2. Charge Reversal

The negative charge on the particles in silica sols is essential to the stability of the sols and to their usefulness in many applications. However, they are not compatible with sols of positively charged particles, as, for instance, those in alumina, titania, and zirconia sols, and when mixed with such sols, they gel or precipitate.

Figure 6-8 shows that in a proper pH range, silica particles can very effectively adsorb metal cations such as Fe^{3+}. Alexander and Iler (1961) showed that polycations of aluminum, zirconium, chromium, titanium and other metals adsorbed even more effectively on silica particles and could reverse the surface charge from negative to positive. In a typical preparation they added one volume of a 12-nm silica sol (Ludox® LS) containing 30% SiO_2 under vigorous stirring to half the volume of a basic aluminum chloride solution containing 0.5 mole of Al and 0.25 mole chloride—see Ch. 3. The resulting clear and stable sol contained 21.2% SiO_2, and 1.7% Al_2O_3, corresponding to a ratio of Al to surface silicon of 1:2, and had a pH of 4.3.

These positively charged, alumina-coated silica sols are dependably stable only when the pH of the sol is kept relatively low at about 4.5, which is a relatively corrosive medium. Mindick and Thompson (1966) developed a method of making stable alumina coated silica sols having a pH ranging from 4.5 to 6.5.

Compared with the Alexander-Iler process, they carefully deionized the silica sol, using strong base and strong acid ion exchange resin to remove destabilizing anions and cations, in two steps with a holding period between the deionization steps from 16 to 24 hours.

In a third process, which is claimed to be less costly and time-consuming than the Mindick-Thompson method, Moore uses basic aluminum acetate to prepare stable, positively charged, alumina coated silica sols, Moore (1971).

Asplund and Otterstedt (1995) studied the adsorption of Al-polycations, primarily the Al_{13}^{7+}-complex discussed in Ch. 3, on silica particles. To solutions of basic Al-chloride (Locron® L, Hoechst) containing varying amounts of alumina, measured as Al_2O_3, they added a fixed volume of 100-nm silica sol (from Nissan, Japan) containing 5% SiO_2 and having a pH of 11.5. Figure 6-34 shows how the pH and zeta potential of the sol and the amount of alumina in the aqueous phase varied with the concentration of alumina in the Locron® solution.

When the concentration of alumina in the Locron® solution was no higher than 1.25 weight % Al_2O_3, based on the weight of silica particles, all the alumina was adsorbed on the sol particles and no alumina could be detected in the aqueous

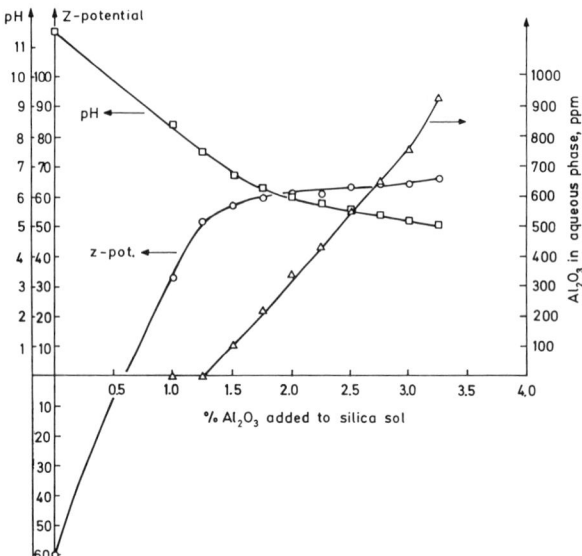

Fig. 6-34. Adsorption of basic Al-chloride ($[Al_{13}(OH)_{24}(H_2O)_{12}]^{7+}Cl$) on 100-nm silica sol. Effect of pH ($\square$), zeta potential (o), and alumina in the aqueous phase (Δ). Asplund and Otterstedt (1995).

phase. At this point the zeta potential was 52 mV, the pH 7.5, and the surface concentration of alumina was 2.7 Al atoms per nm^2. At 2% Al_2O_3 the zeta potential was 61 mV and the surface concentration of alumina was 3.4 Al atoms per nm^2, which appeared to be close to the maximum amount of Al-polycations the silica surface could adsorb. When the concentration of alumina in the aqueous phase was 344 ppm, it corresponded to 38% of the alumina in the system, and 935 ppm Al_2O_3 in the aqueous phase corresponded to 62% of the total amount of alumina in the system.

2.3. Nonionic Reactions of the Silica Surface

The silanol groups of the silica surface, and indeed also the surface hydroxyl groups of many other hydrous metal oxides, can undergo many reactions, several of which are familiar to organic chemists, that adfixes many types of groups, both organic and inorganic, to the surface by covalent bonds. The reactions of the surface are carried out in gas or liquid phase, often under anhydrous conditions, implying that physically bound water must be removed from the surface before reaction can take place—see § 1.3 this section. For a more detailed account of the types of reactions that can be used to modify the particle surface reference is made to Iler (1979). Here there are briefly described some of the more common reactions.

1. Esterification. The surface hydroxyl groups of silica can form esters with alcohols according to the reaction

$$\equiv Si - OH + ROH = \equiv Si - OR + H_2O$$

Esterification of the surface of colloidal silica is most simply achieved by first converting the silica aquasol to an organosol. This can be accomplished by adding a high- boiling water-miscible alcohol, diol or polyol, to the deionized sol and boiling off the water. Complete esterification is possible only under autoclave conditions or high-temperasture reflux.

Stöber et al. (1957) reacted silica with various alcohols at 200°C for 6 h and found that the maximum number of ester groups on 1 nm^2 of silica surface was 4.7 for CH_3O, 3.7 for C_2H_5O, 3.5 for n-C_3H_7O, 3.2 for n-C_4H_9O, and 3.8 for n-$C_8H_{12}O$.

Figure 6-35 shows the silica surface esterified with a highly branched C_{18} alcohol and with n-butanol.

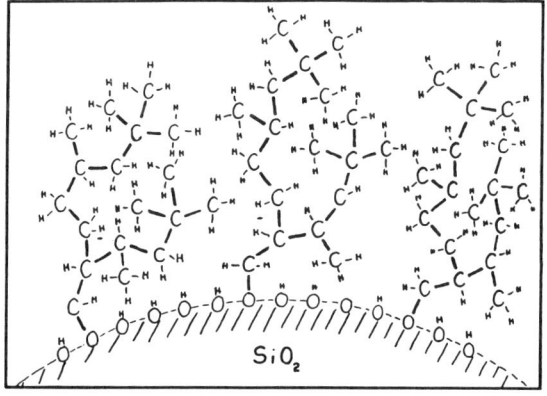

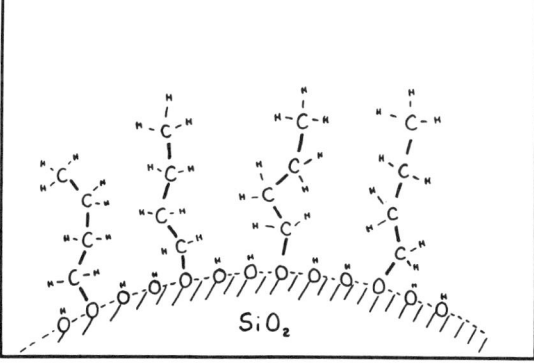

Fig. 6-35. Schematic representation of cross-section of the silica surface esterified with a highly branched C_{18} alcohol and with n-butyl alcohol. Iler (1955). Courtesy Cornell University Press.

Hydrophobic Silica Surface

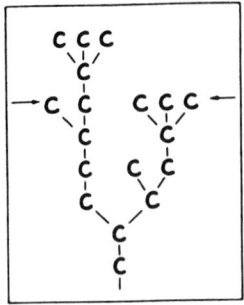

Fig. 6-36. Carbon skeleton of a branched C_{18} alkyl group having a "branch number" of 5. Iler (1955). Courtesy Cornell University Press.

The surface coverage at complete esterification will depend on the size and bulkiness of the alcohol, diol or polyol. Iler (1979) estimated that each alcohol molecule will cover 0.14 n nm^2, where n is the *branch number*—see Fig. 6-36. Thus 5,7,7-trimethyl-2-(1,3,3-trimethylbutyl)-1-octanol in that figure has a branch number 5, and each group covers about 0.7 nm^2.

Esterification and other modifications of the surface of the extremely small silica particles present in alkali silicate solutions—see Ch. 2, §C—will be discussed in §2.4 of this section.

2. Diazotization. Alkoxy groups may also be attached to the silica surface by reacting it with diazomethane;

$$\equiv Si - OH + C_2N_2 = \equiv Si - OCH_3 + N_2$$

This reaction must be carried out in the complete absence of water so as to avoid side reactions.

3. Halogenation. The surface hydroxyl groups may under certain circumstances be replaced by halogens such as fluorine and chlorine. Treatment of the surface of porous glass with a 30% by weight ammonium fluoride solution followed by calcination at 700°C replaces all of the surface silanol groups with fluorine and makes the surface hydrophobic.

Chlorine atoms can be partially substituted for silanol groups by refluxing silica with thionyl chloride

$$\equiv Si - OH + SOCl_2 = \equiv Si - Cl + SO_2 + HCl$$

4. Amination. The chlorinated silica surface is unstable in the presence of ammonia and undergoes the following reaction:

$$\equiv Si - Cl + 2NH_3 = \ \equiv Si - NH_2 + NH_4Cl$$

5. Grignard Reactions. Grignard reactions, which are very useful in organic synthesis, have also been successfully used to modify the surface of silica. Grignard reagents such as methylmagnesium iodide, CH_3MgI, and methyllithium, CH_3Li, undergo specific reactions with many radicals. Silanol groups react according to the equation

$$\equiv Si - OH + CH_3Li = \ \equiv Si - O - Li + CH_4$$

The reaction is quantitative and the amount of methane formed in it has been used to estimate the silanol number, i.e., the number of silanol groups per nm^2.

Grignard reagents also react with a chlorinated silica surface and direct bonds between silicon and carbon are formed:

$$\equiv Si - Cl + CH_3Li = \ \equiv Si - CH_3 + LiCl$$

$$\equiv Si - Cl + C_6H_5Li = \ \equiv Si - C_6H_5 + LiCl$$

6. Reactions with Organic Silicon Compounds. The compounds $(CH_3)_3SiCl$ and $(CH_3)_2SiCl_2$ react selectively and completely with isolated SiOH groups, whereas $(CH_3)_2SiCl_2$ also reacts with some of the geminal groups, i.e., hydrogen bonded adjacent pairs of SiOH—see Fig. 6-6.

Chlorosilanes are used to modify the surface of chromatographic column packings. One widely used surface modification is to introduce octadecyl groups in the silica surface, which can be accomplished by means of octadecyldimethylchlorosilane.

$$\equiv Si - OH + Cl(CH_3)_2 C_{18}H_{37}Si = \ \equiv Si - O - Si(CH_3)_2 - C_{18}H_{37} + HCl$$

2.4. Organic Molecules Hydrogen Bonded to the Silica Surface

Organic molecules hydrogen bonded to small particles of silica constitute only a temporary modification of the particle surface, but we will describe in this section how hydrogen bonding can be used to extract the small particles present in alkali silicate solutions and transfer them into an organic phase so as to form a silicon-rich organosol. Permanent modification of the silica surface can then be accomplished by heating such organolsols under conditions such that the organic hydrogen bonding molecules form esters with the silanol groups of the particle surface.

Kirk (1946) developed an improved process for tanning leather with polysilicic acid made by adding a dilute solution of sodium silicate to dilute acid thus obtaining a 6% SiO_2 sol at pH 2. He found that the silica was adsorbed so rapidly by the hide that it was not evenly distributed, but also discovered that water-miscible organic liquids capable of forming hydrogen bonds retarded the adsorption, giving a uniform tannage.

To study the tanning process, Kirk used a solution of gelatin as a model protein and observed the conditions under which silica precipitated the gelatin. It had long been the practice in the gelatin industry to measure the molecular weight by determining the concentration of sodium chloride required to precipitate gelatin at pH 3-4. Salt solution is simply titrated in until the solution goes suddenly cloudy. The salt dehydrates the gelatin in the solution, removing the water that is hydrogen bonded to the amide linkages so that the gelatin chains can crosslink by hydrogen bonds.

When a hydrogen bonding agent such as diethyl ether or diethylene glycol containing ether oxygens was added, the latter hydrogen bonded to the gelatin, breaking the self bonding and the solution became clear and more salt was needed to precipitate the gelatin. If freshly made silic acid was added to a new sample of gelatin solution, some of the particles were large enough to hydrogen bond with two or more gelatin molecules. The gelatin then acted like a higher molecular weight type and thus required less salt to reach the cloud point. When the silica sol was allowed to age, i.e., allowed to grow to larger particles—see Fig.2-9—and samples were then added to the gelatin solutions, even less salt was needed to reach the cloud point.

2.4.1. Relative Effectiveness of Organic Hydrogen Bonding Compounds. In Kirk's work neither the concentration nor relative proportion of gelatin and silica solution were important, except to ensure that enough material was there to make the precipitate visible at the end point. The method simply balances the precipitating effect of the salt against the solubilizing effect of the hydrogen bonding agent.

Iler (1952) used this method to compare the effectiveness of various hydrogen bonding agents with that of the diethyl ether of diethylene glycol, which he called the standard agent. Thus if another agent such as propyl alcohol was used, the relative concentration required to duplicate the action of the standard agent was an inverse of its activity. He assigned the activity of the standard as 100, and if the test material required twice the concentration to have the same effect, then the activity was assigned the value 50—see Table 6-13. It is evident from the table that in any one group, such as alcohols, the effectiveness depends on the size of the attached hydrocarbon groups and is comparable to its surface activity. However, the hydrocarbon groups must not be so large as to make the compound insoluble in water. The amines at pH about 3 are all present as the amine salts.

Table 6-13. Relative Effectiveness of Hydrogen-bonding Agents, Iler (1955). Courtesy Cornell University Press.

Compounds		Relative Molar Effectiveness
(Standard)	Dimethoxytetraethylene glycol	100
Alcohols	Ethanol	3
	Methanol	6
	Isopropyl alcohol	11
	t-Butyl alcohol	16
Glycols	Ethylene Glycol	0
	Propylene glycol	7
	3-Methyl-1,2-butanediol	18
	Hexamethylene glycol	27
Ketones	Acetone	17
	Methyl ethyl ketone	25
Amides	Formamide	0
	N,N-Dimethylformamide	25
	N,N-Diethylformamide	40
	Acetamide	11
	N,N-Dimethylacetamide	41
	N,N-Diethylacetamide	54
	N-Isobutylacetamide	22
	Urea	7
	Tetramethylurea	44
Primary Amines (as salts)	Methylamine	0
	Cyclohexylamine	25
	2-Ethylhexylamine	32
	m-Toluidine	58
Secondary Amines (as salts)	Dimethylamine	0
	Diethylamine	19
	Piperidine	38
	Dibutylamine	65
	Diamylamine	70
Tertiary Amines (as salts)	Trimethylamine	14
	Pyridine	42
	Quinoline	66
	Cyclohexyldiethylamine	117

2.4.2. Organosols of Colloidal Silica. Kirk also made the observation that when he mixed the silica sol with the hydrogen bonding agent with no gelatin present and saturated the mixture with salt, the organic compound was salted out as a second liquid layer. Moreover, it carried with it most of the silica and he had discovered how to extract colloidal silica from water as a silica organosol. In a typical preparation 20 ml of 5 M SiO_2 solution of sodium silicate, e.g., with ratio 3.3, corresponding to 6 g SiO_2, were added to 180 ml of a cooled (0-5°C), dilute solution of H_2SO_4 with vigorous stirring. The concentration of the acid solution was adjusted so that the pH, after the addition of the sodium silicate solution, was 1.7-2.0 (where silicic acid polymerizes least rapidly—see Section B of Ch. 2) 94 g of cooled tetrahydrofuran was next added to the acid silica sol with moderate to gentle stirring. After 5 minutes 70 g of fine sodium chloride was added with gentle stirring. The mixture was stirred for 20 minutes and transferred to a separatory funnel. After about 30 minutes, the mixture separated into two layers. The upper organic phase, containing most of the organic solvent and the silica was recovered by drawing off the lower aqueous saline layer, Otterstedt et al. (1987).This is the technique that was used in Ch. 2, Section C, to extract the colloidal particles that make up solutions of alkali silicate solutions of different ratios.

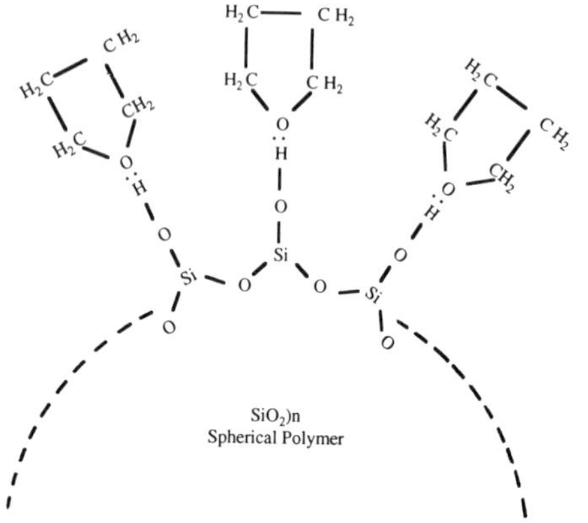

Fig. 6-37. Tetrahydrofuran (H-bonded to spherical polymers in H_2O solution) organophilic-soluble in excess THF. Otterstedt et al. (1987).

Figure 6-37 shows tetrahydrofuran—a small molecule with one oxygen atom that is a strong electron donor—forming a monolayer around the silica particle making it organophilic with its outwardly directed hydrocarbon groups, so that it is salted out into the layer of excess THF.

Normal and tertiary butyl alcohol and triethyl phosphate are powerful hydrogen bonding agents which can be used to form organolsols of colloidal silica. Furthermore, these organosols can be dehydrated by simply boiling off the water. If the organosols are further heated to the boiling point of the hydrogen bonding agent, it will form a stable bond with the silica surface, an ester bond in the case of the alcohols and a silicon-oxygen-phosphorous bond in the case of the triethyl phosphate. Phosphates, e.g., trixylenyl and tritolyl phosphate, are effective plasticizers in PVC compounds, but have a tendency, like many other types of plasticizers, to be lost from the material through migration and evaporation under conditions of high temperature. By bonding plasticizing phosphates to the surface of colloidal silica they can perhaps be converted to highly effective, non-migrating and non-volatile plasticizers for PVC.

Different types of hydrogen bonding agents can extract down to only certain minimum sizes of colloidal silica. No monomer or small cyclic oligomers are extracted from the aqueous brine phase—see Fig. 2-9. Figure 6-38 shows the percent silica extracted from polymerizing acid solutions by four different hydrogen bonding agents. At the beginning, when the colloidal particles are very small and few, very little silica can be extracted. After 40 h, however, triethyl

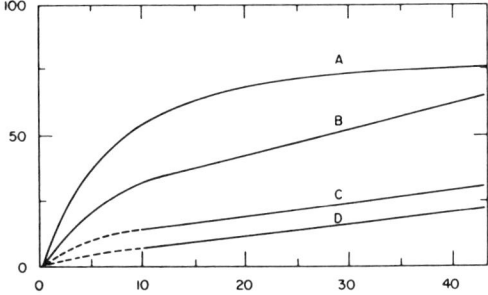

Fig. 6-38. Percent silica extracted from polymerizing silicic acid solution aged for indicated time and saturated with sodium chloride, by (A) triethyl phosphate; (B) tetrahydrofuran; (C) tertiary butyl alcohol, also normal propyl alcohol; (D) acetone. Ordinate: percentage of total silica extracted. Abscissa: age of solution in hours at 25°C. Iler (1980). Courtesy Academic Press.

phosphate and tertiary butylalcohol can extract 75% and 65% of the silica, respectively. It has been reported that if the molecular weight of silica is high, an emulsion or a precipitate is formed instead of a separate liquid phase consisting of hydrogen bonding agent and silica, Iler (1952). Otterstedt et al. (1987), however, found that colloidal particles in the size range 5-15 nm, e.g., Ludox® SM and LS—see Ch. 2, section D—can be extracted from the aqueous brine phase by tetrahydrofuran.

Polyvinylalcohol is a polymeric hydrogen bonding agent that can adsorb on colloidal silica. If one adds just enough polymer to cover the surface, but no excess, all of the PVA hydroxyl groups are turned to the silica surface and outwardly the surface of the particles is all hydrocarbon, making it hydrophobic—see Fig. 6-39. The particles will separate to form a heavy viscous second liquid layer, which may contain up to 40% SiO_2, Iler (1979). If more or less of the PVA is used, no separation takes place; additional chains extend into the water and make the particles hydrophilic.

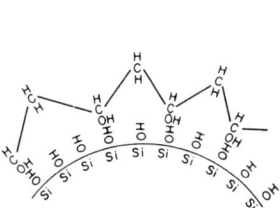

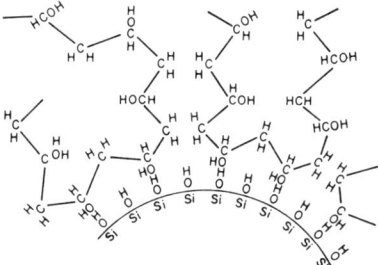

Fig. 6-39a. Silica surface covered with only enough PVA to form a hydrophobic oriented monolayer coating so that coacervation occurs. Iler (1979). Courtesy Academic Press.

Fig. 6-39b. Excess PVA with chain segments bonded to silica surface with remainders of chains extending into solution; no coacervation occurs. Iler (1979). Courtesy Academic Press.

D. Coating of Metal Oxide Surfaces

Applying a uniform, dense and continuos coating to the surface of small particles offers a means to drastically change the chemical nature and properties of the particle surface if the coating procedure is carried out properly. Thus, a great variety of small particles can be coated with a dense, continuous layer of silica. Since the properties and chemical nature of small particles in many applications will be controlled by the chemical nature of the surface, a silica coat will make them behave as silica particles. Similarly, silica, and, indeed, also many other types of particles can be coated with hydrous metal oxides, aluminosilicates, and other substances.

1. Coating of Non-silica Surfaces with Silica

Iler (1959, 1979) developed a method for coating particles, cores, of many different compositions and shapes with a dense, uniform layer of hydrous, amorphous silica—see Fig. 6-40. More recently, this method was studied by Bergna (1994) and Furlong (1994) with special application to coatings on small particles of α-alumina and titanium dioxide, respectively.

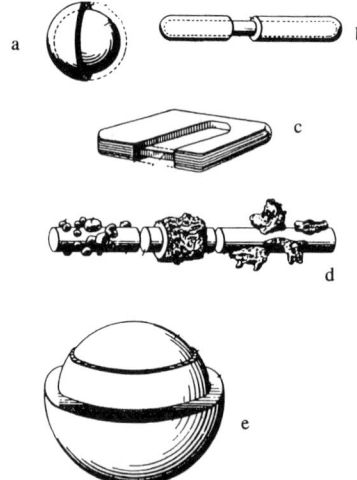

Fig. 6-40. Schematic representations of dense silica coatings formed by the Iler process. Iler (1959).

1.1. General Principles

The first step in the coating procedure is to apply a monomolecular film of silica to the particle surface, which must be water wettable and receptive to silica or be made so by pretreatment.

The second step is to increase the coating thickness by further deposition of silica onto the precursor, monomolecular base coat under such conditions that all silica added in the process will become attached to the coat and will not form secondary particles of silica in the aqueous phase of the dispersion.

The particles to be coated are suspended in an aqueous solution which is supersaturated with silica in the form of monomeric silicic acid, $Si(OH)_4$. As the silica is deposited as a coating, more silica is either formed in the solution or added as supersaturated solution. Silica may thus be present from the beginning as sodium silicate and acid, e.g., HCl, is slowly added to maintain a supersaturated solution of $Si(OH)_4$. The concentration of Na^+ will increase with the coating thickness and care must be taken not to allow the salt concentration to exceed the critical coagulation concentration of silica particles, which is about 0.3M for Na^+ at pH 9-10 and temperatures above 60°C. For thicker coatings the problem of coagulating the particles can be avoided by passing the sodium silicate solution through a cation exchanger to remove sodium ions and adding the resulting solution of polysilicic acid—see Sections B to D of Ch. 2—to the coating solution.

1.2. Cores

The core must be a solid and so finely dispersed that the specific surface is not less than about 1 m^2/g. It can be crystalline or amorphous, inorganic or organic, and may have many different shapes—see Fig. 6-40. Iler (1959) lists the following typical cores, which all are chemically receptive to the attachment of the silica coating:

> Finely divided metal particles such as iron, nickel, and aluminum powders. Metal oxides and hydrated oxides such as aluminum oxide, chromium oxide, iron oxide and nickel hydroxide, titanium dioxide, zirconium oxide, zinc oxide, and cobalt oxide. Metal silicates such as magnesium, aluminum, zinc, lead, chromium, copper, iron, cobalt, and nickel silicates. Natural metal silicates such as the various varieties of clay including kaolin, bentonite, attapulgite, halloysite. Natural fibrous metal silicates such as chrysotile asbestos, palygorskite, amosite, crocidolite, and wollastonite. Plate-like mineral silicates including various varieties of mica such as exfoliated vermiculite, muscovite, phlogophite, biotite, talc, and antigorite. Finely divided synthetic metal silicate products such as fiberglass and rockwool.

Particles of materials which are not receptive to silica can be made so by treating them with a solution of a basic salt of a metal such as aluminum or chromium—see Section B of Ch. 3. The highly charged polycations present in such solutions will attach themselves strongly to the negatively charged, non-receptive-to-silica, core surface so as to form an anchoring layer for the first monomolecular layer of the silica coat—see Fig. 6-40e. Iler (1959) enumerates the following materials which can be used as cores after they have been treated with a solution of a basic metal salt:

Cellulose, paper pulp, cotton fibers, cellulose acetate, rayon, nylon, or the particles in Bakelite emulsions, polystyrene emulsions, and other emulsions of insoluble organic polymers, rubber, such as natural rubber or synthetic latex commonly referred to as conjugated diene elastomers, colored pigments such as copper phthalocyanines, and other organic materials in various physical forms.

1.3. The Skin

If properly carried out, Iler's method will yield a dense uniform skin of hydrated silica on the cores. The development of a dense skin on the cores can be followed by determining the surface area of samples of coated cores during the course of the coating process. A decrease of surface area with increasing amounts of silica added to the particle dispersion indicates that the silica goes onto cores as a dense coat—see Fig. 6-40 a-c. The particles will gain in weight and the specific surface area is measured as square meters per gram. On the other hand, an increase in the specific surface is a tell-tale indication that the skin is porous—see Fig. 6-40d. Iler specifies that the thickness of the skin should not be thinner than about 3 nm and in most applications it will be adequate to have a skin no thicker than 25 nm.

1.4. Rate of Deposition of Monomeric Silica

Although monomeric silicic acid is the material that is deposited onto the core surface, the most practical form of silica for coating cores of different materials is a decationized solution of 3.3 ratio sodium silicate containing about 3% by weight SiO_2. The ultrafine particles in such a solution, about 2 nm—see Section C of Ch. 2—have a high solubility—see Fig. 2-3—and when they are introduced into a dispersion of cores in water of pH 8-10, the aqueous phase becomes strongly supersaturated in monomeric silicic acid with regard to the surface of the relatively large, coated core particles and the skin will grow thicker.

For fundamental studies it may be more desirable to use a solution of monosilicate ions, e.g., made by dissolving reagent-grade sodium metasilicate in dilute NaOH—see Section C of Ch. 2.

The key steps in coating cores with a skin of silica is, as pointed out above, the application of the first monomolecular film of silica on the core surface. Further deposition of silica is always on a silica surface. It is therefore not surprising that the following equation, Iler (1959, 1979), expressing the maximum rate of addition of active silica to the core dispersion, is actually an adaptation of Eqn. 2-15, giving the maximum rate at which active silica can be added in the build-up process, to the coating process:

$$S = \frac{A}{200}(2)^n \qquad (6\text{-}45)$$

where $\quad n = \frac{(T-90)}{10}$

S = maximum parts by weight of active silica which can be added per hour for each part by weight of cores in the system being treated.

T = temperature in degrees C

A = surface area in m^2/g of the cores

If T = 90°C, Eqn. 6-54 reduces to S = 0.005 A, which is the same as Eqn. 2-15 with a k-value of 0.005. (k has the dimension gram active silica added per hour per m^2 of core surface area).

At 90°C 5 mg silica can be added per m^2 per hour, whereas at 30°C the rate would be only 0.078 mg/m^2h. The coating rate at 90°C is 0.45 nm per hour, assuming that the density of the silica coat is 2.2 g/cm^3 and that the coating is uniform.

Iler (1979) indicates that the rate of deposition can be raised by increasing the concentration of a monovalent salt up to just short of about 0.3 molar where silica is flocculated. If active silica is used it may then be possible that not only monomer is deposited on the core surface, but also some of the small particles, between 1.5 and 3.0 nm, present in such solutions of silica. These are then cemented into an impervious coating by simultaneous deposition of monomer and also by closing the micropores by a process similar to sintering.

1.5. Coating of α-alumina with Silica

Bergna et al (1994) studied coating of α-alumina particles, having a specific surface area of 10-13.5 m^2/g and a broad particle size distribution between 0.1 and 0.8 μm, in an aqueous dispersion having a solids content of 10% by weight Al_2O_3. As the silica source they used a solution of monosilicic acid prepared by dissolving reagent grade $Na_2SiO_3.9H_2O$ in 0.1M NaOH and removing the sodium ions by ion exchange at about 5°C at pH below 2.5 so as to minimize polymerization of monosilicic acid in the coating solution before it was used—see Ch. 2, Section B. The calculated concentration of silicic acid in the coating solution was 3 mg SiO_2 per ml. The coating experiment was carried out at 60°C with pH maintained at 8.5 and additions of dilute acid or base as needed. The addition rate of coating solution was about 2 g of SiO_2 per 1000 m^2 per hour, which was claimed to be slow enough to prevent formation of colloidal silica although it was about four times faster than the addition rate calculated by Eqn. 6-54 at 60°C. However, this equation was established by Iler using a solution of active silica, which contains 1.5 to 3.0-nm silica particles, as coating solution. It may therefore be that higher addition rates than those predicted by Eqn 6-54 will work when the coating solution consists of only monosilicic acid and no polymeric silica.

Bergna et al. determined that the specific surface of the particles decreased monotonously with the amount of added silica, thus indicating that the monosilicic acid deposited on the particles as a dense coating. However, the silica is not deposited in complete, continuous layers on the alumina. Instead, the coating will grow thicker in certain places on the alumina surface before complete coverage is obtained. A theoretical monolayer corresponds to 0.8% by weight of silica, but Fig. 6-41 shows that at 0.8% by weight silica on the alumina surface the Si surface coverage, as determined by SIMS, was somewhat less than 30% and the isolelectric point was between pH 4.5 and 5.0, as compared to between 1.7 and 2.0 for a silica surface.

By assuming that the coating grows by random attachment of monomeric silica units to the particle surface with equal preference for bare alumina or previously deposited silica, Bergna could model the growth process. His model predicts that the fraction of bare alumina, equaling 1 minus the fraction of silica on the surface, should decay exponentially with added silica, which is in agreement with the results shown in Fig. 6-41.

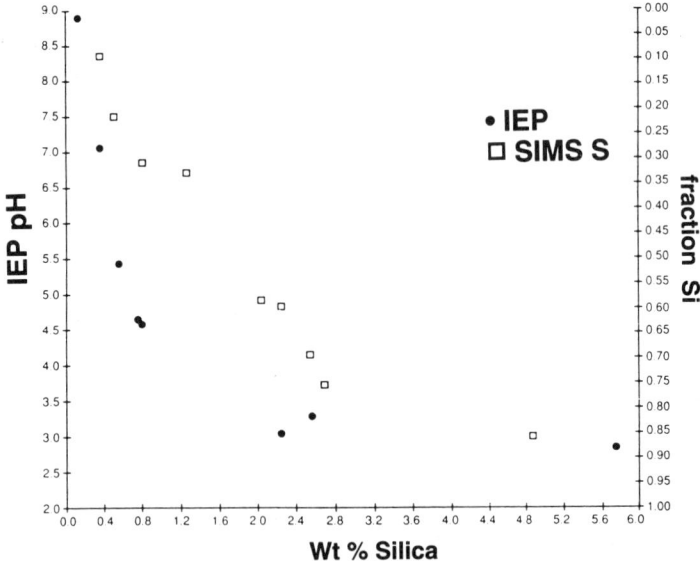

Fig. 6-41. Relationship between isoelectric point, SIMS Si surface coverage, and total weight percent silica on the alumina surface. Firment et al. (1989). Reprinted by permission of John Wiley & Sons, Inc.

1.6. Coating of Titania with Silica

Titania is the most widely used white pigment in paints, plastics and inks since TiO$_2$ particles of suitable size scatter light more cost-effectively than any other material—see Ch. 8. However, the unmodified titania surface is photochemically active and will degrade the organic polymer matrix into which the titania particles are incorporated, e.g., white paint coatings and plastic products. To improve the durability of TiO_2 pigmented polymeric products, the titania particles are therefore often coated with an inert inorganic oxide, usually SiO_2 applied by the process developed by Iler (1959) and described here in §1.1-1.4.

Furlong (1994) examined the Iler dense coating process applied to titania cores, schematically shown in Fig. 6-42.

Sulfuric acid and sodium silicate solutions were added simultaneously but separately to a titania slurry containing 100 g/dm³ of rutile titania particles with a specific surface area of 20 m²/g. The silicate solution was added at a rate of about 0.06 g SiO_2 min⁻¹dm⁻³, i.e., 0.06 g SiO_2 per 2000 m² titania surface area, which is only about one third of the maximum addition rate at 90°C according to Eqn. 6-54, i.e. about 0.17 g SiO_2 min⁻¹ per 2000 m² titania surface area.

For titania particles with a specific surface area of 20 m²/g a theoretical continuous monolayer of silica (0.35 nm thick) corresponds to about 1.6 wt. % SiO_2 on the particles, and Fig. 6.43 shows that at this loading the particle surface is almost silica-like. The isoelectric point decreases with increasing silica loading, but reaches a plateau at somewhat more than 2 wt. % SiO_2.

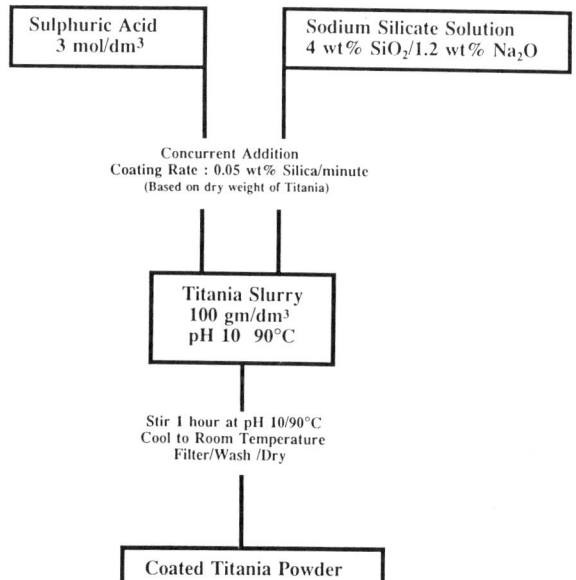

Fig. 6-42. Sequences for the preparation of Iler dense silica (DS) coatings on titania particles. Furlong (1994). Courtesy American Chemical Society.

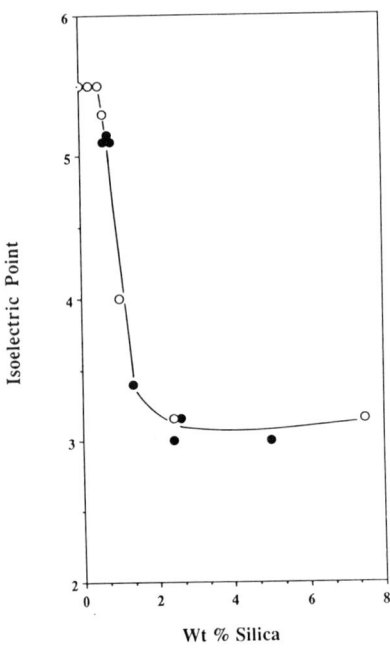

Fig. 6-43. Isoelectric points of DS-coated titanias: O, titania dispersed in aqueous silica; •, dried and coated titania powders. The base titania was 20 m²/g. Furlong (1994). Courtesy of the American Chemical Society.

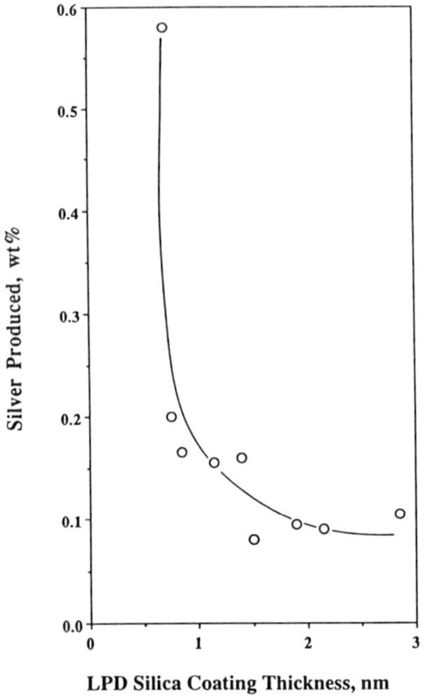

Fig. 6-44. Dependence of the amount of Ag photodeposited on silica-coated titania particles on the thickness of silica coating. Tada et al. (1989). Courtesy of the Chemical Society of Japan.

Figure 6-44 shows that a monolayer of silica, 0.35 nm thick, was not enough to significantly inhibit the photoreduction of aqueous silver ions. Instead, 4-5 monolayers were needed. Furlong (1994) pointed out that "the surface chemistry of silica-coated titania is useful in terms of the inhibition of surface photochemistry, as with pigments, and has potential for control of 'preparative' surface photochemistry".

2. Non-silica Coatings on Various Cores

Non-silica coatings can be applied to cores of many different types, including silica and other hydrous metal oxides. Cores of silica play an especially important role in this coating technology since they are readily available as spherical particles in the size range from 5 to several hundred nanometers.

2.1. General Principles

Just as in the case of coating non-silica cores with a silica coating, the first step is to apply a monomolecular film of the coating material, e.g., a hydrous metal oxide or aluminosilicate, to the core surface, which must be water wettable and receptive to the coating material or made so by pretreatment. Most metal oxides will accept an initial coating from another metal oxide from solution if the pH is close to the point where the metal oxide is precipitated. This is because just before the appearance of a precipitate, there is present polymeric basic metal oxide as subcolloidal particles—see Ch. 3—which cover the surface.

Also as before, the second step is to increase the thickness of the coating by further deposition of the coating material onto the precursor base coat under such conditions that all the coating material added in the process will be added to the coating and not form secondary particles in the aqueous phase of the dispersion.

2.2. Silica Particles Coated with Titania

Greenwood and Otterstedt (1996) developed a method for coating silica cores of diameters from 300-500 nm with up to 500 wt.% TiO_2 (based on the weight of SiO_2) and reversing the charge of the titania coated particles from positive to negative by applying a thin, final coat of silica. The silica cores were made by the Stöber process—see Ch. 2—and had 0.06 or 1.5 aluminosilicate sites per nm^2—see Section C, this chapter. The first step in the coating process was to reverse the charge of the cores by adding them to a rapidly stirred solution containing 0.79 M $TiCl_4$ and 1.37 M HCl. After the addition was completed the TiO_2/SiO_2 weight ratio was 0.03 and the solids content was adjusted to 4 wt.% by dilution with distilled water. Coating was accomplished by heating the dispersion of charge reversed cores to 75°C and adding a freshly made solution of titanium tetrachloride containing 0.79 M $TiCl_4$ and 1.37 M HCl, at a rate of 0.2 mmol TiO_2 per m^2 of core surface area per hour. The pH was maintained at either 1.5 or 2.0 with addition of 3.75 M NaOH solution and the ionic strength was kept low by continuos ultrafiltration through a fluoropolypropylene membrane with a cut-off of 10^5.

Figure 6-45 shows that the coating was more uniform on the cores with 1.5 Al/nm^2 than with 0.6 or no Al/nm^2 and that the particles with the most highly aluminized cores were the least aggregated. Introducing negatively charged aluminosilicate sites in the surface of the cores will facilitate the first step in the coating process, i.e. reversing the charge on the cores by adsorption of polycations of titanium—see Ch. 3—and make for more uniform coatings of titania. At the pH of the first step in the coating process the almost uncharged surface of the unmodified silica particles will have much less affinity for these polycations than the negatively charged surface of the Al-modified cores.

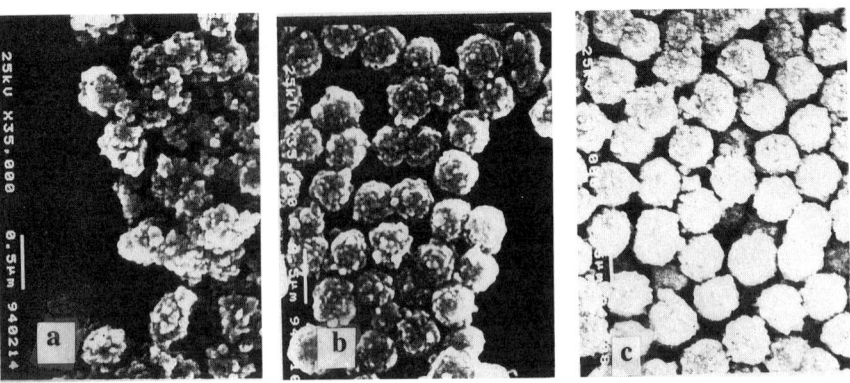

Fig. 6-45. 300-nm Stöber sols, (a) not aluminate-modified, (b) aluminate-modified, 0.6 Al /nm^2, (c) aluminate-modified, 1.5 Al/nm^2. Greenwood and Otterstedt (1996).

Figure 6-46 shows that moderate stirring in addition to the agitation provided by the circulation pump of the microfilitration unit significantly improved the state of dispersion of the coated particles.

It is apparent from Fig. 6-47 that the isoelectric point of silica cores coated with a thick layer of TiO$_2$ is just somewhat lower than that of pure TiO$_2$ particles, whereas the i.e.p. of the charge reversed particles from the first step in the coating process is about one pH unit lower. The titania layer in this case is not thick enough to completely mask the underlying silica surface. X-ray diffraction analysis of the material showed that titania deposited on the cores as anatase. The properties of the composite particles as white pigment will be described in Ch. 7.

It is important to keep the ionic strength low during the coating process so as to prevent particle aggregation and yield a uniform coating. Hsu et al. (1970) solved this problem by coating silica particles of diameters from 0.2 to 1.2 μm made by the Stöber process with titania in a two-stage coating process when the

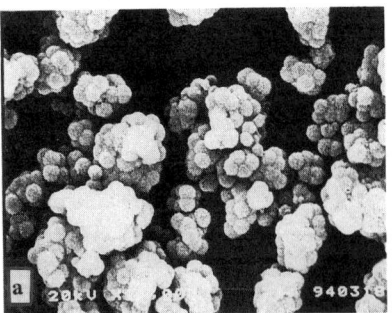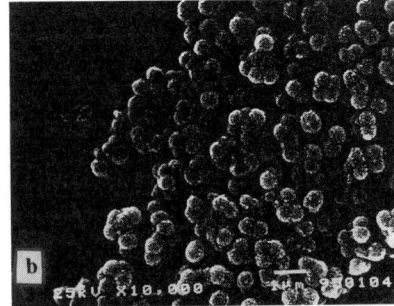

Fig. 6-46. 300-nm Stöber sols (aluminate-modified), 1.5 Al/nm^2, coated with TiO$_2$ under (a) agitation provided by the circulation pump only; (b) moderate stirring and agitation provided by the circulation pump. Greenwood and Otterstedt (1996).

thickest coatings, about 100 wt. % TiO_2 based on the weight of SiO_2 as compared to up to 400 wt. % TiO_2 for Greenwood's composite particles, were applied. In the first step, up to 25 wt. % TiO_2 was deposited by adding a fresh solution of 0.2 M $TiOSO_4$ in 1 M H_2SO_4 to a dispersion of the silica cores under conditions of pH, silica concentration, and addition rate of titanyl sulfate which effected good uniformity of coating. The particles were next separated by filtration, redispersed in pure water and the remaining titanyl sulfate was added.

The coated particles were calcined at different temperatures up to 1000°C. X-ray analysis indicated only anatase below 800°C. At 1000°C rutile began to appear as a minor component.

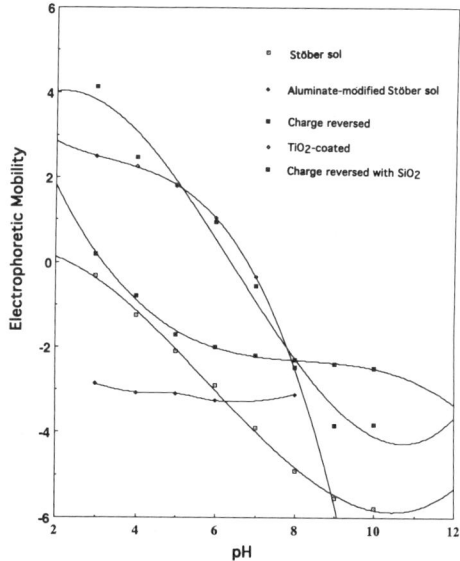

Fig. 6-47. Electrophoretic mobility as a function of pH of 240-nm Stöber sols. (□) unmodified, (♦) is (□) aluminate-modified (1.5 Al/nm²) (◘) is (□) charge reversed with basic aluminum chloride, (○) is (□) coated with TiO_2/nm^2), (■) is (□) charge reversed with 3.3 ratio Na-silicate solution. Greenwood and Otterstedt (1996).

2.3. Composite Particles Coated with Hydrous Metal Oxides

In an attempt to develop new and unique catalytic materials for use in pollution control catalysis Törncrona et al.(1996) studied methods of depositing ceria in a highly dispersed state on the surface of substrate particles—see Fig. 6-48.

The 80 and 240 nm silica cores were made by the build-up process—see Ch. 2, Section D. The sodium content was reduced by passing the sols slowly through a column of strong cation exchange resin in the H^+ form. The cores of 500-nm diameter were made by the Stöber method (Ch. 2, Section B).

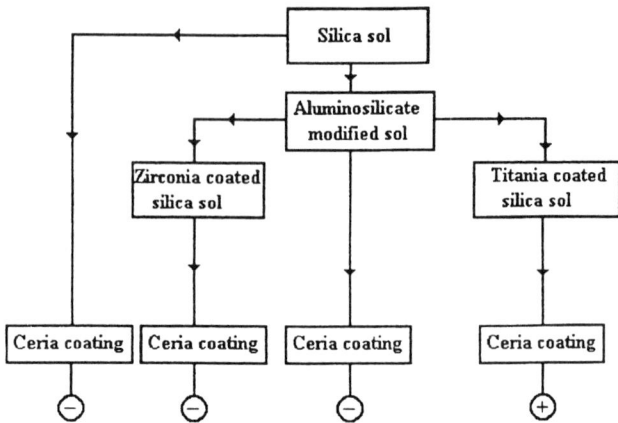

Fig. 6-48. Schematic route for making metal-oxide coated silica sols. Ceria coat applied only to titania-coated silica sol. Törncrona et al. (1996). Courtesy The Royal Socirty of Chemistry

The core surface was aluminized, corresponding to 1.5 aluminosilicate sites per nm^2, by heating the cores with a solution of sodium aluminate—see Section C of this chapter.

Coating of the cores with titania or zirconia was achieved by adding the aluminized core sols slowly (about 20 min.) to titanium chloride solutions, 0.79 M TiCl$_4$ and 1.37 M HCl, and zirconium chloride solutions, 0.64 M ZrCl$_4$, respectively, and heating the solutions at 90°C for 1 h. The coated sols were slowly cooled to room temperature and the pH was raised from about 1.5 to 2.0 by adding an anion exchange resin in the OH$^-$ form. The TEM micrographs in Fig. 6-49 shows that titania deposits as small particles, evenly distributed over the core surface, whereas zirconia goes on as a homogeneous layer.

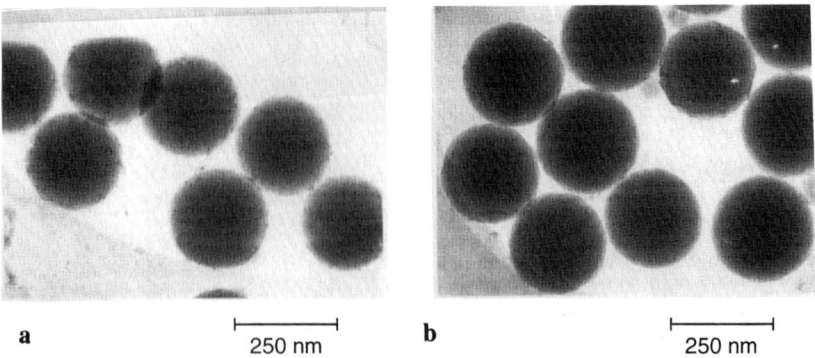

Fig. 6-49. Transmission electron micrograph images of silica particles coated with a thin layer of titania (a) and zirconia (b). Törncrona et al. (1996). Courtesy The Royal Society of Chemistry.

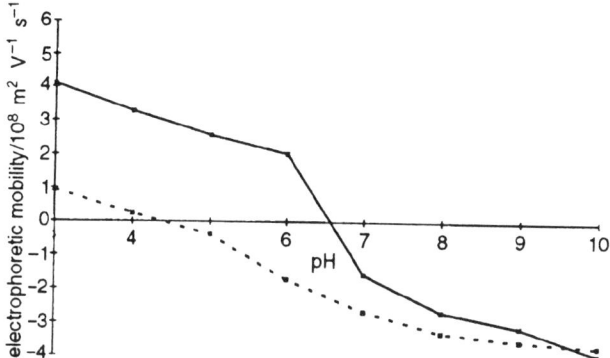

Fig. 6-50. Electrophoretic mobility as a function of pH for silica sols (240 nm diameter) coated with zirconia (—) and titania (- - -). Törncrona et al. (1996). Courtesy The Royal Society of Chemistry.

Figure 6-50 shows that the deposited amount, about 2 atom % titanium and from 0.5 to 3.5 atom % zirconium, as determined by EDS, made the electrophoretic mobilities of the coated particles quite similar to those of particles of titania and zirconia, respectively. The isoelectric point of the titania coated particles was about one pH unit lower than that of pure titanium particles—see also Fig. 6-47.

As Fig. 6-48 shows, only cores with a layer of titania could be coated with ceria. The best results were obtained with a stock solution made by dissolving cerium (IV) sulfate in dilute sulfuric acid at room temperature. To promote the formation of polycations by hydrolysis of Ce^{4+} ions—see Ch. 3—ammonium hydroxide and urea were added. The resulting stock solution was 10mM in $Ce(SO_4)_2$, 50 mM in H_2SO_4, 100 mM in NH_3, and 100 mM in $(NH_2)_2CO$. By adding titania coated sols rapidly to this stock solution and heating it at 80°C for 20 h, ceria was deposited on the particles in an amount corresponding to 5 mass

(a)   (b)

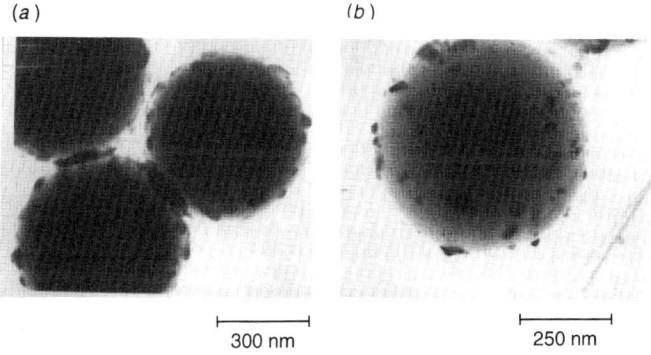

|———| 300 nm |———| 250 nm

Fig. 6-51. TEM images of titania-coated silica particles on which cerium (IV) oxysulfate was deposited after calcination at 550°C (a) and 800°C (b) for 30 min. Törncrona et al. (1996). Courtesy The Royal Society of Chemistry.

% CeO$_2$, based on the total particle weight. Figure 6-51 shows SEM and TEM micrographs of ceria on titania coated silica cores.

The failure to coat unmodified silica particles with ceria was probably due to the fact that at the low pH, about 1.2, the silica surface is weakly positively charged and has little or no affinity for the polycations of cerium (IV). In the case of Al-modified silica cores, aluminum was probably dissolved out from the surface at about pH 2, leaving behind unmodified silica cores. Finally, the much higher solubility of zirconia, compared with titania, at pH 1.2 may be the reason why the zirconia coated cores failed to accept a layer of ceria. The catalytic properties of ceria on SiO$_2$-TiO$_2$ composite will be discussed in Ch. 7.

2.4. Coating Cores with Aluminosilicate

Alexander (1956, 1961) and Bergna (1980) described composite particles with a skin, or a coating, of aluminosilicate. Alexander (1956) claims a process for coating cores with aluminosilicate including:

1. Making a dispersion of particles of 1-100 nm in diameter; particles consist of silicates (e.g., clay particles), silica, or various metal oxides.
2. Adding simultaneously two separate streams of active silica and sodium aluminate with the ratio of Si atoms to Al atoms added between 1:1 and 1:200.
3. Adding streams of active silica and sodium aluminate until the particles are coated with a skin of thickness between 3 and 50 nm.
4. Maintaining the pH between 5 and 12.
5. Maintaining the temperature at 60-100°C.

In 1961 Alexander claimed a stable aquasol of:
1. Negatively charged aluminosilicate particles of diameters from 3-150 nm dispersed in water of pH about 4-10.
2. Si:Al atom ratio 1:1 to 50:1; and a process, which includes:
   a) Adding two separate streams of active silica and sodium aluminate simultaneously to a vigorously agitated body of water of pH from 8-12.
   b) Maintaining the pH between 8-12.
   c) Maintaining the temperature at 50-100°C.

Bergna (1980) also claims a stable aluminate aquasol of:
1. Uniform particles of weight-average diameter of 3-90 nm. (The weight of each particle as measured in micrographs is proportional to the diameter cubed. All particle weights are added up and divided by the number of particles to get the average weight whose cube root is defined as the *weight-average diameter*.)
2. A maximum standard deviation of 0.37d in diameter. A preferred claim is a maximum average deviation of 0.30d.
3. Si:Al atom ratio of 1:1 to 19:1; and a process which includes:
   a) Making a *heel* (see Ch. 2, Section D) of uniform particles of 5-85 nm average diameter; particles consist of aluminosilicate, silica or a variety of refractory oxides.

b) Adding two separate solutions of sodium or potassium silicate and a separate stream of aluminate.

c) Rate of addition is 10 grams of SiO_2 per hour per 1000 m² of surface of the particles on which silica and alumina are being deposited—which is about twice as fast as calculated from Eqn. 6-45 at 90°C.

d) Maintaining the temperature at 50-100°C.

e) Maintaining the pH between 9 and 12 by adding an NH_4^+ or H+ form of ion exchange resin.

f) Removing resin and optionally concentrating the resulting sol.

The three patents outlined above seem to be closely related and one might wonder why Bergna's patent is valid in view of the earlier Alexander patent.

In Bergna's patent (1980) the first claim is specific to an aluminosilicate sol without reference to the process of manufacture. Only the particle size, ratio of Si to Al, and uniformity of particle size are specified.

Alexander (1961) claimed aluminosilicate sols that are clearly defined as to composition, but not limited as to uniformity of particle size. In fact, uniformity is not mentioned.

It would appear that Alexander's claims covered the Bergna sols, but that the latter claimed uniformly sized particles as an improvement over Alexander. It may therefore be that the Bergna claim was allowed because no data was available on the uniformity of particle size of the Alexander sol and it was thus assumed Bergna was an improvement, especially for use in cracking catalysts—see Ch. 7.

Moreover, it is also true that Bergna specified the rates of addition of reagent solutions so as to achieve uniformity. Alexander did not specify in his process claims the rate of addition necessary to produce uniform particles. It may, therefore, have been this failure on the part of Alexander to specify the rates of addition of components that could have led to the conclusion that his sols were not uniform in particle size. Certainly, unless Alexander carried out his process slowly enough, his sols would be non-uniform in particle size.

E. Adsorption of Polymers on the Silica Surface and some Other Surfaces

Most surfaces in nature are negatively charged and interactions between such surfaces and polymers, especially cationic and nonionic polymers, leading to flocculation play an important role in many technical processes; e.g., water treatment, retention of fine material in the papermaking process, and clarification, fining, of various beverages.

Using the silica surface as a model surface, and building on Iler's (1979) survey of the subject, we will review, in this section, the recent literature on interactions between cationic and nonionic polymers and silica.

1. Definitions and Abbreviations

The terms cationic and anionic refer to the pH-range 2-10 in which polymers are usually adsorbed on silica and other materials.

Nonionic Polymers

PEO = polyethylene oxide with repeating unit $(-C_2H_4O-)_n$
PVA = polyvinyl alcohol with repeating unit $(CH_2-CHOH-)_n$
TX 100 = $C_8H_{17}-C_6H_4-(OCH_2CH_2)_{10}OH$
TX 102 = $C_8H_{17}-C_6H_4-(OCH_2CH_2)_{12}OH$
PAM = polyacrylamide; see formula for P(AM-CMA) below

Cationic Polymers

P(AM-CMA) =

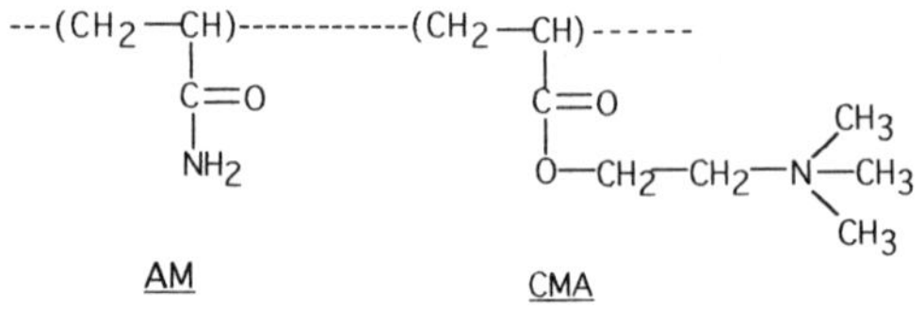

$$---(CH_2-CH)------------(CH_2-CH)------$$

AM   CMA

PEI = polyethylene imine with repeating unit $\left(-(C_2H_4)_2NH_2^+-\right)_n$

PDMDAAC = poly(dimethyl diallylammonium chloride)

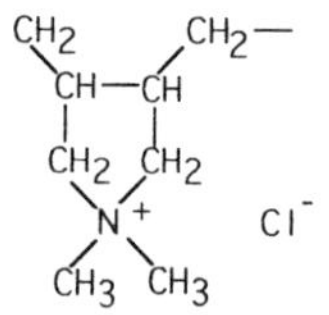

DDTAB = $(C_{12}H_{25}N^+(CH_3)_3)Br^-$

TDTAB = $(C_{14}H_{29}N^+(CH_3)_3)Br^-$

CTAB = $(C_{16}H_{33}N^+(CH_3)_3)Br^-$

Anionic Polymers

P(MS-α-MeSty) = poly(maleic acid-CO-α-methylstyrene)

2. Polymers in General

Synthetic polymers can usually be divided into homopolymers or block copolymers. In homopolymers a small monomer unit is repeated N_p times (N_p is called the degree of polymerization). Block copolymers consist of blocks of one repeating monomer unit succeeded by one or more blocks of other repeating monomer units.

Natural polymers, e.g., proteins and nucleic acids, are often heteropolymers, in which several different monomer units are combined in an irregular pattern. A polyelectrolyte is a homo- or heteropolymer or a block copolymer in which some or all of the monomer units carry an electrical charge.

In dilute solutions interactions between polymer molecules can be neglected. Instead, monomer units along the molecule can interact with each other and the solvent so as to make the polymer chain assume different configurations—Fig. 6-52.

If the polymer-solvent is repulsive, the chain folds back on itself to minimize this interaction and assumes a compact globular shape. On the other

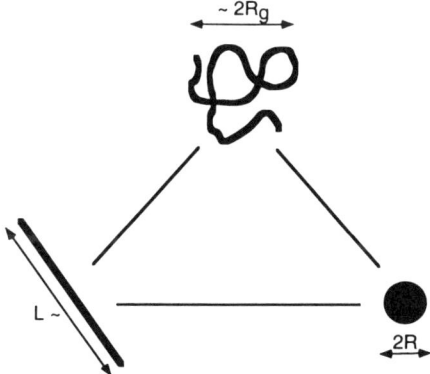

Fig. 6-52. The triangle of Haug, illustrating three extreme types of polymer configuration: the stiff rod ($L \sim N_p$), the compact globule ($R \sim N_p^{0.3}$), and the coil ($R_g \sim N_p^\alpha$ with $0.5 < \alpha < 1$). Adapted from Evans and Wennerström (1994).

hand, if the polymer molecule has a propensity for forming a helical structure, like DNA, it may adopt a stiff linear α-configuration. Most synthetic polymers in solution, however, assume a random coil configuration, which is a much less well-defined structure than a rod or a globule. The size of the coil, as measured by its radius of gyration, R_g, depends on the degree of polymerization, N_p, and the chain extension, which in turn depends on the interaction between monomer units and the solvent.

3. Adsorption of Polymers on Silica

In this chapter colloidal stability, or lack thereof, is discussed in terms of repulsive and attractive interactions between colloidal particles. The key factor determining the stability is an energy barrier separating the well dispersed particles from the collapsed, aggregated system.

Aggregation of colloidal silica, and indeed of most colloidal systems, can be achieved by adding a polymer, in which case one speaks of flocculation, or an oppositely charged ion, e.g., Al^{3+}, in which case one refers to the aggregation as coagulation.

If the heat of adsorption is larger than kT, the adsorption of polymers on the silica surface is usually irreversible. Kinetics and distribution of polymer molecular weights as a rule play a role. Smaller polymers will diffuse faster than large polymers and be the first to reach the surface. However, the adsorption of large polymer molecules is thermodynamically favored and they will in time—hours or days—replace smaller polymers on the surface.

Flocculation may be brought about by different mechanisms, but in the case of colloidal silica two appear to dominate. One is bridging flocculation, which means that a high-molecular-weight polymer is adsorbed on one particle and then is able to collide with another particle and form a bridge across the combined electric double layers of the two particles—see also Ch. 9.

The other mechanism is called patching, which is common with highly charged polyelectrolytes of medium molecular weight (about $5x10^5$). The polyelectrolyte adsorbs on the particle surface, forming positively charged patches that can interact with negatively charged domains on other particles. Patching is more reversible than bridging in the sense that the flocs may more readily reform after having been exposed to shearing—see Ch. 9.

The shape, which may be straight, branched, or crosslinked, and the charge of polymers will affect how they adsorb on particle surfaces. They may adsorb as:

- spheres, like highly branched starch
- random coils with segments bound to the surface
- flat, collapsed configurations
- linked chains with one end attached to the surface and the other end projecting from the surface

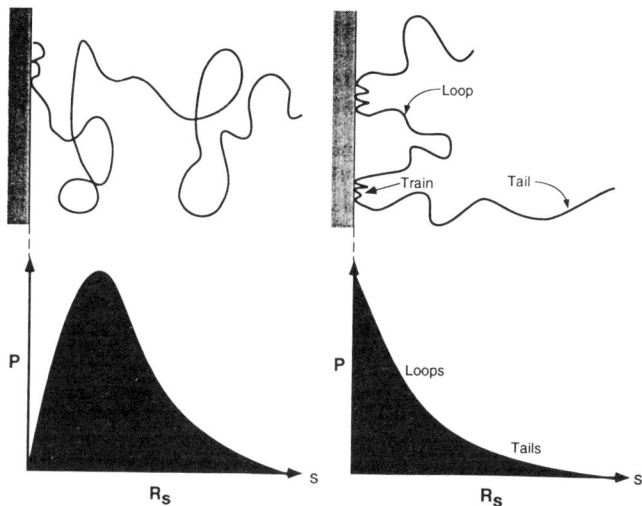

Fig. 6-53. Polymer configurations at a surface. (a) Grafted chain or block copolymer with one highly adsorbing segment. The main chain is assumed to be in a good solvent with no affinity for the surface. Thus, the chains try to avoid the surface. (b) Adsorbing homopolymer, but in a good solvent, indicating a loop, two trains, and the two tails that extend farthest into the solution. Israelachvili (1992). Courtesy Academic Press, London.

According to Iler (1979), PEI is irreversibly bound to silica over a wide pH range. The critical flocculation concentration, c.f.c., depends on the molecular weight at pH above 9 if no salt is present. Since PEI is only weakly charged at pH above 9, and the silica particles are still negatively charged at the c.f.c., flocculation appears to be due to bridging. The silanol groups of the silica surface will hydrogen bond to the oxygen atoms of polyethers and PEO at pH < 7. At low pH the fully hydroxylated silica surface will adsorb a maximum of 7 ethylene oxide groups per nm^2. Iler (1979) also stated that PAM will flocculate silica at pH below 8 and that the effectiveness increases with molecular weight.

PVA will also hydrogen bond to the silica surface at low pH. Maximum adsorption will occur at the isoelectric point of silica, i.e. pH 1.7-2.0. The hydrophobic hydrocarbon groups, which partially cover the surface, will cause aggregation.

The most common types of adsorption are shown in Fig. 6-53.

3.1. Interaction between Silica and Nonionic Polymers, $\tau = 0$

Baudin et al. (1990) studied the interaction between polysaccharides of the type dextran and pullulan and silica sols, having an average diameter of 40 nm and a BET surface area of 32 m^2/g.

In the pH range investigated, 5-8, the two polysaccharides were electrically neutral. Figure 6-54 shows that for the system silica-dextrin, polymer adsorption increases with molecular weight and inclines to level off for very large molecules.

The adsorption was thought to occur through hydrogen bonds between the silanol groups and the hydroxyl groups of the glucose units. Somewhat surprisingly, the maximum adsorption was found to be independent of pH although the number of non-ionized silanol groups increased with decreasing pH. Also, ionic strength, at least not up to 0.5 M NaCl, did not affect the amount dextran adsorbed.

Dilution experiments showed that dextran adsorption was partially reversible, but less at higher molecular weight. Thus, 40% of dextran of molecular weight 10^4 was desorbed on dilution as compared to only 20% of dextran of molecular weight 1.6×10^6. This is unusual behavior for adsorbed polymers, but was thought to be due to the poor affinity of dextran for the silica surface.

Figure 6-55 shows that the thickness of the adsorbed layer increases linearly with molecular weight for pullulans, but not for dextrans. The differences between the two types of polysaccharides was assumed to be due to the fact that dextrans are branched whereas pullulans are linear polymers.

Both dextrans and pullulans were found to be weak flocculants, even at salt concentrations above 0.5 M.

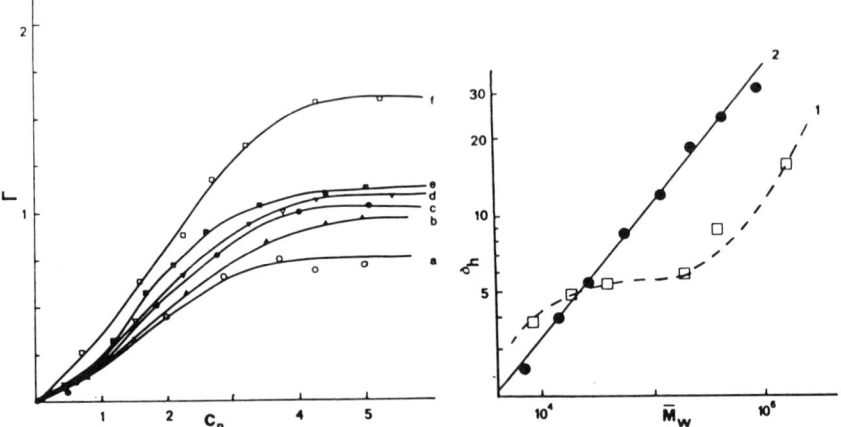

Fig. 6-54. Adsorption isotherms for the system Syton silica-dextran Γ (mg/m²). C_p equilibrium concentration of polymer (g/liter). $M_w = 9900$ (DT 10); (b) $M_w = 17,700$ (S 17); (c) $M_w = 41,500$ (DT 40); (d) $M_w = 185,000$ (S 170); (e) $M_w = 450,000$ (DT 500); and (f) $M_w = 1,630,000$ (DT 2000). Baudin et al. (1990). Courtesy Academic Press.

Fig. 6-55. Plot of the layer thickness δ_h (nm) of dextran and pullulan adsorbed at the water-silica interface versus the polymeric molecular weight. curve 1: □, dextrans; and curve 2: •, pullulans. Baudin et al. (1990). Courtesy Academic Press.

Killmann and Sapuntzjis (1994) and Killmann et al. (1992) measured adsorption isotherms and thickness of the adsorbed layer of PEO of different molecular weights on Stöber sols and polystyrene latices of diameters 106 nm and in the range from 37 to 113 nm (as measured by dynamic light scattering), respectively. The thickness of the adsorbed polymer layer was determined as the difference in hydrodynamic radii between particles with and without adsorbed polymer. The adsorption of polymer on the particles increased with polymer concentration and reached a plateau, corresponding to a plateau thickness of polymer, which depended on the chemical nature of the particle surface and the molecular weight of PEO.

The variation in the thickness of the adsorbed layer, $\bar{d}$, with the amount of adsorbed PEO of different molecular weights is shown in Fig. 6-56 for silica particles and in Fig. 6-57 for polystyrene particles. The thickness increases markedly for both types of particles, but most for the polystyrene particles, as the adsorbed amount of PEO reaches the plateau region of the isotherms. In this region of complete coverage of the particle surface there is strong competition between the polymer molecules for adsorption sites. Compared with lower surface coverages, the random coils can only attach themselves to the surface at relatively few points leaving a larger portion of the polymer chain to extend out into the aqueous phase as loops and tails—see Fig. 6-53. The marked increase in $\bar{d}$ near the plateau region of the adsorption isotherm will therefore be more pronounced for higher molecular weights of the polymer.

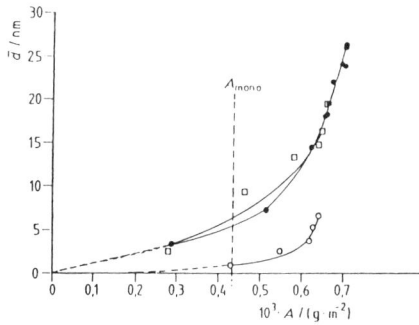

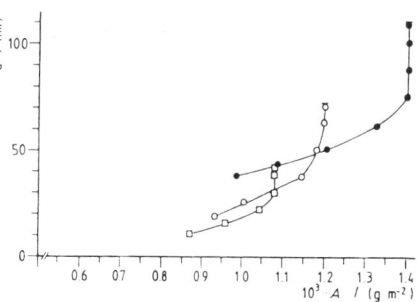

Fig.6-56. Dependence of the adsorbed layer thickness d on the adsorbed amount A of PEO on silica 212 (in H_2O, at 25°C): •, M_w = 996,000; □, M_w = 594,000; O, M_w = 86,000. Killmann and Sapuntzjis (1994). Reprinted from Macromol. Chem. **61**, 42 (1992). with kind permission from Elsevier Science - NL, Sara Burgerhartstraat 25, 1055 KV Amsterdam, The Netherlands.

Fig. 6-57. Dependence of the adsorbed layer thickness d on the adsorbed amount A of PEO on latex 233 (in H_2O, at 25° C): •, M_w = 900,000; O, M_w = 325,000; □, M_w = 160,000. Killmann and Sapuntzjis (1994). Reprinted from Macromol. Chem. **61**, 42 (1992). with kind permission from Elsevier Science - NL, Sara Burgerhartstraat 25, 1055 KV Amsterdam, The Netherlands.

3.2. Comparison between Interactions of Nonionic and Cationic Polymers with Silica

Wang and Audebert (1987) found that P(AM-CMA) with $\tau > 0.15$ flocculated silica sol by a patching mechanism—see §2 of this chapter and Chapter 9—whereas PAM ($\tau = 0$) did not affect the stability of silica sols. In order to investigate the interaction between silica particles and weakly cationic polymers, Wang and Audebert (1987) determined isotherms for the adsorption of weakly charged P(AM-CMA), $0.005 \leq \tau \leq 0.01$, with molecular weights in the range from 1.8×10^5 to 3×10^6, on the surface of silica in the form of silica particles with a wide particle size distribution between 20 to 200 nm. Figure 6-58 shows that the adsorption of P(AM-CMA) reaches a plateau at an equilibrium concentration of polymer in the solution of about 0.5 g/l and that the plateau increases rapidly in height with molecular weight.

Since PAM ($\tau = 0$) does not adsorb on the negatively charged silica surface, it is reasonable to assume that only the cationic sites, and then just a fraction of them, but not the amide groups along the chain, attach to the surface. In the case of a polymer of molecular weight 700 000 and $\tau = 0.0025$, there are only about 25 cationic sites along the chain. A fraction of these stick to the surface, leaving most of the polymer molecule to extend out into the solution as loops and tails. This model of the adsorption explains the large uptake of polymer and makes it plausible that the polymer adsorbs in thick layers, consistent with the results of Killmann and Sapuntzjis (1994) discussed in § 3.1.

P(AM-CMA) of high molecular weight, about 2×10^6, and low cationicity are effective flocculants. Figure 6-59a shows that the flocculation efficiency increases with decreasing cationicity due to increasing thickness of the adsorbed layer, which brings about flocculation by a bridging mechanism—see Ch. 9.

Exposing the flocculated systems to increasing shear rate, achieved by agitation, improved the flocculation yield somewhat for cationicities 0.01 and

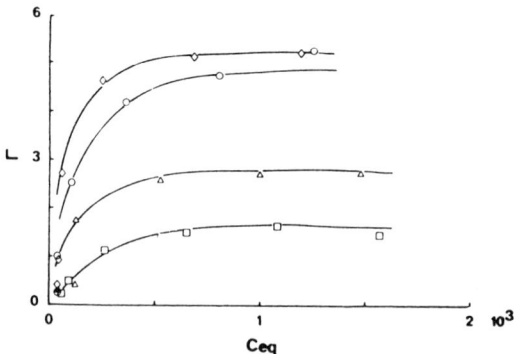

Fig. 6-58. Effect of the molecular weight on the adsorption isotherm. The amount of polymer adsorbed, Γ (mg/m^2), is plotted vs. the polymer concentration in the surrounding medium, Ceq (mg/liter), $\tau = 0.005$. Molecular weights: ($\square$) 1.8×10^5, ($\triangle$) 7.6×10^5, (O) 1.9×10^6, ($\lozenge$) 3×10^6. Wang and Audebert (1987). Courtesy Academic Press.

0.005 due to enhanced particle collision frequency. However, at the lowest cationicity, $\tau = 0.0025$, increasing shear rate broke the interparticle bridging and partially deflocculated the aggregates, Fig. 6-59b.

Kawaguchi et al. (1988) studied the adsorption of fully quaternized PVPP and PEO on silica gel with a surface area of 141 m^2/g at pH 4. The isotherms for both polymers were of high-affinity type and reached a plateau at low equilibrium concentrations of polymer in the solution. The amount of PEO at the plateau did not vary much with molecular weight in the range from 21 x 10^3 to 860 x 10^3, but increased with the amount of KBr present in the solution. The salt solutions are poorer solvents for PEO than pure water and drive the polymer onto the silica surface. The plateau amount of adsorbed PVPP (MW = 550 x 10^3) also increased with KBr concentration, but faster than that of PEO, which was ascribed to the fact that the excluded volume of the polyelectrolyte chains decreased with salt concentration. Figure 6-60 shows that in a competitive situation the adsorbed amount of PVPP decreases monotonically beyond $C_0 = 0.04$ g/100ml, whereas complete adsorption of PEO was found over the entire range of C_0.

It may seem somewhat surprising that cationic PVPP does not interact more strongly with the negative silica surface than nonionic PEO. At pH 4, however, the negative charge on the surface is low—see Fig. 6-30, and strong hydrogen bonding between PEO and Si-OH groups prevail over electrostatic attraction between the cationic pyridinium groups on the PVPP chain and a small number of ionized silanol groups on the surface.

In fact, it was found that at pH 4 adsorbed PVPP molecules were displaced from the surface when PEO was added to the solution. However, at higher pH, e.g., above 8, it might be expected that the situation would be reversed and PVPP would be preferentially adsorbed over PEO.

Scattering experiments provide insight into structures by gaining information about the nature of forces between particles, the geometry of interparticle bonds, the geometry of the bonds and the topology of the aggregates.

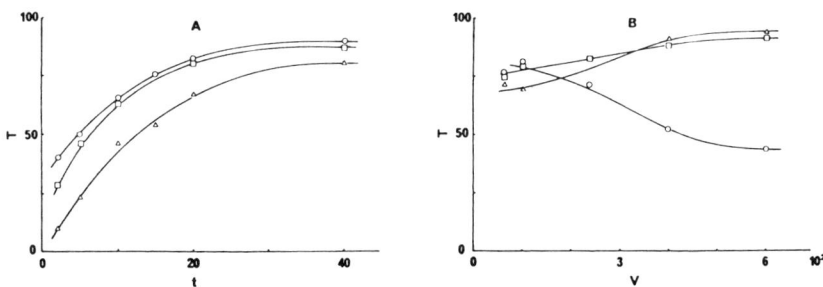

Fig. 6-59. Flocculation according to the jar test method. (A) Effect of the stirring time, t (min), on the transmission, T (%) (speed of stirring, 100 rpm). (O) $\tau = 0.0025$—mol wt 2.4x10^6, ($\square$) $\tau = 0.005$—mol wt 3.0x10^6, (Δ) $\tau = 0.01$—mol wt 2.7x10^6. (B) Effect of the rotation speed , V (rpm) on the transmission, T (%) (stirring time, 20 min); same symbols as in (A). Wang and Audebert (1987). Courtesy Academic Press.

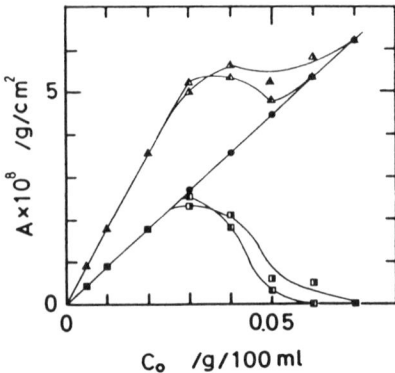

Fig. 6-60. Adsorbed amounts of PEO and PVPP as a function of dosage of PEO (or PVPP) C_0 in competitive adsorption of PVPP and PEO: (•) adsorbed amounts of PEO-12 and PEO-860; (□) adsorbed amount of PVPP and (▲) total adsorbed amount of PVPP and PEO in PVPP/PEO-12 mixture; (■) adsorbed amount of PVPP and total (▲) adsorbed amount of PVPP and PEO in PVPP/PEO-860 mixture. Kawaguchi et al. (1988). Courtesy Academic Press.

Scattering intensities, $I(Q)$, are functions of Q, which is inversely related to interparticle distance and has the dimension Å^{-1}. Wong et al. (1988) used small-angle neutron scattering to study the flocculation behavior of 40-nm silica particles with P(AM-CMA), having cationicities from 0.01 to 1 and molecular weights from 0.05×10^6 to 1.7×10^6, and PEO with molecular weights between 0.6×10^6 and much above 5×10^6, as flocculants.

Figure 6-61 shows the scattering intensities of flocs made with P(AM-CMA) of different cationicities. At high Q-values (short distances between particles) interparticle interference reveal that flocs made with polymers of different cationicities have quite different structures. A strong short-range order peak is observed for flocs made with low charge density polymers, whereas the curves representing flocs with highly charged polymers change directly from the high-Q scattering into the slow monotonic increase at low Q's. At lower Q's the structure of the flocs varies markedly with the cationicity of the polymer and at the lowest Q's, corresponding to interparticle distances of about 5-10 particle diameters, the flocs scattered strongly, indicating that their structure must be heterogeneous.

Since the charge on the silica particles depends on pH—see Fig. 6-30—the electrostatic interactions between P(AM-CMA) and the particles, and therefore also the light scattering intensity, vary with pH, as is shown in Fig. 6-62.

At low pH, the repulsion between the silica particles is quite weak and they can approach each other closely. The trace of a peak at high Q's in Fig. 6-61 suggests that the particles are on the average quite close. The charge on the particle surface increases with pH, resulting in stronger repulsions between the particles. The shift in the peak of the scattering intensity curve toward smaller Q's indicates that the average separation between the particles increases with pH and surface charge.

Figure 6-63 compares the scattering curves for P(AM-CMA) ($\tau = 0.01$) and PEO at pH 7-8 and low ionic strength, conditions under which the two polymers might be expected to flocculate silica particles in a similar manner. The figure suggests that the structure of the flocs produced by the two polymers are quite similar. The main difference is that the structure of the PEO flocs depends more strongly on pH than those of low-charge P(AM-CMA), and that PEO must have higher molecular weight than P(AM-CMA) to bring about flocculation.

Moudgil and Somasundaran (1984) showed that nonionic PAM of molecular weight up to 1.2×10^6 did not adsorb on silica particles 90% of which were smaller than 10 μm and had a surface area of 280 m^2/g in the pH range 2.6-9.8, whereas P(AM-CMA) adsorbed with a maximum amount at pH 9.8.

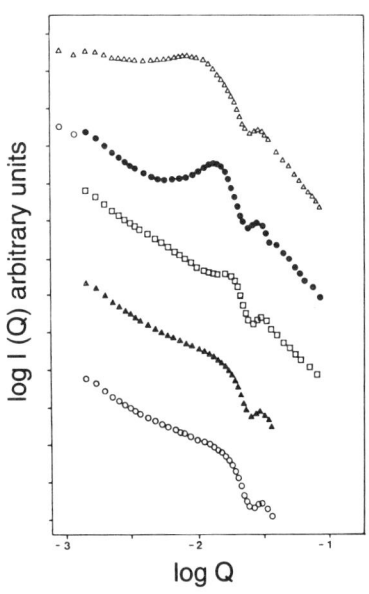

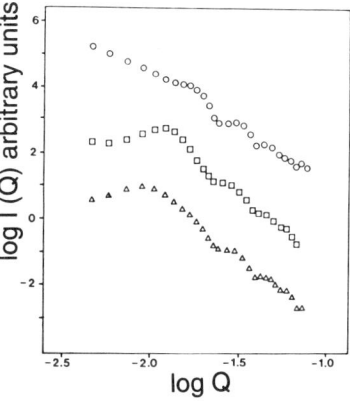

Fig. 6-61. Comparison of spectra of flocs made with polyacryamide of molecular weights ca. 1×10^6 and cationicities of 1, 5, 13, 30, and 100% (from top to bottom); spectra are shifted vertically with respect to one another. Short-range order between the silica particles is apparent from the depression and peak at intermediate Q ($0.008 < Q < 0.011 Å^{-1}$) for cationicities of 1-13%. Wong et al. (1988). Courtesy Academic Press.

Fig. 6-62. pH effects on structure: flocs made with AM-CMA 5%+, M 1.7×10^6 at pH 2.7, 7.4, and 9.9 (from top to bottom). As the pH rises, the silica acquires a larger and larger charge; the result is that the peak shifts to lower Q and the depression becomes more pronounced. Wong et al. (1988). Courtesy Academic Press.

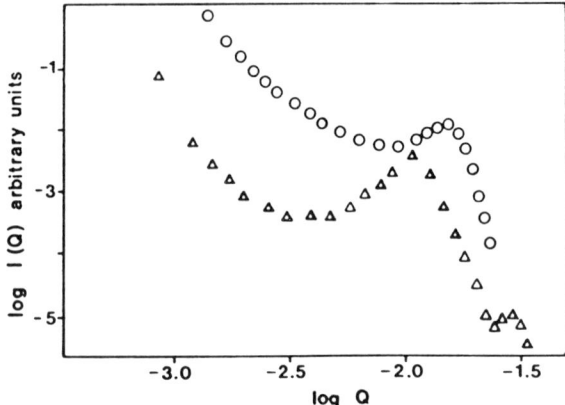

Fig. 6-63. Spectrum of silica flocculated by PEO, 4×10^6 (circles), and for comparison, that of a floc made with AM-CMA 1%+, $M3\times10^6$ (triangles): similar short-range order with large-scale heterogeneities is found for both kinds of flocs, despite a different flocculation mechanism for each. For a lower-molecular-weight PEO (0.6×10^6), no short-range order peak is found. Wong et al. (1988). Courtesy Academic Press.

Partyka et al. (1993) used microcalorimetry and more conventional techniques to study the interactions between silica (microporous silica of specific surface area of 22 m^2/g) and nonionic, cationic, and zwitterionic amphoteric surfactants.

The nonionic surfactants were oxyethylenic octylphenols of the general formula C_8H_{17}-C_6H_4-$(OCH_2CH_2)_nOH$ with n in the range from 9 to 40. For n < 20, they adsorbed on silica with an isotherm of sigmoidal shape—see Fig. 6-64, and reached a plateau whose level increased with temperature. The sigmoidal shape disappeared when the polar chain length increased, i.e., when n increased beyond 20.

Figures 6-65 and 6-66 show the adsorption isotherm and the differential molar enthalpies of adsorption of a nonionic surfactant with n = 9-10 on silica versus the coverage, Θ. At low coverages, Θ < 0.15, the differential molar enthalpies, $\Delta_{1,2}h_2$, are exothermic, corresponding to the beginning of the isotherm, where the adsorption of individual molecules are observed—step 1 in Fig. 6-66. At Θ > 0.15, $\Delta_{1,2}h_2$ becomes endothermic and the adsorption isotherm changes drastically. In the second step of Fig. 6-66, partial micellization begins to take place, and increases with Θ—step 3 in Fig. 6-66—and is essentially complete at Θ = 0.8—step 4 in Fig. 6-66. The maximum value of $\Delta_{1,2}h_2$, attained at Θ = 0.5, corresponds roughly to the micellization enthalpy.

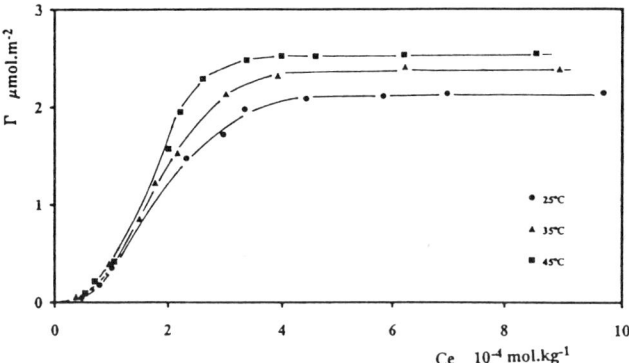

Fig. 6-64. Adsorption isotherms of TX102 onto silica showing the effect of temperature: ●, 25°C; Δ, 35°C; ■, 45°C. Partyka et al. (1993). Reprinted from Colloids and Surfaces A: Physicochemical and Engineering Aspects, **76**, 267 (1993) with kind permission from Elsevier Science - NL, Sara Burgerhartstraat 25, 1055 KV Amsterdam, The Netherlands.

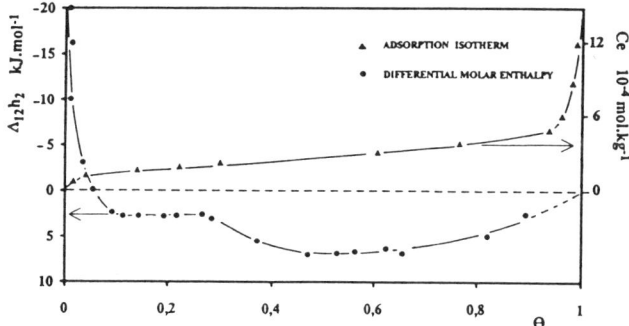

Fig. 6-65. Adsorption isotherm (Δ) and differential molar enthalpies of displacement (●) of TX100 onto silica at 20°C. Partyka et al. (1993). Reprinted from Colloids and Surfaces A: Physicochemical and Engineering Aspects, **76**, 267 (1993) with kind permission from Elsevier Science - NL, Sara Burgerhartstraat 25, 1055 KV Amsterdam, The Netherlands.

The presence of salt will, like an increase in temperature, cause dehydration of the silica surface and increase surface-surfactant and surfactant-surfactant interactions. The plateau in the isotherms will therefore increase with salt concentration (and temperature).

Cationic surfactants of the monoalkyl trimethylammonium bromide group, $C_nH_{2n+1}N^+(CH_3Br^-)$, DDTAB:n=12, TDTAB:n=14, and CTAB:n=16, adsorbed strongly in the beginning of the isotherm, shown in Fig. 6-67 for the case of TDTAB. The initial adsorption increases with the length of the monoalkyl chain. The figure also shows the change in pH (a) and electrophoretic mobility along the isotherm (b).

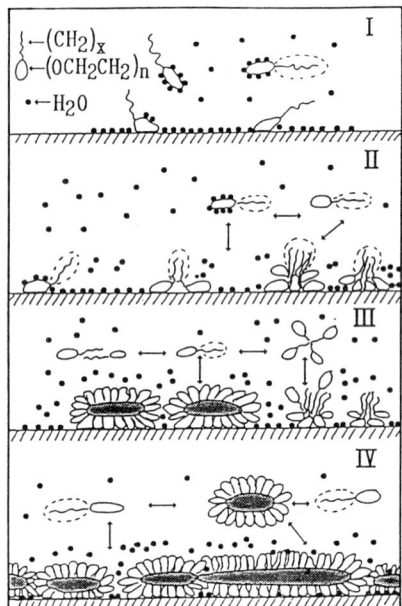

Fig. 6-66. Schematic representation of the sequential steps of adsorption of non-ionic surfactants onto a hydrophilic surface. Partyka et al. (1993). Reprinted from Colloids and Surfaces A: Physicochemical and Engineering Aspects, **76**, 267 (1993) with kind permission from Elsevier Science - NL, Sara Burgerhartstraat 25, 1055 KV Amsterdam, The Netherlands.

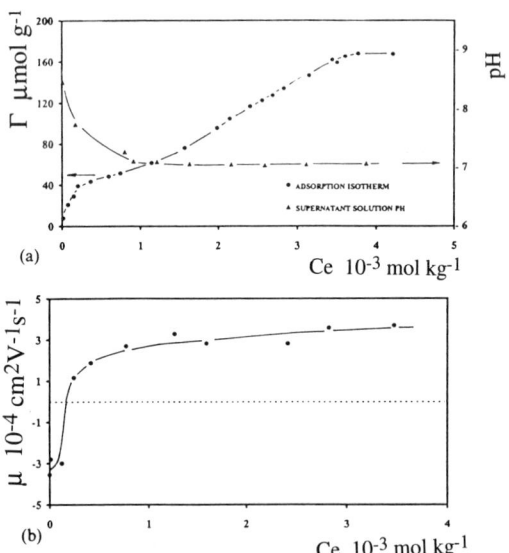

Fig. 6-67a & b. Adsorption isotherm of TDTAB onto silica (●) and pH values of the supernatant solutions (Δ) at 25°C. (b) electrophoretic mobilities of silica particles vs. equilibrium concentrations of TDTAB at 25°C. Partyka et al. (1993). Reprinted from Colloids and Surfaces A: Physicochemical and Engineering Aspects, **76**, 267 (1993) with kind permission from Elsevier Science - NL, Sara Burgerhartstraat 25, 1055 KV Amsterdam, The Netherlands.

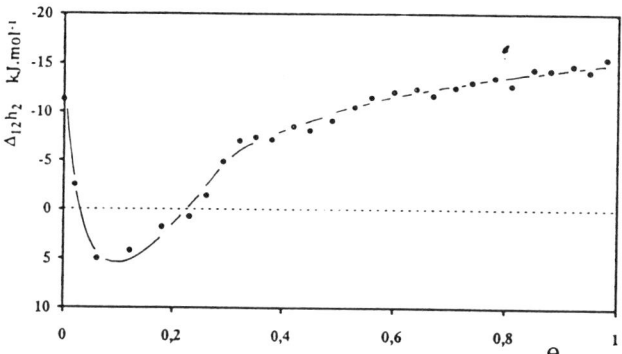

Fig. 6-68. Differential molar enthalpies of displacement of TDTAB onto silica at 25°C. Partyka et al. (1993). Reprinted from Colloids and Surfaces A: Physicochemical and Engineering Aspects, **76**, 267 (1993) with kind permission from Elsevier Science - NL, Sara Burgerhartstraat 25, 1055 KV Amsterdam, The Netherlands.

The positively charged surfactant molecules are at first adsorbed individually on the negatively charged silica surface. The negative charge on the particles is neutralized at a surface coverage of about 15%. The change in the differential enthalpies of adsorption along the isotherms is shown in Fig. 6-68.

At the early stage of the adsorption process the differential molar enthalpy is exothermic, but becomes endothermic for Θ's in the region from about 0.04 to 0.25, which Partyka et al. (1993) attributed to a displacement of water by surfactant molecules or sodium ions on the surface. However, as more and more molecules are adsorbed on the surface, lateral alkyl group-alkyl group interactions give an increasing exothermic contribution to the differential molar enthalpy. Lengthening the mono-alkyl chain will make the enthalpy of adsorption more exothermic.

3.3. Interaction between Silica and Cationic Polymers

Mabire et al. (1984) tested cationic polymers, P(AM-CMA) with cationicities in the range from < 0.05% to 100% and molecular weights between a few thousand and a few million as flocculants for silica particles with a wide particle size distribution around a mean size of 125 nm and a specific surface area of 14 m^2/g at pH about 7.

Flocculation is a dynamic process taking place when particles or flocs of various sizes collide. When the particle or cluster size is less than 0.5 μm, particle aggregation is controlled primarily by Brownian motion and by mechanical agitation for larger sizes. The floc size varies considerably during a flocculation experiment. Figure 6-69 shows that although the aggregate size increases rapidly in the very beginning, during the first 20 seconds, it then decreases monotonously to a plateau.

It is reasonable to assume that the floc size increases with stronger interaction between the particles and less agitation. Fig. 6-69 then shows that at a

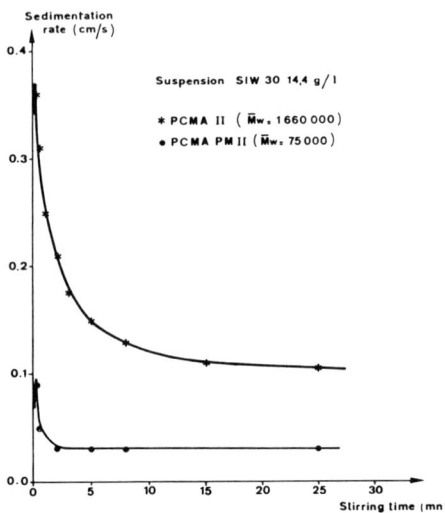

Fig. 6-69. Evolution of the sedimentation rate with the stirring time. Silica suspension Si W 30, 14.4 g/liter, stirring 100 rpm, (∗) Poly CMA M_w=1.6x106· (•) Poly CMA M_w= 75, 000. Mabire et al. (1984). Courtesy Academic Press.

given stirring rate, the floc size, as measured by sedimentation rate, increases with the molecular weight of the flocculant, suggesting that the interaction between the particles increases with the size of the polymer (at a given cationicity).

Figure 6-70 shows how the zeta potential of the particles and the turbidity of the supernatant liquid—an indication of the flocculation efficiency—varies with the amount of flocculant added. For a given value of the cationicity, τ, the

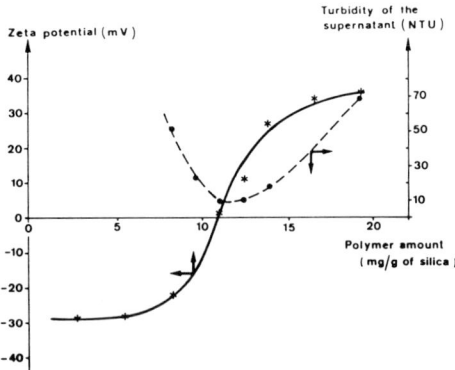

Fig. 6-70. Evolution of ζ-potential (∗) and supernatant turbidity (•) vs. the flocculant amount. Suspension Si W 30; 1.8 mg/liter, stirring 100 rpm for 2 min, Poly CMA M_w=3.4x106· Mabire et al. (1984). Courtesy Academic Press.

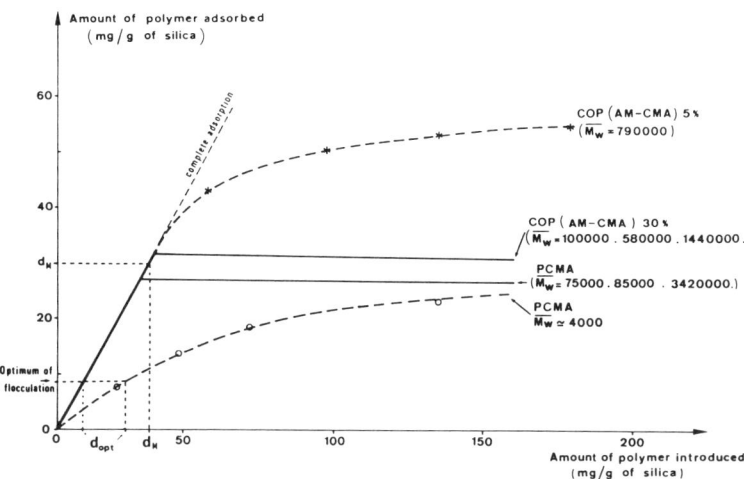

Fig. 6-71. Adsorption isotherms of some polymers on silica Si W 30. Mabire et al. (1984). Courtesy Academic Press.

slope of the curve, as well as the amount of polymer needed to reach zero zeta potential, $\zeta = 0$, is independent of the molecular weight within a wide range of M_w, e.g., 75,000 to 2.4×10^6 and with $\tau = 1$. Furthermore, the amount of flocculant added at $\zeta = 0$ is also the amount required for optimum flocculation.

Mabire et al. (1984) also noted that the adsorption of flocculant on the silica surface is fast, strong, and irreversible when the amount of polymer (weight of polymer per weight of silica) does not exceed the limit d_M—see Fig. 6-71—corresponding to complete coverage of the surface, i.e. $\Theta = 1$.

The amount of polymer needed for optimal flocculation, d_{opt}, is always less than 0.5 d_M. For $\tau = 1$ it is 0.3 d_M and for $\tau = 0.05$ it is 0.45 d_m. The fact that optimal flocculation occurred at $\Theta < 0.5$, which is optimal coverage for bridging flocculation—see Ch. 9—suggested to Mabire et al. (1984) that patching was the dominant mechanism rather than bridging when highly charged polyelectrolytes were used as flocculants. Bridging, on the other hand, is the important mechanism for nonionic polymers and weakly charged polyelectrolytes.

Buchhammer et al. (1993,1995) studied the formation of non-stoichiometric, cationic complexes between an anionic polyelectrolyte, P(MS-α-MeSty), and cationic polyelectrolytes, PDMDAAC and PEI, respectively, and the

Table 6-14. Characteristics of adsorbed particles at pH 5.8. Buchhammer et al. (1995). Reprinted from Colloids and Surfaces A: Physicochemical and Engineering Aspects, **95**, 299 (1995) with kind permission from Elsevier Science - NL, Sara Burgerhartstraat 25, 1055 KV Amsterdam, The Netherlands.

Particle	BET surface area (m^2/g)	Specific surface charge (C/g)	Surface charge density (C/m^2)	Zeta potential (mv)
Silica	650	-0.278	-0.0004	-28
Glass beads	1	-0.115	-0.1100	-52

adsorption of such complexes on silica gel and glass beads at pH of about 5. The characteristics of the adsorbents are shown in Table 6-14.

Both polycations were branched, but PDMDAAC had a higher molecular weight, ~ 78,000, and basic strength (due to the quaternary ammonium groups) than PEI, ~ 4000. The polyanion, P(MS-αMeSty), was linear with a stiff backbone chain of styrene groups and had a molecular weight of about 24,000. The adsorption isotherms of positively charged polyelectrolyte complexes, PEC, with a molar ratio of polyanion to polycation of about 0.6 on silica (a) and glass beads (b) are shown in Fig. 6-72.

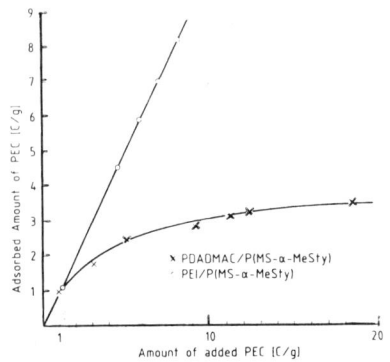

Fig. 6-72a. Adsorption isotherms of PEC on silicic acid. Buchhammer et al. (1995). Reprinted from Colloids and Surfaces A: Physicochemical and Engineering Aspects, **95**, 299 (1995) with kind permission from Elsevier Science - NL, Sara Burgerhartstraat 25, 1055 KV Amsterdam, The Netherlands.

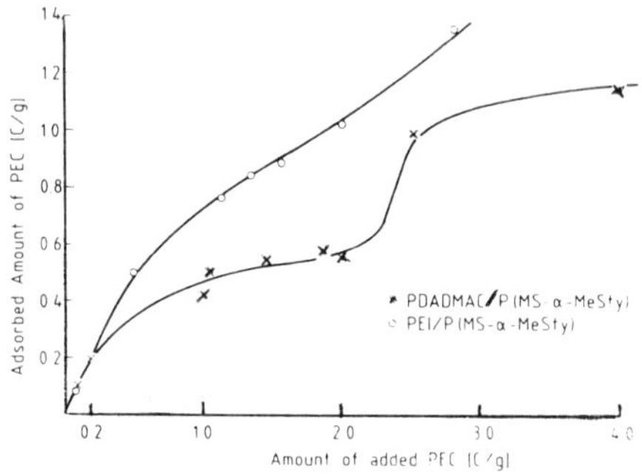

Fig. 6-72b. Adsorption isotherms of PEC on glass pearls. Buchhammer et al. (1995). Reprinted from Colloids and Surfaces A: Physicochemical and Engineering Aspects, **95**, 299 (1995) with kind permission from Elsevier Science - NL, Sara Burgerhartstraat 25, 1055 KV Amsterdam, The Netherlands.

More PEC was adsorbed on the silica particles than on the glass beads due to the higher specific surface area of the former. The amount of adsorbed PEI complex increased monotonously with PEC on both adsorbents, whereas the PDADMAC complex reached a plateau. The behavior of the latter complex was interpreted to be the consequence of the fact that it had a more defined size and compact structure than the PEI complex and that therefore less of it was needed to cover the surface. The very high surface charge density of the glass beads allowed the formation of a second adsorption layer; hence the second plateau in the isotherm in Fig. 6-72b.

Denoyel et al. (1990) also used microcalorimetry to study the homogeneity of the surface, the adsorption energy, and the fraction of polymer units (P(AM-CMA)) adsorbed on the surface of precipitated silica particles with a specific surface area of 32 m^2/g and well-dispersed, plate-like particles of montmorillonite (200x200x1 nm) with a specific surface area of 800 m^2/g at pH ≤ 7.

The surfaces of montmorillonite and silica appear to be very different with regard to their affinity for CMA. Figure 6-73 shows a high-affinity isotherm for

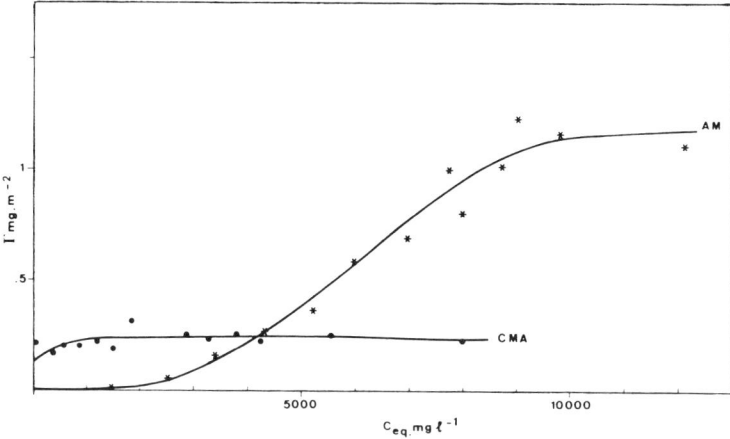

Fig. 6-73. Adsorption isotherms of CMA and AM on montmorillonite. Denoyel et al. (1990). Courtesy Academic Press.

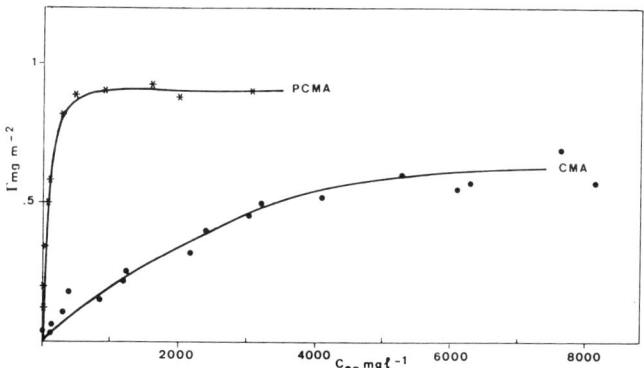

Fig. 6-74. Adsorption isotherms of CMA and PCMA on silica. Denoyel et al. (1990). Courtesy Academic Press.

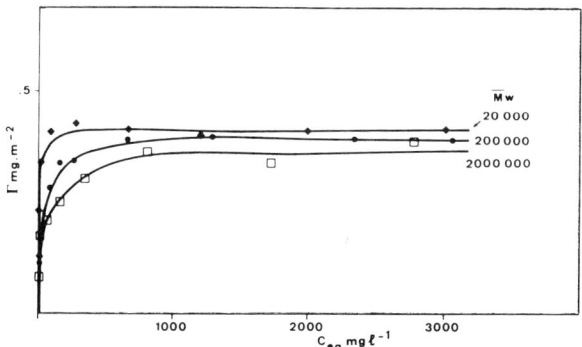

Fig. 6-75. Adsorption isotherms of various PCMA on montmorillonite. Denoyel et al. (1990). Courtesy Academic Press.

the adsorption of CMA on montmorillonite whereas Fig. 6-74 shows a relatively low-affinity isotherm for CMA on silica.

On the other hand, Figures 6-74 and 6-75 show that the adsorption of P(AM-CMA) on both montmorillonite and silica follows high-affinity isotherms.

Figure 6-76 illustrates that molar integral enthalpy of displacement of $Na(H_2O)_6^+$ by CMA is almost constant over the Θ-range. The amount of adsorbed CMA at the plateau in Fig. 6-73 corresponds roughly to the amount of Na^+ adsorbed in the double layer at the surface. Since the surface areas of the CMA molecule and the $Na(H_2O)_6^+$ ion are approximately equal, 1.3 and 1.44 nm^2, respectively, every Na^+ ion in the double layer is exchanged for a specifically adsorbed CMA molecule—see § A of this chapter. The energy of displacing hydrated Na^+ ions by CMA molecules is 10 kJ/mol, corresponding to about 4kT per molecule CMA.

The enthalpy of exchanging hydrated Na^+ ions for P(AM-CMA) depends on the molecular weight of the polymer and decreases from 5.7 kJ/mol for $M_w = 2 \times 10^4$ to 1.4 kJ/mol for $M_w = 2 \times 10^6$—see Fig. 6-76. From the adsorption isotherms in Fig. 6-75 and from the known cationicity of the polymer, Denoyel et al. (1990) calculated that the enthalpy of exchange per CMA unit, ΔH/CMA unit, was constant (4kT per unit) and practically independent of coverage, indicating that the chain conformation is also independent of surface coverage. The interaction between the polymer and the surface is strong. The polymer is strongly attached to the surface with little leeway for rearrangement. The fraction of polymer units effectively adsorbed decreases with molecular weight from 0.55 for $M_w = 2 \times 10^4$ to 0.13 for $M_w = 2 \times 10^6$.

The effect of cationicity on the enthalpy of displacement and adsorption behavior of P(AM-CMA) with a molecular weight of 5×10^5 is shown in Table 6-15.

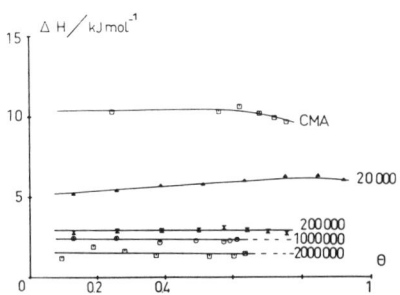

Fig. 6-76. Enthalpies of displacement of solvent (water with 1.2×10^{-3} NaCl) from montmorillonite by monomer CMA and by PCMA of various molecular weights. Denoyel et al. (1990). Courtesy Academic Press.

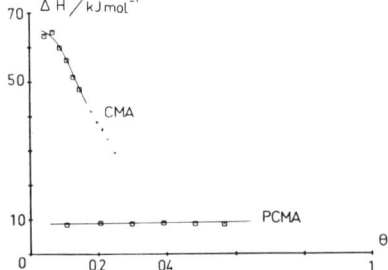

Fig. 6-77. Enthalpies of displacement of water from silica by monomer CMA and by PCMA of various molecular weights. Denoyel et al. (1990). Courtesy Academic Press.

τ %	Γ (mg/m^2)	Average M_w of a segment	ΔH (kJ/mol)	ΔH (kJ/mol CMA)	No. of (segment/ nm^2)	No. of CMA (unit/nm^2)	No. of absorbed CMA/nm^2
0	0.80	72	very small	-	6.8	-	-
1	3.40	72	0.1	10	28.1	0.28	0.28
5	1.90	77.5	0.45	9	14.5	0.73	0.65
13	1.20	86.9	0.95	7.3	8.2	1.07	0.77
23	0.65	99	1.8	7.8	4.0	0.92	0.72
30	0.55	107.6	1.8	6.0	2.9	0.89	0.53
100	0.40	193.5	2.4	2.4	1.2	1.20	0.29
CMA	0.25	193.5	10	10	0.74	0.74	0.74

Table 6-15. Effect of Polymer Cationicity on the Enthalpies of Displacement and Adsorption Behavior. Denoyel et al. (1990). Courtesy Academic Press.

The most striking result is that the number of adsorbed CMA units per nm^2 rapidly increases to a maximum of about 0.80 at 7-15%. At higher cationicities, electrostatic repulsions and chain stiffening decrease the density of cationic groups on the surface.

Figure 6-77 shows that the enthalpy of displacement of Na$^+$ ions and water molecules from the silica surface by CMA decreases rapidly with increasing Θ. The high value at low coverage corresponds to the adsorption of cationic CMA molecules on the few SiO$_2^-$- sites (0.3 charges per nm^2) present on the surface at the conditions pH $\leq$ 7 and the salinity in the experiments. P(AM-CMA), on the other hand, adsorbed on the silica surface with a constant enthalpy of displacement over the whole Θ-range, indicating that the small number of energetic surface sites are saturated by an excess of cationic units on the polymer. Since the number of such sites is so small, there is always an excess of cationic sites and the enthalpy of displacement is practically independent of molecular weight, contrary to the situation for montmorillonite.

However, the number of energetic sites increases rapidly with pH and at higher pH, e.g., higher than 9, the silica surface can be expected to become more similar to the montmorillonite surface with regard to the adsorption behavior of P(AM-CMA).

Bleier and Goddard (1980) used optical and microelectrophoretic methods to study the flocculation of silica dispersions, with particle sizes of 14 nm and 80 nm, respectively, by cationic polyelectrolytes of various cationicity. The cationic polyelectrolytes were PEI, quaternary nitrogen substituted cellulose ethers, and poly(N-methyl-4-vinylpyridinium iodide) with molecular weights in the range from 1×10^5 to 2.5×10^5.

Figure 6-78a shows the general flocculation behaviors of silica (80-nm particles) with poly(N-methyl-4-vinylpyridinium iodide)—100% cationicity—and the ζ-potential of the system (b). The flocculation zone is bounded by the critical flocculation concentration (c.f.c.) and the critical stabilization concentration (c.s.c.). At the optimal flocculation concentration (o.f.c.) the system is negatively charged and has a ζ-potential of about -20mv. The breadth of the flocculation

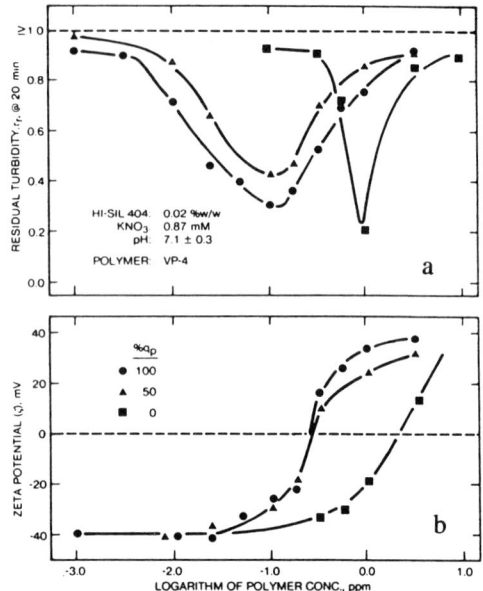

Fig. 6-78a & b. Effect of degree of quaternization, q_p, of VP-4 on the composite stability and ζ-potential profiles of HI-SIL 404. Bleier and Goddard (1980). Reprinted from Colloids and Surfaces **1**, 407 (1980) with kind permission from Elsevier Science - NL, Sara Burgerhartstraat 25, 1055 KV Amsterdam, The Netherlands.

zone, i.e., the ratio of c.s.c. to c.f.c., is a qualitative measure of the sensitivity of the silica dispersion to the presence of flocculant.

The critical flocculation concentration decreases and the width of the flocculation zone increases with increasing cationicity of the polyelectrolyte.

Bleier and Goddard (1980) concluded that optimum flocculation of a silica dispersion occurs when significantly less than 50% of the surface is covered by polyelectrolyte. The negative charge on the silica particles is thus only partially neutralized, implying that the flocculation takes place by the mechanism of patching—see Ch. 9. Complete charge neutralization is not required for optimal flocculation/sedimentation, and it is not likely that bridging is the major flocculation mechanism.

Törncrona et al. (1995) found that the negative charge on 22-nm silica particles (Ludox ®TM) could be reversed by slowly adding the silica sol to an agitated solution of a polyelectrolyte with high cationicity and a molecular weight of 50,000. The amount of adsorbed polymer per m^2 of silica was 2.0 mg—see also Ch. 7.

4. Summary and Conclusions

Adsorption of nonionic polymers on silica occurs by hydrogen bonding and is favored by low pH, at which the degree of ionization of the silanol groups is low, and by high molecular weight. The adsorbed layer can be quite thick, up to 40% of the particle diameter. The bonding to the surface is usually not very strong and dilution may cause desorption. In the case of polydisperse polymers, the smaller molecules, having faster diffusion rates, will adsorb first, but will be replaced later by slower moving, thermodynamically favored, larger molecules.

Nonionic polymers flocculate colloidal silica and other colloidal particles to which they can adsorb by hydrogen bonding, by a bridging mechanism. Flocculation is aided by electrolytes, which compress the diffuse layer—see Section A —and facilitate bridging of the particles by polymer molecules.

In the case of cationic polymers, the maximum amount of adsorbed polymer and the thickness of the adsorbed layer increase with decreasing cationicity (τ). Polymers with high τ assume a flat configuration on the silica surface and optimally flocculate anionic colloids by the patching mechanism when the surface coverage is about 30%. As $\tau \to 0$, the surface coverage for optimal flocculation increases to about 50% and the bridging mechanism plays an increasingly important role—see Ch. 9.

Table 6-16 summarizes the influence of the cationicity of P(AM-CMA) on the flocculation behavior of silica particles.

Table 6-16. Influence of Copolymer Cationicity on Flocculation Behavior. Wang and Audebert (1987).Courtesy Academic Press.

Property	$\tau > 0.15$	$\tau < 0.05$
Kinetics of adsorption	quick (minutes)	slow (hours)
Shape of adsorption isotherm	quick saturation of the surface	progressive saturation of the surface leading to quasi plateau
Total amount of flocculant adsorbed	Γ_{max} relatively small (≈ 1 mg/m^2), practically no dependence of mol. wt.	Γ_{max} increases with mol. wt. and when τ decreases (for mol. wt. 3×10^6, $\tau = 0.01$)
Layer thickness δ_H	slightly dependent on molecular weight	high values of δ_H which increases when M increases and τ decreases
Adsorption reversibility	adsorption practically irreversible except for oligomers	possible desorption
Electrokinetic potential	ζ became positive by addition of excess flocculant	ζ tends to zero by addition of an excess of flocculant
Optimum of flocculation	practically no molecular weight dependent ($\tau = 1$, dose opt. ≈ 0.5 mg/m^2 silica)	decreases when mol. wt. increases (i.e., $\tau = 0.01$; mol wt $\approx 3 \times 10^6$, tap water; dose opt. ≈ 1.4 mg/m^2 silica)
Mechanical properties of the flocs	good shear resistance; large floc volume (i.e., $\approx 16 \text{cm}^3$/g silica Syton W 30 in flocculation by tube rotation)	sticky mud (i.e., $\tau = 0.01$ mol wt $\approx 3 \times 10^6$, tap water; floc volume $\approx 4 \text{cm}^3$/g silica Syton W30)

CHAPTER 6

References

Alexander, G. B., U. S. Patent 2,913,419 (1956).

Alexander, G. B., U. S. Patent 2,974,108 (1961).

Alexander, G. B., and Iler, R. K., U. S. Patent 2,897,797 (1959).

Alexander, G. B., and Iler, R. K., U. S. Patent 3,007,878 (1961).

Asplund, T. and Otterstedt, J-E., unpublished work (1995).

Allen, L. H., and Matijevic, E., J. Colloid and Interface Sci., **31**, 287 (1969).

Allen, L. H., and Matejevic, E., J. Colloid and Interface Sci., **33**, 420 (1970).

Baudin, I., Ricard, I., and Audebert, R., J. Colloid Interface Sci. **138**, 324 (1990).

Bergna, H. E., U. S. Patent 4,217,240 (1980).

Bergna, H. E., *The Colloid Chemistry of Silica.* Advances in Chemistry Series 234, American Chemical Society, Washington, D. C. (1994).

Bleier, A., and Goddard, E.D., Colloids and Surfaces, **1**, 407 (1980).

Brinker, C. J., and Scherer, G. W. *Sol-Gel Science*, Academic Press, San Diego, 1990.

Buchhammer, H-M., Kramer, G., and Lunkwitz, K., Colloids and Surfaces A: Physicochemical and Engineering Aspects, **95**, 299 (1995).

Buchhammer, H-M., Petzhold, G., and Lunkwitz, K., Colloids and Surfaces A: Physicochemical and Engineering Aspects, **76**,81 (1993).

Cattania, M. G., Ardizzone, S., Bianchi, C. L., and Carella, S., Colloids Surf., **76**, 233 (1993).

De Boer, R. K., Angew. Chem., **75**, 383 (1985).

De Boer, R. K., and Vleeskens, J. M. K., Ned. Akad. Wet. Ser. B, **61**, 2 (1958).

Denoyel, R., Darand, G., Lafuma, F., and Audebert, R., J. Colloid Interface Sci., **139**, 281 (1990).

Evans, D. F., and Wennerström, H., *The Colloidal Domain*, VCH Publishers (1994).

Everett, D.H., *Basic Principles of Colloid Science*, Royal Society of Chemistry, London (1988).

Firment, L.E., Bergna, H.E., Swartzfager, D.G., Bierstedt, P.E., and van Kavelaar, M. L., Surface Interface Anal., **14**, 46 (1989).

Furlong, D. N.,*The Colloid Chemistry of Silica.* Advances in Chemistry Series, 234. American Chemical Society, Washington, D. C. (1994).

Garg, A., and Matijevic, E., Langmuir, **4**, 38 (1988).

Greenwood, P., and Otterstedt, J-E., unpublished work (1996).

Hingston, F. J., and Raupach, M., Austral. J. Soil Res., **5**, 295 (1967).

Hohl, H., and Stumm, W., J. Colloid Interface Sci., **43**, 409 (1976).

Hsu, W. P., Rönnquist, L., and Matijevic, E., Langmuir **8**, 31 (1988).

Iler, R. K., J. Phys. Chem., **56**, 673 (1952).

Iler, R. K., *The Colloid Chemistry of Silica and Silicates*, Cornell University Press, NY (1955).

Iler, R. K., U. S. Patent 2,885,366 (1956).

Iler, R. K., U. S. Patent 2,885,366 (1959).

Iler, R. K., J. Colloid Interface Sci., **43**, 399 (1973).

Iler, R. K., J. Colloid Interface Sci., **55**, 25 (1976).

Iler, R. K., *The Chemistry of Silica,* John Wiley, New York (1979).

Iler, R.K., J. Colloid Interface Sci., **75**, 138 (1980).

Israelachvili, J., *Intermolecular and Surface Forces*, 2nd ed., Academic Press, London, 291 (1992).

James, R. O. and Parks, G. A., *Characterization of Aqueous Colloids by their Electrical Double Layer and Intrinsic Surface Chemical Properties. Surface and Colloid Science.* E. Matijevic (ed.), Plenum Press, New York (1987).

Kawaguchi, M., Kawaguchi, H., and Takahashi, A., J. Colloid Interface Sci., **124**, 57 (1988).

Killmann, E. and Sapuntzjis, P., Colloids and Surfaces A: Physicochemical and Engineering Aspects, **86**, 229 (1994).

Killmann, E., Sapuntzjis, P., and Maier, P., Macromol. Chem., Macromol. Symp., **61**, 42 (1992).

Kondo, S., Igarashi, M. and Nakai, K., Colloids and Surfaces, **63**, 33 (1992).

James, R. O., and Healy, T. W., J. Colloid Interface Sci., **40**, 42 (1972).

Kirk, J. S., U. S. Patents 2,408,654-2,408,656 (1946).

Lekki, J., and Lakowski, J., Trans. IMM, **80**, 174 (1971).

Liu, J. C., and Huang, C. P., Langmuir, **8**, 1851 (1992).

MacMobility Program, Department of Mathematics, University of Melbourne, Parkville, Victoria 3052, Australia.

Matijevic, E., and Allen, L. H., Sci. Technol., **3**, 264 (1969).

Matijevic, E., and Scheiner, P., J. Colloid Interface Sci., **63**, 509 (1978).

Mabire, F., Audebert, R., and Quivoron, C., J. Colloid Interface Sci., **97**, 120 (1984).

Milonjic, S. L., Colloids and Surfaces, **63**, 113 (1992).

Mindick, M., and Thompson, A. C., U. S. Patent 3,252,917 (1966).

Moore, E. P., U. S. Patent 3,620,978 (1971).

Moudgil, M. and Somasundaran, P., Fundamentals of Adsorption, Proc. of the Engineering Foundation Conference, Schloss Elman, Fed. Rep. Ger., 355 (1984).

Oestreicher, C. A., and McGlashan, D. W., AMME Annual Meeting, San Francisco, (1972).

Otterstedt, J-E., unpublished work (1995).

Otterstedt, O. E., Otterstedt, J-E., Ekdahl, J., Backman, J., and Anderson, C-H., J. Appl. Polym. Sci., **34**, 2575 (1987).

Park, S. W., and Huang, C. P., J. Colloid Interface Sci., **117**,431 (1987)

Parks, G. A., Chem. Rev., **65**, 177 (1965).

Partyka, S., Lindheimer, M., and Faucompre, B., Colloids and Surfaces A: Physicochemical and Engineering Aspects, **76**, 267 (1993).

Pulfer, K., Schindler, P. W., Westall, J. C., and Grauer, R., J. Colloid Interface Sci., **101**, 554 (1984).

Schindler, P. W., and Stumm, W., *Aquatic Surface Chemistry: Chemical Processes at the Particle-Water Interface.* Stumm, W.(ed.) John Wiley, New York (1987).

Schindler, P. W., Fürst, B., Dick, R., and Wolf, P. U., J. Colloid Interface Sci., **55**, 469 (1976).

Shaw, D. H., *Introduction to Colloid and Surface Chemistry*, Butterworths, London (1970).

Stöber, W., Bauer, G., and Thomas, K., Ann. Chem., **604**, 104 (1957).

Stumm, W., Kummert, R., and Sigg, L., Croatica Chemica Acta, **53**, 292 (1980).

Tada, H., Saitoh, Y., Miyata, K., Kawahara, H., J. Jpn. Soc. Colour Mater., **62**, 399 (1989).

Törncrona, A. and Otterstedt, J-E., unpublished work (1990).

Törncrona, A., Löwendahl, L., Otterstedt, J-E., and Jansson, K., J. Mater. Chem., **6**, 213 (1996).

Törncrona, A., Sterte, J., and Otterstedt, J-E.,J. Mater. Chem., **5**, 121 (1995).

Vordonis, L., Akratopolu, A., Koutsoukos, P. G., and Lycourghiotis, A., *Preparation of Catalyst IV*, Louvain-la-Neuve, 309, Elsevier, Amsterdam (1987).

Wang, T.K., and Audebert, R., J. Colloid Interface Sci., **119**, 459 (1987).

Westall, J., and Hohl, H., Advan. Colloid Interface Sci., **2**, 265 (1980).

Wong, K., Cabane, B., and Duplessix, R., J. Colloid Interface Sci., **123**, 466 (1988).

Zhuravlev, L. T. Colloids and Surfaces A: Physicochemical and Engineering Aspects, **74**, 71 (1993).

Introduction

Many of the useful properties of catalysts arise from reactions between the solid surfaces of the catalysts and various gaseous reactants. Catalysts of large specific surface areas are often desired and they can be obtained by the use of porous bodies. The large surface areas are associated with the internal surfaces of microporous systems which consist of networks of a great many very small pores.

The terms "macroporous" and "microporous" substances suggest that porous substances may be divided into at least two subgroups. It is generally agreed that porous materials such as sponge and pumice can be classified as macroporous materials and, as long as the pores can be seen with the naked eye, there will be no difference of opinion about the classification. In 1927 Scott Russell, distinguishing between macropores and micropores in building stone, fixed the borderline in that case at a pore diameter of 0.005 mm. For catalysts, however, the division line is fixed at about 50 nm, i.e., 100 times smaller than the figure fixed by Russell. This brings paper, fabrics, cloth, building stone, coke, and wood mainly into the group of macroporous materials; and catalysts and also other adsorbents, mainly into the group of microporous materials. This does not mean that the first group is devoid of micropores or that the second group does not contain macropores, but that the characteristic properties of these materials are mainly given by macropores in the first case and by micropores in the second.

Many technically important materials are made from small particles or consist of small particles. Materials having a porous structure are often made of small particles. Porous structures, particularly of inorganic materials, are thus made by packing small particles of different form and shape in different ways. Densely packed spherical particles will thus give rise to a regular pore structure with a narrow pore size distribution and having an average pore size which is directly proportional to the diameter of the particles and a specific surface area that is inversely proportional to the particle diameter. In the case of structures of loose random packing, i.e., structures with coordination numbers less than 6, the inverse relationship between pore size and specific surface area is not valid and it is possible to make porous structures characterized by a combination of high specific surface area and large pores, the pore size distribution of which, however, is broad compared with structures of densely packed spheres. Rod-like, fibrillar particles, can be randomly packed to a "pick-up-sticks" structure characterized by a broad pore-size distribution and an average pore size that is of the same order of magnitude as the diameter of the rods and a specific surface area that is inversely proportional to the particle diameter.

It is thus possible to use small particles to make porous structures of well-defined pore structure by varying shape, size, and packing density of the particles in a controllable manner. Such structures are used in catalysts, catalyst supports, adsorbents, chromatographic column packings, and many other technically important materials.

1. Structure of Catalysts

The most widely used industrial and environmental catalysts are inorganic solids. Clays and zeolites, discussed in Chapters 4 and 5, are such materials. In the molecular-sized pores of the well-defined structures of zeolites and some clays, so-called pillared clays, catalysis takes place in an almost solvent-like environment. Zeolites therefore provide a transition between solutions and the more typical catalytic solids, which catalyze reactions on their surfaces.

These more typical catalytic solids include metals, metal oxides, and metal sulfides, sometimes in combination with each other. It is obvious that efficient solid catalysts must have high specific surface area. Since catalytic reactions take place on the surface, reaction rates increase in proportion to the surface area, provided that diffusional constraints can be neglected. The optimal form of a catalyst is therefore frequently a porous solid of high internal surface area. The only other way to achieve a high surface area per unit volume reactor is to use the solid in the form of very small particles. However, from a practical point of view this is not realistic. The pressure drop over the reactor would be impossibly high and the fine particles may be entrained in the product stream and foul up pipes and pumps.

Industrial catalysts have specific surface areas of up to several hundred square meters per gram. Many metal oxides can be prepared in the form of porous high-surface-area solids. The porous structure is achieved by small particles of the metal oxide arranging themselves into a structure of high mechanical strength combined with high surface area and suitable pore size. Many important catalytic materials, i.e., the noble metals, however, cannot be made in the form of sturdy, high-surface-area, porous structures. Instead, they can be deposited in the form of exceedingly small particles on the surface of strong, high-surface-area, porous supports made of small particles of metal oxides such as alumina and silica. This type of important catalysts, among them those used in catalytic converters for autos, thus consist of a catalytically active surface, e.g., noble metals, metal oxides or metal sulfides, in the form of minute particles, deposited onto the surface of a porous support. In some cases, e.g., automotive catalytic converters, the support is in turn deposited on the surface of a carrier which may consist of a ceramic or metallic monolith having a large number of channels in a honeycomb structure.

Table 7-1. Matrix of catalyst components			
Support + carrier ------>	1. Conventional	2. Improved	3. Improved
Catalytically active substances			
1. Conventional	γ-alumina + a) Pt & Rh: catalytic converters for cars b) Pt, Pd: oxidation catalysts for VOC's c) Pt+Re: reforming catalysts d) MoS$_2$: hydrotreatment catalysts	hydrothermally treated γ-alumina + a) Pt & Rh: catalytic converters for cars b) Pt, Pd: oxidation catalysts for VOC's c) MoS$_2$+ CoS: hydrotreatment catalysts d) Pt + reforming catalysts	fibrillar boehmite + a) Pt, Pd: oxidation catalysts b) MoS$_2$ + CoS: hydro-treatment catalysts c) conventional zeolites: catalytic converters Spherical silica + a) Pt, Pd: oxidation catalysts
2. Improved	γ-alumina + Pd:Pt (80:20) oxidation catalysts for VOC's	hydrothermally treated γ-alumina + Pd:Pt (80:20) oxidation catalysts for VOC's	fibrillar boehmite or spherical silica + Pd:Pt (80:20): oxidation catalysts
3. New	γ-alumina + colloidal zeolites: catalytic converters for cars		fibrillar boehmite + colloidal zeolites: catalytic converters for cars

Table 7-1 shows a matrix of catalyst components and indicates how conventional, improved and new catalytic substances can be combined with conventional, improved and new supports to make new and improved catalysts for cleaning exhausts from vehicles, combustion of volatile organic compounds, and different processes for upgrading oil. Matrix element 1,1 shows that conventional catalysts for cleaning exhausts from cars, combustion of volatile organic compounds, hydrotreatment, reforming and alkylation consist of noble metals, metallic oxides or metal sulfides on porous supports of γ-alumina. These catalysts are used as reference catalysts when the new or improved catalysts in the other matrix elements are discussed.

2. Catalysts with Supports on Carriers - Catalysts for Pollution Control

Supports of silica and especially alumina, so-called washcoats, on ceramic, which is the most common material, and metallic monoliths, are now primarily used in catalytic converters for cars and in catalysts for complete oxidation, i.e. combustion, of volatile organic solvents. However, washcoats of alumina and silica on other carriers, e.g., fibers, are also being studied with the objective of developing catalysts that, in addition to removing gaseous pollutants, also can eliminate particulate pollutants, e.g., soot, from diesel-powered vehicles.

2.1. Monolithic Carriers

Figure 7-1 shows schematically how the channel walls of monoliths of ceramic or metallic materials have been covered with washcoats of small particles of different shapes and sizes. To the left is a washcoat of boehmite alumina, the most frequently used support today, in the form of corpuscular particles. To the right is a washcoat, also of boehmite alumina, but in the form of rods that are packed to a "pick-up-sticks"-structure. In the middle is a washcoat of uniformly-sized silica spheres packed to a porous structure of uniform pores.

The methods for preparing dispersions of washcoat particles described below are all laboratory methods, but similar methods are being used to make such dispersions on an industrial scale.

2.1.1. Preparation of Dispersions of Conventional Boehmite. Dispersions of alumina can typically be made by adding 300 g Disperal®, which is a spray-dried powder of boehmite from Condea in Germany, to a vigorously stirred solution of 9.7 g HCl (37%) in 690 g water. The alumina dispersion was agitated for 10 minutes and then centrifuged at 2500 rpm for 1 h. The supernatant phase, consisting of about 30-nm aggregates of much smaller, about 4-nm, primary particles, was used in the preparation of washcoats and catalysts, Larsson et al. (1987).

2.1.2. Preparation of Dispersions of Fibrillar Boehmite. Fibrillar boehmite can be prepared by autoclaving a solution of basic aluminum chloride, OH:Al = 2.5 and containing 2 wt. % Al_2O_3, at 160° C for 10 h. Agitation was accomplished by rotating the autoclave, Evaldsson et al. (1989) and Sterte and Otterstedt (1986); see also Ch. 3 §B.

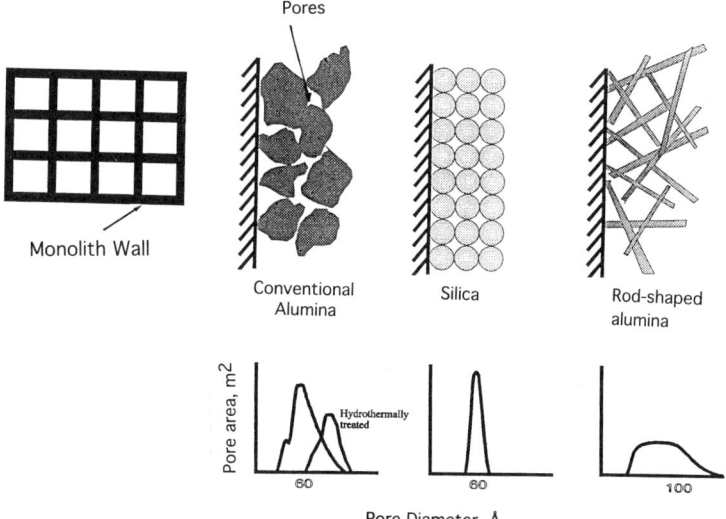

Fig. 7-1. Washcoats of particles of different shapes on monolith walls. Otterstedt (1995).

After the heat treatment, the Teflon®-lined autoclave was cooled under tap water. Gelation was brought about by raising the pH of the solution from about 1.8 to about 9.0 by adding 1.5M ammonia solution. Chloride ions were removed by washing the gel with dilute ammonia solution, pH 10, until chloride ions could no longer be detected with AgNO₃. The gel was re-dispersed by lowering the pH to 4-5 with 1M solution of acetic acid. The resulting sol was used to prepare washcoats and catalysts.

2.1.3. Preparation of Dispersions of Silica. In a typical preparation silica sol (Ludox® TM, 49.5 wt. % SiO_2) was first passed at 3 cm/min. through a column filled with Dowex 50W-X8 (a strong acid cation-exchange resin), 20-50 mesh, in the ammonium form. The pH of the effluent was raised to 10 by addition of ammonium hydroxide solution. The resulting sol solution was stored overnight to permit sodium ions occluded within the silica particles to diffuse out into the solution. The sequence of exchange and storing at pH 10 overnight was repeated until the sodium content was reduced to the desired value—about 0.05 wt. % Na_2O, Axelsson et al. (1988).

2.1.4. Application of Washcoats on Monoliths. Axelsson et al.(1988) applied washcoats by cutting monolith samples of 15-mm lengths with a square cross-section containing 81 square channels from a commercial honey-comb structure of cordierite, Axelsson and Otterstedt (1988). The corners were trimmed off, resulting in a cross-section with 69 channels.

Washcoat was deposited on the monolith samples by repeated immersions in colloidal solutions containing up to 30 wt.% silica or alumina. The immersion time was about 1 second. Excess colloidal solution was removed by gently blowing air through the channels. The samples were dried with hot air at 130 °C for 5 min. About three immersions were required to deposit enough washcoat to give the sample a specific surface area equal to that of commercial catalysts, i.e., about 20 m²/g. All samples were finally heated in air at 50 °C for 1.5 h, whereby boehmite was converted to γ-alumina.

2.1.5. Hydrothermal Treatment of Washcoats of γ-Alumina. Löwendahl and Otterstedt (1990, 1991) treated monoliths with washcoats of γ-alumina (originally boehmite) hydrothermally at 550-1000 °C for various times, 0.5-4.0 h, in a furnace while adding water at a rate of 1.5 cm³/min. Two hours at 815 °C turned out to give the washcoat pore structure an optimal combination of specific surface area and pore-size distribution.

2.2. Fibrous Carriers

Catalysts with washcoats on monolithic carriers work very well as catalytic converters in gasoline-powered vehicles. They exhibit low pressure-drops and reduce emissions of hydrocarbons, CO, and NO by about 90%. However, monolithic catalysts do not work equally well when it comes to cleaning the exhausts from diesel-powered vehicles, which, in addition to gaseous pollutants, also contain particles of soot. Ordinary monolithic converters cannot trap the soot particles, and they pass unaffected through the catalyst. Special monoliths can be used where every second channel is plugged so that the exhaust gases are forced

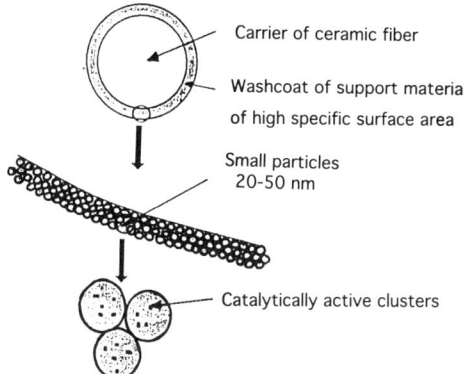

Carrier of ceramic fiber

Washcoat of support material
of high specific surface area

Small particles
20-50 nm

Catalytically active clusters

Fig. 7-2. Washcoat of support on carrier of ceramic fiber. Törncrona et al. (1995).

through the porous walls, which act as a soot filter for the open channels. However, after a certain time the pores of the channel walls become plugged with soot, which must be burned off periodically. Monolithic soot traps of this kind are being tested in various places, but, so far, they have not gained wide acceptance, mainly because of the problem with regeneration of the soot-capturing activity.

Törncrona et al. (1995, 1997a) developed a concept for combining catalytic functions with a soot-capturing function which involves providing thin fibers, 3-10 μm, of, e.g., silica, with washcoats of alumina or silica; see Fig. 7-2. Catalytically-active substances, e.g., noble metals, are deposited on the washcoat. Soot particles are caught on the fiber structure, which acts as a filter, and, as they come into contact with the catalyst particles on the washcoat, are continually combusted. Gaseous pollutants are also eliminated or reduced in number by the catalyst particles.

2.2.1. Pretreatment of Silica Fibers. Törncrona et al. (1995) utilized electrostatic attraction between oppositely charged surfaces to build up a porous washcoat on silica fibers. The surface of silica fibers can assume a negative charge if it contains silanol groups and if the pH is above 2—see Ch. 6 §C. During the production of silica fibers they are exposed to temperatures higher than 1500° C, which leads to almost complete dehydroxylation of the fiber surface. Moreover, spinning aids are oftentimes used and remains of such chemicals can interfere with adsorption of particles on the fiber surface. Commercial silica fibers of diameter 2 μm were therefore heated at 500° C for 1 h to burn off organic material. The fiber surface was re-hydroxylated by depositing monomeric silicic acid from small silica particles on the fibers in an autoclave by the process called Ostwald ripening—see Ch. 2 §A. A quantity of 5 g of calcined fibers was placed in a 300-ml Teflon®-lined stainless steel autoclave with 3 g of 7-nm silica sol, Ludox® SM containing 30% SiO_2, and 250 ml water of pH 9. The autoclave was heated to 200° C and kept at this temperature for 16 hours, whereby the 8-nm Ludox® SM particles were dissolved into monomeric silicic acid which deposited on the surface of the silica fibers and formed a thin layer of amorphous silica, the surface of which contained about 5 silanol groups per square nanometer; see Ch. 6 § C.

2.2.2. Charge Reversal of Silica Particles and Fibers. Application of multilayers of silica particles on glass plates has been described by Iler (1980) and on silica fibers by Törncrona et al. (1995). Amorphous silica fibers were charge-reversed by putting 1 g of 2-μm fibers in 100 ml of an aqueous solution containing 0.4 wt. % of a cationic polymer, e.g., Berocell® 6100 from Berol Akzo-Nobel, Sweden, or 0.4 wt. % (Al_2O_3) of basic aluminum chloride, e.g., Locron® L from Hoechst, Germany. Surplus charge-reversing agent was rinsed out with 100 ml of distilled water and the pH was adjusted to 8.0 in the case of cationic polymer and to 4.0 in the case of basic aluminum chloride. Silica particles of 22-nm size, e.g., Ludox® TM from DuPont, USA, were charge-reversed by slowly adding 200 ml of 0.4 wt. % silica sol to 200 ml of a stirred solution of cationic polymer containing 0.2 g polymer. About 2.0 mg polymer was adsorbed per square meter of silica surface.

2.2.3. Build-up of Washcoats of Spherical Silica and Fibrillar Alumina on Silica Fibers. The various steps of the coating processas described in § 2.2.1, 2.2.2, and here, are shown in Fig. 7-3. Pretreated fibers, 5.0 g, were immersed in 400 ml of 0.4 wt. % solution of charge-reversing agent. The surplus of charge-reversing agent was rinsed out with 400 ml of distilled water and pH was adjusted to that of the solution of charge-reversing agent. The charge-reversed fibers were next put into 400 ml of 0.1 wt. % silica sol. Surplus silica sol was rinsed out with 400 ml of distilled water. The silica fibers with one layer of adsorbed silica particles were put in a glass column and 200 ml of charge-reversed silica sol, containing 0.2 wt. % SiO_2, were passed over the fibers in the column at a rate of 10 cm/min. Surplus sol in the fibers was rinsed out with distilled water. A quantity of 200 ml of anionic sol, containing 0.2 wt. % SiO_2, was then passed over the fibers at the same rate. This sequence was repeated twice. The coated fibers were hydrothermally heated at 750° C in 100% steam for 24 hours to fixate the porous wash-coat on the fiber surface.

According to a simplified coating procedure, the silica fibers were first charge-reversed by using a solution containing 0.4 wt. % cationic polymer and were then coated with silica particles in one step. The silica content of the sol was raised to 2.0 wt. % SiO_2 in order to ascertain that the specific surface area of the coated fibers was adequate, i.e., about 20 m^2/g.

Particles of fibrillar boehmite can also be deposited on silica fibers by a one-step procedure at pH 4.0. Pretreated fibers, 5.0 g, were immersed in 200 ml of boehmite sol containing 2.0 wt. % Al_2O_3. The surplus sol particles were rinsed out with distilled water. It is not necessary to charge-reverse the silica fibers since the particles of fibrillar boehmite are positively charged. In order to fixate the washcoat of fibrillar boehmite on the fiber surface, the coated fibers were calcined in air at 550° C for 90 minutes.

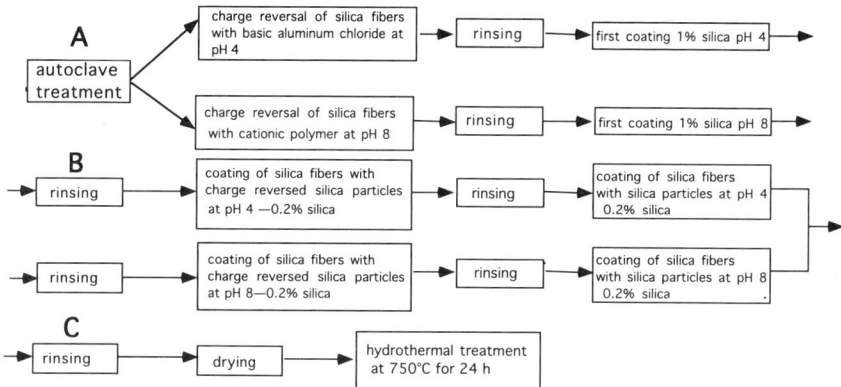

Fig. 7-3. Schematic diagram for preparing washcoats on fibers. Törncrona et al. (1995). Courtesy of The Royal Society of Chemistry.

2.2.4. Fiber Bundles with Washcoats of Fibrillar Boehmite. In the previous sections it was described how a washcoat of spherical silica or fibrillar boehmite could be applied to the individual, well-separated fibers in a wad of quartz fibers. Törncrona et al. (1997a) used similar methods to apply washcoats of fibrillar boehmite in the individual strands of yarn of 10-μm Almax® α-alumina fibers (Mitsui Mining Co. Ltd., Japan).

Fibrillar alumina was prepared by autoclaving a basic aluminum chloride solution (0.8 mol/dm^3 Al) at 160° C for 24 h—see §2.1.2 this chapter.

The yarn bundle of α-alumina fibers was calcined at 600° C for 2 h to remove any organic residue. The surface charge of the fibers was reversed by immersing the yarn in a solution containing the potassium salt of polyvinylsulfate, [-CH$_2$CH(OSO$_3$K)-]$_{1500}$ at pH 4. Adsorption of 0.80 mg polyvinylsulfate per m^2 of fiber surface reversed the positive charge on the surface to a high negative charge corresponding to an electrophoretic mobility of -3 (mms^{-1})/(v cm^{-1})—see §A Ch. 6. The charge-reversed strands of the fiber bundle were next coated with particles of fibrillar alumina by a process similar to the one described in §2.2.3 and outlined in Fig. 7-3. Four coatings increased the specific surface area of the α-alumina fibers from 0.09 m^{2}g^{-1} to 0.7 m^{2}g^{-1}. The SEM micrographs in Fig. 7-4 show that four coatings completely covered the coarse fiber surface, Fig.7-4a, with a porous, fairly even layer of fibrillar boehmite, Fig. 7-4b. All fiber bundle strands were coated with a uniform layer of fibrillar boehmite.

Woven structures of yarns of fibers, e.g., of α-alumina, washcoated with fibrillar boehmite or spherical silica particles may be interesting catalyst supports. Deposition of catalytically active materials, e.g., noble metals and/or oxides of transition metals, on the particles of the washcoat may yield effective catalytic filters for capturing and combusting soot particles from diesel-powered vehicles— see also § 2.4.6 below.

This method can also be used to coat ceramic fibers in general with many different types of particles. Thus, fibers coated with zirconia particles may provide

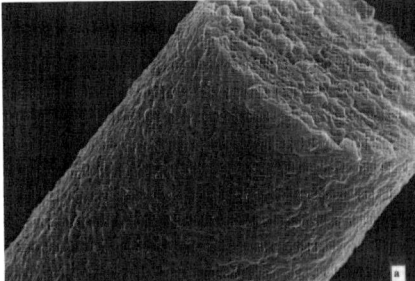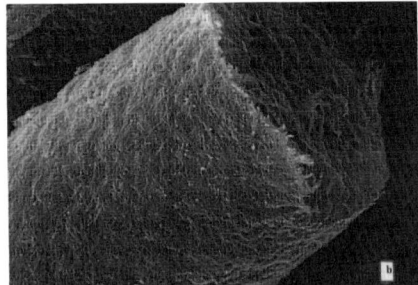

Fig. 7-4. SEM micrographs of Almax® α-alumina fibers (diameter = 10μm) uncoated (a), and coated with fibrillar boehmite (b). Törncrona et al. (1997a). Reprinted from Journal of the European Ceramic Society, **17** 1459 (1997) with kind permission from Elsevier Science - NL, Sara Burgerhartstraat 25, 1055 KV Amsterdam, The Netherlands.

a suitable interfacial bond between the fibers and the matrix in long-fiber reinforced ceramic matrix composites (CMC).

2.3. Structure of Washcoats of Alumina and Silica on Carriers

The structure of the washcoat is very important to the performance of catalysts. The washcoat must thus have a favorable combination of specific surface area and pore size to ascertain high activity and good selectivity of the catalyst. Furthermore, the structure must be stable at the reaction conditions in order to prevent loss of expensive, catalytically active substances by being buried in collapsing pores.

2.3.1. Washcoats of Alumina and Silica on Monoliths.
Löwendahl and Otterstedt (1990, 1991) "pre-collapsed" the finest pores by heating washcoats of alumina on monoliths of cordierite at 550 and 814° C for 2 h and at 750, 814, and 877° C in 100% steam for 2 h. Figure 7-5 shows that the maximum of the pore

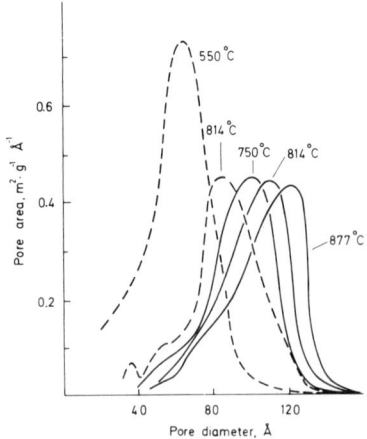

Fig.7-5. Pore area distributions of dry-calcined (----) and hydrothermally treated (——) boehmite applied as washcoat on monoliths. Time of treatment was 2 h. Löwendahl and Otterstedt (1990). Reprinted from Applied Catalysis, **59**, 89 (1990) with kind permission from Elsevier Science - NL, Sara Burgerhartstraat 25, 1055 KV Amsterdam, The Netherlands.

area distribution at 60 Å for samples that were first thermally treated at 550° C, moved to larger pore diameters when the temperature of the hydrothermal treatment increased. At 814° C the maximum occurs at 110 Å whereas thermal treatment at the same temperature only shifts the maximum to 85 Å. Comparatively large molecules, such as xylene, can diffuse faster in this structure than in one whose pore area distribution is centered around 60 Å. Hydrothermal treatment of washcoats of corpuscular boehmite thus opens up the pore structure so that transport of large molecules is facilitated while maintaining a large specific surface area. Deposition of catalytically-active substances, e.g., noble metals, on the surface of the support after it has been hydrothermally treated minimizes the risk of deactivating the catalyst by burying these expensive metals in the smallest pores as the support is exposed to severe conditions of temperature and moisture in use.

In Fig. 7-6 are shown x-ray diffractograms of corpuscular boehmite (Disperal®) which has been subjected to different thermal and hydrothermal treatments, Löwendahl and Otterstedt (1990). Heating Disperal® in air at 550° C gave a diffractogram that is identical with that which in the literature is called "low-temperature" γ-alumina. Higher temperatures brought about transitions to "high-temperature" γ-alumina, and at 1000° C θ-alumina was formed.

Hydrothermal treatment gave sharper and more well-defined diffractograms, which indicates a better-developed crystalline structure, than thermal treatment at the same temperature. It thus appears that moist heat gives rise to a more ordered crystalline structure than dry heat, or, in other words, hydrothermal treatment at a certain temperature converts alumina to a transition form that thermal treatment can only achieve at a higher temperature. It may also be that hydrothermally treated alumina contains more hydroxyl groups and a different distribution between Brønsted and Lewis acid sites on the surface of the washcoat and may, for these reasons, provide a different surrounding for catalytic metals than thermally treated alumina.

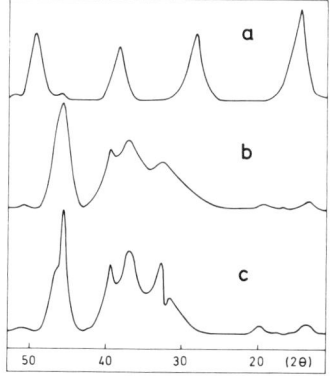

Fig.7-6. X-ray diffractometer curves of alumina powders from boehmite (Disperal®). (a) Disperal® powder; (b) dry-calcined at 814° C for 2h; (c) hydrothermal treatment at 814° C for 2 h. Löwendahl and Otterstedt (1990). Reprinted from Applied Catalysis, **59**, 89 (1990).with kind permission from Elsevier Science - NL, Sara Burgerhartstraat 25, 1055 KV Amsterdam, The Netherlands.

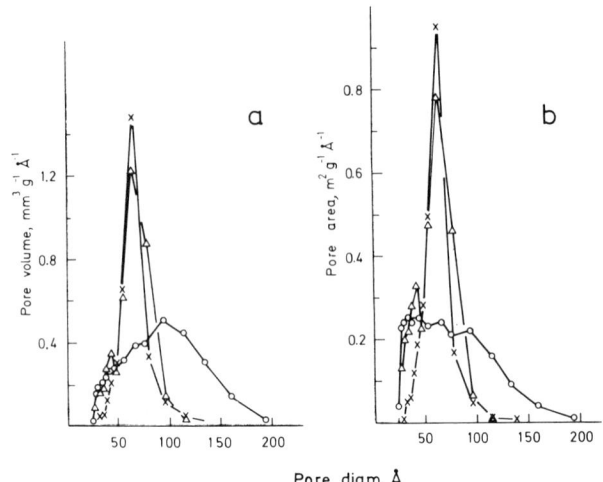

Fig. 7-7. Pore-size distributions of wash-coats. (o) fibrillar alumina; (Δ) corpuscular alumina; (x) silica particles. Evaldsson et al. (1989). Reprinted from Applied Catalysis, **55**, 123 (1989) with kind permission from Elsevier Science - NL, Sara Burgerhartstraat 25, 1055 KV Amsterdam, The Netherlands.

Transition Electron Microscopy, TEM, studies showed that the main difference between thermally and hydrothermally treated alumina was that hydrothermally treated alumina contained larger crystallites, Skoglundh et al. (1992).

In Fig. 7-7 pore size and pore area distributions of corpuscular boehmite (the most common washcoat material in commercial catalytic converters), fibrillar boehmite, and de-cationized silica are compared, Evaldsson et al. (1989). Washcoats of corpuscular boehmite and silica both have pore sizes distributed around a maximum of 60 Å, but the alumina structure has more area and volume in pores smaller than 50 Å and broader distributions than the silica structure. Washcoats of fibrillar alumina have by comparison much broader distributions of surface area and volume and therefore provide a much more open pore structure.

Figure 7-8 shows that the hydrothermal stability of a porous silica structure depends strongly on the alkali content, Axelsson et al. (1988). Hydrothermal treatment at 750° C and 100% steam for 24 h caused the silica sample containing 0.05 wt. % Na_2O to lose only 17% of its original specific surface area of 125 m^2/g, whereas the drop in specific surface area was 55% for the sample containing 0.61 wt. % Na_2O. The same hydrothermal treatment of corpuscular γ-alumina reduced the initial specific surface area of 213 m^2/g by 38% to 132 m^2/g, which is only insignificantly higher than the specific surface area of low-sodium silica. From the point of view of providing catalytically active substances with support area under severe hydrothermal conditions, low-sodium silica is therefore comparable with γ-alumina.

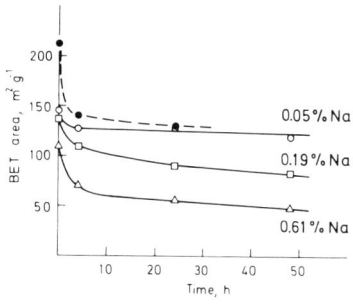

Fig. 7-8. Influence of hydrothermal treatment on alumina (----) and silica (——) applied as washcoat. No platinum deposited. Axelsson et al. (1988). Reprinted from Applied Catalysis, **44**, 251 (1988) with kind permission from Elsevier Science - NL, Sara Burgerhartstraat 25, 1055 KV Amsterdam, The Netherlands.

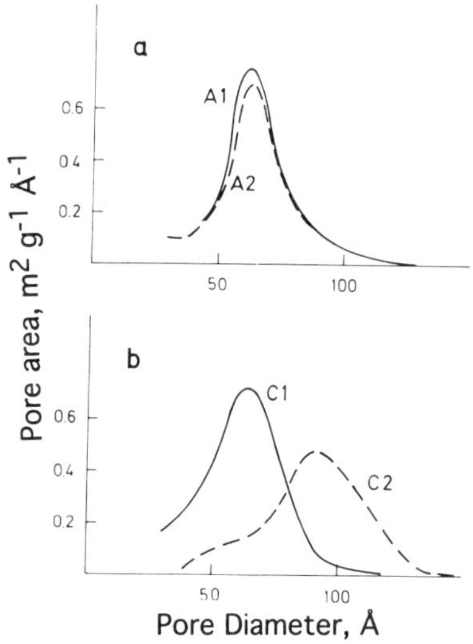

Fig. 7-9. Pore area distributions for (a) silica and (b) alumina washcoats before and after deactivation at 750°C for 24 h. Axelsson et al. (1988). Reprinted from Applied Catalysis, **44**, 251 (1988) with kind permission from Elsevier Science - NL, Sara Burgerhartstraat 25, 1055 KV Amsterdam, The Netherlands.

Figure 7-9 shows that the pore-area distributions of washcoats of alumina and silica are quite different and that they also respond differently to hydrothermal treatment. The washcoat of low-sodium silica had a fairly narrow distribution centered around a pore diameter of about 60 Å (sample A1), which did not change on hydrothermal deactivation (sample A2). When the silica washcoat contained more sodium the effect of hydrothermal treatment became more noticeable, but was still small. The washcoat of alumina had a much broader pore area distribution (sample C1), which was flattened, broadened, and shifted towards larger pore diameters on deactivation (sample C2); cf. Fig. 7-5. The smaller pores of the washcoat disappeared and larger ones grew.

2.3.2. Washcoats of Alumina and Silica on Fibers. The pore volume distribution of of a washcoat of 22-nm silica particles on 2-μm silica fibers before and after hydrothermal treatment at 750°C for 24 h is shown in Fig. 7-10. The distribution is

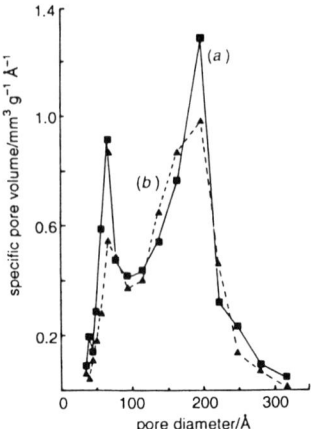

Fig. 7-10. Pore-volume distributions of 2-μm amorphous silica fibers coated with 22-nm silica particles: (a) before and (b) after hydrothermal treatment at 750° C for 24 h. Törncrona et al. (1985). Courtesy the Royal Society of Chemistry.

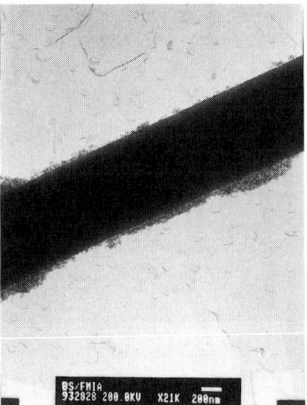 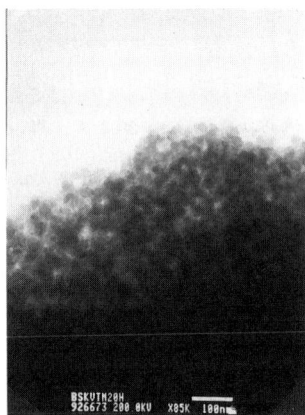

Fig. 7-11. TEM micrographs of uneven layers of 22-nm silica particles adsorbed on the surface of 2-mm amorphous silica fibers hydrothermally treated at 750° C for 24 h. Törncrona et al. (1995). Courtesy the Royal Society of Chemistry.

bimodal with one peak at 60 Å, which arises from densely-packed silica particles, and another at 700 Å, which is thought to come from pores between small aggregates of silica particles. The particles are thus not regularly densely packed on the fiber surface, which can also be seen in Fig. 7-11. On the other hand, the silica particles have been laid down in a fairly even layer, which increased the specific surface area of the fibers to 41 m^2/g before hydrothermal treatment. The pore size and its distribution can be changed by varying the size of the silica particles.

Hydrothermal treatment of silica fibers with washcoats of silica particles at 750° C in 100% steam for 24 h sintered the particles somewhat and reduced the specific surface from 41 to 32 m²/g. The moderate changes of the pore volume distribution during the hydrothermal treatment is due to the fact that the washcoat consisted of low-sodium silica particles. Such washcoats are thus characterized by high hydrothermal stability. Figure 7-12 shows that the pore volume distribution of a washcoat of fibrillar γ-alumina on silica fibers, which was calcined at 550° C for 90 minutes, is broader than the pore size distribution of a washcoat of silica particles on the same type of fibers. Figure 7-13 shows that the pore structure consists of randomly packed rods, which make up a so-called "pick-up-sticks" structure.

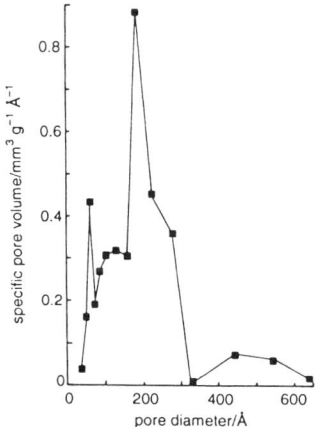

Fig.7-12. Pore-volume distribution of silica fibers coated with fibrillar alumina particles after calcination at 550° C for 90 min. Törncrona et al. (1995). Courtesy the Royal Society of Chemistry.

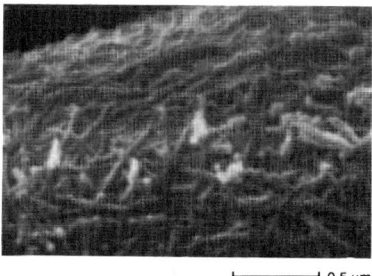

Fig. 7-13. SEM micrograph of a silica fiber coated with particles of fibrillar alumina after thermal treatment at 550° C for 90 min. Törncrona et al. (1995). Courtesy the Royal Society of Chemistry.

This structure is characterized by a broad pore size distribution with an average pore size of the same order of magnitude as the diameter of the rods. The main peak of the distribution occurs at 200 Å whereas the average diameter of the rods is about 400 Å (their length is about 9000 Å). During calcination at 550° C the specific surface area decreased from 35 to 28 m^2/g.

2.4. Catalysts with Supports of Alumina and Silica on Carriers

Catalysts for pollution control, i.e., catalysts for cleaning of automobile exhausts and complete oxidation, e.g., combustion of volatile organic compounds in, for instance, ventilation air from different processes, consist of catalytically active substances, frequently noble metals or transition metal oxides, deposited on the surface of washcoated carriers.

2.4.1. Deposition of Catalytically-active Substances on Supports. Two methods are commonly used for depositing, say, platinum and rhodium on washcoats, Evaldsson et al. (1989). In the first method, called "impregnation to incipient wetness", the channels of a washcoated monolith are first filled with a solution of e.g., chloroplatinic acid and the water is next removed by evaporation; say at 100° C for 1 hour. The second method utilizes the fact that the surfaces of alumina and silica possess ion exchange characteristics. The surface of silica is negatively charged at pH > 2 and is therefore a cation exchanger in this pH range. To attach themselves to a silica surface, ions of noble metals must therefore be present as cations, e.g., in the case of platinum as tetramine platinum ions, $Pt(NH_3)_4^{++}$. An alumina surface, in contrast, is positively charged at pH < 8 and is therefore an anion exchanger in this pH range. Anions of noble metals are therefore required; e.g., again in the case of platinum as chloroplatinic ions. Deposition of noble metals on the support surface is effected by circulating appropriate solutions of noble metal ions through the channels of the monolith whereby the complex noble metal ions attach themselves to the surface of the washcoat.

Catalytically active substances such as metal oxides or mixtures of such oxides can be introduced into washcoated monoliths by an impregnation-calcination procedure, Lin et al. (1995) and Skoglundh et al. (1994). Oxides of transition metals can thus be inserted by immersing a sample of washcoated monolith in an aqueous solution of the appropriate mixture of, e.g., metal nitrates; typically 2.5 M with respect to the sum of the different metal ions. Surplus solution is removed by blowing air at room temperature through the monolith channels.

Drying is accomplished by blowing hot air through the channels. In a typical experiment this procedure was repeated six to seven times to obtain 6 g of transition metal oxides per 100 g of catalyst after calcination at 800° C for 2 hours. The specific surface area of wash-coated monolith decreased by about 40% after deposition of the transition metal oxides to a typical value of about 11 m^2/g for a "ready-to-use" catalyst.

2.4.2. Testing of Catalysts. In work by Skoglundh et al. (1991) the reactor used for catalyst testing consisted of a vertical stainless steel tube, 900-mm long and with an inner diameter of 16 mm, encased in a tubular furnace. Each catalyst was typically first exposed to oxidizing conditions in an air flow of 500 cm^3/min at 500° C for 30 minutes and then to reducing conditions in a hydrogen flow of 200 cm^3/min for 1 hour at 450° C.

The light-off temperature, T_{50}, was defined as the temperature of the inlet gas at which the conversion of the reactant was 50%. The T_{50} temperature is a measure of the temperature at which the catalyst ignites a reaction. Low values of T_{50} of catalysts are desirable so as to reduce so-called cold-start emissions, i.e., emissions that occur because the catalyst has not become warm enough to ignite the various emission-reducing reactions. The space velocity during the determination of T_{50} was normally 144,000/h, determined at 20° C.

The efficiency of the catalysts was determined by measuring the conversion of a reactant at different space velocities in the range 100,000-300,000/h, within a reaction temperature range of 145-500 °C.

2.4.3. Palladium and Platinum on Thermally and Hydrothermally Treated Wash-Coats of Alumina. Figure 7-14 shows that the T_{50}'s for the oxidation of xylene over Pd:Pt supported on hydrothermally-treated alumina were lower over the whole range of Pd:Pt mol-percent ratios than over the same catalysts on thermally-treated alumina, Skoglundh et al. (1991) The light-off temperature had a minimum at a Pd:Pt ratio of 80:20, but it was less pronounced for catalysts with hydrothermally-treated supports. Furthermore, the minimum in T_{50} on hydrothermally-treated alumina was only about 17 °C higher than for particles of pure platinum. It is also noteworthy that T_{50} for pure palladium on hydrothermally-treated alumina was as low as for pure platinum on thermally-treated alumina. Pd-Pt particles with a Pd:Pt mole-% ratio of 80:20 on hydrothermally-treated alumina supports are effective and economically attractive catalysts for complete oxidation of volatile organic compounds.

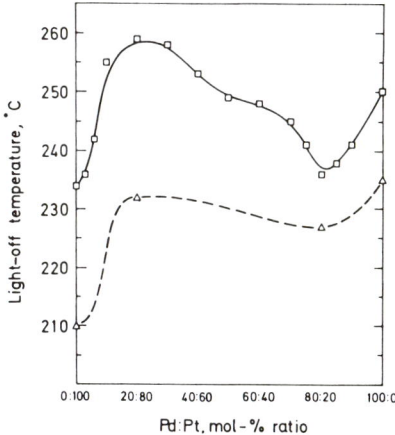

Fig. 7-14. Effect of composition of noble metals on T_{50} for oxidation (SV=144,000 h^{-1}) of 220 ppm xylene over catalysts with thermally ($\square$) and hydrothermally ($\triangle$) treated washcoats. [Pt +Pd] = 10 µmol per gram catalyst. Skoglundh et al. (1991). Reprinted from Applied Catalysis, **77**,9 (1991) with kind permission from Elsevier Science - NL, Sara Burgerhartstraat 25, 1055 KV Amsterdam, The Netherlands.

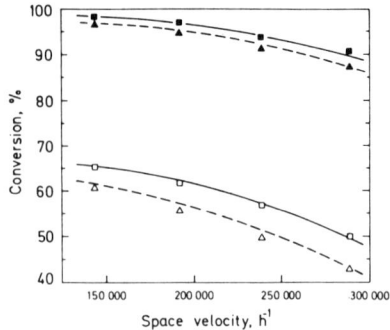

Fig. 7-15. Effect of gas flow rate on conversion at 290° C for the catalytic oxidation of 600 ppm xylene (unfilled symbols) and 600 ppm carbon monoxide (filled symbols), respectively, over thermally (□) and hydrothermally (Δ) treated washcoats. Pd:Pt = 80:20 mol %, [Pd+Pt] = 10 μmol per gram catalyst. Skoglundh et al. (1991). Reprinted from Applied Catalysis, **77**,9 (1991). with kind permission from Elsevier Science - NL, Sara Burgerhartstraat 25, 1055 KV Amsterdam, The Netherlands.

Hydrothermal treatment of alumina supports decreases the surface area and shifts the pore-size distribution to larger pores—cf. Figs. 7-7 and 7-8. It may, therefore, be expected that larger molecules diffuse more rapidly in a hydro-thermally-treated support structure and can reach the available catalyst particles in the structure more readily. Figure 7-15 shows that the conversion of xylene at a given space velocity is higher over catalysts on hydrothermally-treated supports and that the difference, as expected, increases with space velocity. The oxidation of carbon monoxide, which is a smaller molecule than xylene, was less sensitive to the pore structure of the support.

2.4.4. Efficiency of Catalysts with Washcoats of Alumina and Silica.

Figure 7-16 shows the conversion of xylene versus temperature at three different space velocities, 90,000, 123,000, and 169,000/h, for catalysts with washcoats of fibrillar alumina (F3), 0.052 wt. % Pt, corpuscular alumina (D1), 0.046 wt. % Pt, and spherical silica (K), 0.039 wt. % Pt, Evaldsson et al. (1989). A comparison of the effect of the nature of the washcoat on the efficiency of platinum as an oxidation catalyst for xylene is obscured by the fact that the platinum content is not the same in the three catalysts. The efficiency is highest for the F catalyst, even when it is expressed as mg Pt per square meter of washcoat, and lowest for the D1 catalyst.

Assuming that the conversion of xylene is not very sensitive to the platinum content within the studied range of platinum concentrations, then the fibrillar boehmite washcoat is the most effective support for platinum. Figure 7-7 shows that catalyst F3 has most of its pore area in pores larger than 50 Å whereas catalyst D1 has a significant fraction of its pores in pores smaller than 50 Å. Assuming that platinum is uniformly distributed over the surface of the catalysts, more platinum is, therefore, readily accessible to the fairly large xylene molecules in the washcoat of fibrillar alumina than in corpuscular alumina. Catalyst K, having the lowest platinum content, is also more efficient than catalyst D1, which is consistent with the fact that the silica washcoat has relatively more pore area in pores larger than 50 Å than does the washcoat of corpuscular alumina.

2.4.5. Metal Oxides on Washcoats of Alumina and Silica. Lin et al. (1995) studied the catalytic activity of mixtures of metal oxides with either copper oxide or cobalt oxide as the main component (see also §2.4.2). Not unexpectedly both oxide mixtures were quite good oxidation catalysts. The mixture based on Cu, $La_{0.45}$ $Sr_{0.15}$ Cu $Ce_{0.35}$ $Zr_{0.05}$ Ox, on a washcoat of fibrillar γ-alumina thus had a light-off temperature of 365° C for complete oxidation of propene, and the corresponding cobalt mixture had a T_{50} of 376° C for the same reaction. Addition of small amounts of Pt + Rh, e.g., 0.5mg per gram catalyst, lowered T_{50} for oxidation of propene over the Cu-based catalyst to 268° C, which is a very significant reduction, and over the cobalt-based catalyst to 218° C, which is a remarkable decrease.

The type of washcoat material also affected the light-off temperature for complete oxidation of xylene, which varied from 354° C for the cobalt mixture on silica to 375° C for the same mixture on corpuscular boehmite. For the Cu-mixture T_{50} was comparatively high, 408° C. Both mixtures had low-activity for reducing NO to N_2 at the stoichiometric ratio between oxygen and propene. The Cu-version on fibrillar boehmite showed an NO-conversion of 15% at 500° C, whereas the cobalt version had practically no activity at all.

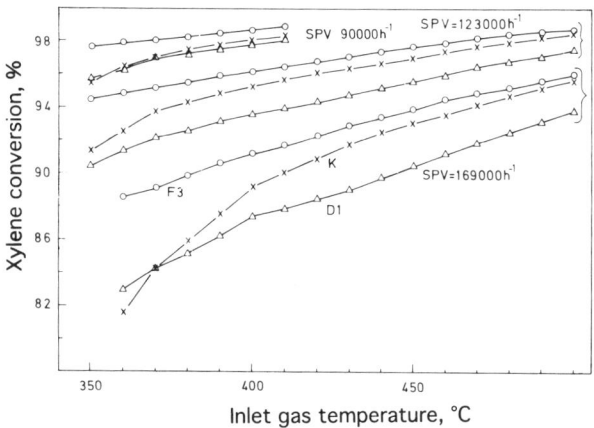

Fig. 7-16. Dependence of xylene conversion on temperature and space velocity. F3, fibrillar alumina; D1, corpuscular alumina; K, spherical silica. Evaldsson et al. (1989). Reprinted from Applied Catalysis, **55**, 123 (1989) with kind permission from Elsevier Science - NL, Sara Burgerhartstraat 25, 1055 KV Amsterdam, The Netherlands.

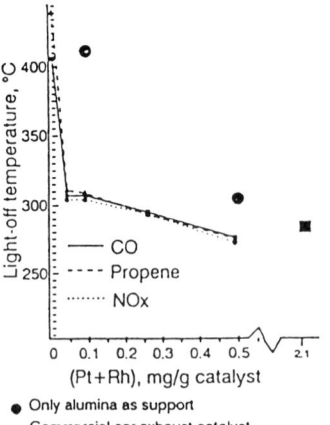

Only alumina as support ●
Commercial car exhaust catalyst ■

Fig. 7-17. Light-off temperatures for promoted $La_{0.45}$ $Sr_{0.15}$ $Co_{1.0}$ $Ce_{0.35}$ $Zr_{0.05}$ O_x as car exhaust catalyst at $S=1$ and $SV= 90,000$ h^{-1}, applied on corpuscular alumina washcoat. Lin et al. (1995). Reprinted from Applied Catalysis B: Environmental, **6**, 237 (1995). with kind permission from Elsevier Science - NL, Sara Burgerhartstraat 25, 1055 KV Amsterdam, The Netherlands.

Addition of small amounts of Pt + Rh, up to 0.49 mg per gram of catalyst, brought about very significant, sometimes almost dramatic, improvements of the NO conversion and oxidation activity for all combinations of the two types of oxide mixtures with the three kinds of washcoats. Figure 7-17 shows that the cobalt-based mixture on corpuscular alumina with 0.49 mg Pt + Rh gave the same light-off temperature for NO-reduction, about 285° C, as 2.1 mg Pt + Rh on corpuscular alumina in commercial catalytic converters. Already very small amounts of Pt + Rh, <0.1 mg per gram catalyst, caused large decreases of T_{50} for the reduction of NO.

The cobalt-based mixture of metal oxides is thus a much more effective support for the noble metals than corpuscular boehmite, especially at low concentrations of the noble metals. One reason may be that Rh deposited directly on γ-alumina will be partially deactivated by forming mixed oxides with aluminum, Wan and Dettling (1987).

Another reason for the high activity for NO reduction of combinations of mixtures of metal oxides with small amounts of noble metals may be that these metals may promote formation of H_2, e.g., by the watergas shift reaction. Hydrogen, adsorbed dissociatively on the noble metals, spills over to the metal oxides and reduces some of them, such as CeO_2, to a lower oxidation state, which in some manner participates in the reduction of NO to N_2, Lin et al. (1995).

2.4.6. Platinum on Fibers with Washcoats of Silica. Figure 7-18 shows the variation with temperature of the activity for oxidation of propene over Pt on a washcoat of 22-nm silica particles deposited on 2-μm silica fibers, Törncrona and Otterstedt (1995a). The light-off temperature is about 200° C, which is about the same as for a commercial Pt-based oxidation catalyst at the same space velocity, 100,000/h. Figure 7-19 shows that T_{50} for the oxidation of CO is about 170° C.

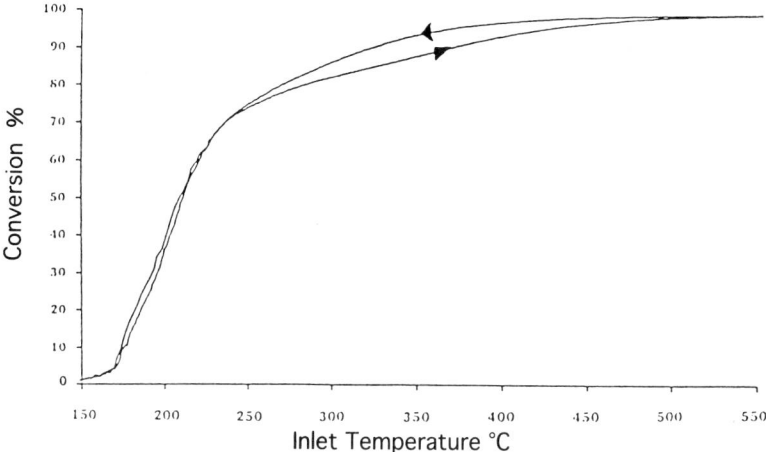

Fig. 7-18. Conversion of propene as a function of inlet temperature for a fibrous catalyst (1 mg Pt/g catalyst) at space velocity 100,000 h[-1] with a washcoat of 22-nm silica particles. Törncrona et al. (1995a).

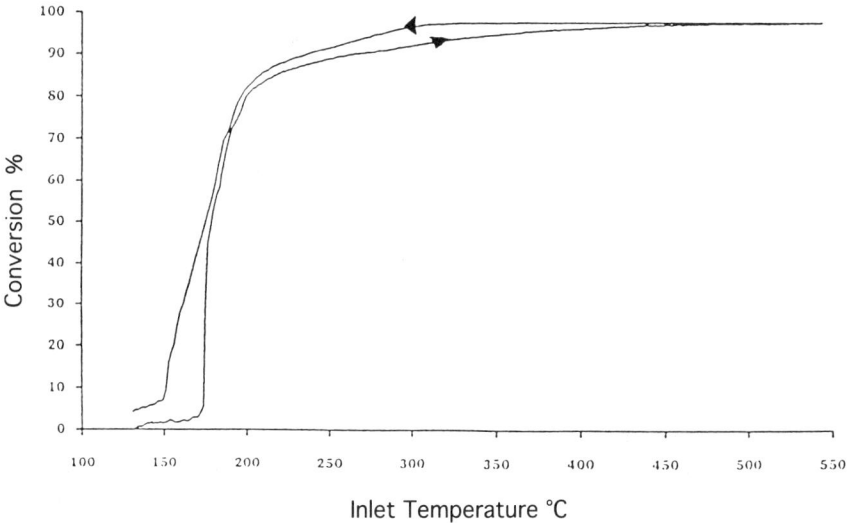

Inlet Temperature °C

Fig. 7-19. Conversion of CO as a function of inlet temperature for a fibrous catalyst (1 mg Pt/g catalyst) with a washcoat of 22-nm silica particles at space velocity 100,000 h[-1]. Törncrona et al. (1995a).

Törncrona et al. (1995) also investigated the extent to which the oxidation of propene and carbon monoxide was limited by mass transport by measuring the conversion of each reactant at different space velocities in the range 30,000-150,000/h at the constant temperature of 350° C. Figure 7-20 shows that the

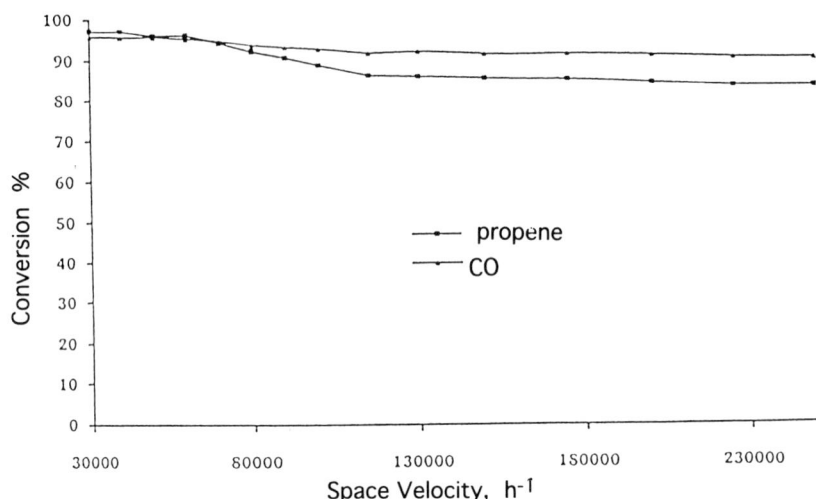

Fig. 7-20. Conversion propene and carbon monoxide as functions of space velocity at 350° C for a fibrous catalyst with a washcoat of 22-nm silica particles containing 1 mg Pt per gram catalyst. Törncrona et al. (1995a).

oxidation of CO was practically independent of space velocity whereas the oxidation of propene decreased somewhat with space velocity up to 120,000/h, but thereafter remained constant.

2.4.7. Reduction of NO_x over Zeolite ZSM-5 under Oxidizing Conditions.

Catalytic converters for gasoline-powered vehicles have a structure as shown in Fig. 7-1. The catalysts, rhodium, usually in combination with platinum, are deposited on the washcoat of boehmite alumina by the methods described in §2.4.1 above. Provided the air-to-fuel ratio in the car engine is close to the stoichiometric ratio required for completely combusting the hydrocarbon fuel, such catalysts, commonly called three-way catalysts, efficiently reduce NO_x and oxidize unburned fuel and CO in the car exhaust to nitrogen and water and CO_2, respectively.

Diesel engines and so-called lean-burn engines, however, operate at air-to-fuel ratios considerably higher than the stoichiometric ratio and three-way catalysts cannot reduce NO_x under these conditions, although they still function as efficient catalysts for combusting hydrocarbons and CO. There is, therefore, a need to develop catalytic converters that can remove harmful emissions from the exhaust from diesel-powered and lean-burn vehicles; especially by being able to reduce NO_x to N_2 under overall oxidizing conditions.

It was shown by Iwamoto (1981) that copper ion exchanged zeolite ZSM-5 can reduce NO_x to N_2 in an oxygen-rich atmosphere. In §4.3 of Ch. 5 it was shown that colloidal zeolite ZSM-5 could be prepared with $SiO_2:Al_2O_3$ ratios in the range from 40 to 80. Eriksson et al. (1997) prepared samples of monolith-based catalysts with conventional and colloidal zeolite ZSM-5 of different $SiO_2:Al_2O_3$ ratios and containing Cu^{2+} as the counter ion, incorporated in the washcoat of corpuscular alumina—see § 2.3 above.

The NO_x reducing and hydrocarbon combusting capabilities of the catalysts were evaluated using an oxidizing reactant gas mixture, composed of 400 ppm NO, 800 ppm propene, and 10% oxygen in argon as carrier gas and with a space velocity of 100 000 h^{-1}, by the method described in §2.4.2 above. Figure 7-21 shows that the conversion of NO_x to N_2 is about 50% higher over colloidal zeolite ZSM-5 than over the conventional zeolite ZSM-5 and has a maximum between ratios 50 and 60 for both types of zeolites. Moreover, the light-off temperatures T_{50} for both the NO_x reduction and the hydrocarbon oxidation were 50 to 60° C

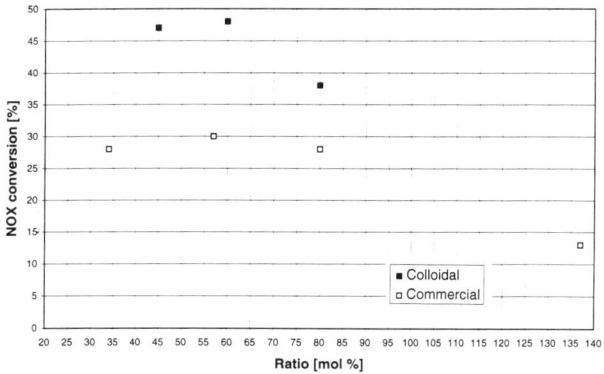

Fig. 7-21. Reduction of NO_x over Cu-ion exchanged colloidal and commercial zeolite ZSM-5. Eriksson et al. (1997).

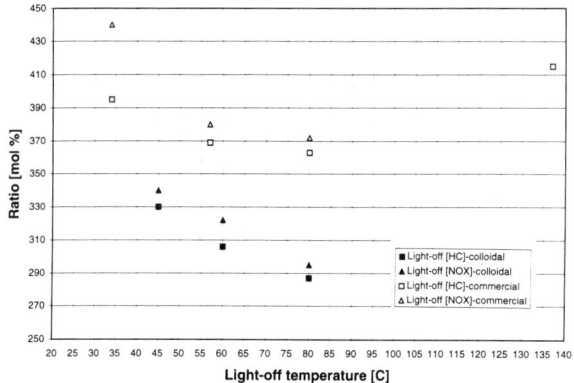

Fig. 7-22. Light-off temperature of NO and propene over Cu-ion exchanged colloidal and commercial zeolite ZSM-5. Eriksson et al. (1997).

lower over the colloidal zeolites, Fig. 7-22, making the colloidal version of zeolite ZSM-5 a more efficient catalyst than conventional ZSM-5 for these reactions. The reason for the higher efficiency of colloidal ZSM-5 may be their higher external surface area or the shorter diffusion paths in and out of these zeolites. During the short residence time in the reactor, about 35 milliseconds at a space velocity of 100,000 h-1, much of the interior of the large conventional ZSM-5 crystals may not be accessible to the reactant molecules.

2.4.8. Catalysis with Low Light-off Temperature, T_{50}. About 80% of the emissions from cars equipped with catalytic converters stem from the first three minutes of driving before the catalyst has been heated to its light-off temperature. These cold-start emissions can be reduced or even eliminated by several or a combination of several methods. Thus, the catalyst can be placed closer to the engine so as to faster reach its light-off temperature. Fast catalyst light-off can also be achieved by electrical heating of the catalyst. Any of these methods would be more effective, the lower the light-off temperature of the catalyst is.

Törncrona et al. (1997b) prepared catalysts by intermittently impregnating alumina washcoats supported on a monolith of cordierite with aqueous solutions of one or two of Co^{+2}, Ce^{3+}, $PtCl_6^{2-}$, $PdCl_4^{2-}$ ions or complexes with subsequent drying, calcination and reductions; see § 2.4.1, § 2.4.2 and §2.4.5 above. After the last step in the catalyst preparation—the reduction step,—the catalyst was placed in the reactor and the temperature was raised to 550° C while the reactant gas mixture, which could be either oxidizing or reducing, flowed through the reactor. The catalyst was stabilized in the reactant gas mixture for about 90 min., after which time the catalyst was cooled to about 80° C either in an atmosphere of 10% O_2 balanced with N_2 in a step called the pre-oxidation step, or in an atmosphere of 4% H_2 balanced with N_2 in a step called the pre-reduction step. The light-off temperatures for CO and propene (HC) over catalysts containing Pt or Pd promoted

Table 7-2. Nominal compositions and light-off temperatures (T_{50}) of CO and C_3H_6 in synthetic car exhaust, determined after preoxidation in 10 vol. % O_2/N_2 at 550° C and 4 vol. % H_2/N_2 at 550° C, respectively, using a net oxizing feed (S = 1.17). Törncrona et al. (1997b). Reprinted from Applied Catalysis B: Environmental, **14**, 131 (1997) with kind permission from Elsevier Science - NL, Sara Burgerhartstraat 25, 1055 KV Amsterdam, The Netherlands.

Sample	Metal oxide content (mg)	Precious metal content (mg)	Light-off temperatures (° C) Net oxidizing feed (S = 1.17)			
			Preoxidized		Prereduced	
			$T_{50(CO)}$	$T_{50(propene)}$	$T_{50(CO)}$	$T_{50(propene)}$
LTP1	0	2.0(Pt)	311	307	312	308
LTP2	0	1.09(Pt)	245	243	260	257
LTP3	40(Ce)	2.0(Pt)	246	246	220	220
LTP4	40(Ce)	1.09(Pt)	257	257	261	254
LTP5	40(Co)	2.0(Pt)	238	238	165	175
LTP6	40(Co)	1.09(Pd)	245	245	170	177
LTP7	40(Ce)	0	545	540	550	550
LTP8	40(Co)	0	340	365	185	200

Ce calculated as CeO_2.   Co calculated as Co_3O_4.   Molar amounts of Pt and Pd the same with cerium or cobalt oxides, and either pre-oxidized or pre-reduced according to the procedure just described, were measured. Table 7-2 shows the results for a slightly oxidizing reactant gas mixture, i.e., one containing somewhat more oxygen than is required for complete oxidation of CO and propene to water and CO_2 at a space velocity of 90,000 h^{-1}.

Compared with Pt and Pd, catalyst samples LTP1 and LTP2, respectively, the same metals promoted with cerium oxide, catalyst samples LTP3 and LTP4, and cobalt oxide, catalyst samples LTP5 and LTP6, achieved considerable reductions of the light-off temperatures for CO and propene. In particular, the light-off temperatures for CO and propene were lowered by 130-150° C over pre-reduced catalysts promoted with cobalt oxide. In fact pre-reduced cobalt oxide alone on an alumina washcoat, catalyst sample LTP8, was almost as effective a catalyst as pre-reduced Pt and Pd promoted with cobalt oxide, catalyst samples LTP5 and LTP8, and much more effective than pre-reduced Pt and Pd, catalyst samples LTP1 and LTP2.

X-ray photoelectron spectroscopic investigation of the pre-oxidized and pre-reduced catalysts indicated differences in their surface composition, but it was not clear how these differences correlated with variations in catalytic behavior.

Törncrona et al. (1996) washcoated cordierite monoliths with silica-titania composite particles on which cerium or cobalt oxide had been deposited—see § D of Ch. 6. Deposition of Pt or Pd on these washcoats also resulted in catalysts with remarkably reduced light-off temperatures for CO and propene.

3. Catalysts for Hydroprocessing of Oil

Removal of sulfur, nitrogen, oxygen, and metals from oil by catalytic reactions in so-called hydroprocessing or hydrotreatment processes has been of great importance ever since oil began to be used as a source of energy. Oil and oil products must be purified since most of the catalysts used in downstream refining of oil are negatively affected by sulfur and metals. Catalysts for reforming, a hydro-treatment process that raises the octane number of the naphtha fraction, are thus sensitive to poisoning by sulfur.

The last twenty years have seen increased demands to reduce emissions of sulfur from motor vehicles and industries using oil as the energy source. Also, the specifications for the sulfur content of fuel oils have continually become stricter, particularly in the United States and Japan where fuel oils today may contain no more than 0.3 weight percent sulfur compared to 0.6-0.8 wt. % in Sweden and 1-3% in Europe in general. These stricter specifications have had the effect that the importance of catalytic hydroprocessing has increased in the refining industry and that improved efficiency of the process as well as of the catalysts is being demanded. Figure 7-23 indicates that hydrotreatment catalysts consist of a support-carrier, on the surface of which catalytically active substances are deposited.

By far the most commonly-used support material is γ-alumina, which is formed by thermal treatment of the starting material boehmite. Other materials can also be used with γ-alumina when special functions are required of the support. In

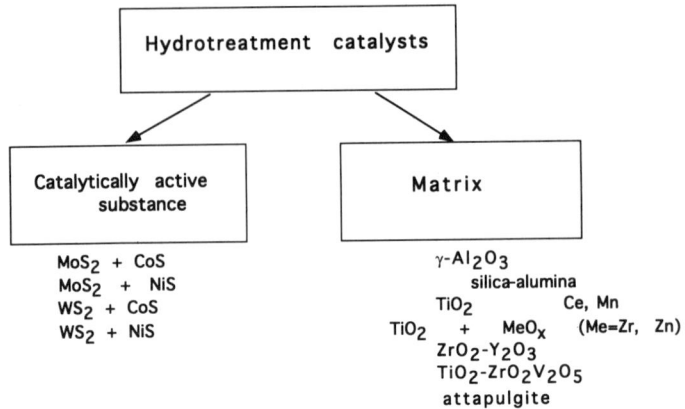

Fig. 7-23. Build-up of hydrotreatment catalysts.

some situations, it is desired that the hydrotreating catalyst, in addition to removing sulfur, also should bring about a certain amount of cracking of the oil. This can be achieved if the support is made of, or contains silica-alumina, the surface of which contains strongly acid aluminosilicate sites which catalyze cracking of hydrocarbons—see §4.2.3 below and §C and §D of Ch. 6.

The most widely used catalytic material is a mixture of molybdenum sulfide and either nickel or cobalt sulfide. The mixture of catalytically active substances is introduced into the support as a mixture of corresponding metal salts, e.g., chlorides or nitrates—see §2.4.5 and §2.4.8 above. During calcination in air the salts are decomposed to oxides, which will be transformed to sulfides in the reactor. The catalyst usually contains 2-5 % cobalt or nickel and 7-15 % molybdenum.

The alumina support of catalysts used for desulfurization of distillates has a pore structure with a pore size distribution that is typically centered around 70-80 Å. The heavier oil fraction to be desulfurized, the heavier molecules it will contain. Metals such as vanadium and nickel, and also to some extent sulfur, are contained in the heaviest molecules. Fast diffusion of heavy oil into the catalyst, where it can undergo desulfurization and demetalization, respectively, requires that the pore structure of the support can accommodate these very large molecules. Hydrotreatment of heavy oils thus requires catalysts with pore size distributions that are quite different from those of catalysts for desulfurization of light oils.

3.1. Hydrotreatment Catalysts with different Pore Structures

Catalysts for desulfurization of distillates have a pore structure characterized by a pore size distribution centered around about 70 Å and a specific surface area of 200-250 m^2/g. The alumina support is usually produced by extruding a well-peptized, viscous dispersion of corpuscular boehmite in water to about 5-mm-long extrudates of 0.8-3 mm diameter. After drying and calcining (see Section 2.1.4), the extrudates are impregnated with aqueous solutions of molybdenum and cobalt or nickel salts, e.g., chlorides or nitrates.

The metals are first converted to oxides by calcining in air at about 550° C and then to sulfides by sulfiding in the reactor. Catalyst supports with pore structures designed to desulfurize and demetallize heavy oils can be made from the supports of γ-alumina just described by heating them at about 1050° C for 2 hours, Oleck et al. (1976). The γ-alumina will then be transformed into a mixture of the crystalline phases θ- and α-alumina.

During this phase transition the specific surface area decreases to between 40 and 100 m²/g and the pore structure is enlarged so that about 65 % of the pore volume consists of pores with diameters in the range from 180-300 Å. The average pore size of the catalyst has thus become significantly larger, allowing large molecules to rapidly diffuse in and out of the catalyst, but this happens at the expense of specific surface area.

Another way of achieving rapid material transport without sacrificing too much of the catalyst's specific surface and, as a consequence, activity, is to make support structures with bimodal pore-size distributions. There would be a natural network of large, about 500-1000 Å, pores for rapid transport, but it would not provide much surface area for catalysis. This would be complemented by numerous small pores, about 80-100 Å, which provide the surfaces needed for desulfurization and metal deposition. Such two-pore catalysts can be made by mixing into the dispersion of well-peptized, corpuscular boehmite, cellulose fibers or microcrystalline cellulose. During calcination of the extrudates the organic material will burn off and leave behind large pores, Tischer (1981). Figure 7-24 shows bimodal pore-size distributions that have been obtained in alumina supports by admixing with Avicel®, a type of microcrystalline cellulose.

Two-pore catalysts can also be manufactured by pressing many microspheres of alumina, e.g., γ-alumina, lightly together so that the spaces between the microspheres become the macropores and the pores inside the microspheres are the micropores.

Support structures with large average pore size in combination with high specific surface area can also be accomplished by using rod-like particles of boehmite in a "pick-up-sticks" structure, Ying et al. (1995)—see also § 2.3.1.

Fibrillar boehmite was made by autoclaving a basic aluminum chloride solution at 160° C for 24 hours. A well-peptized dispersion of these particles was extruded as pellets, which were impregnated with molybdenum and nickel. Figure 7-25 shows the pore-size distributions of the finished catalyst (catalyst A) and a commercial catalyst from Shell Oil Co. (catalyst B). The specific surface areas of the two catalysts were 138 and 158 m²/g, respectively. Figure 7-26 shows that the catalyst with a support of fibrillar alumina demetallized a heavy oil more effectively than the commercial catalyst with smaller average pore size, whereas their desulfurization efficiencies were about equal.

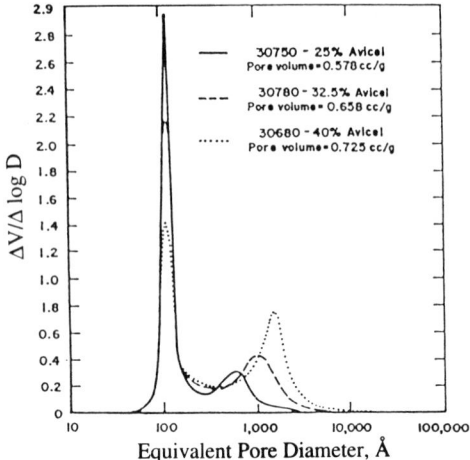

Fig. 7-24. Bimodal supports prepared by Avicel® addition and calcined at 1000°C/3 h. Tischer (1981). Courtesy Academic Press.

4. Fluid Catalytic Cracking Catalysts, FCCC

Fluid Catalytic Cracking, FCC, is the most efficient process for producing gasoline and the catalytic cracker is the core of modern high-conversion refineries. The FCC process was originally designed to convert gas oil from the atmospheric and vacuum distillation towers, but today FCC units also convert atmospheric and vacuum residues to gasoline and other distillates. This section is patterned after the excellent review of FCC cracking catalysts by Scherzer (1989).

Fluid catalysts are small spherical particles ranging from 40-150 μm in diameter. The active component of the modern cracking catalyst is a zeolite, usually zeolite Y, incorporated into a matrix which can be considered to be a

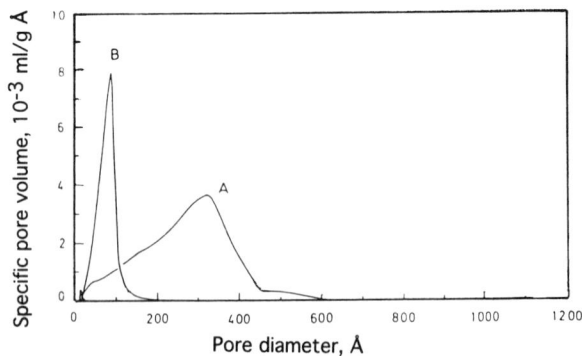

Fig. 7-25. Desorption pore diameter distributions for catalysts A and B. Adapted from Ying et al. (1995). Courtesy American Chemical Society.

support/carrier for the zeolite. The matrix support imparts mechanical strength to the catalyst, provides a certain activity for pre-cracking of heavy oils and dilutes the activity of zeolite, which is too high to allow the catalyst to be made entirely of zeolite. The matrix may also contain different additives to provide the catalyst with special functions and properties. The composition of fluid catalytic cracking catalysts is shown in Figure 7-27.

4.1. Zeolites

The most frequently used zeolite in FCCC is ion-exchanged zeolite Y, which is normally first made in the sodium form, NaY. In the sodium form, however, zeolite Y is not catalytically active, but must be ion-exchanged with rare earths

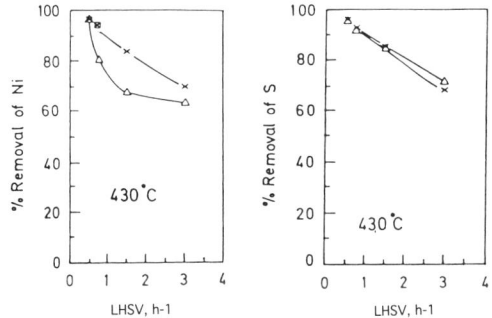

Fig. 7-26. Removal of a) nickel; b) sulfur; from residual oil by hydroprocessing at 430° C and LHSV in the range 0.5-3.0 h^{-1}. Catalysts A (x), B(Δ). Adapted from Ying et al. (1995). Courtesy American Chemical Society

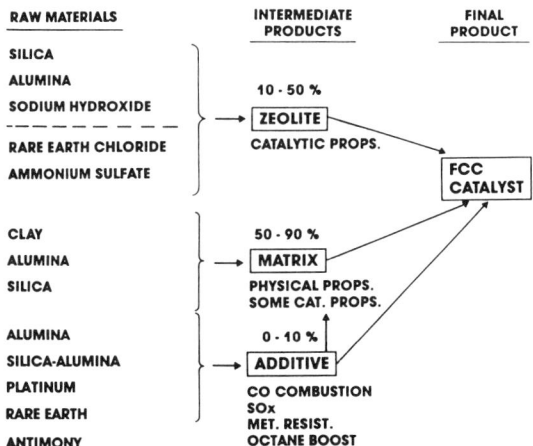

Fig. 7-27. The composition of FCC catalysts. Scherzer (1989). Courtesy Marcel Dekker.

and/or ammonium ions, in order to convert the zeolite into a catalytically active and hydrothermally stable form.

4.1.1. REY—Rare Earth Zeolite Y. REY is made from NaY with a silica-to-alumina ratio, $SiO_2:Al_2O_3$, of about 5.0 by ion exchanging it with solutions of rare earth ions. Commercial rare earth salts are mixtures of lanthanum and cerium with smaller amounts of neodymium and praseodymium. After the ion exchange, the zeolite still contains 3-4% Na_2O, which is too much to ensure high hydrothermal stability. To reduce the sodium content further the rare earth exchanged zeolite is calcined at 450-750° C and then again ion exchanged with a rare earth salt solution.

After the second ion exchange the REY, actually CREY, calcined REY, contains up to 16 % rare earth oxide and less than 1 % Na_2O, Scherzer (1972) and Moscou and DeBest (1970). REY zeolites are active and gasoline-selective cracking catalysts, but yield gasoline of rather low octane number, primarily because the high total acidity of the zeolite enhances the rate of hydrogen transfer, leading to a high rate of conversion of olefins into paraffins and of aromatics into condensed polycyclics.

4.1.2. USY—Ultrastable Y. USY zeolite can be prepared by first ion exchanging Na-Y zeolite with an ammonium salt solution in order to reduce the sodium content to 3-4 % Na_2O. The partially ammonium-exchanged zeolite is next calcined at about 760° C in steam to stabilize the zeolite and move the remaining sodium ions into ion exchange positions, Maher and McDaniel (1968). Ammonium exchange of the calcined zeolite removes most of the remaining sodium ions so that in its final form the zeolite contains less than 1% Na_2O and has a unit cell size of about 24.45 Å compared with about 24.70 Å for REY zeolite and the starting NaY zeolite. The stabilization in the calcination step involves de-aluminization of the framework, i.e., aluminum moves from the framework into the zeolite pores where it forms "extra framework alumina", EFA. It is true that this will increase the silica-to-alumina ratio of the framework, but the ratio for the zeolite as a whole will remain unchanged. By leaching USY zeolite with an acid solution, Moscou et al. (1970), or an acid ion exchange resin, Scherzer and Albers (1985), EFA and additional sodium can be removed. The leaching shall preferably lead to silica-to-alumina ratio of between 6 and 12 and a sodium content of about 0.1%.

USY zeolite has lower activity than REY zeolite, but gives a higher yield of high-octane gasoline since it enhances the olefin fraction of the gasoline. The activities of USY and acid-leached USY zeolites can be enhanced by rare earth exchange, Scherzer (1984).

4.1.3. High-silica Zeolite Y, HSY. Acid-leached USY can be said to be a type of HSY. Another type of HSY can be prepared by extracting aluminum from the framework of zeolite Y and replacing it with silicon without forming extra framework alumina in the pores. Zeolite LZ-210 is one example of this type of HSY zeolite, which is prepared by treating a partially ammonium-exchanged zeolite Y with a solution of $(NH_4)_2SiF_6$ at controlled pH, Skeels and Breck (1986). Silicon from $(NH_4)_2SiF_6$ replaces aluminum in the framework and the resulting $(NH_4)_3AlF_6$ is removed from the zeolite by washing. Depending on the preparation conditions, the silica-to-alumina ratio varies from 6 to 15. The hydrothermal stability of the HSY zeolites is very high and they also have high selectivity for high-octane gasoline.

4.1.4. ZSM-5, Pentasil. The zeolite designated ZSM-5 belongs to the family of silica-rich zeolites, and can be prepared by reacting organic quaternary ammonium hydroxides, in place of NaOH, with silica, e.g., in the form of a silica sol; and alumina, e.g., in the form of sodium aluminate, at about 150°C, Argauer and Landolt (1972). In the final step of the preparation, the calcination, the organic cation is decomposed and Na, H-ZSM-5 is formed—see Ch. 5.

ZSM-5 zeolites have been prepared with silica-to-alumina ratios from about 20 to higher than 8000. At the highest ratios the silicon content approaches that of pure silica as in silicalite, which is isostructural with ZSM-5. Electron microprobe analysis and XPS have shown that alumina is unevenly distributed, especially in larger crystals, > 5 μm, von Balmoos, (1981). Aluminum is often enriched at the surface as compared to the bulk of the crystal.

The octane-enhancing ability of ZSM-5 is a result of its unique shape-selective properties. Cracking catalysts containing ZSM-5 together with some form of zeolite Y will crack straight-chained hydrocarbons in the gasoline fraction, which have low octane numbers, in the ZSM-5 crystals, whereas branched hydrocarbons, which have high octane numbers, cannot penetrate into the pores of ZSM-5 because of their size and bulk. ZSM-5 will therefore enrich the high-octane components of the gasoline fraction.

4.2. Matrices

The composition of matrices of zeolitic FCC catalysts is similar to that of pre-zeolite catalysts, which consisted of synthetic silica-alumina gels and acid-leached clays.

In zeolitic cracking catalysts the zeolite, i.e., the catalytically active substance, is embedded in the matrix. The composition of the matrix and catalyst preparation conditions are chosen to impart desirable physical and catalytic properties to the catalyst particles.

4.2.1. Functions

Binder. The matrix binds the zeolite crystals in the microspherical catalyst particles into a structure that is hard enough to survive collisions with other particles and the reactor wall and able to withstand severe hydrothermal conditions without deterioration.

Diffusivity. The reactants diffuse through the matrix and undergo cracking in the zeolite crystals to products, which also must be able to diffuse through the matrix. For efficient transfer of reactants into and products out of the catalyst particles, the matrix must therefore have a pore structure that allows fast diffusion of hydrocarbon molecules. A seminal paper on diffusion in zeolite structures is that by van den Broeke (1995).

Dilution. Catalyst particles consisting entirely of zeolite crystals would be much too active and cause overcracking to undesirable products. The matrix therefore dilutes the zeolite component and moderates its cracking activity.

Sink for Sodium and other Metals. The matrix may act as a sink for ions of sodium and other metals, e.g., vanadium, nickel, and iron, which detrimentally affect the activity and selectivity of zeolites. By solid-phase ion exchange, sodium ions can diffuse from the zeolite crystals into the matrix, thus improving the thermal and hydrothermal stability of the zeolite.

4.2.2. Classification. Catalyst matrices may be classified by chemical composition, the origin of the components (synthetic, semi-synthetic, or natural), by their catalytic capacity (low, medium, or high activity), by some physical or chemical property (e.g., low or high specific surface area, low or high density).

Matrices usually contain two or more components. The binder is a necessary component in the catalyst and usually consists of amorphous silica, alumina, silica-alumina, or silica-magnesia. Clay, usually kaolin, halloysite, or montmorillonite, is often another component and it may be modified by chemical or thermal treatment. The mechanical strength of the catalyst particles is generally improved by the clay. Some catalysts contain other inorganic oxides such as TiO_2, ZrO_2, P_2O_5, B_2O_3, etc. Matrices consisting of a synthetic component (binder) and a natural component (clay) are called semi-synthetic matrices.

4.2.3. Synthetic Components

i. Amorphous Silica

Silica from a silica sol can be used as an effective, low-activity binder in the preparation of spray-dried FCC catalysts. The most common silica source is in the form of an acid sol, containing 3-5 % SiO_2, which can be prepared by acidifying a 3.3 ratio sodium silicate solution to pH 3—see Ch. 2.

Freshly prepared silica sol is most effective as a binder since the particles are still small and have had no time to form microgel, let alone gel. However, such sols are unstable and will gel in time. An increase in pH will also cause rapid gelling—see Ch. 6.

Amorphous silica binders matrices have low catalytic activity owing to their Brönsted acidity. Due to Ostwald ripening, their specific surface area will rapidly decrease under hydrothermal conditions—see Ch. 2. On the other hand, silica sols in combination with clays can form matrices that can have high surface areas and relatively high pore volumes, > 0.3 cm^3/g, with a high proportion, about 50 %, of pore volume in pores with diameters in the range 500-1000Å. This is beneficial primarily in cracking heavy feedstocks, e.g., resids.

ii. Alumina

Alumina is an important component of many zeolitic catalysts, which can serve several functions: (a) increase the catalytic activity of the matrix, (b) improve the mechanical strength of the catalyst, (c) trap metals present in heavy feed stocks, (d) improve the hydrothermal stability of the catalyst.

Pseudoboehmite, one of several commercially available aluminas, is the alumina most frequently used in catalyst formulations. It is usually dispersed or peptized in water with an acid—see Ch. 3—and mixed with the other catalyst components, after which the mixture is spray dried.

Pseudoboehmite is often used together with clays and other components in the catalyst matrix formulation. Addition of small amounts of ammonia-stabilized silica sol to the mixture of alumina and clay will enhance the attrition resistance of the catalyst, Lim and Stamires (1978).

iii. Silica-Alumina Gel

Before the advent of zeolitic cracking catalysts, synthetic silica-alumina gels were used as cracking catalysts. Much of the knowledge accumulated during that time has been used to formulate and design matrices containing amorphous silica-alumina as binder.

Synthetic silica-alumina gels can be prepared in several ways. One method is to react a silica sol with alumina by adding alumina, either in the form of a dispersion of an oxide-hydrate or as a solution of an aluminum salt, to the silica sol, Lim and Stamires (1978). Another way is by co-precipitation of a silica-alumina gel. A solution of sodium silicate, usually of $SiO_2:Al_2O_3$ ratio 3.3, is reacted with a solution of one or more aluminum salts under controlled conditions, Seese et al. (1980). In this case the gel will contain sodium salt, which must be removed by washing, either before or after the gel is mixed with zeolite and clay, to prevent the damaging effect of sodium on the thermal and hydrothermal stability of the zeolite.

The acidity of the gel depends on the alumina content, which varies between 10 and 30 % Al_2O_3. The Brønsted acidity reaches a maximum in silica-alumina gels with 30 % Al_2O_3; however, see §4.4 below.

The structure of silica-alumina gels is made up of aggregates of small spherical particles. The specific surface area may vary from 100-600 m^2/g, depending on factors such as preparation conditions, type of packing of the gel particles, particle size, thermal-hydrothermal treatment of the gel, etc., that is, the same factors that affect the pore volume and pore-size distribution of the gel; see Ch. 10. Calcination in the presence of steam decreases the surface area and pore volume, but increases the average pore size—see, e.g., Fig. 7-9b. The ion exchange capacity of silica-alumina gels is due to the presence of aluminosilicate sites on the gel surface—see Ch. 6 §C. Silica-alumina gels in an ion-exchanged form, as with rare earths, has been used as a matrix component in zeolitic cracking catalysts, which increases both catalytic activity and formation of coke and gas, Alafandi and Stamires (1981).

4.2.4. Natural Components

i. Natural Clays. Clays, in the form of aqueous dispersions of fine particles, which can be modified by chemical or thermal treatment, are used in all commercial catalyst formulations. Natural clays have low catalytic activity, but they affect the physical properties of the catalyst. Catalysts containing clays are in general harder, denser, and have better attrition resistance than catalysts with synthetic matrices. Clays also serve as fillers, since they are usually less expensive than other matrix components—see Ch. 4.

Kaolinite and halloysite are common clays in catalyst formulations. They have a small average particle size with a large fraction of the particles having a particle size < 2 μm, and are usually purified of elements such as iron, which has a detrimental effect on catalyst performance—see also Ch. 4 and Ch. 9. The small particle size and high degree of dispersion decrease catalyst attrition. Figure 7-28

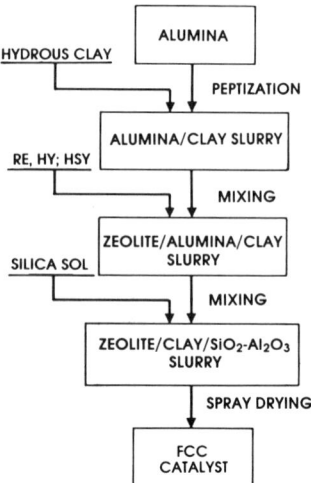

Fig. 7-28. Simplified diagram for FCC catalyst manufacturing process with silica-alumina sol derived matrix. Scherzer (1989). Courtesy Marcel Dekker.

shows a flow chart for preparing an FCC catalyst with a semi-synthetic matrix consisting of kaolin clay and silica-alumina gel as binder, Scherzer (1989).

ii. Chemically Modified Clays. Treatment of natural clays such as kaolinite, halloysite, and montmorillonite with acids enhances their catalytic activity, and such materials were actually the first to be used as cracking catalysts. Mild leaching of raw clay with a mineral acid, e.g., dilute sulfuric acid, removes most impurities such as iron, which decreases catalyst selectivity, forming more coke and gas at the expense of gasoline.

Pillared clays are other chemically modified forms of clays that have catalytic activity. Montmorillonite pillared or crosslinked with polycations of aluminum or zirconium thus shows good activity for cracking resids—see also § 4.4 below and Ch. 4.

iii. Chemically and Thermally Modified Clays. In Chapter 4 it was shown that kaolinite will undergo the following phase transitions on heating:

$$
\text{kaolinite} \xrightarrow[-H_2O]{500\text{-}600°C} \text{metakaolin} \xrightarrow{950°C} \text{spinel} \xrightarrow{>1000°C} \text{mullite}
$$

Converting kaolin to a mixture of spinel and mullite by calcination at about 1000° C and then removing some silica from the spinel by leaching the mixture with caustic results in a matrix material with good hydrothermal stability and activity for cracking heavy oils, Lussier and Surland (1989).

4.3. Octane-enhancing Catalysts

The first octane-enhancing catalysts appeared in the United States in the mid-1970's. About one-third of the gasoline sold in the United States and Europe comes from FCC units, and any increase in the octane number of FCC gasoline is therefore important to the refiner. Consequently, since the mid-1980's there has been a clear trend away from chiefly gasoline-producing catalysts toward high-octane and high-liquid-yield oriented FCC catalysts. Figure 7-29 shows that octane-enhancing FCC catalysts, like gasoline-oriented catalysts in general, consist of two components: zeolite and matrix, Scherzer (1989).

Some octane catalysts contain a third component: an octane-enhancing additive, e.g., zeolite ZSM-5 (see §4.1.4), which is either incorporated into the catalyst matrix or added as separate particles. Although the zeolite and matrix components both affect the octane rating of the FCC gasoline, it is the zeolite that plays the major role.

Fluid cracking catalyst octane catalysts can be classified according to their composition in the following manner:

1. catalysts with an octane-boosting zeolite Y in a catalytically inert matrix
2. catalysts with an octane-boosting zeolite Y in a catalytically active matrix—so-called octane-resid catalysts
3. catalysts with an octane-boosting or a conventional rare earth-exchanged zeolite Y and an octane-enhancing additive such as zeolite ZSM-5.

4.4. Catalysts for Cracking Heavy Oils

Cracking of heavy oils, such as resids, exposes the cracking catalyst to much more severe conditions than does cracking of gas oils. The catalyst must meet the following requirements in order to endure these conditions:

1. resistance to poisoning by metals; nickel and vanadium in particular
2. hydrothermal stability at high regeneration temperatures
3. low yields of coke and gas, especially in an environment with a high metal concentration
4. good activity for cracking very large molecules.

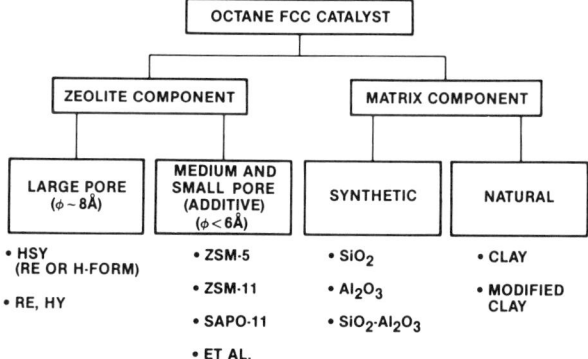

Fig. 7-29. Composition of octane FCC catalysts. Scherzer (1989). Courtesy Marcel Dekker.

Otterstedt et al. (1991) prepared a catalyst for cracking heavy oils by incorporating different ion-exchanged forms of zeolite Y in silica-alumina matrices having silica-to-alumina ratios in the range 10:90-70:30, i.e., at the opposite end of the weight ratio range typical for conventional silica-alumina matrices, which have ratios from 90:10 to 75:25—see §4.2.3.3 above. A silica sol was first prepared by deionizing a 3.3 ratio sodium silicate solution containing about 5 % SiO_2 to pH 2.5. To this sol was then added, with stirring, colloidal boehmite (Disperal) in amounts calculated to give silica-alumina matrices of desired ratios. The various types of zeolite Y, CREY, REUSY, USY, and LZ, were added to the silica-alumina sols under vigorous stirring and the resulting slurries were spray dried to give particles in the range 20-100 µm. Prior to testing, all catalysts were heated with 100 % steam at 790° C for 18 hours in order to simulate the aging of the catalysts in the regenerator of an FCC unit.

Figure 7-30, a plot of gasoline yields from cracking a fraction of Wilmington crude, boiling in the range 445–541° C, over CREY catalysts, against the matrix composition, shows that the gasoline yield has a maximum for silica-to-alumina weight ratios in the range 10:90–30:70, the range of ratios in which the deactivated surface area has a minimum—see Fig. 7-31.

In Ch. 6 silica surfaces containing aluminosilicate sites were discussed and the conclusion was reached that the silica surface can accomodate a maximum of about two such sites per square nanometer. Similarly, an alumina surface can also accomodate a maximum of about two aluminosilicate sites per square nanometer.

Matrices with silica-to-alumina ratios in the range 10:90-30:70 consist essentially of a porous alumina structure. There is more than enough silica to form

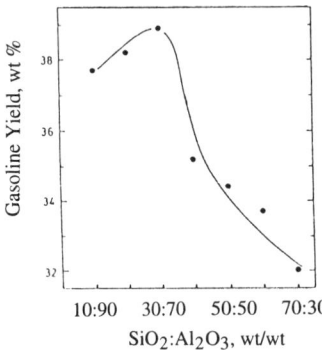

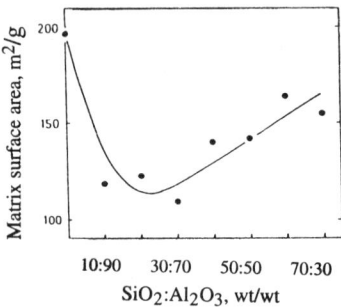

Fig. 7-30. Gasoline yield from cracking heavy oil over catalyst containing 17% CREY in matrices with different silica-to-alumina ratios. Otterstedt et al. (1991). Reprinted from Applied Catalysis, **70**, 43 (1991) with kind permission from Elsevier Science - NL, Sara Burgerhartstraat 25, 1055 KV Amsterdam, The Netherlands.

Fig. 7-31. Effect of matrix silica-to-alumina weight ratio on matrix surface area after deactivation at 790°C for 18 h in 100% steam. Otterstedt et al. (1991). Reprinted from Applied Catalysis, **70**, 43 (1991) with kind permission from Elsevier Science - NL, Sara Burgerhartstraat 25, 1055 KV Amsterdam, The Netherlands.

two aluminosilicate sites per square nanometer on the alumina surface. Surplus silica will form catalytically relatively inactive debris in the alumina structure and hence contribute only little to the formation of coke and gas at the expense of gasoline.

It will, however, contribute to the surface area of the matrix, and the minimum of the matrix surface area in Fig. 7-31 corresponds to a minimum amount of surplus silica in the matrix structure. Conventional silica-alumina matrices, with silica-to-alumina ratios between 90:10 to 75:25, on the other hand, consist mainly of a porous silica structure. Surplus alumina, i.e., alumina that has not been utilized to form aluminosilicate sites on the silica surface, about two per nm^2, appear as debris in the matrix structure. This alumina debris, however, is relatively active catalytically and contributes to the formation of coke and gas at the expense of gasoline.

In another type of catalyst, Sterte and Otterstedt (1988) used a matrix of alumina-pillared montmorillonite. Alumina pillars were prepared from basic aluminum chloride solutions which were hydrothermally treated at 120 and 140° C. The catalysts containing 20 % REY in admixture with alumina-montmorillonite complexes were prepared by dispersing the zeolite in slurries of the respective alumina montmorillonites.

A reference catalyst containing 20 % REY in a kaolin-binder matrix was also prepared, designated ACH-K-REY. All catalysts were dried at 60° C in air and were ground in a ball mill. The 40 to 100-mm fraction was steam heated at 750° C in a muffle furnace for 18 hours. Table 7-3 shows that the three catalysts with matrices of alumina-pillared montmorillonite are more active than the reference catalyst, ACH-K-REY, but that only the catalyst ZAMC-140, i.e., with pillars from a basic aluminum chloride solution autoclaved at 140° C, exhibited considerably higher conversion compared with the reference catalyst ACH-K-REY, namely 72.1 % vs. 57.2 %, as well as a higher gasoline yield of 30.4 % vs. 27.6 %.

Table 7-3.MAT results for alumina-montmorillonites containing REY. Adapted from Sterte and Otterstedt (1988). Reprinted from Applied Catalysis. **38**, 131 (1988) with kind permission from Elsevier Science - NL, Sara Burgerhartstraat 25, 1055 KV Amsterdam, The Netherlands.						
Catalyst	Conversion wt %	Gasoline wt %	Coke wt %	Gas wt%	LCO wt %	T °C
ACH-K-REY	57.2	27.6	12.7	17.0	15.1	560
ZAMC	62.3	26.6	15.3	20.4	9.9	560
ZAMC-120	63.1	27.9	15.6	19.6	11.1	560
ZAMC-140	72.1	30.4	19.6	22.1	12.7	560

5. Catalysts for the Chemical Industry

In the preceding Sections 3 and 4, it was described how small particles play an important role in the preparation of hydroprocessing and fluid cracking catalysts, which are widely used in the petroleum refining industry. Small particle technology can also be used in the preparation and improvement of catalyst properties for many processes in industrial chemistry.

A recent example of importance is the new Du Pont process, which became operational in 1996, for converting n-butane to tetrahydrofuran, THF, over vanadium phosphorous oxide and carbon supported Pd/Rh catalysts. This process minimizes byproduct formation, and combines significant environmental improvements with increased economy and product quality.

The success of this process rests on three innovations: 1) the use of a circulating fluidized bed reactor for the partial oxidation of n-butane; 2) the development by Bergna (1987), using small particle technology, of attrition-resistant particles; and 3) the development of a highly selective catalyst for the hydrogenation of maleic acid to THF under rather mild conditions in a bubble column reactor.

Vanadium phosphorous oxides, with $(VO)_2P_2O_7$ as the predominant species, are the preferred catalysts for n-butane oxidation. Because surface oxygen of these catalysts participates in the selective oxidation of n-butane to maleic anhydride, this oxygen must be restored periodically by re-oxidation of the catalyst.

In order to be used in fluidized bed processes, small particles of vanadium phosphorous oxide have to be cemented together to obtain porous microspheres or amphora-shaped strong agglomerates with high attrition resistance.

Conventional fluid catalysts, i.e., fluid catalytic cracking catalysts, derive their high mechanical strength from the matrix, which constitutes from 50 to 70% of the catalyst, and in which the active catalytic material, i.e., zeolites, is embedded. High attrition resistance has thus been gained at the expense of catalytic activity since the matrix, often a combination of kaolin clay with a silica binder, is virtually inert catalytically. It turns out , however, that this is not a problem in the case of fluid cracking catalysts since the amount of zeolite that can be incorporated into the mechanically strong matrix is sufficient to provide the catalyst with high catalytic activity. This may not be true for other catalytic materials, e.g., vanadium phosphorous oxides.

A conventional approach to impart attrition resistance to a catalyst grain is to embed small particles of the active catalyst in a continuous framework or skeleton made of a hard and relatively inert material like colloidal silica. In this case, the percentage of hard materials required to impart sufficient attrition resistance to the catalyst composite particle is so high (~50%) that it may affect the activity and/or selectivity of the catalyst.

Satisfactory attrition resistance can be conferred to the porous grains of the active catalysts with a much smaller amount of a cementing, inert material (~10 wt. %) uniformly distributed within a thin layer in the peripheral zone of the porous catalyst grains. This cementing materials can be either subcolloidal silica or extremely small-particle-size colloidal silica (< 5 nm diameter).

Figure 7-32 shows schematically that in the new Du Pont catalyst the vanadium phosphorous oxide active particles are cemented in the peripheral zone of porous microspheres by polysilicic acid filling the spaces between active particles. Microporosity in the cemented periphery is high enough to provide passage of reactants into and products out of the interior of the catalyst. By concentrating the silica binder in a peripheral zone, high attrition resistance can be achieved with minimum dilution of the catalytic material. Porous catalyst microspheres of 45 to 150 μm diameter were prepared by spray drying slurries of 0.5 - 1 μm vanadium phosphorous oxide particles and polysilicic acid (see § D of Chapters 2 and 6) of a SiO_2 concentration not exceeding about 6 wt. %, and calcining the spray-dried porous microspheres at a temperature which is not deleterious to the catalytic properties, but high enough to improve the cementation of the particles that constitute the porous microspheres. A temperature range of 300° to 400° C is typical for this calcination.

During the spray drying operation the extremely small particles of polysilicic acid, about 2 nm, migrate to the particle surface where they form the silica-rich, hard shell. The total concentrations of silica correspond to at most 10% of the catalyst weight.

The attrition resistance of catalysts made according to Bergna's method is so high that the catalyst particles can be used for many months without loss of activity, Bergna (1987).

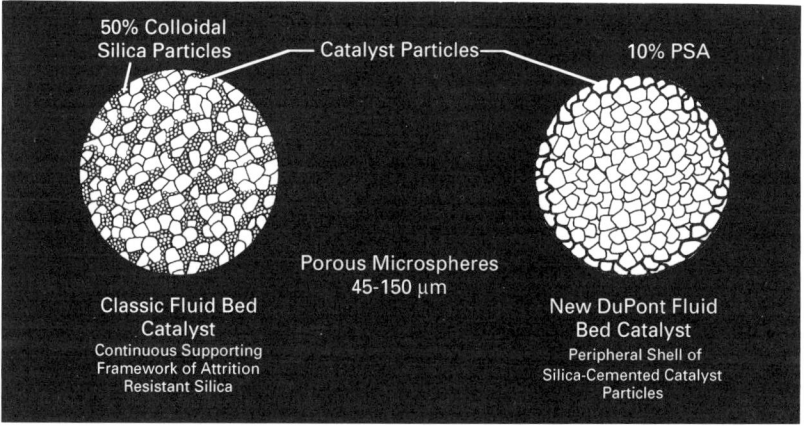

Fig. 7-32. Fluid bed catalysts attrition resistance: two concepts. On the right a cross-section of a vanadyl phosphate microporous microsphere containing 10% amorphous silica. Courtesy H. E. Bergna.

References

Alafandi, H., and Stamires, D., U.S. Patent 4,259,210 (1981).
Argauer, R. J., and Landolt, G. R., U. S. Patent 3,702,886 (1972).

Axelsson, I. M., Löwendahl, L., and Otterstedt, J.-E., Applied Catalysis, **44**, 251 (1988).

Bergna, H., U.S. Patent 4,677,084 (1987).

Eriksson, L., Löwendahl, L., Gevert, B., Törncrona, A., and Otterstedt, J-E., Applied Catalysis. (submitted for publication, 1997).

Evaldsson, L., Löwendahl, L., and Otterstedt,J.-E., Applied Catalysis, **55**, 123 (1989).

Iler, R. K., J. Colloid Interface Sci., **75**, 138 (1980).

Iwamoto, M., Yahiro, H., Tanda, K., Mizuno, N., Mine, Y., and Kagawa, S. J. Phys. Chem., **95**, 3727 (1991).

Larsson, L. B., Löwendahl, L., and Otterstedt, J.-E., *Catalysis and Automotive Pollution Control*. Eds. A. Crucq and A. Frennet, Elsevier Science, Amsterdam (1987).

Lim, J., and Stamires, D., U.S. Patent 4,086,187 (1978).

Lin, P.-Y., Skoglundh, M., Löwendahl, L., Otterstedt, J-E., Dahl, L., Jansson, K., and Nygren, M., Applied Catalysis B:. Environmental, **6**, 237 (1995).

Löwendahl, L., and Otterstedt, J.-E., Applied Catalysis, **59**, 89 (1990).

Lussier, R.J., and Surland, G. J., U.S. Patent 4,836,913 (1989).

Maher, P. K., and McDaniel, C. V., U. S. Patent 3,402,996 (1968).

Moscou, L., and DeBest, E. H., Neth. Appl. **68**, 17, 207 (1970).

Oleck, S. M., Slein, T. R., Sherry, H. S., and Milstein, D., U.S. Patent 3,931,052 (1976).

Otterstedt, J.-E., and Löwendahl,L., Swedish Patent 8901066-4, Nr. 464392 (1991).

Otterstedt, J.-E., Törncrona, A., Löwendahl, L., and Sterte, J., Swedish Patent. 470573 (1994).

Otterstedt, J.-E., Zhu, Y.-M., and Sterte, J., Applied Catalysis. **70**, 43 (1991).

Otterstedt, J-E. unpublished work (1995).

Scherzer, J. and Albers, E. W., U. S. Patent 3,676,368 (1972).

Scherzer, J. U. S. Patent 4,477,336 (1984).

Scherzer, J., and Humphries,A. P., U.S. Patent 4,512,961 (1985).

Scherzer, J., Catal. Rev.-Sci. Eng, **31**(3), 215 (1989).

Seese, M. A., Albers, E. W., and Magee, J. S., U.S. Patent 4,226,743 (1980).

Skeels, G. W., and Breck,D. W., U. S. Patent 4,610,856 (1986).

Skoglundh, M., Löwendahl, L., Menon, P. G., Stenbom, B., Jacobs, J. P., van Kessel, O., and Brongersma, H. H., Catalysis Letters, **13**, 27 (1992).

Skoglundh, M., Löwendahl, L., Jansson, K., Dahl, L., and Nygren, M., Applied Catalysis, B: Environmental, **3**, 259 (1994).

Skoglundh, M., Löwendahl, L., and Otterstedt, J.-E., Applied Catalysis, **77**,9 (1991).

Sterte, J. and Otterstedt, J.-E., Mat. Res. Bull., **21**, 1159 (1986).

Sterte, J., and Otterstedt, J.-E., Applied Catalysis. **38**, 131 (1988).

Tischer, R.E., Journal of Catalysis. **72**, 255 (1981).

Törncrona, A., Sterte, J., and J.-E. Otterstedt. J. Mater. Chem., **5**, 121 (1995).

Törncrona, A., and Otterstedt, J-E., unpublished work (1995a)

Törncrona, A., Löwendahl, L., Otterstedt, J-E., and Jansson, K., J. Mater. Chem., **6**, 213 (1996).

Törncrona, A., Löwendahl, L., Otterstedt, J-E., Brandt, J., Journal of the European Ceramic Society, **17**, 1459 (1997a).

Törncrona, A., Skoglundh, M., Fridell, E., Thormählen, P., and Jobson, E., Applied Catalysis B: Environmental, **14**, 131 (1997b).

Van den Broeke, L. J. P., A.I.Ch.E. Journal, **41**, 2399 (1995).

Von Balmoos, R., and Meier, W. M., Nature, **289**, 782 (1981).

Wan, C. Z., and Dettling, J. C., *Catalysis and Automotive Pollution Control*. Ed. A. Crucq and A. Frennet, Elsevier Science, Amsterdam (1987).

Ying, Z.-S., Gevert, B., Sterte, J., and Otterstedt, J-E., Ind. Eng. Chem. Res., **34**, No. 5, 1566 (1995).

CHAPTER 8. PIGMENTS

The Dry Color Manufacturers Association (USA) gives the following definition of a pigment: "A pigment is a colored, black, white, or fluorescent particulate organic or inorganic solid, which is usually insoluble in, and essentially physically and chemically unaffected by, the vehicle or substrate into which it is incorporated." —Pigments Handbook (1988).

Dyestuffs, on the other hand, are often molecularly dispersed, that is soluble in the carrying substrate, and may bleed into other contacting media in which they are soluble.

Pigments in the form of small particles will change appearance by selective absorption or by scattering of light. They are usually finely dispersed so as to be an integral part of decorative, protective and functional coatings, e.g. automotive finishes, printing inks, industrial coatings, marine paints, traffic paints, and exterior and interior house paints.

1. Classification

The following classification of pigments is based on coloristic and chemical considerations and forms the basis of this chapter:

• White pigments: TiO_2, ZnS, ZnO
• Extender pigments: clays, calcium carbonate
• Black pigments: carbon black, iron oxide
• Colored pigments:
      colored inorganic pigments: iron, chrome, and cadmium compounds
      colored organic pigments: azo compounds, phthalocyanines
• Specialty pigments:
      Luster pigments
      Luminescent pigments: ZnS
      Magnetic pigments: γ-Fe_2O_3, Fe_3O_4, CrO_2, metal particles
      Transparent pigments: TiO_2, Fe oxides
      Anticorrosive pigments: phosphates of Zn, Al, and Cr.

2. Economic Aspects

World production of inorganic pigments—colored, white, black, and others—in 1987 was about 4.5 million tons with consumption broken down as follows in Table 8-1.

The sales of inorganic pigments in 1989 added up to 13 billion U. S. dollars. Prices range from about \$0.1-0.2/kg for extender pigments such as clays and calcium carbonate to about \$1.5-2.0/kg for TiO_2 and about \$4-6/kg for enamel grade carbon black for automotive uses.

Table 8.1. World Production of Inorganic Pigments in 1987. Heine and Völz (1992). Courtesy VCH Publishers, Weinheim.	
Titanium dioxide	69%
Synthetic iron oxides	11%
Carbon black pigments	9%
Lithopone	5%
Chromates	3%
Zinc oxide	1%
Chromium oxide	<1%
Mixed metal oxide pigments	<1%
Others	<2%

Table 8-2. Major Pigments in the U.S., Europe, and Japan ranked in order of value. Otterstedt (1992).

Type	Color Index	Amount (10⁶ kg)	Value (10⁹ USD)	Price (USD/kg)
1 Diarylide	Pigment Yellow 12	20.8	0.28	13.5
2 Phthalocyanine	Pigment Blue 15:3	11.5	0.19	16.8
3 Quinacridone	Pigment Violet 19	3.19	0.19	58.5
4 Lithol Rubine	Pigment Red 57:1	12.2	0.18	14.6
5 Alkali Blue	Pigment Blue 19 & 61	9.23	0.11	11.4
6 Barium Lithol Red	Pigment Red 49:1	8.35	0.09	10.8
7 Phthalocyanine Green	Pigment Green 7	3.82	0.09	23.1
8 Red Lake C	Pigment Red 53:1	6.79	0.08	11.8
9 Diarylide Yellow	Pigment Yellow 14	5.87	0.07	11.9
10 Phthalocyanine	Pigment Blue 15:4	4.22	0.06	15.0
	Totals	86.0	1.34	15.5 avg

The world production of colored organic pigments in 1987 was 1.75 x 10^5 tons, Zollinger (1987), and sales amounted to about 4 billion dollars. Table 8-2 ranks the ten major organic pigments used in the United States, Europe, and Japan in 1992, Otterstedt (1992). A recent extensive book on all aspects of industrial organic pigments is that by Herbst and Hunger (1997).

3. Properties

Inorganic pigments are mostly single-component particles of oxides, hydrous oxides, silicates, sulfates, or carbonates of various metals. However, small particles with a composite structure, which may have interesting pigment properties, were discussed in Chapters 3 and 6, and will also be discussed in this chapter.

3.1. Chemical Properties

The chemical nature of the surface of particles making up inorganic compounds were discussed in Chapter 6. Selected chemical properties of some inorganic pigments are shown in Table 8-3, Kittel (1974).

Table 8-3. Chemical Properties of some Inorganic Pigments. Kittel (1974).

Pigment type	Moisture content DIN 53193 wt. %	pH DIN 53200	Water soluble components DIN 53197 wt. %	Acidity or alkalinity number DIN 53202
Titanium dioxide, rutile (surface treated)	0.4-0.5	5.5-8,5	0.1-0.2	alk. 3-8
Titanium dioxide, anatase (untreated)	0.2-0.3	7.0-7.5	0.4-0.5	alk. 11-15
Zinc oxide	0.1-0.2	~7	≤0.2	
Iron oxide, red		5.5-7.0	0.1-1.2	
Iron oxide, yellow		3.5-4.5	0.3-0.4	

Table 8-4. Chemical Properties of Major Organic Pigments. Kittel (1974). Courtesy W.A. Colomb in der H. Heenemann Gmbh.					
Pigment type	pH DIN 53200	Fastness (scale according to DIN 53230: 0=best, 5= worst)			
		Water	Alkali	Acid	Ethanol
Monoazo	5.0-8.5	0	0	0	0-5
Diazo	5.5-7.0	0	0	0	0-1
Phthalocyanine, blue	6.5-7.5	0	0	0	0
Phthalocyanine, green	5.0-7.5	0	0	0	0-1
Polycyclic	3.0-9.5	0	0	0	0

Most organic pigments are azo, monoazo, or diazo compounds, phthalocyanine compounds, or polycyclic compounds. Some of their chemical properties are shown in Table 8-4, Kittel (1974).

3.2. Physical Properties

From a practical and applications point of view the following properties are the most important:

       Density   Particle size and shape

       Surface area   Particle size distribution

       Oil number   Hardness

       Texture

Density and surface area, and possibly hardness, are the only physical properties that can be directly determined for pigment powders. The other physical properties depend on the degree of dispersion of the particles in a dispersion medium.

Density. Inorganic pigments and extender pigments range in density from 1.8 to 8 g/cm^3. The density of titania (rutile) is 4.0 - 4.1 g/cm^3 and that of carbon black is 1.8 g/cm^3.

Surface area. The BET method, which is the standard method for determining the specfic surface of small particles, was outlined in Chapter 1.

The surface area is inversely proportional to density and size of the pigment particles and ranges from about 20 to 1000 m^2/g for carbon blacks, from about 10-100 m^2/g for organic pigments, and from 5 to 50 m^2/g for inorganic pigments.

For complete dispersion of a pigment in a vehicle, e.g., a binder, the entire surface area of the pigment must be wetted by the vehicle. One kg of a pigment with a specific surface area of 50 m^2/g, typically one of the major organic pigments, has a total surface area of 5 hectares (more than 10 acres!).

Oil number. The oil number can be used to estimate roughly the amount of binder required to completely disperse a pigment powder. It depends, in an unspecified manner, on the particle size, i.e. the surface area, and the interaction between the particle surface and a binder-vehicle. The oil number can be determined by the British Standard Test B5.3483:1962 or by DIN 53199. It is the amount of linseed oil that must be ground into 100 g pigment powder on a plate using a palette knife until the mixture just smears.

Texture. This term is used to describe the ease of dispersing a pigment powder in a vehicle-binder. Carr (1967) claims that the following texture rating can be assigned to carbon blacks and organic pigments in linseed oil

$$\text{Texture rating} = S/y$$

where S = specific surface area (g/m^2) and y = oil number (g oil/100g pigment).

The value of S/y is 1.0 for many carbon blacks and organic pigments. For others it ranges from 0.2 to 2.0. Carr interpreted the variation as follows:
- If S/y > 1, breakdown of the pigment aggregates is incomplete and the texture is harder than normal.
- If S/y = 1, breakdown of the pigment aggregates is complete and the pigment is completely dispersed and the texture is normal.
- If S/y < 1, breakdown of the pigment aggregates is complete and some of the primary particles or crystals have been fractured so as to expose fresh surfaces. Such pigments are said to have softer texture than normal.

Particle size, shape and distribution. The concepts of particle size and particle shape described in Table 8-5 and Fig. 8-1 correspond to those used in the internationally accepted classification of pigment particles, Heine and Völz (1992).

Table 8-5. Definitions of Particles and associatied Terms. Heine and Völz (1992). Courtesy VCH Publishers, Weinheim.	
Term	Definition
Particle	individual unit of a pigment that can have any shape or structure
Primary or individual particles	particles recognizable as such by appropriate physical methods (e.g. by optical or electron microscopy)
Aggregate	assembly of primary particles that have grown together are are aligned side by side; the total surface area is less than the sum of the surface areas of the primary particles
Agglomerate	assembly of primary particles (e.g., joined together at the corners and edges), and/or aggregates whose total surface area does not differ appreciably from the sum of the individual surface areas
Flocculate	agglomerate present in a suspension (e.g., in pigment-binder systems), which can be disintegrated by low shear forces

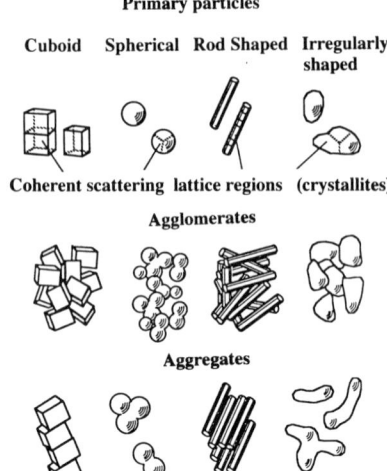

Fig. 8-1. Primary particles, agglomerates, and aggregates. Heine and Völz (1992). Courtesy of VCH Publishers, Weinheim.

Care must be exercised when using the terms particle size and particle size distribution since these properties of small particles can be defined in several ways as shown in Table 8-6.

Table 8-6. Particle size, particle size distribution, and characteristic quantities. Heine and Völz (1992). Courtesy VCH Publishers, Weinheim.	
Term	Definition
Particle size	geometrical value characterizing the spatial state of a particle
Particle diameter D_{eff}	diameter of a spherical particle or characteristic dimension of a regularly shaped particle
Equivalent diameter D	diameter of a particle that is considered as a sphere
Particle surface area S_T	surface area of a particle: a distinction is made between the internal and external surface areas
Particle volume V_T	volume of a particle: a distinction is made between effective volume (excluding cavities) and apparent volume (including cavities)
Particle mass m_T	mass of a particle
Particle density ρ_T	density of a particle
Particle size distribution	statistical representation of the particle size of a particulate material
Distribution density	gives the relative amount of a particulate material in relation to a given particle size diameter. Density distribution functions must always be normalized.
Cumulative distribution	normalized sum of particles that have a diameter less than a given particle size parameter
Fractions and class	a fraction is a group of particles that lies between two set values of the chosen particle size parameter that limits the class
Mean value and other similar parameters	the mean values of particle size parameters can be expressed in many ways; some values are used frequently in practice
Distribution spread	parameter for characterizing the nonuniformity of the particle size

The average ultimate particle size of most commercial pigments ranges from 0.01 to 1.0 μm (10 to 1000 nm). Extenders and some pigments may have average diameters of as much as 50 μm.

Hardness. The abrasiveness of a pigment is not identical to its intrinsic hardness, i.e., the hardness of the primary particles. Consequently, the Mohs hardness scale is not a useful indication of the abrasiveness of a pigment.

3.3. Optical Properties

Desired optical properties of pigments are color strength, hiding power, transparency, and brightness. Some of these properties, e.g., transparency and hiding power, are contradictory, and a careful selection of pigments is required so as to obtain optimal optical properties in a given application.

The basic color of a pigment is determined by its chemical constitution, but the hue, and also transparency and brightness, are modified by particle size and shape, and differences in refractive index between the pigment and the medium in which it is dispersed.

According to the Mie theory, the color strength and hiding power (scattering strength) are functions of the particle size, refractive index, and the absorption and scattering coefficients, Mie (1908).

Interaction between a light beam of wavelength λ and a particle of diameter d causes part of the light to be absorbed and another part to be scattered.

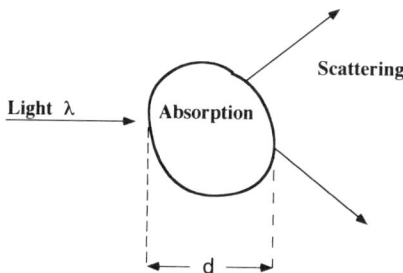

Fig. 8-2. Absorption and scattering of light by a small particle.

Absorption and scattering both depend on the particle size, but in different ways. In the so-called four-flux formulation, an extinction index, H_{ext}, is defined as follows:

$$H_{ext} = K + S \qquad (8\text{-}1)$$

where K = Kubelka-Munk absorption coefficient and S = scattering coefficient, both for unit concentration and thickness

Moreover

$$H_{ext} = \frac{3f}{2d} Q_{ext} \qquad (8\text{-}2)$$

where Q_{ext} = extinction efficiency, f = volume fraction of particles
(pigment) in the vehicle (e.g., binder), and d = particle diameter and

$$Q_{ext} = \frac{4C_{ext}}{pd^2} \tag{8-3}$$

where C_{ext} = extinction cross-section

$$\frac{\pi d^2}{4} = \text{geometric cross-section}$$

Since extinction = absorption + scattering, Q_{ext} can be expressed analogously to
H_{ext}:

$$Q_{ext} = Q_{abs} + Q_{sc} \tag{8-4}$$

If Eqn. 8-4 is inserted in Eq. 8-2, we obtain

$$H_{ext} = \frac{3f}{2}\frac{Q_{abs}}{d} + \frac{3f}{2}\frac{Q_{sc}}{d} \tag{8-5}$$

where now $\frac{3f}{2}\frac{Q_{abs}}{d} = K \tag{8-7}$

$$\frac{3f}{2}\frac{Q_{sc}}{d} = S \tag{8-8}$$

Both Q_{abs} and Q_{sc} depend on the refractive index $m = m_1 + im_2$, but
whereas Q_{sc} primarily depends on the real part, i.e. m_1, Q_{abs} depends mainly on
the imaginary part, i.e. m_2.

Models may be used to calculate Q_{ext}, Q_{abs}, and Q_{sc} as functions of d.
Figures 8-3 to 8-5 show how they vary with $\alpha = \frac{\pi}{\lambda} d$ for different absorption
indices, Kerker (1969).

The absorption index m is defined as

$$m = \frac{\text{refraction index of particle}}{\text{refraction index of dispersion medium}} \tag{8-9}$$

Brandreth and Otterstedt (1996) used the curve with absorption index m =
1.28 - 1.37i in Fig. 8-4 to calculate Q_{abs}/d as a function of d for light of
wavelength 500 nm. The reason for choosing this particular curve is that the
imaginary part of m, i.e. m_2 =1.37, is large, which is typical of a material with
high light absorbance; e.g. an organic pigment. Curve 1 in Fig. 8-6 shows that the
normalized Q_{abs}/d, which is roughly equivalent to the color strength first slowly
decreases as d increases to about 80 nm, but then rapidly decreases as d increases
from about 80 nm to about 300 nm. At d = 400 nm the color strength is only

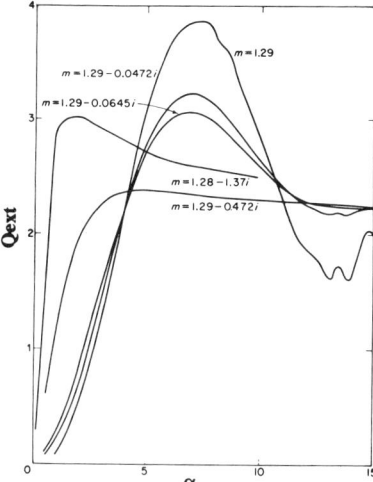

Fig. 8-3. Extinction efficiency for various absorption indices plotted against α. Kerker (1969). Courtesy of Academic Press.

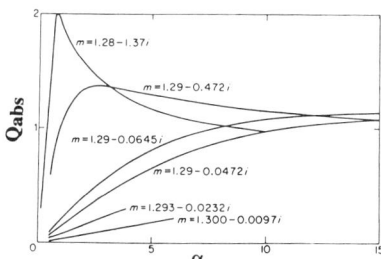

Fig. 8-4. Scattering efficiency for totally reflecting sphere (m = ∞) plotted against α. Kerker (1969). Courtesy of Academic Press.

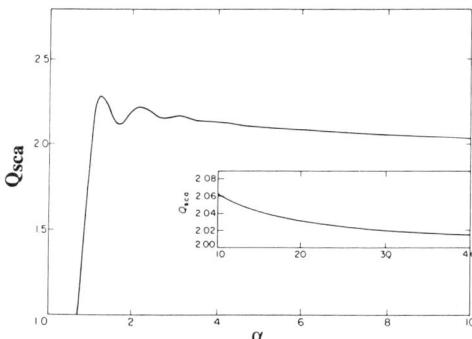

Fig. 8-5. Absorption efficiency for various absorption indices plotted against α. Kerker (1969). Courtesy of Academic Press.

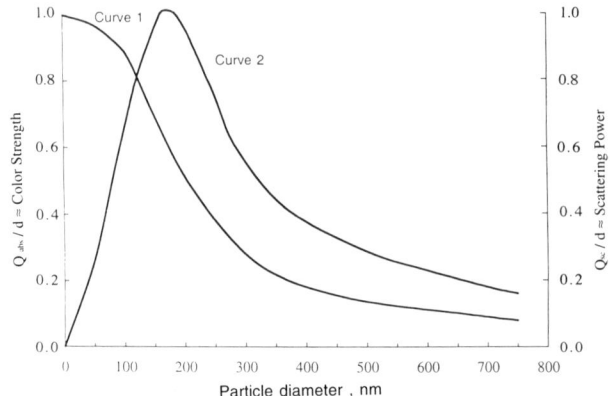

Fig. 8-6. Variation of color strength, Q_{abs}/d, (curve 1) and scattering power, Q_{sc}/d, (curve 2) with size of pigment particles. Otterstedt and Brandreth (1996).

about 25% of the color strength at very small d's. It is important to note that the higher the absorbance of the pigment, the larger $|m_2|$ is and the faster Q_{abs} will increase in Fig. 8-4.

Expressed differently, one could say that Q_{abs} is squeezed against the y-axis. The effect on Qabs/d in Fig. 8-6 will be even stronger. Curve 1 will be "pushed" against the y-axis and the rapid decline in color strength with increasing d will begin at smaller d's. This means that the higher the absorption coefficient, or color strength, of a pigment, the more finely the pigment will have to be dispersed so as to attain its full or potential, color strength.

Similarly, Otterstedt and Brandreth (1996) used the curve in Fig. 8-5 to calculate Q_{sc}/d, the scattering, as a function of d. Curve 2 in Fig. 8-6 shows that the scattering, which is roughly equivalent to the hiding power of the pigment, increases with d to a maximum at d ~ 180 nm, which is smaller than half the wavelength of the light (500 nm) that was used in the calculations. A rule of thumb is that the scattering maximum occurs for particles with a size corresponding to half the wavelength of the scattered light.

3.4. Light, Weather, and Heat Stability

The ability to withstand degradation by light and weather depends primarily on the chemical composition, particle size, structure, and concentration of the pigment in the medium in which the pigment is dispersed. Testing of these properties is generally carried out by open-air exposure and accelerated weathering under controlled conditions in special devices.

The thermal stability of pigments also depends on the chemical composition, structure and type of binder used. In tests for evaluating the heat stability of pigments in a given binder two temperatures are usually noted: the highest temperature to which the pigment can be heated without changing its color and the temperature at which a small, but in general still acceptable change in color tone can be observed.

3.5. Comparison between the Properties of Organic and Inorganic Pigments

Inorganic and organic pigments differ in regard to the following properties:

Color strength. Organic pigments generally have higher absorption coefficients and color strength.

Hiding power. The absorption index is usually larger for inorganic pigments, making them scatter light more efficiently than organic pigments.

Density. Inorganic pigments as a rule have higher densities.

Solubility. Inorganic pigments are generally insoluble in organic solvents and show less tendency to bleed out into an organic medium, e.g. a polymeric binder.

Heat resistance. Organic pigments are more heat sensitive than inorganic pigments.

Brightness. Organic pigments are usually brighter than inorganic pigments. They scatter light less efficiently.

4. White Pigments

White pigments are small particles of some of the very few compounds that have refractive indices of about 2 or higher, shown in Table 8-7.

Titanium dioxide is the principal white pigment because of its light scattering power, which is superior to that of any other white pigment. This section will therefore primarily deal with titanium dioxide, but also briefly will describe zinc sulfide and zinc oxide.

Table 8-7. Characteristics of some White Hiding Pigments. Schiek (1992). Reprinted by permission of John Wiley & Sons, Inc.

Pigment	Chemical Composition	Refractive Index[a]	Average particle size (μm)
titanium dioxide, anatase	TiO_2	2.55	0.2
titanium dioxide, rutile	TiO_2	2.70	0.2-0.3
zinc oxide	ZnO	2.01	0.2-0.35
zinc sulfide	ZnS	2.37	0.2-0.3
lithopone	28% ZnS, 72% $BaSO_4$	1.84	0.2-0.3
lead carbonate, basic white lead	$2PbCO_3 \cdot Pb(OH)_2$	2.0	1.0
lead sulfate, basic	$PbSO_4 \cdot PbO$	1.93-2.02	0.8
antimony oxide	Sb_2O_3	2.1	1.0

[a] A common refractive index for a paint vehicle is 1.6

4.1. Titanium Dioxide

Titanium dioxide occurs in nature in the crystalline forms rutile, anatase, and brookite. Rutile and anatase are manufactured in large quantities, which are primarily used as pigments, but also as catalysts and in ceramics.

Properties. Pure TiO_2 is chemically stable and non-toxic. Anatase and rutile, the two commercially important forms, scatter light more efficiently than any other known material. When produced in a very fine particle size, the outstanding light scattering efficiency gives TiO_2 its extraordinary hiding power and bright white color.

TiO_2 pigments strongly absorb UV light, resulting in the formation of excited species that can oxidatively degrade organic polymeric binders in which titanium dioxide is incorporated and thus catalyze the degradation of, e.g., paint films—see also Ch. 6 § D.

Table 8-8 shows some typical properties of anatase and rutile.

In 1994 the total worldwide production capacity was 3.77×10^6 metric tons, of which 54% employed the chloride process. The chloride process will probably continue to displace the older sulfate process because it is a continuous process and is less intensive with respect to capital investment, labor and energy consumption. Furthermore, users, in particular those incorporating advanced finishing technology, shift to chloride-process pigment because of its superior performance.

Surface and surface coating. TiO_2 particles made by the sulfate process have a layer of phosphate on their surface, which makes them readily disperse in water. Such TiO_2 is used as an opacifier in paper where photochemical reactivity does not matter.

In the chloride process $AlCl_3$ is added as a co-oxidant to promote the formation of rutile and reduce particle aggregation. Chloride process pigment therefore contains about 1% alumina, which has been enriched at the surface of the particles, Braun (1992). Dispersability can be improved by coating the pigment particles with up to 5% of a layer of precipitated hydrous alumina. This coating, however, does not significantly reduce the photoreactivity of the TiO_2. In §D of Ch. 6 it was stated that it required 4-5 monolayers of SiO_2, corresponding to 6.5-8% SiO_2 in the form of a dense coating on the particles to make the TiO_2 particles durable and non-photoreactive—see Fig. 6-44. Durable

Table 8-8. Typical Pigment Properties of Anatase and Rutile TiO_2. Schiek (1992). Reprinted by permission of John Wiley & Sons, Inc.		
Property	Anatase	Rutile
density, g/cm^3	3.8-4.1	3.9-4.2
refractive index	2.55	2.76
oil absorption, g oil/100 g pigment	18-30	16-48
tinting strength, Reynolds method	1200-1300	1650-1900
particle size (av), μm	0.3	0.2-0.3

pigment can also be obtained by coating the particles with other metal oxides, e.g., co-precipitated hydrous alumina-zirconia, Howard (1977), or zirconium-tin oxides, Matsunaga (1983).

Coating also affects the acid-base characteristics of the particle surface, Lloyd et al. (1992). Silica coating increases surface acidity and promotes the adsorption of basic molecules, whereas alumina coating makes the surface basic and enhances the adsorption of acidic species.

Uses. Of the 3.06 million metric tons of TiO_2 used in 1992, 51% was used in coatings, 19% in plastics, 14% in paper, and the balance of 8% in several different applications such as elastomers, ceramics, cosmetics, and foods.

Composite structures. Greenwood et al. (1998) prepared a series of composite pigments consisting of silica cores of different diameters coated with varying amounts of anatase TiO_2—see § D of Ch. 6. When the deposition of titania on the cores was complete, the charge on the SiO_2-TiO_2 composite particle was reversed to negative by running the dispersion of particles into a solution of 3.3 ratio sodium waterglass. Table 8-9 shows the composition and some of the properties of the composite particles.

The composite SiO_2-TiO_2-SiO_2 particles and commercial rutile TiO_2 from Hoechst, supplied as an aqueous dispersion with a solids content of about 50%, were incorporated into polyacrylate films at various concentrations and the reflectance of the pigments was measured. Figure 8-7 shows that the smallest cores, 30 nm, with the thickest layer of anatase TiO_2, 400 wt % (based on core weight) came closest to the reflectance of pure rutile TiO_2.

In Fig. 8-7 the reflectance was plotted against the total (SiO_2 + TiO_2) weight fraction of pigment in the films. If instead, as in Fig. 8-8, the reflectance is plotted against the weight fraction of TiO_2 in the films, the reflectance of the anatase TiO_2 in the composite particles moves much closer to that of pure rutile TiO_2, in particular at pigment concentrations higher than 8-10% TiO_2.

Moreover, using Eqn. 8-9, Judd and Wyszecki (1963), to calculate the surface reflection R for light traveling through a material of refractive index n_1 and striking the surface of a material with refractive index n_2, one can estimate that the reflectance of anatase is only about 80% of that of rutile. Rutile TiO_2, instead of anatase TiO_2, in the composite particles would therefore increase the reflectance of every point on the composite pigment curves in Fig. 8-8 by about

Table 8-9. Composition and Properties if SiO_2-TiO_2 Composite Particles. Greenwood et al. (1998)			
Core diameter mm	Amount of TiO_2 on core, wt %	Electrophoretic mobility, 10^8 m^2/vs	Particle diameter (DLS), nm
300	150	-4.29	540
300	325	-4.21	556
300	400	-4.05	608
407	82	-3.98	659
407	170	-3.98	655
407	233	-4.05	630
500	82	-3.27	695
500	138	-3.12	660

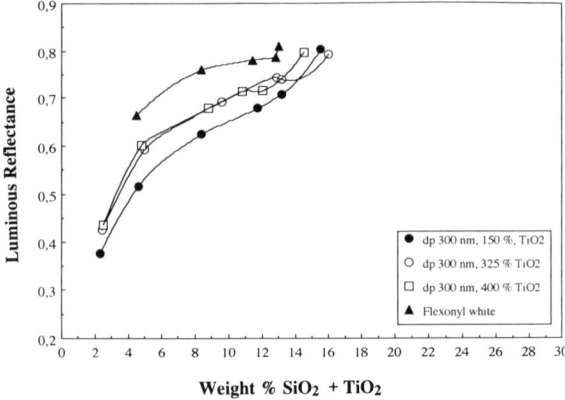

Fig. 8-7. Luminous reflectance of 300-nm SiO_2 cores coated with TiO_2. Greenwood et al. (1998).

25%. Relatively thick layers (400 wt. %) of rutile TiO_2 deposited on 300 nm silica cores would therefore reflect light more effectively than solid particles of rutile SiO_2.

$$R = \left[\frac{n_2 - n_1}{n_2 + n_1}\right]^2 \qquad (8-9)$$

Particles of TiO_2 sandwiched between SiO_2 can be made with very high uniformity of size. The thickness of the outer layer of SiO_2 can be carefully controlled so as to give very durable and easily dispersible pigments.

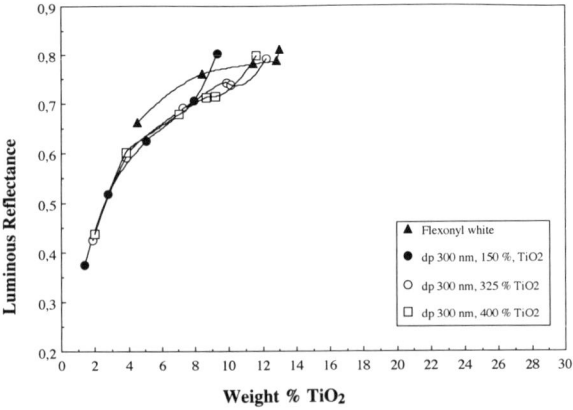

Fig. 8-8. Luminous reflectance of 300-nm SiO_2 cores coated with TiO_2; See also Table 8-9. Greenwood et al. (1998).

4.2. Zinc Oxide

Properties. Zinc oxide is a fine white powder that becomes yellow at temperatures above 300° C. It is an amphoteric oxide that reacts with acids and dissolves in alkaline solution with formation of zincates. The refractive index and particle size of ZnO are given in Table 8-7. Its density and oil absorption value are 5.6 g/cm^3 and 10-25 g oil/100 g ZnO, respectively.

Manufacture. Most of the ZnO used today is produced from sphalerite, ZnS, as the starting ore by either the direct, US process, or the indirect, French process. In the US process a compound containing ZnO is reduced with coal to zinc vapor, which is oxidized to ZnO powder. The French process takes place in two steps. Zinc metal is first vaporized and the vapor is then oxidized with the ZnO powder being collected.

Uses. ZnO is much inferior to TiO$_2$ in hiding power and is therefore used as a white pigment where some of its other properties are valued. It can, for instance, act as a fungistat in exterior oil paints. ZnO absorbs UV radiation and will therefore protect organic polymeric binders from photodegradation and reduce chalking.

The largest use of ZnO is as an activator for vulcanization accelerators in natural and synthetic rubber.

4.3. Zinc Sulfide and Lithopone

Lithopone is a mixed pigment which originally consisted of 28-30 wt. % ZnS and 70-72 wt % BaSO$_4$. Products with up to 60 wt. % ZnS have been developed.

Properties. The refractive index and particle size of ZnS are shown in Table 8-7. The density and oil absorption value are 4.0 g/cm^3 and 13 g oil/100 g ZnS, respectively.

Clean white pigments should not absorb light at all in the visible region, i.e., between 480 and 800 nm. ZnS and BaSO$_4$ meet this condition and do not exhibit the yellowish undertone associated with rutile TiO$_2$.

Manufacture. The raw materials for making lithopone are ZnSO$_4$, which can be obtained from different sources, and BaS, which is produced by reducing naturally occurring barite with coke.

The reaction of equimolar amounts of ZnSO$_4$ and BaS yields a white coprecipitate with the theoretical composition 29.4 wt. % ZnS and 70.6 wt. % BaSO$_4$. Coprecipitates with higher ratios of ZnS, e.g., 62.5 wt. % ZnS and 37.5 wt. % BaSO$_4$, can be obtained by the following precipitation reaction:

$$ZnSO_4 + 3ZnCl_2 + 4BaS = 4ZnS + BaSO_4 + 3 BaCl_2$$

Uses. Although the refractive index of ZnS is higher than that of ZnO, its hiding power and opacifying ability is still inferior to that of TiO$_2$. Lithopone has only 20-25% of the opacity of TiO$_2$.

Nevertheless, ZnS and lithopone are used as pigments in primers, plastic masses, putties, and emulsion paints when a clean white color is required.

Lithopone absorbs strongly in the UV region and is therefore useful as a white pigment in UV-cured paint systems. Like zinc compounds in general, ZnS has a fungicidal and algicidal action and will prevent attack by fungi or algae on paints.

5. Extender Pigments

Extenders—also called mineral fillers—are mostly inorganic solids of different chemical composition. The particle size can be fine or coarse and the particles may have many different forms. Extenders are added to coatings such as lacquers and paints to improve properties, e.g., rheological and mechanical properties, and product economics. They are generally white or somewhat grayish and differ from regular white pigments primarily by having much less hiding power.

The principal mineral fillers are clays—kaolin, bentonite, talc, and mica—along with calcium carbonate, barite, alumina, and silica.

Extender pigments are produced either directly from natural minerals, or by precipitation from aqueous solution.

Clays have been described in some detail in Ch. 4, silicas in Ch. 2, and aluminas in Ch. 3. Several important extender pigments are also discussed in Ch. 9 and in the filler section of Ch. 10.

6. Black Pigments—Carbon Black

The term carbon black refers to a group of well-defined, industrially manufactured products made under carefully controlled conditions. As a black pigment, carbon black absorbs light strongly over the whole visible spectrum and is therefore used for jet black colors and for darkening other pigment compositions. The largest use of carbon black, however, is as a reinforcing filler in rubber—see below and the section on fillers in Ch. 10.

6.1. Properties

Carbon black is a solid consisting of spherical particles showing a significant degree of two-dimensional symmetry. It does not have the third dimensional symmetry of graphite and diamond and is therefore a defect solid. The key physicochemical properties of carbon blacks, which vary between different manufacturing processes, are structure, particle size, surface area, and chemical composition.

Crystallite and particle structure. The crystalline regions in carbon black consist of well-developed graphite platelets stacked roughly parallel to one another, but random in orientation with respect to adjacent layers. These regions are 1.5-2.0 nm wide and 1.2-1.5 nm high, corresponding to 4 to 5 layer planes per crystallite containing about 375 carbon atoms. A particle of a carbon black with a surface area of 100 m^2/g contains over 4000 crystallites ordered in a concentric layer plane arrangement.

Morphology. Particles, aggregates, and agglomerates are terms used to describe structures of increasing size and complexity in carbon black, and indeed in many other particulate solids. These three terms were defined and described in Fig. 8-1 and Table 8-4. The carbon black particles are roughly spherical elements that are joined in the aggregate structures. Aggregates are the primary dispersible elements of carbon black and agglomerates are undispersed, but dispersible clusters of aggregates held together by van der Waals forces or binders.

The performance of carbon black as a pigment and as a reinforcing filler depends to a large extent on the size and shape of the aggregates. In Fig. 8-9 the electron micrograph shows that there is a very wide range in aggregate size, but within each aggregate the particles seem to be about the same size, and that the size of the aggregates is directly related to the size of the particles.

The aggregates can also have many different shapes: from compact grape-like clusters to open dendritic or branched assemblages and to fibrous arrangements.

The concept of structure was introduced about fifty years ago to account for and explain the difference between qualities of carbon blacks that had very similar surfaces areas but very different effects on the properties, e.g., viscosity and hardness, of the materials in which they were incorporated. Structure has to do with fluffiness and the term is used to describe the relative void volume properties of carbon blacks of the same surface area. Open dendritic and fibrous aggregates, which have the same surface area, have a high structure whereas compact clusters have a low structure.

The structure of carbon blacks can be measured by a void volume determination such as the dibutyl absorption test (analogous to the oil absorption test for determining the oil number, described in § 3.2 this chapter). Dibutyl phthalate is added to dry carbon black while mechanically working the mixture in a device that can also measure the mixing torque. The torque is at first very small, but increases rapidly when the mixture begins to form a coherent mass.

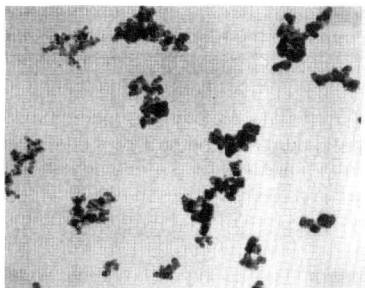

Fig. 8-9. Electron micrograph of reinforcing-grade of N399 tread black (100,000 x). Dannenberg and Paquin (1992). Reprinted with permission of John Wiley & Sons.

Type	Carbon	Hydrogen	Oxygen	Sulfur	Ash	Volatile
Table 8-10. Chemical Composition of Carbon Blacks, %. Dannenberg and Paquin (1992). Reprinted by permission of John Wiley & Sons, Inc.						
rubber-grade furnace	97.3-99.3	0.20-0.40	0.20-1.20	0.20-1.20	0.10-1.00	0.60-1.50
medium thermal	99.4	0.30-0.50	0.00-0.12	0.00-0.25	0.20-0.38	
acetylene	99.8	0.05-0.10	0.10-0.15	0.02-0.05	0.00	<0.40

The amount of DBP required to achieve a standard torque is taken as the relative structure index; low DBP absorption capacity implies a low structure and vice versa.

High structure carbon blacks give lower bulk densities and lead to high binder demand in paint formulations and to high modulus in vulcanized rubber.

Chemical composition. Table 8-9 shows that hydrogen and oxygen are the two most important non-carbon elements of carbon black, but also that some types also contain sulfur, ash and volatiles.

The oxygen is present as C_xO_y complexes on the surface of the aggregates whereas hydrogen and sulfur also are present in the interior of the aggregates. The volatiles originate from the functional groups attached to the carbon black layer planes—see Fig. 8-10.

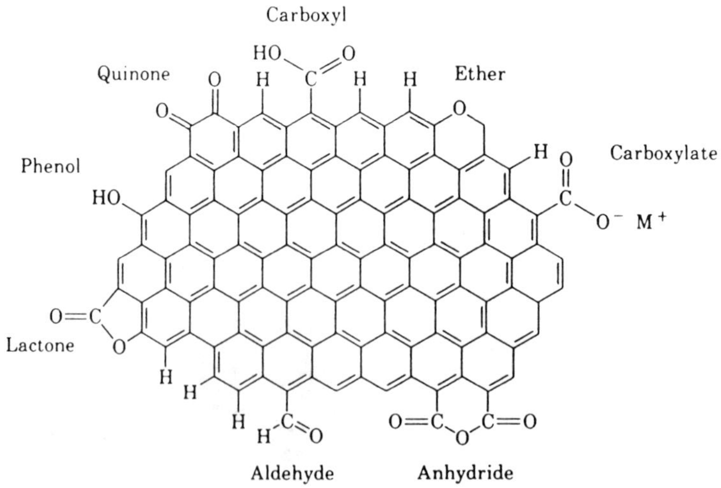

Fig. 8-10. Aromatic layer plane with functional side groups. Dannenberg and Paquin (1992). Reprinted with permission of John Wiley & Sons.

6.2. Formation Mechanisms

Figure 8-11 shows schematically the various steps in the formation of carbon black, Medalia et al. (1976).

The various steps in the sequence are not well understood. However, the morphology of the primary particles of carbon black suggests that the first nuclei of pyrolyzed hydrocarbons condense from the gas phase. Further carbon layers or their precursors are then adsorbed at the surface of the growing particles with an orientation that is always parallel to the existing surface. In high-structure carbon blacks, several particles collided and joined while they were growing. These loose arrangements were formed into aggregates by further carbon deposition. Ionic species such as K^+ and Ca^{2+} have a strong influence on the morphology of the aggregates and their surface area during the formation of carbon black, which indicates that ionic mechanisms are involved in the nucleation and aggregate formation steps, Frianf and Thorley (1961).

At temperatures above 1200°C and beginning at the particle surface the carbon layers rearrange to a graphitic structure. Graphite crystallites are formed at 3000°C and the carbon black particles assume polyhedral shapes.

6.3. Manufacture

The various manufacturing processes for making carbon blacks are summarized in Table 8-11, Medalia et al. (1976).

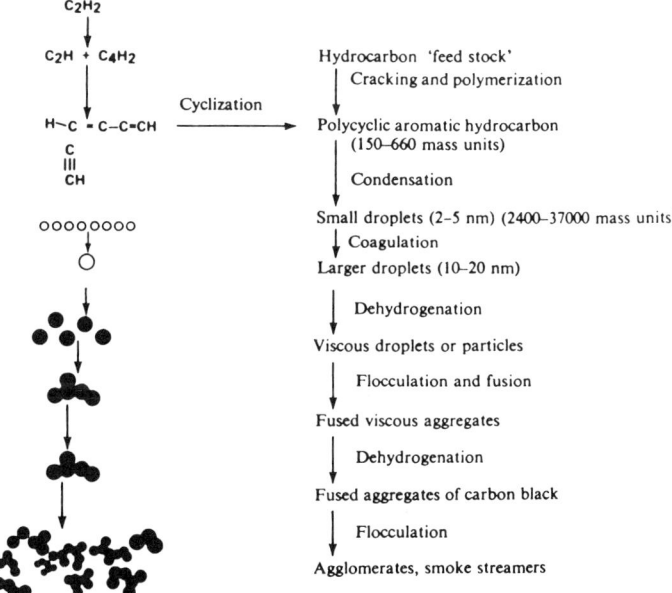

Fig. 8-11. Schematic mechanism of carbon black formation.Medalia et al. (1976). Courtesy American Chemical Society

Table 8-11. Manufacturing Processes for Carbon Black. Medalia et al. (1976). Courtesy American Chemical Society.

Process or type	Process feature	Raw material
Furnace	Partial combustion and cracking in refractory furnace	Petroleum residue, natural gas
Impingement (channel or roller)	Diffusion flame impingement on cold surface of channel iron or of steel roller	Natural gas, town gas, coal tar
Thermal	Thermal cracking in super-heated refractory chamber	Natural gas
Lampblack	Partial combustion in open pans	Coal tar, petroleum residue
Acetylene	Thermal cracking	Acetylene

The furnace process is the most important process and about 95% of the carbon blacks used in the world are made by this process. Table 8-10 shows that the acetylene process yields a very pure carbon black. The lampblack process is the oldest carbon black process still in use. The ancient Egyptians and Chinese employed methods similar to modern processes collecting the lampblack by deposition on a cool surface.

6.4. Uses

The particle sizes obtained by different manufacturing processes and required by various applications are summarized in Fig. 8-12.

In 1982 about 4.3×10^6 metric tons of carbon black were produced worldwide. Approximately 96% of this amount was used as a reinforcing filler in rubber, of which 65-70% was used in the tire industry and 25-30% in the mechanical rubber goods industry. Only 5-6% of all the carbon black was used for pigmentation purposes.

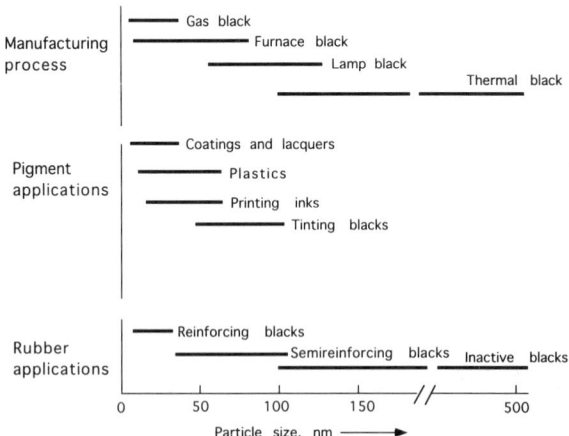

Fig. 8-12. Mean particle size and typical applications of various carbon blacks. Voll and Kleinschmit (1992). Courtesy of VCH Publishers, Weinheim.

Table 8-12. Types and Applications of Special Pigment Grades of Carbon Blacks
Dannenberg and Paquin (1992). Reprinted by permission of John Wiley & Sons, Inc.

Normal grades			
Type	Surface area m^2/g	DBPA [a] ml/100 g	Volatile content %	Uses
high color	230-560	50-120	2	high jetness for alkyd and acrylic enamels, lacquers, and plastics.
medium color	220	70-120	1-1.5	medium jetness and good dispersion for paints and plastics; ultraviolet and weathering protection for plastics.
regular color	80-140	60-114	1-1.5	for general pigment applications in inks, paints, plastics, and paper; gives ultraviolet protection in plastics, high tint, jetness, gloss, and dispersibility in inks and paints.
	46	60	1.0	good tinting strength, blue tone, low viscosity; used in gravure and carbon paper inks, paints, and plastics.
	45-85	73-100	1.0	main use in inks; standard and offset news inks.
low color	25-42	64-120	1.0	excellent tinting black-blue tone; used for inks-gravure, one-time carbon paper inks; also for paints, sealants, plastics, and cements.
thermal blacks	7-15	30-35	<0.5	tinting-blue tone; plastics and utility paints.
lamp blacks	20-95	100-160	0.4-0.9	paints for tinting blue-tone.
Surface oxidized grades			
high color	400-600	105-121	8.0-9.5	used for maximum jetness in lacquers, coatings, plastics, fibers, record disks.
medium color, long flow	138	55-60	5	used in lithographic, letterpress, carbon paper, and typewriter ribbon inks; high jetness.
medium color, long flow	96	70	2.5	used for gloss printing and carbon paper inks; excellent jetness. dispersibility; tinting strength, and gloss in paints.
low color	30-40	48-93	3.5	used for tinting where flooding is a problem; easy dispersion.

a Dibutyl phthalate absorption

Tables 8-12 and 8-13 show applications of principal rubber grade and special pigment grade carbon blacks, respectively.

Table 8-13. Application of Principal Rubber-grade Carbon Blacks. Dannenberg and Paquin (1992). Reprinted by permission of John Wiley & Sons, Inc.		
Designation	General rubber properties	Typical uses
N110, N121	high abrasion resistance	special tire treads, airplane, off-road racing
N220, N299, N234	high abrasion resistance good processing	passenger, off-road, special service tire treads
N399, N347, N375, N330	high abrasion resistance, easy processing, good abrasion resistance	standard tire treads, rail pads, solid wheels, mats, tire belt, sidewall, carcass retread compounds
N326	low modulus, good tear strength, good fatigue, good flex cracking resistance	tire belt, carcass, sidewall compounds, bushings, weather strips, hoses
N550	high modulus, high hardness, low die swell, smooth extrusion	tire innerliners, carcass, sidewall, innertubes, hose, extruded goods, v-belts
N650	high modulus, high hardness, low die swell, smooth extrusion	tire innerliners, carcass, belt, sidewall compounds, sheeting
N660	high modulus, high hardness, low dieswell, smooth extrusion	carcass, sidewall, bead compounds, innerliners, seals, cable jackets, hose, soling, EPDM compounds
N762	high elongation and resilience, low compression set	mechanical goods, footwear, innertubes, innerliners, mats

7. Colored Pigments

White pigments have the characteristic that the scattering coefficient can vary but is very large over the whole spectrum of visible light whereas for black pigments the absorption coefficient is very large and constant over the entire visible region—see § 3.3 this chapter. In contrast, for colored pigments, the absorption and scattering coefficients depend strongly on the wavelength and may assume very high values at certain wavelengths and very low values at others. For very brilliant and transparent colors the absorption coefficient should be large in a narrow wavelength region whereas the scattering coefficient should be small over the whole visible spectrum. Characteristically, such colors may be obtained with some very finely dispersed organic pigments.

At the same particle size, inorganic pigments, compared with organic pigments, generally have lower color strength due to smaller absorption coefficients and duller colors, but higher hiding power due to larger scattering coefficients.

7.1. Inorganic Colored Pigments

Colored inorganic pigments can be classified according to their chemical composition as in Table 8-14.

Table 8-14. Classification of Inorganic Colored Pigments. Heine and Völz (1992). Courtesy VCH Publishers, Weinheim.

Chemical class	Green	Blue-green	Blue	Violet	Red	Orange	Yellow	Yellow-green
Oxides and oxide-hydroxides					iron oxide red	iron oxide brown, mixed brown spinel and corundum phases	iron oxide yellow	
Chromium oxide pigments	chromium oxide	chromium oxide hydrate green						
Mixed metal oxide pigments phases	cobalt green and blue	cobalt green and blue	cobalt green and blue				nickel rutile yellow, chromium rutile yellow	bismuth vanadate
Sulfide and sulfo-selenide pigments				cadmium sulfoselenide, cadmium mercury sulfide			cadmium sulfide	cadmium zinc sulfide
Chromate pigments	chrome green				molybdate red	chrome orange	chrome yellow, zinc yellow, alkaline earth chromates	
Ultra-marine pigments	ultramarine green, blue, violet, and red							
Iron blue pigments			iron blue					
Others			manganese blue	cobalt, manganese violet			naples yellow	

They can be classified according to their constituent components as follows:
- Single component pigments: most of the pigment classes in Table 8-14 are of this type.
- Mixed metal oxide pigments: different metal oxides have been mixed, often by means of grinding in the dry state.
- Substrate pigments: one component, usually a pigment, has been deposited onto substrate particles, which could be another pigment or an extender, preferably by wet chemistry methods.
- Core pigments: a pigment composition is deposited on core particles, which could consist of inexpensive materials such as extenders, by precipitation or wet mixing of the components. In this way functional

pigments can be made of expensive pigment compositions at a tolerable cost.

Small particle technology, that is, methods and techniques for making small particles of different metal compounds and for applying coatings of various compositions to small particles, as described in Chapters 2, 3, and 6, can be used advantageously to make single-component, mixed-oxide substrate, and core pigments and to improve the quality and properties of such pigments.

In this section only the most economically important colored inorganic pigments, namely pigments of iron oxide, chrome oxide, and mixed metal oxides, are described. For more detailed information on these pigments and for information on the other classes of colored pigments in Table 8-14, reference is made to Kirk Othmer's Encyclopedia of Chemical Technology (1992), Ullmann's Encyclopedia of Industrial Chemistry (1992), and Paul (1995).

7.1.1. Iron Oxide Pigments. These pigments are the most important members of the group of colored inorganic pigments and are either naturally occurring or synthetic materials.

Properties. Natural and synthetic iron oxide pigments are well-defined crystalline materials; Ullmann (1992):

1) α-FeOOH, goethite [1310-14-1], has a diaspore structure whose color changes from green-yellow to brown-yellow with increasing particle size, where the number in brackets is the CAS registry number.

2) γ-FeOOH, lepidocrocite [12022-37-6], has a boehmite structure with color changes from yellow to orange with increasing particle size.

3) α-Fe$_2$O$_3$, hematite [1317-60-8], has a corundum structure with color changes from light red to dark violet with increasing particle size.

4) γ-Fe$_2$O$_3$, maghemite [12134-66-6], has a spinel superstructure, is ferrimagnetic, and has a brown color.

5) Fe$_3$O$_4$, magnetite [1309-38-2], also has a spinel structure, and has a black color.

The stability and inertness of iron oxide pigments under various conditions are so good that no surface treatment, e.g., by coating, is needed to improve these properties. Synthetic iron oxides are chemically purer and more uniform in particle size, i.e., a narrower particle-size distribution, than their natural counterparts.

Manufacture. High-quality iron oxide pigments with controlled particle size and shape and particle-size distribution can be made by the following types of processes:

1) Solid state reactions at elevated temperatures, usually at 500-1000 °C, in an oxidizing atmosphere (red, black, and brown).

2) Precipitation and hydrolysis of iron salt solutions—see Ch. 3 (yellow, red, orange, and black).

3) Laux process, in which iron reduces aromatic nitro compounds in the presence of Fe(II) chloride or Al chloride (black, yellow, red).

Table 8-15. Reaction Equations for the Production of Iron Oxide Pigments. Buxbaum & Printzen (1992). Courtesy VCH Publishers, Weinheim.

Color	Reaction	Process
Red	$6FeSO_4 \cdot xH_2O + 1.5 O_2 = Fe_2O_3 + 2Fe_2(SO_4)_3 + 6H_2O$	Copperas process
	$2Fe_2(SO_4)_3 = 2Fe_2O_3 + 6SO_3$	
	$2Fe_3O_4 + 0.5O_2 = 3Fe_2O_3$	calcination
	$2FeOOH = Fe_2O_3 + H_2O$	calcination
	$2FeCl_2 + 2H_2O + 0.5 O_2 = Fe_2O_3 + 4HCl$	Ruthner process
	$2FeSO_4 + 0.5O_2 + 4NaOH = Fe_2O_3 + 2Na_2SO_4 + 2 H_2O$	precipitation
Yellow	$2FeSO_4 + 4NaOH + 0.5O_2 = 2\alpha\text{-}FeOOH + 2Na_2SO_4 + H_2O$	precipitation
	$2Fe + 2H_2SO_4 = 2FeSO_4 + 2H_2$	
	$2Fe_2SO_4 + 0.5O_2 + 3H_2O = 2\alpha\text{-}FeOOH + 2H_2SO_4$	Penniman process
	$2Fe + 0.5 O_2 + 3H_2O = 2\alpha\text{-}FeOOH + 2H_2$	
	$2Fe + C_6H_5NO_2 + 2H_2O = 2\alpha\text{-}FeOOH + C_6H_5NH_2$	Laux process
Orange	$2FeSO_4 + 4NaOH + 0.5O_2 = 2\gamma\text{-}FeOOH + 2Na_2SO_4 + H_2O$	precipitation
Black	$3FeSO_4 + 6NaOH + 0.5O_2 = Fe_3O_4 + 3Na_2SO_4 + 3H_2O$	1-step precipitation
	$2FeOOH + FeSO_4 + 2NaOH = Fe_3O_4 + Na_2SO_4 + 2H_2O$	2-step precipitation
	$9Fe + 4C_6H_5NO_2 + 4H_2O = 3Fe_3O_4 + 4C_6H_5NH_2$	Laux process
	$3Fe_2O_3 + H_2 = 2Fe_3O_4 + H_2O$	reduction
Brown	$2Fe_3O_4 + 0.5O_2 = 3\gamma\text{-}Fe_2O_3$	calcination
	$3Fe_3O_4 + Fe_2O_3 + MnO_2 + 0.5O_2 = (Fe_{11}, Mn)O_{18}$	calcination

The equations for the reactions involved in these different process types are shown in Table 8-15, Ullmann(1992).

In 1982 the production of synthetic iron oxide pigments was estimated to be between 500,000 to 600,000 metric tons, compared to about 100,000 metric tons of natural oxides.

Uses. Synthetic iron oxides have good tinting strength (color strength) and excellent hiding power. Their lightfastness and resistance to alkalis make them very useful in demanding applications. The principal areas of use for iron oxide pigments are shown in Table 8-16.

Table 8-16. Main Areas of Use for Natural and Synthetic Iron Oxide Pigments. Buxbaum and Printzen (1992). Courtesy VCH Publishers, Weinheim

Use	Amount, %		
	Europe	U.S.A	World
Coloring construction materials	64	37	60
Paints and coatings	30	48	29
Plastics and rubber	4	14	6
Miscellaneous	2	1	5

7.1.2. Chrome Oxide Pigments Chromium oxide pigments, or chromium oxide green pigments, are synthetic materials. Natural mineral deposits of chromium oxide are not known.

Properties. Chromium oxide is a dense, crystalline material, density 5.2 g/cm^3, of the corundum type. It is quite hard—abut 9 on the Moh scale—which makes it a good grinding material, but too abrasive for many pigment applications.

The particle size depends on the manufacturing process, but it is distributed around mean values in the range 0.5-0.6 μm.

The refractive index of chromium oxide is quite high, 2.5, which is almost as high as that of rutile TiO_2, 2.7. Chromium oxide pigments are green with an olive green tint. Small particles are lighter green with a yellowish hue, whereas larger particles are darker, with a bluish tint.

Chromium oxide pigments are very inert materials with outstanding lightfastness and excellent resistance to acids, alkalis, and high temperatures.

Manufacture. Chromium oxide pigments are produced by reduction of alkali dichromate with sulfur or carbon:

$$Na_2Cr_2O_7 + S = Cr_2O_3 + Na_2SO_4$$

or by thermal decomposition of ammonium dichromate. In the industrial process a mixture of ammonium sulfate or chloride is calcined:

$$Na_2Cr_2O_7 \cdot 2H_2O + (NH_4)_2SO_4 = Cr_2O_3 + Na_2SO_4 + 6H_2O + N_2$$

Ullmann (1992) estimates that 37,000 metric tons of chromium oxide, including 18,000 metric tons as pigments, were used worldwide in 1990.

Uses. The outstanding stability and weathering properties of chromium oxide green pigments make them very useful as colorants for roofing granules, cement, concrete, and outdoor industrial coatings, and ceramics.

7.1.3. Mixed Metal Oxide Pigments. Mixed metal oxides refer to pigments that crystallize in a stable oxide lattice. The color is due to colored cations incorporated in the lattice. The number of possible mixed metal oxides obtainable from different substituent elements and oxide lattices is very large, but only a limited number of structures, principally spinel, rutile, and hematite structures, have the stability and optical properties necessary for pigment applications. The important mixed oxide spinel pigments are shown in Table 8-17 which also shows the colors of such types.

Nickel rutile yellow [8007-18-9] with the approximate composition $Ti_{0.8}Sb_{0.1}Ni_{0.05}O_2$ is a light lemon-yellow pigment. Chromium rutile yellow [68186-90-3] with the approximate composition $Ti_{0.9}Sb_{0.05}Cr_{0.05}O_2$ has a color that varies with particle size from light to medium ochre.

Mixed metal oxide pigments have good thermal and chemical stability and also have a high refractive index.

Trivial name	Formula	CAS registry no.	C.I. name and no.	Color	Crystal structure
Cobalt blue	$CoAl_2O_4$	[1345-16-0]	Pigment Blue 28:77 346	reddish blue	spinel
	$Co(Al, Cr)_2O_4$	[68187-11-1]	Pigment Blue 36:77 343	greenish blue	spinel
Cobalt green	$(Co,Ni, Zn)_2 TiO_4$	[68186-85-6]	Pigment Green 50:77 377	green	inverse spinel
Zinc iron brown	$ZnFe_2O_4$		Pigment Yellow 119:77 496	light to medium brown	spinel
Spinel black	$Cu(Cr, Mn)_2O_4$	[68186-91-4]	Pigment Black 28:77 428	black	spinel intermediate between normal and inverse spinel
	$Cu(Fe, Cr)_2O_4$	[55353-02-1]	Pigment Black 22: 77 429		

Table 8-17. Mixed-phase Spinel Pigments. Mansmann, Räde and Wilhelm (1992). Courtesy VCH Publishers, Weinheim

Manufacture. Mixed metal oxide pigments are usually the products of solid state reactions between finely divided and intimately mixed reactants at 800-1400° C. Intimate mixing of the components is accomplished by stirring together aqueous dispersions of very small particles of oxides, hydroxides, carbonates, or nitrates.

Ullmann (1992) estimates that the world consumption of mixed metal oxide pigments, excluding the former Eastern block and China, was about 1100 metric tons in 1990.

Uses. Mixed metal oxide pigments are especially suitable for coatings that have to meet high standards of lightfastness and resistance, e.g., baking enamels, coil coatings, and powder coatings.

An important use for rutile yellow is in coil coating of aluminum and steel for the building industry.

7.2. Organic Colored Pigments

In the introduction to this chapter it was stated that pigments, including organic coated pigments, are much less soluble in various media than organic dyestuffs. Small particles of organic pigments are crystalline in nature and consist of individual organic molecules held together by cohesive forces, which are so strong that the pigment is insoluble or very sparingly soluble in most organic solvents, including water, and organic polymeric binders. Inorganic pigments are generally even more insoluble than organic pigments, but here the pigment particles do not consist of individual molecules; rather they are genuine chemical compounds in which the atoms or molecules are bound to one another by strong covalent or ionic forces. Dyestuffs are also crystalline materials consisting of individual molecules, in many cases quite similar to those making up organic pigments, but the molecules are held together by much weaker forces than in pigments. Solvents, plasticizers, and polymers can readily separate, i.e. dissolve, the molecules of many organic dyes. Dyes will therefore bleed into such media whereas pigments will not.

Table 8-18, Lincke (1969), shows the principal classes and colors of organic pigments.

There is no white organic pigment, and organic black pigments are rare and have poor lightfastness compared to carbon black.

Small particle technology, according to the scope of this book, is primarily concerned with small particles of inorganic materials. Inorganic pigments, white, colored, and extenders, fall well within this scope and consist of small particles discussed at length in chapters 2, 3, 5, and 6. However, organic pigments, and also carbon blacks, are quite different from the small particles treated in those chapters, not only in their chemical nature, but also in how they are prepared. Methods for producing organic pigments will therefore not be described here and the properties and uses of the most important classes of organic pigments will only be briefly summarized.

Nevertheless, this section will end with a description of the mating of small particle technology with organic chemistry to produce new unique pigments of composite structure.

Table 8-18. Classification of the Most Important Organic Pigments. Lincke (1969). Courtesy Curt Vincentz Verlag

Pigment class	Yellow	Red	Violet	Blue	Green
Azo	Hansa yellow Benzidine yellow Nickel azo yellow Azo condensation yellow	Toluidine red Lithol red Naphthol red			
Phthalocyanine				Phthalo-cyanine blue	Phthalo-cyanine green
Polycyclics	Anthra pyrimidine Isoindoline	Thioindigo Perylene Quinacri-done	Isoviolan-throne	Indanthrene blue	

7.2.1. Azo Pigments. These pigments contain the azo group -N=N- as the chromophore, of which there may be only one (monoazo pigments), two (diazo), or three(triazo). The nomenclature of azo pigments is the same as that of azo dyes. Yellow, red, and orange are the commercially most important colors, although azo pigments can be made in any color.

Monoazo pigments have yellow and orange colors and are made by coupling diazotized substituted anilines with an activated methylene group in linear or cyclic compounds.

Hansa yellow pigments and β-napthol pigments are the two commercially most important types of monoazo pigments. C.I. Pigment Yellow 1, 11680 [2512-29-0] is a representative pigment of the Hansa Yellow type. (C.I. stands for Color Index, 11680 is the C.I. constitution number of the pigment, and the number in brackets is the C.A.S. number.)

C.I. Pigment Yellow 1, 11680 [2512-29-0]

C.I. Pigment Red 3, 12120 [2425-85-6] is a good representative of monoazopigments of the β-naphthol type and is one of the most important organic pigments.

C.I. Pigment Red 3, 12120 [2425-85-6]

Hansa yellow and β-napthol monoazo pigments have good lightfastness and resistance to weathering, but moderate to poor solvent and bleed resistance.

There are two main types of diazo-pigments, diarylide pigments and bis(N-acetoacetarylide) pigments. C.I. Pigment Yellow-12, 21090 [6358-85-6] is a very important diarylide pigment and one of the most important organic pigments.

C.I. Pigment Yellow-12, 21090 [6358-85-6]

Pigment Yellow-16, 20040 [5979-28-2] is the bis(N-acetoacetarylide) pigment that has the greatest industrial importance.

C.I. Pigment Yellow-16, 20040 [5979-28-2]

Diazo pigments have larger molecular size and contain more conjugated double bonds than monoazo pigments and thus have higher solvent resistance and tinting strength but lower lightfastness.

A third type of monoazo pigments is azo pigment lakes, which are based on soluble dyes having sulfonic acid groups. Insoluble pigments are formed by precipitating the soluble dyes with metal ions such as Ca^{2+}, Sr^{2+}, Ba^{2+}, or Mn^{3+}.

C.I. Pigment Red 57:1, 15850:1 [5289-04-9] is a lake pigment and one of the most important organic pigments, widely used in printing ink formulations.

C.I. Pigment Red 57:1, 15850:1 [5289-04-9]

It has a brilliant magenta color, good lightfastness, and good solvent resistance.

7.2.2. Phthalocyanine Pigments. Phthalocyanines constitute the most important group of organic pigments and are characterized by high color intensity, excellent lightfastness and resistance to bleed and chemicals, extreme heat stability, and exceptionally high tinting strength, but are restricted to blue and green colors. Phthalocyanines are suitable for all areas of use.

C.I. Pigment Blue 15, 147-14-8 Copper Phthalocyanine Blue

7.2.3. Quinacridone Pigments. Quinacridones are polycyclic pigments consisting of five six-membered rings in a linear arrangement. They are insoluble and do not bleed in most organic solvents and application media. Their lightfastness, weathering stability, and chemical stability are high. The colors range from orange, maroon, scarlet, red, magenta to violet. C.I. Pigment Violet 19, 46500 [1047-16-1], β-modification, has a violet shade and high tinting strength.

C.I. Pigment Violet 19, 46500 [1047-16-1]

This pigment is particularly suitable for adjusting red shades in automotive finishes and can also be used in metallic finishes.

7.2.4. Perylene Pigments. Perylenes are also polycyclic pigments. Their colors are various shades of red. The lightfastness and bleed fastness are very good and the thermal stability is high. C.I. Pigment Red 179, 71130 [5521-31-3] is a maroon perylene.

$$CH_3—N \overset{O}{\underset{O}{\bigcirc}} \cdots \overset{O}{\underset{O}{\bigcirc}} N—CH_3$$

C.I. Pigment Red 179, 71130 [5521-31-3] Perylene Maroon

Perylenes are primarily used to pigment plastics, but are also used to pigment high-quality stoving enamels.

7.2.5. Composite Pigments. Many important pigment properties such as color strength, transparency, hiding power, etc., depend on the size of the pigment particles: cf. Fig.8-6 and the discussion in § 3.3 of this chapter. One of the problems in conventional pigment technology is to make pigment of an optimal particle size for a given application. Organic pigments, for instance, are of molecular size when they are formed by reaction in solution, but grow rapidly to crystals, which may be quite small, e.g., smaller than 10 nm in size. However, pigment crystals which are formed very close to one another, attract each other by van der Waals forces and form aggregates. Further processing in the pigment plant may cause the aggregates to agglomerate into larger particles. Clever process technology may prevent untoward increase in pigment particle size, but it is generally necessary to subject the pigment particles to extensive grinding, which is energy consuming, so as to make the particles fine enough for use.

Otterstedt et al. (1995) approached the problem from the opposite direction. Instead of expending energy on making relatively large particles smaller, they used cores of the proper size as templates to make composite pigment particles of uniform size and shape. Their composite pigment is shown in Fig. 8-13.

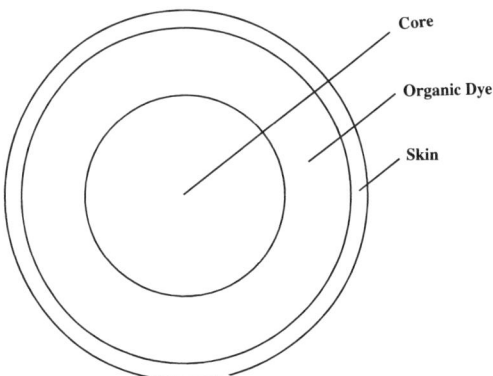

Fig. 8-13. Composite pigment. Otterstedt et al. (1995).

Suitable core particles are silica particles in the size range from 8 to 100 nm, the surface of which could be modified with aluminum silicate—see § C of Chapter 6, so as to make the surface charge negative over a wide range. The negative charge on the particles was reversed by running the silica sol into a solution of basic aluminum chloride—see § C Ch. 6. An anionic dye of high color strength and lightfastness was next deposited on the positively charged core particles. At this step in the preparation of the composite pigment the system consisted of a dispersion of discrete particles of uniform size coated with a layer of insolubilized dye. In order to improve bleed resistance and to equalize the chemical nature of the particle surface, i.e., make it chemically the same regardless of the type of dye that was used to coat the cores, a film of an optically transparent and insoluble material, e.g. silica, was deposited onto the dye layer composite pigment. Composite pigments of this type combine the best properties of organic dyes and pigments. They have the brilliance and transparency of dyes, and the bleed resistance, heat resistance, and lightfastness of pigments.

8. Specialty Pigments

Specialty pigments are pigments that in addition to color effects, and sometimes instead of such effects, provide special properties such as magnetism, luster, and corrosion resistance.

The most important types of pigment along with their properties and uses will be briefly summarized. Special attention will be given to those pigments that can be prepared, modified, or treated by the methods described in this book. For more detailed information on specialty pigments, reference is made to Ullmann's *Encyclopedia of Industrial Chemistry* (1992), as well as Schiek (1992), and Paul (1996).

8.1. Magnetic Pigments

Fine particles of the following materials can be used as magnetic pigments:
1. Iron oxide. Needle-shaped particles of γ-Fe_2O_3 and Fe_3O_4.
2. Cobalt containing iron oxide. Iron oxide is either doped with Co, 2-5 wt % Co, or small particles of γ-Fe_2O_3 or Fe_3O_4 are coated with cobalt hydroxide, 2-4 wt % Co.
3. Chromium dioxide. Small needles of CrO_2.
4. Metallic iron. Small needle-shaped particles of metallic iron with particle size in the range 0.25-0.35 μm and a length to width ratio of about 10:1.
5. Barium ferrite. $BaFe_{12}O_{19}$ in the form of small hexagonal platelets.
Properties and uses of magnetic pigments of iron oxide, cobalt-containing iron oxide, and metallic iron are shown in Table 8-19.

The shape of the pigment particles is extremely important, since the noise level of magnetic tapes decreases with decreasing particle size.

Table 8-19. Some Quality Requirements for Iron Oxide and Metallic Iron Magnetic Pigments. Leitner and Kathrein (1992). Courtesy VCH Publishers, Weinheim.

Field of application	Pigment type	Approximate particle length, μm	Specific surface area, m^2/g	Coercive field strength, H_c, kA/m	Saturation magnetization, M_s, $mT \cdot m^3/kg$	M_r/M_s
Computer tapes	γ-Fe_2O_3	0.60	13-17	23-25	86-90	0.80-0.85
Studio radio tapes	γ-Fe_2O_3	0.40	17-20	23-27	85-92	0.80-0.85
IEC I compact cassettes standard (iron oxide operating point)	γ-Fe_2O_3	0.35	20-25	27-30	87-92	0.80-0.90
high grade	Co-γ-Fe_2O_3	0.30	25-37	29-32	92-98	0.80-0.90
IEC II compact cassettes CrO_2 operating point)	Co-γ-Fe_2O_3 Co-Fe_3O_4	0.30	30-40	52-57	94-98	0.85-0.92
IEC IV compact cassettes (metal operating point)	metallic iron	0.35	35-40	88-95	130-160	0.85-0.90
Digital audio (R-DAT)*	metallic iron	0.25	50-60	115-127	130-160	0.85-0.90
1/2-inch Video	Co-γ-Fe_2O_3 Co-Fe_3O_4	0.30	25-40	52-57	94-98	0.80-0.90
Super-VHS video	Co-γ-Fe_2O_3	0.20	45-50	64-72	94-96	0.80-0.85
8-mm Video	metallic iron	0.25	50-60	115-127	130-160	0.85-0.90

*R-DAT: rotary digital audio tape

Iron oxide particles are stabilized with a protective coating of such materials as silicates, phosphates, and chromates.

8.2. Luster Pigments

Nacreous or pearlescent pigments are used in paints, plastics, inks, and cosmetics where brilliance and luster are required in addition to color. The original pearlescent pigments were made from fish scales and tissue.

In Fig. 8-15 the optical properties of conventional and luster pigments, i.e. nacreous or pearlescent pigments and metallic effect pigments, are shown and compared. Conventional pigments interact with light by absorption and/or scattering—Fig. 8-15A. In metallic effect pigments small metal flakes, usually aluminum or copper-zinc bronze, act as mirrors that reflect most of the incident light—Fig. 8-15B. In natural pearls, layers of high refractive index $CaCO_3$ alternate with layers of low refractive index material—Fig. 8-15C. The small, high-refractive-index platelets of nacreous pigments simulate the luster of pearls when they are oriented in a parallel alignment in a matrix of low refractive index, e.g., a paint binder or a plastic—Fig.8-15D.

The dominant types of nacreous pigments are small composite particles with a mica core. The transparent mica platelets, about 5-100 μm across and 0.3-0.6 μm thick, are coated with layers of metal oxides, usually titania, chromia, and

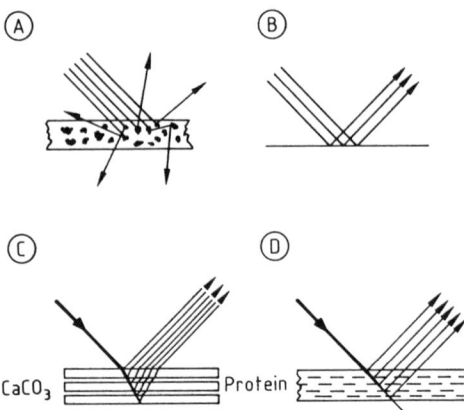

Fig. 8-14. Optical properties of conventional and luster pigments. A) Conventional pigment that absorbs and scatters light; B) Metal effect pigment with complete regular reflection; C) Natural pearl composed of alternating layers of protein and $CaCO_3$; D) Nacreous pigment: the pearl is simulated by parallel orientation of the pigment platelets. Frans and Härtner (1992). Courtesy VCH Publishers, Weinheim.

iron oxide, or a combination of such oxides, of various thickness, which give different colors—see Fig. 8-15.

Mica-based nacreous pigments are analogous to other composite particles discussed in this book, e.g., the titania on silica cores in Figs. 6-45 and 6-46, ceria on silica-titania cores in Fig. 6-51, and chromia, iron oxide and cobalt oxide on silica-titania cores in §2.4.8 of Ch. 7. The methods of small particle technology can therefore be used to develop new kinds of luster pigments, e.g., combination pigments which combine the properties of nacreous and colored pigments. Mica platelets can thus be coated with inorganic colorants, e.g., yellow, orange, and

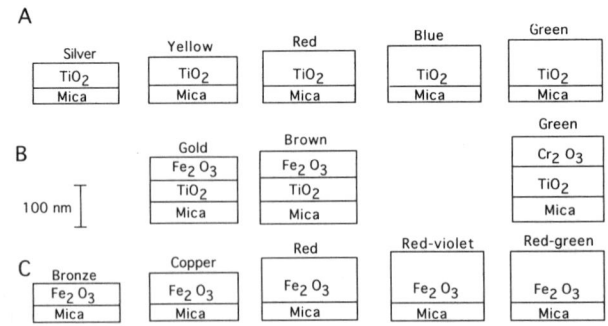

Fig. 8-15. Upper half of metal oxide - mica pigments. Increasing layer thickness of metal oxide causes different interference colors in reflection. Combination with absorption colorants (e.g., Fe_2O_3) produces metallic effects. (A) Interference colors; (B) Combination pigments; (C) Metallic colors. Frans and Härtner (1992). Courtesy VCH Publishers, Weinheim.

red iron (III) oxides and hydrous oxides. Similarly, organic colorants may be applied to mica platelets by using the technique developed by Otterstedt et al. (1995), outlined in § 7.2 of this chapter.

8.3. Transparent Pigments

The transparency of pigments depends on the difference in refractive index of the pigment, n_1, and the medium, n_2, in which the pigment is incorporated, and the particle size of the pigment. In a given pigment-binder system (n_1-n_2) is constant and the only way to affect the transparency of the pigment is by changing its particle size.

The Mie theory can be used to describe the scattering behavior of conventional pigment particles which have a mean size of, say, 200 nm, corresponding to a radius a=100 nm—see § 3.3 of this chapter. In the visible spectrum, $400 \leq \lambda \leq 700$ nm, the bandwidth a/λ is $0.14 \leq a/\lambda \leq 0.25$. The Mie theory is applicable in this bandwidth region and predicts that the scattering efficiency of pigment particles decreases rapidly with particle size. Also, the scattering is intense, more prominent in the foreward direction, and does not depend strongly on the wavelength.

Thus, pigment particles scatter less and less light as they get smaller, i.e. they become more and more transparent. Transparent pigments, e.g., micronized or ultrafine TiO_2, have a particle size in the range $10 \leq 2a \leq 50$ nm, corresponding to a bandwidth range of $0.007 \leq a/\lambda \leq 0.053$. In this bandwidth range the scattering of light is given by the Rayleigh equation

$$I_s = \frac{8\pi^4 a^6}{\lambda^4 r^2} \cdot \left|\frac{m^2-1}{m^2+2}\right|^2 \cdot \left(1+\cos^2\theta\right) \cdot I_o \qquad (8\text{-}10)$$

where a = radius of the scattering particle,
   λ = wavelength of scattered light,
   r = distance from particle to where intensity is measured,
   m = concentration of particles
   θ = angle between incident light and scattered light
   I = intensity of incident light

The Rayleigh equation can be used to explain the "frost" effect of ultrafine TiO_2 in metallic automotive coatings. This effect will impart bluish or yellowish hues to the lacquer, depending on the angle of observation—see Fig. 8-17. The intensity of the scattered light is inversely proportional to the fourth power of the wavelength. Blue light, $\lambda \sim 425$ nm, will scatter five to six times stronger than red light, $\lambda \sim 650$ nm. Hence the blue hue of the scattered light and the yellowish hue of the reflected light.

The most important transparent pigments consist of iron oxide, yellow and red, and titanium dioxide. The primary particles of transparent iron oxide are

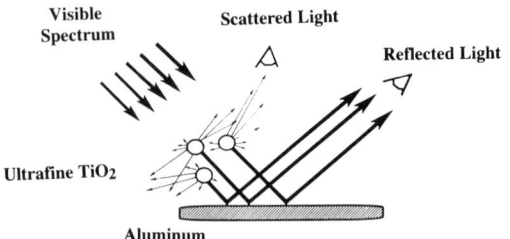

Fig.8-16. Schematic illustration of the frost effect of ultrafine titanium dioxide particles in metallic layers. Kettler and Richter (1992). Courtesy Curt Vincentz Verlag.

needle shaped with an average length of 50-100 nm, a width of 10-20 nm, and a thickness of 2-5 nm.

Transparent pigments are primarily used in metallic paints and lacquers. They also have high absorption of UV and are therefore used to color plastic bottles and films for packaging of UV-sensitive foods (transparent iron oxides), and as UV-screens in sun tan lotions (transparent TiO_2).

In Section D of Chapter 6 it was mentioned that many types of polymeric binders, pigmented with TiO_2, must be protected from photoinduced degradation brought about by catalytic action of the base TiO_2 surface. This is even more the case with transparent TiO_2, since the surface area of the ultrafine particles exposed to the binder is so much higher. One way of protecting the binder is to coat the TiO_2 particles with a dense layer of silica. The amount of SiO_2 needed to shield the binder from the surface of conventional TiO_2 particles with a mean particle size of about 250 nm is 6-7 wt % SiO_2, based on the weight of TiO_2, corresponding to 4-5 monolayers of SiO_2 on the particle surface—see Section D of Chapter 6. Assuming that the surface of transparent TiO_2 also must be covered with 4-5 monolayers of SiO_2, 60-70 wt % SiO_2 would be required, since the particles are ten times smaller than those of conventional TiO_2 pigment.

References

Braun, J. H., Baidins, A., and Marganski, E., Progress in Organic Coatings, **20**, 105 (1992).

Buxbaum, C., and Printzen, H., Ullmann's *Encyclopedia of Industrial Chemistry*, vol. A20, 297, VCH Publishers, Weinheim (1992).

Carr, W., *Pigments Handbook*, vol. III, John Wiley & Sons, N.Y., 11 (1973).

Dannenberg, E. M., and Paquin, L., Kirk-Othmer *Encyclopedia of Chemical Technology*, vol. 4, 1037, John Wiley and Sons, N.Y. (1992).

Franz, K-D., and Härtner, H., Ullmann's *Encyclopedia of Industrial Chemistry*, vol. A-20, 331, VCH Publishers, Weinheim (1992).

Frianf, G. F., and Thorley, B., U. S. Patent 3,010,794 (1961).

Greenwood, P., Otterstedt, J-E., and Niklasson, G. A., to be published (1998).

Heine, H., and Völz, H. G., Ullmann's *Encyclopedia of Industrial Chemistry*, vol. A-20, 245, VCH Publishers, Weinheim (1992).

Herbst, W., and Hunger, K., *Industrial Organic Pigments*, 2nd ed., John Wiley, N.Y. (1997).

Howard, P. B., U. S. Patent 4,052,223 (1977).

Judd, D: B., and Wyszecki B., Color in Business, *Science and Industry*, John Wiley and Sons, N.Y. (1963).

Kerker, M., *The Scattering of Light, Academic Press*, N. Y. (1969).

Kettler, H., and Richter, G., Farbe und Lack, **98**, 93 (1992).

Kittel, H., *Lehrbuch der Lacke und Beschichtungen*, Band II, W.A. Colomb in der H. Heenemann Gmbh., Berlin, 29 (1974).

Leitner, L., and Kathrein, H., Ullmann's *Encyclopedia of Industrial Chemistry*, vol. A-20, 33, VCH Publishers, Weinheim (1992).

Lincke, G., Farbe und Lack, **75**, 632 (1969).

Lloyd, T. B., Li, J., Fowkes, F. M., Brand, J. R., and Dizikes, L. J., Journal of Coatings Technology, **64**, 91 (1992).

Mansmann, M., Räde, D., and Wilhelm, V., Ullmann's *Encyclopedia of Industrial Chemistry*, vol. A20, 309, VCH Publishers, Weinheim (1992).

Matsunaga, M., Usami, T., Okada, H., and Futumato, F., U. S. Patent 4,405,376 (1983).

Medalia, A. I., and Irwin, D., *Characterization of Powder Surfaces*, Ed. C. D. Parfitt and K. S. W. Sing, Academic Press, London (1976).

Mie, G., Ann. Phys. **25**, 377 (1908).

Otterstedt, J-E., and Brandreth, D. A., unpublished work (1996).

Otterstedt, J-E., Otterstedt, O., and Sterte, J., European Patent 0593493 (1995).

Otterstedt, O., and Gustavsson, S. R., M. Sc. Thesis, Report No. 92-16, Department of Industrial Management and Economics, Chalmers University of Technology, Gothenburg, Sweden (1992).

Paul, S., *Surface Coatings*, 2nd ed., John Wiley & Sons, N.Y. (1996).

Schiek, R. G., Kirk-Othmer *Encyclopedia of Chemical Technology*, vol. 17, 877, John Wiley and Sons, N. Y. (1992).

Voll, M. D., and Kleinschmid, P., Ullmann's *Encyclopedia of Industrial Chemistry*, vol. A-5, 154, VCH Publishers, Weinheim (1992).

CHAPTER 9. SMALL PARTICLES IN PAPER

The word paper is derived from *papyrus*, a sheet made in ancient times by pressing together very thin strips of an Egyptian reed, *cyperus papyrus*. The modern material, paper, consists of sheet materials that are comprised of bonded, flexible, cellulose fibers which, while very short, 0.5-4 mm, are about 100 times as long as they are wide. Small particle fillers or pigments, in the form of clays or other inorganic materials are used to give paper improved properties, e.g. opacity, brightness and printability, or to improve the economics of the papermaking process. In this chapter we will focus on the use of small particles as process aids to improve retention and dewatering on paper machines.

1. Paper and Paperboard

The types of fibers and filler used in various paper and boards are shown in Table 9-1, which also shows that retention and dewatering aids are used in all the grades, Eklund and Lindström (1991).

2. Fillers

Mineral fillers in the form of small particles are used in paper for various reasons. There has always been the economic incentive to substitute low-cost fillers and extenders for some high-cost fibers in paper, but there is also the incentive to improve several of the properties of paper. The use of fillers increases opacity and brightness of the paper and also improves printability by making printing ink absorption more uniform, gives higher gloss after calendering and leads to better "feel" and dimensional stability.

The disadvantages of using fillers in paper are reduced mechanical strength, caused by the filler particles interfering with the hydrogen bonding between the cellulose fibers, heavier paper, greater wear on the wire of the paper machine, and higher content of fine material in the circulating water system. Table 9-2 shows the types and amounts of fillers used in different grades of paper.

Pigments, which are also small particles of inorganic materials, are used to improve the optical properties of paper and are usually more expensive than

	Fiber Type	Filler/Pigment Type	Filler percentage	Retention and Dewatering Aids
Standard Newsprint	100% TMP[3]	possibly specialty pigments	-	x
SC-offset[1]	50% TMP 25% bleached softwood kraft pulp	clay	25%	x
Fine Paper	35% bleached softwood kraft pulp 50% bleached hardwood kraft pulp	CaCO$_3$ or clay	15%	x
LWC[2]	50% TMP 50% bleached softwood kraft pulp	(only from broke)		x
Sack Paper	100% unbleached softwood kraft pulp	-	-	x
Liner	100% unbleached softwood kraft pulp	-	-	x
Fluting Media	100% NSSC[4]	-	-	x
Liquid Board	79% CTMP[5] 30% bleached soft and hardwood kraft pulps	-		x
Tissue	30 bleached softwood kraft pulp 30% bleached hardwood kraft pulp 40% bleached softwood sulphite	-		x

Table 9-1. Composition of different types of papers and boards, Eklund and Lindström (1991). Courtesy D.T. Paper Science.

1 supercalendered-offset. 2 light-weight coated. 3 thermomechanical pulp. 4 neutral sulfite semi-chemical pulp. 5 chemi-thermo mechanical pulp

cellulose fibers. Pigments are also often made synthetically while fillers are ground minerals.

2.1. Types of Fillers

The most common types of fillers are:

- kaolin or clay ($Al_2O_3 \cdot 2\ SiO_2 \cdot 2\ H_2O$)

- talc ($3\ MgO \cdot H_2O$)

- calcium carbonate ($CaCO_3$)

- gypsum ($CaSO_4 \cdot 2H_2O$)

- mica ($3\ Al_2O_3 \cdot K_2O \cdot 6\ SiO_2 \cdot 2\ H_2O$)

of which the most important, kaolin, was discussed in Ch. 4.

The properties of some commonly used fillers are shown in Table 9-3.

Table 9-2. Fillers and pigments used in paper, Eklund and Lindström (1991). Courtesy D.T. Paper Science

Paper Grade	Content	Type of filler
Newsprint	0-10%	clay, talc, special pigment
Magazine paper (uncoated, SC)	20-30%	clay, talc
Fine paper	0-25%	clay, talc, chalk, TiO_2
Wrapping paper	0-10%	clay, talc, chalk, TiO_2

Table 9-3. Properties of some commonly used fillers, Eklund and Lindström (1991). Courtesy D.T. Paper Science.

Filler	Brightness (ISO, %)	Specific surface (m^2/g)	Particles < $2\mu m$, %	Density (kg/m^3)	Refractive index, n
Kaolin clay	76-82	5-8	12-45	2.6	1.55
EEC Grade C	81	8	45	-	-
EEC Grade M	79	7.5	30	-	-
Calcium carbonate ($CaCO_3$)	80-96	2-6	36-60	2.7-2.8	1.56
Omya DX50 (chalk)	83	-	50	-	-
Omya Hydrocarb 60M (ground limestone)	96	5.4	-	-	-
Talc ($3MgO \cdot 4SiO_2 \cdot H_2O$)	81-96	6-7	20-25	-	1.57
Finntalc P20	81	6.5	20	-	-

Calcium carbonate is used as a collective term for the minerals chalk, limestone and marble, and the synthetic material, precipitated calcium carbonate, PCC.

Chalk consists primarily of pure $CaCO_3$ with no more than 5 mole per cent Mg in the crystal lattice. The mineral is known for its whiteness, purity, porosity, and brittleness. Its basic particles exist in two size ranges, 0.5-4 μm and 10-100 μm. The smaller particles arise from chalk-bearing algae and plankton, the second from shellfish. The ratio of fine to coarse material varies from site to site. Chalk is softer than limestone because of its microcrystalline character.

Limestone was formed by precipitation and sedimentation in sea water and often contains fossils from shell-fish. Under the influence of high sediment pressures and high temperatures from the center of the earth (500-700°C), limestone and dolomite are transformed into marble, a coarse recrystallized form of rock.

PCC, precipitated calcium carbonate, is used in applications where very high purity and fine particle size calcium carbonate is required. The development of uniform crystal-size PCC in the late 1980's enabled paper mills to make better-quality paper at a lower cost in an environmentally friendly process, *Financial World* (1993).

Compared with kaolin clay, calcium carbonate is more soluble and sensitive to attack by acids and acidic materials, e.g., alum, which restricts its use as a filler in papermaking.

2.2. Pigments

In this context pigments are man-made, mostly inorganic materials, which improve the optical properties of paper, but are more expensive than filler. They may be considered specialty fillers. The most frequently used are:

• titanium dioxide (used for certain papers as the main pigment) (TiO_2)

• hydrated aluminum oxide ($Al(OH)_3$, crystalline)

• sodium/calcium aluminosilicates

• calcined clay ($Al_2O_3 \cdot SiO_2$)

• barium sulfate (often the main pigment when it is used) ($BaSO_4$)

• diatomaceous earth (SiO_2)

• plastic pigment

• zinc oxide (ZnO) etc.

Their basic properties are shown in Table 9-4.

Table 9-4. Properties of some pigments used in papermaking, Eklund and Lindström (1991). Courtesy D.T. Paper Science.

Pigment/filler	Brightness (ISO, %)	Specific surface (m^2/g)	Particles < 2mm (%)	Density (kg/m^3)	Refractive index, n
Calcinated clay	92.5	16-17	90	2.7	-
Silicates	97-98	40-130	-	1.8-2.1	1.50-1.55
NaAl-silicate	97	130	-	2.1	1.51
Titanium dioxide	97	8-25	$D_{50} = 0.15$	3.9-4.2	2.55-2.72
Anatase	97	8-10	-	3.9	2.55
Rutile	97	-	-	4.2	2.72
$Al(OH)_3$	95	6-8	99	2.42	1.57

3. Papermaking

Different paper machines have various configurations at the wet end of the machine, but Fig. 9-1 shows schematically a representative setup. In the mixing chest fibers and paper chemicals are mixed to an aqueous slurry, the furnish, containing about 0.5-2% fiber. Some of the chemicals may be added at a later stage, e.g., to the machine chest or before, or into a pump. From the head box, the furnish is filtered on a wire screen, where the fibers adhere weakly to one another. When more water is removed from the mat formed on the screen by suction, the sheet becomes stronger, but is still relatively weak. When the sheet is dried it becomes still stronger, and becomes the material known as paper. Modern paper machines produce an endless paper sheet, up to 10 m wide, at a speed of over 20 m/s, i.e., one hectare (more than two acres) every 50 seconds. The machine is more than 100 m long and produces about 250,000 metric tons per year.

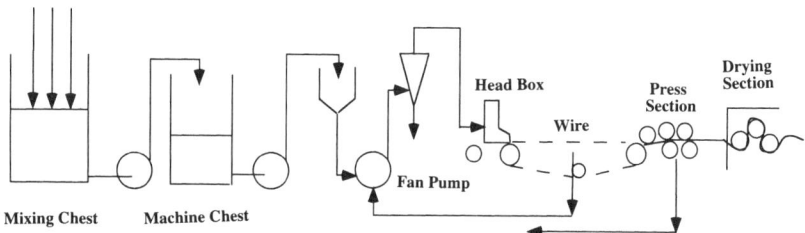

Fig. 9-1. The wet end of a paper machine.

Environmental and economic pressures have reduced water usage in paper production in the last 30-40 years from 80-90 m³ per metric ton to less than 10 m³/ton. During the last decade many efforts have been made to reduce the use of water even more with the ultimate objective of achieving a paper mill that is 100% closed.

The Problem

The achievement of a closed or nearly closed paper mill with respect to water usage is intimately related to the retention of fiber fines and chemicals and other additives in the furnish on the wire. Poor retention will cause the fines and other small particles to go through the wire with the water and make reuse of the *back water* difficult or impossible; see Fig. 9-1.

The nature of the problem can be understood if one first considers the composition of the furnish:

1. Fibers, 0.5-4 mm long and 50-400 μm thick, but also including fragmented fibers, so-called "fines".

2. Chemicals, which are either compounds giving special properties to the paper or aiding the process.

    a) dry-strength agents

        purpose: to achieve high strength in the dried paper sheet.

        products: starches with a grain size of 50-80 μm, polyacrylamide

        (PAM) with a molar mass between 100,000 and 500,000, and others.

    b) wet-strength resins

purpose: to impart good strength to the wet sheet.

products: polyamide-polyamine-epichlorohydrin resin (PAMAM-EPI), urea formaldehyde resin (UF), and melamine-formaldehyde resin (MF); all of colloidal size, and others.

c) sizing agents

purpose: to make the paper less hydrophilic

products: allyl ketene dimers (AKD) as a colloidal dispersion.

alkenyl succinic acid anhydride (ASA) as an oil or an emulsion, rosin as a colloidal dispersion.

3. Fillers, small mineral particles in the size range less than 2 μm.

4. Retention aids; see below.

The nature of the problem is further illustrated by Fig. 9-2 showing the dimensions in the wet end and Fig. 9-3, which compares the size of the holes in the wire with the sizes of the cellulose fibers, fines, filler particles and the various chemical additives.

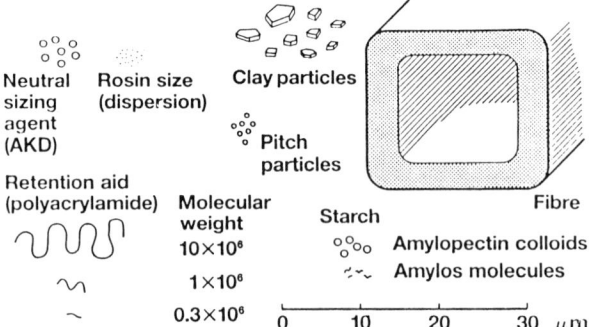

Fig. 9-2. Dimensions in the wet end of the papermaking process.

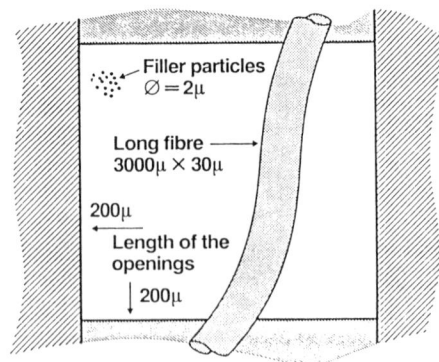

Fig. 9-3. Small particles on the wire in the papermaking process.

The difficulty in retention is further aggravated by the fact that all the particles of the furnish are negatively charged and therefore have no bonding to each other to form aggregates large enough not to pass through the holes of the wire.

The obvious solution to the problem is therefore to put into the system particles or additives of opposite charge to cause agglomeration of the paper components to larger clumps that cannot go thorough the wire. This is accomplished by so-called retention aids.

4. Retention

The term retention refers to the holding back of the components of the stock during dewatering. The fibers are retained on the wire whereas fillers, fines, and additives of colloidal size may be washed through the mat formed on the screen. Retention is accomplished by a combination of mechanical means, i.e., filtration, and the physico-chemical mechanism of agglomeration or flocculation.

Mechanical retention during sheet formation on the wire may be considered a filtration process. The fibers in the stock, which are 500-4000 μm long and 20-100 μm thick, are captured on the wire and form a three-dimensional network consisting of 2-100 layers of fibers on the wire. As the layers form, they capture progressively smaller fibers and other colloidal particles in the stock suspension, making the pore structure gradually finer with the largest pores on the wire side and the finest on the top side.

Mechanical retention is least efficient in the beginning of the sheet formation and, although it becomes more effective as more layers form, it cannot retain a satisfactorily high proportion of the finest components of the stock. The losses for newsprint are typically about 50%.

By adding special chemicals, retention aids, to the stock, the fines and other colloidal components can be made to flocculate or aggregate into agglomerates too large to go through the wire.

Retention aids may consist of either one component or two components. They can act by changing the electrostatic repulsion forces between colloidal particles or affect the stability of colloids by adsorbing on two or more particles causing them to form larger aggregates.

Retention Mechanisms

Although good retention is most likely attained by the joint action of more than one mechanism and a given retention aid may act by several mechanisms, it is still useful to distinguish between some principal types of aggregation mechanisms, which are shown in Table 9-5.

There are no sharp distinctions between the terms coagulation and flocculation, but here *coagulation* denotes aggregation by the action of low-molecular weight electrolytes whereas *flocculation* means aggregation brought about by polymers, which can be natural or synthetic.

Table 9-5. Survey of some aggregation mechanisms, Eklund and Lindström (1991). Courtesy D.T. Paper Science.				
	Mechanism	Floc shear resistance	Effect of increased content of dissolved anionic wood polymers	Effect of increased electrolyte content
1	Charge neutralization (coagulation)	-	- -	+
2	Heterocoagulation	+	- -	-
3	Patching	+	- -	-
4	Bridging -adsorption flocculation	++	-	++
	-sensitizing flocculation	++	-	++
5	Complex flocculation -ion complex	++	(-) (+)	- -
	-non-ionic complex	++	(-) (+)	(-) (+)

-, -- = retention decrease   +, ++ = retention increase   + = increase
++ = strong increase

In high-speed modern paper machines the floc is subjected to high shear and the effect of the shear resistance of the floc on retention is indicated in Table 9-5. The trend toward reduced water usage in the production of paper increases the amounts of soluble anionic wood polymers and electrolytes in the stock, which will also affect the retention on the wire.

1. Charge neutralization: Electrolytes are the simplest kind of coagulants that can be used to improve retention. They act fully in accord with the DLVO (Deryagin-Landau-Verwey-Overbeek) theory by screening the charges and compressing the electrolytic double layer of the negatively charged particles of the stock, thus allowing attractive forces to come into play and aggregate the particles—Fig. 9-4. Aggregation by charge neutralization is a fairly slow process which has lost some of its importance as the speed of paper machines has become ever faster.

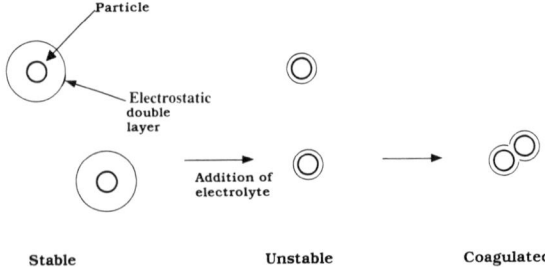

Fig. 9-4. Coagulation of particles by addition of electrolyte which cause charge neutralization.

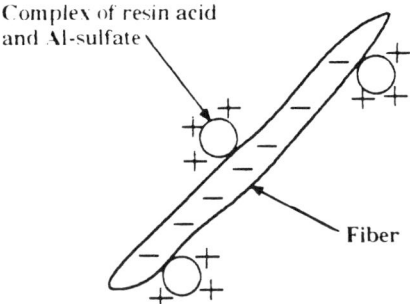

Fig. 9-5. Heterocoagulation using resin acid complexes.

2. Heterocoagulation: This mechanism involves adsorption of oppositely charged particles, e.g., complexes of resin acids and aluminum sulfate, on the surfaces of fibers and filler particles. Heterocoagulation is sensitive to soluble anionic wood polymers and electrolytes, with which cationic sizing particles, e.g., the complex shown in Fig. 9-5, preferentially interact.

3. Patch flocculation - Patching: Patching resembles charge neutralization, but is different. In this mechanism cationic polymers are strongly adsorbed in a flat configuration on the negative surfaces of the particles, on which they form cationic patches—see Fig. 9-6.

Adsorption leads to partial charge neutralization and electrostatic attraction between oppositely charged patches on different particles leads to flocculation. If the cationic polymer is small and the patches are smaller in size than the thickness of the electrolytic double layer—which depends on the concentration of electrolytes—aggregation will take place by the mechanism of charge neutralization.

A characteristic difference between charge neutralization and patching is that the rate of coagulation for the former mechanism increases with electrolyte concentration. Once an optimal electrolyte concentration has been attained, however, the rate of flocculation by patching will decrease with electrolyte content due to the fact that the electrolyte cations will force the polymer from the particle surface.

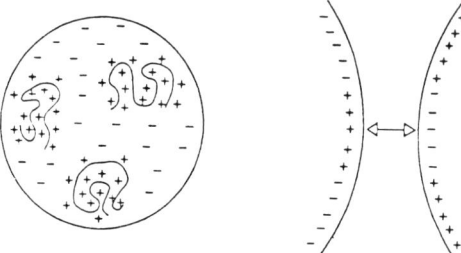

Fig. 9-6. Patch flocculation using cationic polymers.

Relatively short-chained cationic polymers of average molar mass and high charge density are suitable for patch flocculation. Modified polyethylene imines, polyamines, and polyamide-amine-epichlorohydrin resins are in this category.

4. Bridging: In this mechanism flocculation is accomplished by long-chain, i.e., high molar mass, polymers forming binding bridges between particles, as seen in Fig. 9-7.

For effective bridging to occur, it is very important that the polymers adsorbing on the surface of the particles form loops and tails that protrude into the solution. To what extent this happens depends on the type of polymers, contact time, and properties of the surface of the particles to be flocculated. Suitable polymers are weakly charged or non-ionic, i.e., high molecular-mass polyacrylamide and polyethylene oxide.

Flocs made by bridging are large but fairly easily broken by shearing, which may tear the bridging polymers and retard the process of re-flocculation.

Bridging can be two types: adsorption flocculation and sensitizing flocculation.

Adsorption flocculation: Electrostatic and/or Van der Waals forces make the polymer adsorb on two or more particles.

Sensitizing flocculation: Polymers causing this type of flocculation are either non-ionic or have the same charge as the particles they are to flocculate.

Fig. 9-7. Flocculation using high-molar-mass polymers to form bridging.

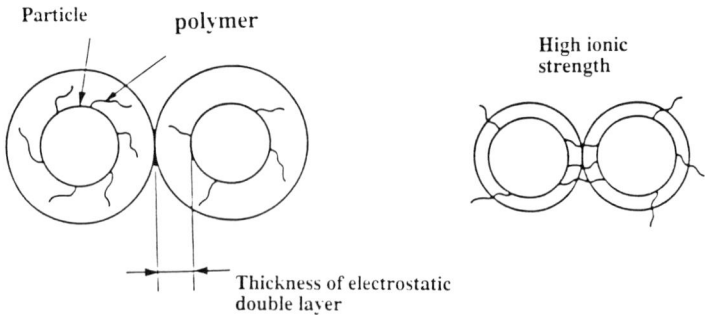

Particle   polymer   High ionic
                                                                 strength

Thickness of electrostatic
double layer

Low ionic

strength

Fig. 9-8. Sensitizing flocculation using long-chain polymers which can link through the electrostatic double layer.

The closest approach of charged particles in solution is determined by the thickness of the electrostatic double layer. Charged particles can only come sufficiently close together so as to be bridged by non-ionic polymers or polymers of the same charge if the surface potential of the particles is reduced or the electrostatic double layer is compressed, e.g., by addition of electrolytes. Flocculation will occur when the effective chain length is greater than twice the thickness of the electrostatic double layer. See Fig. 9-8.

5. *Complex flocculation:* Flocculation by any of the four mechanisms described above can be accomplished by only one flocculant or retention aid. Much more effective flocculation and retention can be achieved by using combinations of retention aids. The most common combinations are between oppositely charged retention aids, which can form complexes of varying strength with each other. It is, however, also possible to use combinations of non-ionic retention aids that can form complexes by hydrogen bonding.

Complex flocculation by electrostatic interaction: Complex formation that leads to good retention may occur between oppositely charged polymers or between charged polymers and oppositely charged microparticles; the latter systems, which may be called dual micro-particulate retention aids, are of great industrial importance and will be described in detail in the next section. Figure 9-9 shows how complexes and flocs are formed by the interaction of oppositely charged polymers.

A relatively short-chain, highly charged cationic polymer forms patches on the particles and aggregates them to a weak floc. A long-chain, anionic polymer is next added to the primary floc, which strengthens it by forming complexes with the cationic patches on the particles.

Given enough time, the cationic polymer may penetrate into the cellulose fibers and steadily decrease the number of cationic sites on the fiber surface. In order to achieve strong complex formation and high retention only a short time should elapse before addition of the anionic polymer. If the time is very short,

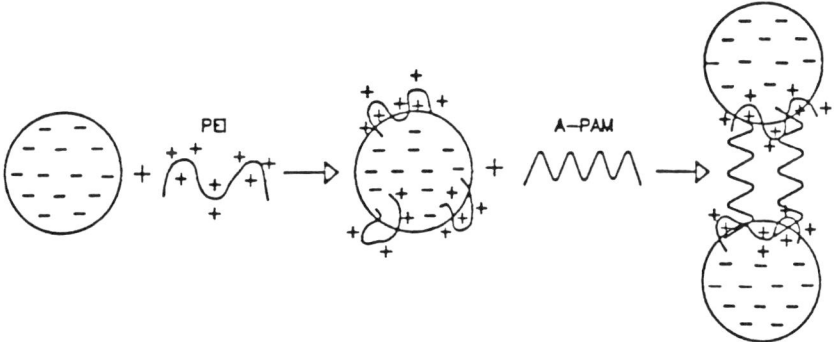

Fig. 9-9. Combination of polyelectrolytes with different charge. Primary flocs are formed with the help of patching, after which primary flocs link together by bridging. Eklund and Lindström (1991). Courtesy D.T. Paper Science.

much of the cationic polymer may not have had time to adsorb on the fibers and filler particles and may instead crosslink with the anionic polymer to a network that may occlude fines and colloidal particles and thus actually improve retention.

Complex formation by non-ionic interaction: Polyethylene oxide and phenolic resins are examples of retention aids which form hydrogen bonded complexes where interaction occurs between the oxygens in the PE-oxide and the phenolic hydroxyl groups of the phenolic resin.

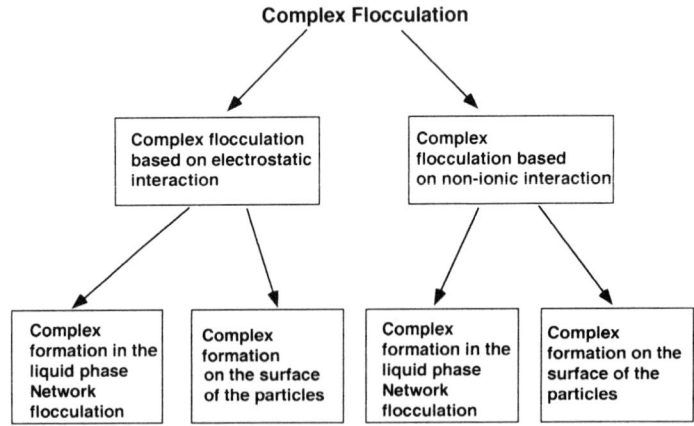

If an excess of cationic polymer is used or does not have sufficient time to adsorb on the fibers and the filler particles, then a large part of the cationic polyelectrolyte remains in the liquid phase when the anionic compound is added. Crosslinking will then take place and form a network in which small particles can be enclosed. Complex flocculation leading to network formation may be an effective mechanism for improving retention.

Figure 9-10 summarizes the concept of complex flocculation.

Complex Flocculation

| Complex flocculation based on electrostatic interaction | Complex flocculation based on non-ionic interaction |

| Complex formation in the liquid phase Network flocculation | Complex formation on the surface of the particles | Complex formation in the liquid phase Network flocculation | Complex formation on the surface of the particles |

Fig. 9-10. A schematic diagram to summarize the concept of complex flocculation.

5. Retention Aids

Retention of fillers and fine materials in the furnish for papermaking machines is accomplished by the addition of natural and synthetic polymers—in the last fifteen years often in combination with colloidal particles of natural or synthetic origin.

Some commonly used polymers are shown in the following non-exhaustive listing of one-component retention aids.

1. Polyethylene imines, PEI, are strongly cationic and strongly branched polymers with a molar mass between 100,000 and 1,000,000.

$$H_2N-(CH_2CH_2N)_x^--(CH_2CH_2NH-)_y$$

$$\begin{array}{l} H_2C \\ \quad \backslash NH-(CH_2-CH_2-NH)_nH \quad \text{OR} \\ H_2C \end{array}$$

POLYETHYLENE IMINE

$$\begin{array}{l} CH_2 \\ | \\ CH_2 \\ | \\ NH_2 \end{array}$$

2. Polyethylene amines, contain secondary amine groups and are linear, strongly cationic polymers with a molar mass of about 100,000.

$$-CH_2-CH_2-\underset{\underset{H}{|}}{N}-$$

3. Polyacrylamides, PAM, are non-ionic polymers with a molar mass of about 1,000,000.

$$\left[\begin{array}{c} CH_2-CH-CH_2-CH-CH_2 \\ \quad\; | \qquad\qquad | \\ \quad\; C=O \qquad\;\; C=O \\ \quad\; | \qquad\qquad | \\ \quad\; NH_2 \qquad\;\; NH_2 \end{array}\right]_n$$

4. Cationic polyacrylamides, contain tertiary amine groups which can be quaternized. Molar mass is about 1,000,000.

$$\begin{array}{l} CH-CH_2 \\ | \\ C-NH-CH_2-N{\overset{\displaystyle\diagup CH_3}{\diagdown CH_3}} \\ \| \\ O \end{array}$$

5. Anionic polyacryamides, A-PAM, can be synthesized by co-polymerizing acrylamide with acrylic acid. It contains anionic carboxyl groups and has a molar mass of about 1,000,000.

$$\begin{array}{l} CH-CH2 \\ | \;\diagup O \\ C \\ \;\diagdown O \end{array}$$

6. Polyethylene oxide, PEO, is non-ionic and has a molar mass of about 1,000,000.

$$\left[O-(CH_2)_2-O-(CH_2)_2-O-(CH_2)_2-O\right]_n$$

7. Cationic starch

Dual Retention Aid Systems

Cationic natural and synthetic polymers have long been used to improve retention of fines and fillers on the wire of paper machines. Such polymers, i.e. cationic starch or cationic polyacrylamide, produce a high degree of flocculation in the furnish. This floc, however, is not very strong and is easily broken and re-dispersed by hydraulic shear. Furthermore, when long-chain polymers are used, chain rupture and rearrangement of the polymer fragments on the particle surfaces may occur. Nevertheless, single-component retention aids improve the first-pass retention, though not to the same degree as dual retention aid systems.

Such systems have been used in the paper industry for many years. Component one, a cationic polymer, usually of the patching type, is first added to the furnish, followed by the addition of the second component, an anionic polymer of the bridging type. Figure 9-11 schematically compares single component and dual retention aid systems. The application of retention aids has been optimized in the sense that the retention maximum in the figure corresponds to a zeta potential of value zero, i.e., the charges on the positive components in the system exactly balance the charges on the negative component, which may be difficult to accomplish in an actual situation. When an optimal amount of cationic component, in this case cationic starch, in the single-component system, is added, the furnish system has no charge and flocculation and retention are maximized. In the dual system cationic starch has to be present in the furnish so as to reach zero ζ–potential after the given amount of the second component, an anionic polymer, has been added. Thus, the maximum in flocculation and retention is not only higher than for the single component system, but it also occurs at larger dosages of cationic starch, which is beneficial since starch is not only a retention aid, but is also an additive that increases the dry strength of paper.

Moore (1976) studied charge relationships of dual polymer retention aids in furnishes containing bleached kraft softwood and TiO_2 as pigment. After adjusting the pH of the slurry to the desired pH of 4.5-5.0 with HCl, alum was

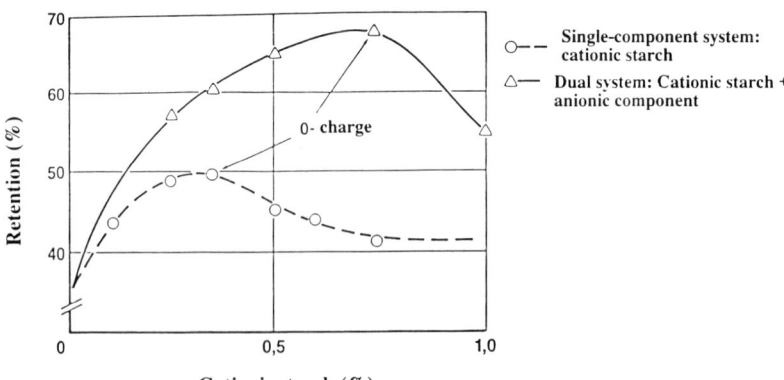

Fig. 9-11. Single-component and dual retention aid systems. Andersson and Larson (1984). Courtesy Arbor Publications.

added, the pH readjusted with dilute NaOH, and the mixture was stirred for one minute. The cationic polymer was added and the suspension was stirred for one minute before the anionic polymer was added and the final slurry was vigorously stirred for an additional minute. The point of charge neutralization and the retention were measured. Figure 9-12 indicates that the mechanism for obtaining maximum retention with dual systems involves more than just balancing charges.

It is to be expected that the retention maximum occurs at lower loading levels of cationic polymer (with 30% cationic substitution) when the furnish contains alum since alum contributes to the cationic charge of the system. However, the peak of maximum retention and the ζ-potential zero value do not coincide. Moore (1976) concluded that the dual systems he studied achieved high retention and produced a shear-resistant floc. Complex formation involving bridging structures appears to overwhelm charge neutralization as a flocculation mechanism. It may be possible to optimize the amounts of cationic and anionic components so that the retention maximum occurs at the ζ-potential equal to zero, but this may be difficult to achieve in practice.

In the last 10-15 years a special kind of dual retention aid, a microparticle-containing flocculant system, often referred to as microparticulate retention/dewatering aid, was developed. The commonly used commercial systems comprise colloidal silica in combination with cationic starch or cationic synthetic polymers, Andersson and Larson (1984) and Andersson and Lindgren (1996), the so-called Compozil® system, or sodium montmorillonite together with cationic polyacrylamide, Langley and Litchfield (1986) and Wågberg et al. (1995), the Hydrocol® system. In both systems, the cationic polymer is added first and the extensive flocs then formed are broken down and partially re-dispersed by high-shear forces. The anionic microparticles are added just before the paper is formed and cause final flocculation of the furnish. The two systems are quite different in composition, but they share the following characteristics

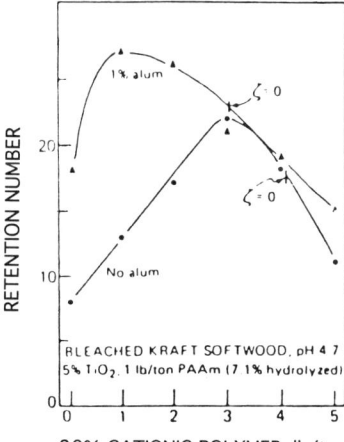

Fig. 9-12. The effect of alum on the optimum ratio of cationic to anionic polymers for retention . Moore (1976). Courtesy TAPPI.

which distinguish them from conventional, polymer-based, dual retention aids, Lindström et al. (1989):

- strong, reversible flocculation
- more effective dewatering in the wire and press sections
- formation on the wire yields sheets of higher porosity and permeability.

The Compozil® system, Eklund and Lindström (1991), and Hydrocol® system, Wågberg et al. (1995), were recently studied to elucidate the mechanism responsible for their unique properties. See Table 9-6.

Compozil® system: Andersson and Lindgren (1996) used a Britt Dynamic Drainage Jar to investigate the retention effects of combinations of various types of anionic silica with either cationic starch or polyacrylamides of different charge density. The furnish consisted of a 60/40 mixture of fully bleached birch and pine sulfate pulps with 30% (based on total solids) chalk as the filler. The solids content and pH of the furnish were 0.5% and 8.1, respectively. The polyacrylamides had charge densities between 2-25% cationicity, corresponding to between 0.25 and 3.0 meq/g, and a molecular weight of 5×10^6. The cationic starch had a degree of cationic substitution of 0.4, corresponding to 0.25 meq/g.

Table 9-6. Dual retention/dewatering/dry strength-agent systems, Eklund and Lindström (1991). Courtesy D.T. Paper Science.

Components in preferential addition sequence	Flocculation mechanism
Aluminum salt + A-PAM	Patching/bridging
PAE + A-PAM	Patching/bridging
PPE + A-PAM	Patching/bridging
PEI/modified PEI + A-PAM	Patching/bridging
Cationic starch + A-PAM	Patching/bridging
Phenolic resin + PEO	Network flocculation
Act. Montmorillonite + A-PAM	Network flocculation
Cationic starch + Colloidal silica	Patching/bridging
Cationic PAM + Act. Montmorillionite	Patching/bridging
Trimethylol amine + PVOH	Hybrid mechanism

Table 9-7. Properties of anionic colloidal silica, ACS. Andersson and Lindgren (1996). Courtesy Arbor Publications.

	S.A. (m²/g)	Size (nm)	S (%)	A.Size (nm)	L (nm)	W (nm)	L/W	C.D. (meq/g)
ACS1	897	3.0	43	-	-	-	-	0.86
ACS2	830	3.3	31	-	-	-	-	0.80
ACS3	890	3.1	20	-	-	-	-	0.86
ACS4	619	4.4	98	4.5	-	-	-	0.59
ACS5	545	5.0	47	13	26	5.4	5	0.52
ACS6	500	5.4	31	17	36	5.2	7	0.48
ACS7	280	10.0	98	-	-	-	-	0.27

The properties of anionic colloidal silicas, ACS, used are shown in Table 9-7 where:

S.A. is the surface area measured by Sears titration (see Ch. 1).

Size is the primary particle size calculated from the surface area (Eqn. 1-23).

S is defined as the percentage of silica in the dispersed phase and can be obtained from viscosity measurements. A high S-value indicates well-dispersed, non-aggregated colloidal particles, whereas a low-value suggests that the primary particles have formed microaggregates, perhaps linear structures containing up to 7-8 primary particles.

A. Size is the average size of the microaggregates as determined by dynamic light scattering, DLS.

L is the length of the microaggregates as determined by DLS and viscosity measurements

W is the width of the microaggregates determined by DLS and viscosity measurements

C.D. is the calculated charge density of the ACS at pH 8 in meq/g (see Ch. 6).

Figure 9-13 shows that for the system cationic starch-ACS and for a given amount of silica added, retention increases with the charge density (i.e., surface area) of the ACS.

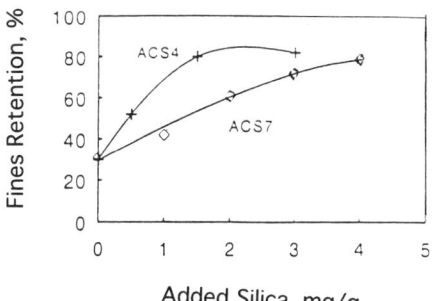

Fig. 9-13. Effect of charge density on retention. Andersson and Lindgren (1996). Courtesy Arbor Publications.

For a given cationic starch the charge density of the silica will thus determine the position of the maximum retention peak, but will not affect the height of the peak. The structure, or degree of microaggregation, of the silica having 5-nm primary particles does not appear to affect the retention, Fig. 9-14a, whereas structure is very important for high retention when the primary particle size of silica is 3 nm, Fig. 9-14b.

Thus, there appears to be a critical size of the primary particles, somewhere between 3 and 5 nm, below which non-structured, or non-aggregated silica improves retention much less than structured ACS. Andersson et al. speculate that this critical size is approximately the same as the distance between

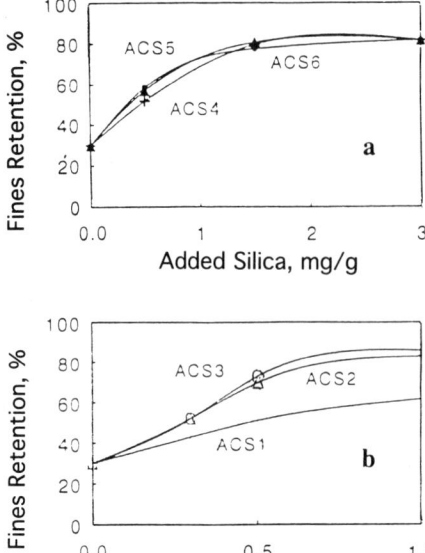

Fig. 9-14. Effect of silica structure on retention in the system cationic starch - ACS: (a) 5-nm primary particle size. (b) 3-nm particle size. Andersson and Lindgren (1996). Courtesy Arbor Publications.

the tuft of branches in the amylopectin molecule, which is the main constituent of potato starch (see Fig. 9-15).

Very small silica particles may therefore be able to penetrate into the amylopectin molecule, so that their cationic charge is screened from other cationic particles in the furnish and they lose their flocculation efficiency.

Figure 9-16 shows that compared with the system cationic starch-ACS, in the system cationic polyacrylamide, CPAM-ACS, retention is much more sensitive to the structure of the silica particles.

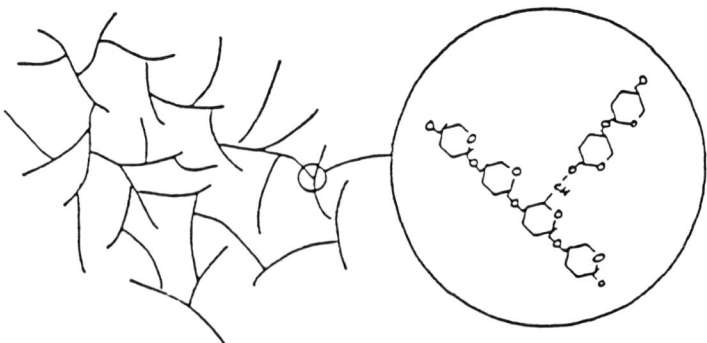

Fig. 9-15. Amylopectin. linear parts with a length of 20-30 glucose units built up to a tree-like structure through branching on carbon 6. Eklund and Lindström (1991). Courtesy D.T. Paper Science.

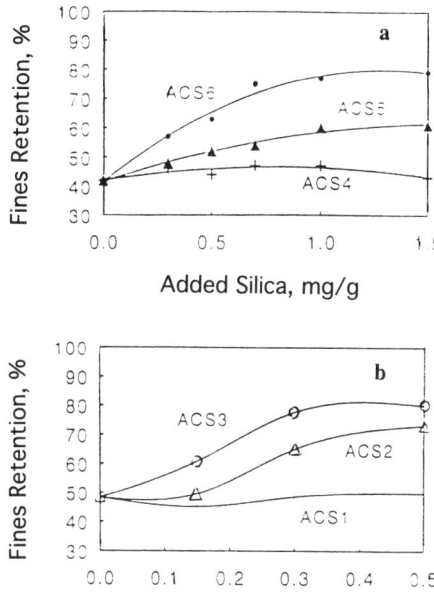

Fig. 9-16. Effect of silica structure on retention in the system CPAM-ACS (a) 5-nm particles.(b) 3-nm particles. Andersson and Lindgren (1996). Courtesy Arbor Publications.

Very small particles, neither 5-nm ACS4, Fig.9-16a, nor 3-nm ACS1, Fig. 9-16b, had any effect on the retention. High retentions were obtained only when at least one dimension of the silica particles was quite large. Thus Table 9-7 shows that ACS5 had a length of 26 nm, but was very much less effective than ACS6, which had a length of 36 nm.

Figure 9-16 also shows that, just as for the system cationic starch-ACS (Fig.9-14), the charge density of the ACS affects the position of the retention maximum peak, but not its height. It requires three times more ACS6 of primary particle size 5 nm than ACS3 of primary particle size 3 nm to reach maximum retention, but this has the same value for the two ACS types.

Andersson and Lindgren (1996) also studied the effect of the cationicity of the CPAM on retention and found that a cationicity of 10 mole % (1.2 meq/g) gave maximum retention, thus supporting the hypothesis that electrostatic interactions between the two components overwhelm other types of interaction, e.g., hydrogen bonding.

They also used their data to construct a model for the system CPAM-ACS, shown in Fig. 9-17 for a constant dosage of CPAM of 0.8 mg/g. Each curve shows the predicted retention for an ACS with a constant S-number, and, as expected, maximum retention increases with increasing structure, or degree of microaggregation, of the ACS. Also, ACS with low structure gave more pronounced retention maxima than more highly structured ACS, which was interpreted to indicate that charge neutralization was the dominant mechanism

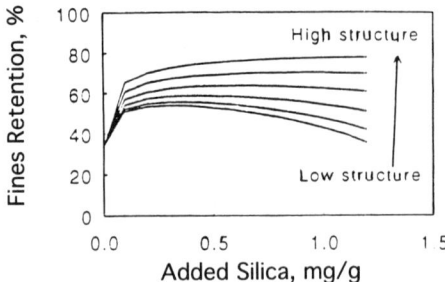

Fig. 9-17. Retention model for the CPAM-ACS system Andersson and Lindgren (1996).
Courtesy Arbor Publications.

when ACS of small particle size was used. The flatter maxima of the upper curves suggest that another mechanism, microparticle bridging, comes into play when large-particle-size ACS was used.

For both the cationic starch-ACS and CPAM-ACS systems the main flocculation mechanisms were assumed to be electrostatic interactions, e.g., charge neutralization and bridging.. For bridging the relatively dense and compact amylopectin molecules in starch, relatively small ACS particles are suitable provided their size is not smaller than a critical value of about 5 nm (see Fig. 9-18a). Cationic PAM has a more open structure than amylopectin and large, extended ACS particles are required to fit between the loops and bridge CPAM adsorbed on fibers and filler particles (see Fig. 9-18b).

Hydrocol® system: Wågberg et al.(1986) measured the adsorption of cationic polymer on cellulose fibers and of sodium montmorillonite on polymer-pre-saturated cellulose fibers, and using a Fiber Optic Flocculation Sensor (FOFS)

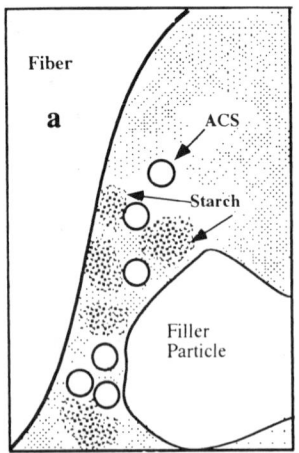
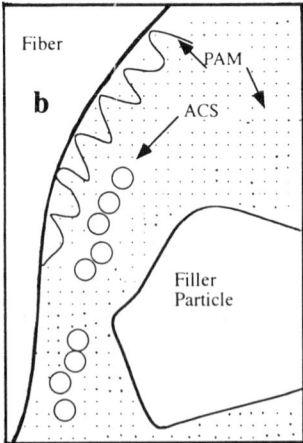

Fig. 9-18. Bridges of ACS microparticles. (a) cationic starch - ACS. (b) cationic polyacrylamide - ACS, Andersson and Lindgren (1996). Courtesy Arbor Publications.

instrument, studied the flocculation behavior of the system cellulose fiber-cationic polyacrylamide-sodium montmorillonite. The cellulose fibers were from a fully blended kraft softwood pulp.

The cationic polyacrylamide, CPAM, had a molecular weight of 5.8 x 10^6 and charge density of 1.8 meq/g. The cationic polymer used for presaturating cellulose fibers was poly-dimethyldiallyl ammonium chloride, poly-DMDAAC, with a molecular weight of 1.2 x10^6 and a charge density of 6.19 meq/g. The investigators assumed that the mechanism of the Hydrocol® system involved bridges by montmorillonite particles between loops and tails of adsorbed polymer molecules on adjacent fibers and set out to prove that assumption by doing a series of experiments.

La Mer's bridging model (1963) for flocculation of colloidal systems by polymers predicts that the maximum rate of flocculation is attained at a surface coverage of the particles by polymers of 50%.

Figure 9-19 shows the flocculation index for the single-component system CPAM and the dual system CPAM + montmorillonite. According to the investigators a polymer dosage of between 500 to 1000 mg/g is close to 50% surface coverage of the fibers. At low levels of polymer addition polymer bridging is probably the principal flocculation mechanism in the two systems, and their flocculation indices do not differ much in this region. However, the flocculation index curves for the two systems diverge at higher additions of CPAM and Wågberg et al. suggest that a new bridging mechanism, bridging by montmorillonite particles between loops and tails of adsorbed polymer on adjacent fibers, now comes into play.

They argue that the microparticle bridging would be promoted by the extended conformation of the polymer present at large dosages of polymer, whereas polymer bridging according to La Mer's model would reach a maximum plateau at a surface coverage of 50%. The results of other experiments involving the single component and dual systems and cellulose fibers precovered with poly-DMDAAC also supported the microparticle bridging model.

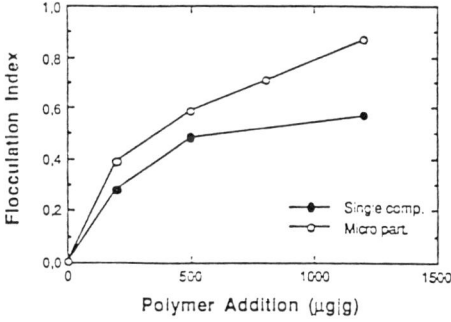

Fig. 9-19. Flocculation of the cellulosic fibers, measured as flocculation index, caused by the single-component system and the microparticle system. The polymer added was cationic poly-acrylamide and the amount of montmorillonite was 2 mg/g fiber. The fiber concentration was 2 g/l. Wågberg et al.(1995). Courtesy TAPPI.

In the last set of experiments they measured the adsorption of Na-montmorillonite on cellulose fibers precovered with CPAM, wherein excess CPAM was removed before the montmorillonite was brought into contact with the fibers. Figure 9-20 shows that the adsorption of Na-montmorillonite increases rapidly at first, but that there is a change in slope at about 500 µg CPAM/g, which resembles the behavior of the flocculation index curve in Fig. 9-19, indicating that it is the interaction between preadsorbed polymer molecules on adjacent fibers and montmorillonite particles that causes flocculation in the microparticle-based system.

From these results Wågberg et al. (1986) concluded that the good retention obtained by the CPAM-montmorillonite system, i.e., the Hydrocol® system, was caused by a mechanism involving bridging by microparticles, i.e., montmorillonite particles, of loops and tails of adsorbed polymers on adjacent fibers.

In the Hydrocol® system, montmorillonite particles do what highly structured, i.e., microaggregated, ACS particles do in the Compozil® systems consisting of CPAM and anionic colloidal silica (see Fig. 9-18b).

Nalco System: Nalco developed a microparticle retention system consisting of a cationic starch in combination with an anionic mixture of a high molecular weight water soluble anionic polymer and colloidal silica, Johnson (1991).

The degree of cationic substitution on the starch ranges between 0.02 to 0.15 and most preferably between 0.025 to 0.10. The particle size of the dispersed silica particles, usually in the form of a colloidal silica sol, should fall in the range between 2-25 nm. In patent example 1, Johnson used a silica sol with a particle size of 4 nm.

Many types of water soluble anionic polymers can be used, such as polyacrylamide containing about 30 mole percent acrylic acid and 70 mole percent acrylamide monomer, but the molecular weight should fall in the range between $5 - 25 \times 10^6$.

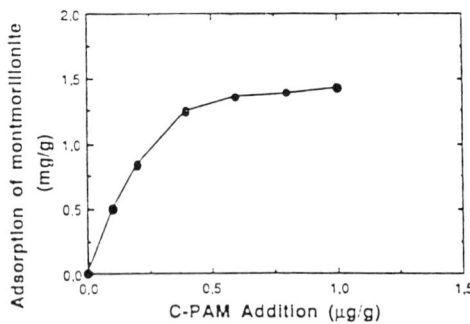

Fig. 9-20. Adsorption of montmorillonite on cellulosic fibers precovered with C-PAM. The adsorption time for the C-PAM was 3 min and the concentration was 2 g/l. During the subsequent montmorillonite adsorption experiments the fiber concentration was 20 g/l and the adsorption time was 5 min. The montmorillonite addition was kept constant at 2 mg/g fiber. Wågberg et al (1995). Courtesy TAPPI.

The anionic combination of anionic polymers and silica sol is preferably added to the papermaking stock after the addition of cationic starch. It is added in an amount so as to accomplish electroneutralization.

The most preferred results were obtained when the starch to polymer weight ratio ranged from 50:1 to 5:1, the polymer to silica weight ratio ranged between 8:1 to 1:1, and the starch to silica weight ratio ranged from 75:1 to 30:1.

References

Andersson, K. and Larsson, H., Nordisk Cellulosa, Nr. 1, 57 (1984).

Andersson, K. and Lindgren, E., Nordic Pulp and Paper Research Journal, No. 1, 15 (1996).

Eklund, D. and Lindström, T., *Paper Chemistry*, D.T. Paper Science Publications, Grankulla (1991).

Financial World, June 8, 42 (1993).

Johnson, K. A., European Patent Specification 234513B1(1991).

La Mer, V.K. and Healy, T.W., *Reviews of Pure and Applied Chemistry*, **13**, 112 (1963).

Langley, J.G. and Litchfield, E., 1986 Papermaker's Conference/ Tappi Proceedings, 89 (1986).

Lindström, T., Hallgren, H. and Hedborg, F., Nordic Pulp and Paper Research Journal, 4:2, 99 (1989).

Moore, E.E. (1976), Tappi, **59**, Nr. 6, 120.

Wågberg, L., Björklund, M., Åsell, I. and Swerin, A., 1986 Papermaker's Conference/ Tappi Proceedings, 71 (1986).

CHAPTER 10. VARIOUS APPLICATIONS OF SMALL PARTICLES

A. Functions

B. Applications

Various Applications of Small Particles

In Chapter 1 it was pointed out that many important technical materials are made of or contained small particles. Some of the appli7)ich the authors have more extensive experience were described in Chapters 7-9. Obviously, there are many other important applications of small particles and a selection of these, to some extent based on the interest and experience of the authors, are discussed in this chapter.

    Table 10-1 shows functions that can be achieved by using small particles and their their utilization in various applications. In Section A of this chapter small particles functions are described with each function exemplified in a few applications. The importance of small particles in selected applications is treated in Section B.

Table 10-1. Functions - Applications Chart

Functions	Catalysts	Ceramics & Refractories	Chemicals & Specialties	Coatings and Finishes	Drugs & Cosmetics	Electronics	Metals	Paints, Inks, and Colors	Paper	Plastics, Polymers, & Resins	Textiles
Adsorbancy	X		X		X				X		
Binding-Bonding	X	X		X	X	X		X	X	X	X
Dispersing		X	X	X	X		X	X	X	X	X
Emulsifying			X	X	X				X	X	
Flocculating		X	X	X					X		
Frictionizing				X					X		X
Mordanting		X		X		X		X	X	X	X
Reactivity	X		X		X	X		X	X	X	X
Reinforcing	X	X		X					X		
Soil Resistance			X	X		X		X		X	X
Static Electricity Reduction		X	X			X		X	X	X	X
Suspending		X	X	X	X	X		X	X	X	
Thickening		X	X	X	X			X	X	X	
Thixotropy		X	X	X	X			X	X	X	
Viscosity Control		X	X	X	X			X	X	X	
Wetting			X				X	X		X	X

A. FUNCTIONS

Many of the functions described here can be achieved by other means than small particles. However, when a certain function depends on the presence of a high surface area of a special chemical nature, this function can probably most effectively be brought about by the use of small particles. Some of the functions discussed in this section are well described in the scientific and technical literature; others are not, and there are some which are somewhat speculative in nature and should be considered as suggestions for further research or application studies.

1. Adsorbency

Small charged particles dispersed in water will adsorb other dispersed or dissolved species of opposite charge and remain dispersed if their charge is reversed; see Section C of Chapter 6. Thus, Ludox® AM, 12-nm alumina-modified silica sol from Du Pont, will adsorb 0.53 mmol/g, or about 1.5

molecules/nm^2, of Astrazon® Rot BBL, a cationic dye of molecular weight 530 from Bayer, at pH 10-11. In Section E of Chapter 6 and in Chapter 7 it was shown that 22-nm silica particles reversed their charge by adsorbing 2 mg of Berocell® 6100, a cationic polymer with the repeat unit $[CH_2CHOHCH_2N(CH_3)_2]_n^+$ and a molecular weight of 50000 g/mol from AkzoNobel (Sweden), per m^2 surface area at pH 8.0.

Similarly, fibrillar boehmite alumina—see Chapter 3—with a typical specific surface area of 275 m^2/g will adsorb from 25 to 100% of its weight of anionic organic dyes, which form a monomolecular layer on the alumina fibrils, at pH below 5.0. Polyvalent anions and fluoride ions will also adsorb strongly on the alumina particles at pH below 5.0, the latter ions probably through the formation and subsequent adsorption of AlF_6^{3-}.

Small particles can also be used as adsorbents when converted to macroscopic bodies of different shapes, for instance granules or extrudates. The wash-coats on ceramic monoliths, shown in Fig. 7-1 and used as catalyst supports, are examples of such macroscopic structures of small particles. Figure 7-25 shows the pore size distribution of extrudates of fibrillar boehmite particles. Such extrudates typically have a pore volume of about 0.5cc/g. The pore size and pore area distributions of wash-coats of corpuscular boehmite, fibrillar boehmite and decationized silica are shown in Fig. 7-7. Structures of corpuscular boehmite and silica have pore sizes distributed around a maximum of about 60Å but the alumina structure has more area and volume in pores smaller than 50Å and broader distributions than the silica structure. Structures of fibrillar alumina have by comparison much broader distributions of surface area and volume and therefore provide a much more open pore structure.

Small particles and structures of small particles thus offer opportunities for use in purification and separation processes for either liquid or gas streams. By careful selection of small particles, i.e., choosing particles of a suitable shape and with surfaces modified in an appropriate way, adsorbents with high selectivities for a wide range of chemical compounds and structures can be obtained

1.1. Adsorbents - Silica Gels and Powders. Molecular sieve zeolites and silica gel are the most widely used and effective adsorbents and can adsorb a large number of substances, so called adsorbates. Their adsorption capacity depends on the nature of the adsorbate, the concentration of the adsorbate in the carrier medium, and the temperature of the adsorbent. In Fig. 10-1 the capacities for adsorption of water in zeolite 4A—see Chapter 5—and narrow-pore silica gel are compared.

Molecular sieve is a much more effective desiccant for air with a low moisture content, say below 1g H_2O/m^3, than silica gel, whereas at higher moisture contents, say above 10 g H_2O/m^3, the reverse is true. Also, an increase in adsorbent temperature causes the silica gel to lose significantly more moisture adsorption capacity than the molecular sieve.

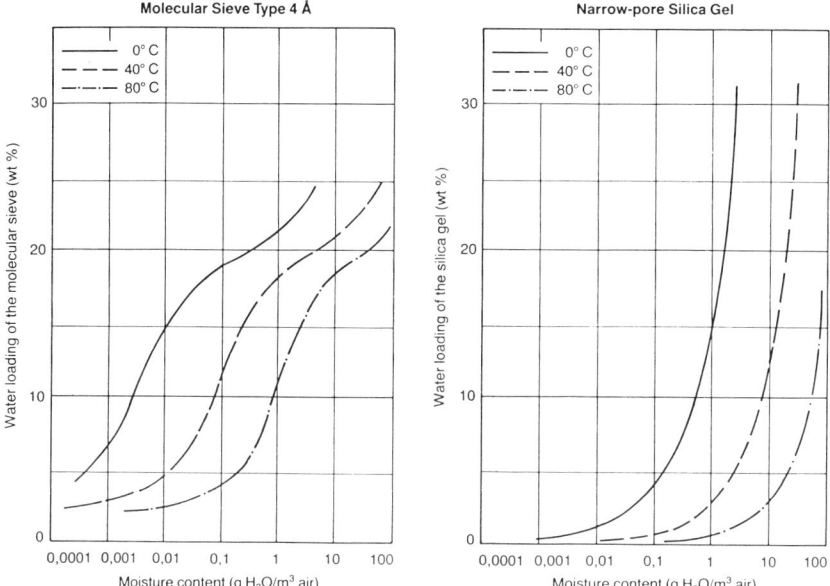

Fig. 10-1. Adsorption capacity of Zeolite 4A and silica gel. Courtesy W.R. Grace Co.

The structure, properties, and applications of molecular sieve zeolites were discussed in Chapter 5. Silica gel belongs to the group of synthetic silica products called *silica powders*, which also include precipitated silica and fumed silica. A comparative overview of the most important properties of silica powders is shown in Table 10-2.

More detailed information concerning the manufacture, properties and applications of silica powders than is given here can be found in the monumental monographs by Iler (1979), Vail (1952), and by Bergna and Roberts (1993).

1.1.1. Gelling of Silica. Figure 2-8 shows that in a system of growing particles of silica the final result will depend on the conditions of pH and electrolyte condition. At pH above 8 and at low salt concentrations, stable silica sols will result, whereas at low pH and/or high electrolyte concentrations, gel will form.

Effect of pH - Hydroxyl Ion Concentration. Figure 6-26 shows that the gel time of silica sols, including sols of extremely small particle size obtained by decationizing alkali silicate solutions, first decreases with pH to a minimum between 6 and 7, but then increases and reaches a region of very high stability between pH 8 and 10. In Chapter 2 it was explained that hydroxyl ions play a key role in silica systems. They catalyze both the polymerization and dissolution of silica. The decrease in gel time, with increasing pH thus reflects the catalyzing action of hydroxyl ions in forming siloxane bridges between the particles; see

Table 10-2. Properties of Synthetic Silicas (courtesy Degussa)				
Property	Fumed or Pyrogenic Silica	Precipitated Silica	Silica Gel	
			Xerogel	Aerogel
1	BET Surface Area m^2/g	50 to 600	30 to 800	250 to 1000	250 to 400
2	Average primary particle size nm	5 to 50	5 to 100	3 to 20	3 to 20
3	Size of agglomerates or aggregates	1)	1 to 40	1 to 20	3 to 20
4	Density g/cm^3	2.2	1.9 to 2.1 2)	2.0	2.0
5	Tamped vol. $cm^3/100$ g	1000 to 5000	200 to 2000	100 to 200	800 to 2000
6	Loss on drying (2 h at 105°C) %	<2.5	3 to 7	3 to 6	3 to 5
7	Loss on ignition (2 h at 1000°C) %	1 to 3	3 to 7	3 to 6	3 to 5
8	pH	4	5 to 9	3 to 8	2 to 5
9	Predominant pore size Å	not porous @ < 300 m^2/g	>300	20 to 200	>250
10	Dibutyl phthalate absorption $cm^3/100$ g	250 to 300	175 to 320	100 to 350	200 to 350
11	Pore size distribution	1)	very wide	narrow	narrow
12	Fraction of internal surface area3)	0	small	very high	high
13	Structure of aggregates and agglomerates	chain-like aggregates	moderately aggregated almost spherical particles	very strongly agglomerated particles	clearly agglomerated porous particles
14	Thickening effect	very pronounced	exists	definitely exists	exists

1) Not applicable; 2) Depends on water content; 3) Estimated by comparing surface areas obtained by BET and SEM.

Fig. 10-2. However, hydroxyl ions also ionize the silanol groups on the surface of the silica particles, thus providing the surface with a negative electrostatic charge which rapidly increases with pH. Above pH 8, and at low salt concentrations, the electrostatic repulsion between the particles is so strong that they cannot come sufficiently close together for siloxane bridges to form and the sol will not gel.

Effect of Salt. Figure 10-3 shows that when the negatively charged surface groups on the colloidal particles are screened by layers of cations from neutral salts, the electrostatic repulsion between the particles is substantially reduced. Above a certain salt concentration, which depends on the type of salt, the electrostatic repulsion between the particles will have become so decreased that the particles can come close enough together for siloxane bridges to form, and they will aggregate, gel, or coagulate after a certain time, which depends on pH and, of course, also on the concentration of particles.

Generally speaking, the larger the size of the hydrated cation of the neutral salt, the more efficiently it can screen the charged particle surface and the shorter the gel time. The effect is particularly noticeable with large organic cations such as guanidine, tetramethylammonium-, and tetraethanol-ammonium ions.

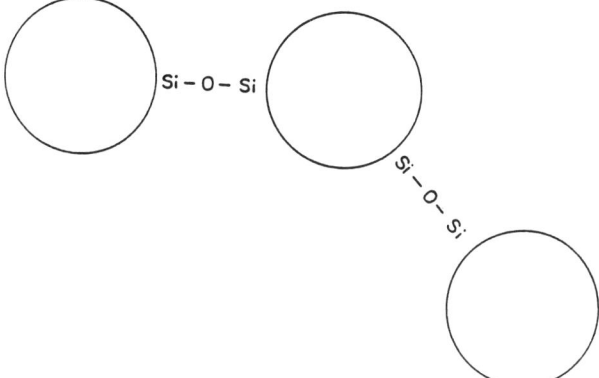

Fig. 10-2. Siloxane bridges between silica particles.

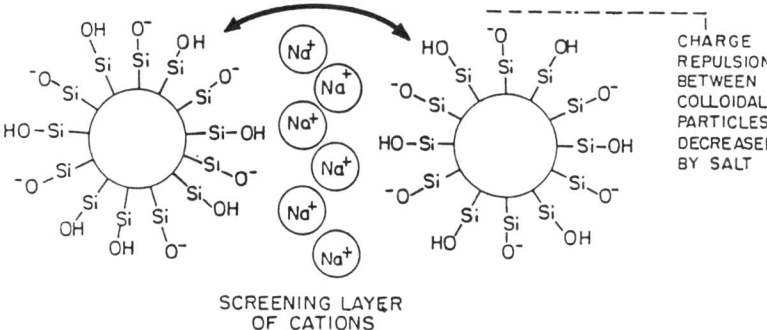

Fig. 10-3. Screening and flocculation of charged silica particles by cations.

Anions of neutral salts affect the gel time and gelling behavior of colloidal silica in an indirect way. For the cations of a neutral salt to surround and screen the negatively charged silica particles, they must be able to move away from the charge field of their own anions. The higher this charge field is the less likely are the cations to concentrate around the silica particles, and the less effective the salt will be in screening and bringing about gelling.

Salts containing highly charged anions such as sulfate ions are less effective screening agents than those containing monovalent anions such as chloride. Acetate salts are better screening agents than the corresponding chloride salts.

Effect of Temperature. The activation energy of the polymerization of silica, i.e., the formation siloxane bonds, will probably be different on the acid side and the basic side. However, according to Iler (1979), it is about 15 Kcal mole^{-1} in the pH region 4 to10. This means that the gel time will decrease by a factor of about 2 for every 10° C change in temperature.

1.1.2. Structure of Silica Gel. Iler (1979) described a silica gel as a coherent, rigid three-dimensioned network of contiguous particles of colloidal silica.

Silica gels are usually characterized by the properties shown in Table 10-2. However, their structure, giving rise to the properties in the table, can best be described qualitatively in terms of the coalescence between the particle packing and the size of the ultimate particles making up the gel.

When the particles in a silica sol gel, siloxane bridges are formed between particles in contact with one another. The necks between the particles represent surfaces of negative curvature and thus areas of very low solubility, see Fig. 2-3. Silicic acid, supplied by areas of positive curvature and/or deliberately added to the gel under formation, will therefore deposit in the area of contact between the particles and thicken the necks, i.e., increase the coalescence and the mechanical strength of the gel, see Fig. 10-4.

The packing of the particles in a gel can be represented by the coordination number, which is defined as the number of other particles that are in contact with a selected particle. Figure 10-5 shows the coordination number

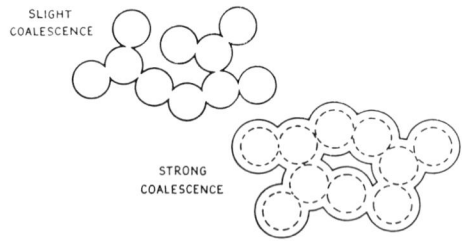

Fig.10-4. Reinforcement of gel structure by deposition of silica. Iler (1955). Courtesy Cornell University Press.

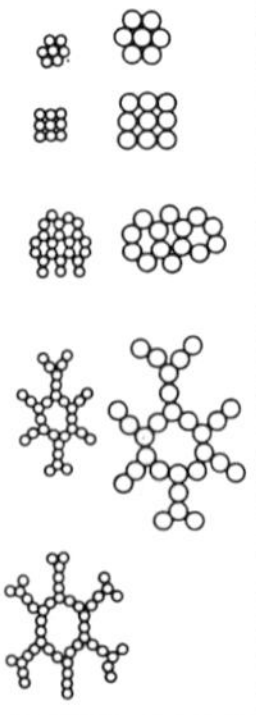

COORDINATION NUMBER			VOLUME %		PORE VOL.
			SOLID	PORES	cm³ g⁻¹
12			74.5	25.5	0.155
6			52	48	0.42
3			5	95	8.6
3-2-3			1.3	98.7	35
3-2-2-3			0.83	99.17	54

Fig. 10-5. Coordination number for various packing geometries and corresponding volume fractions and pore volumes. Iler (1979). Reprinted by permission of John Wiley & Sons, Inc.

for various packing geometries and the corresponding volume fractions of solid and pores, and pore volume. Densely packed gels have structures in which the particles are in close packed hexagonal, randomly close packed (not shown in Fig. 10-5), or close packed cubic configurations with coordination numbers 12, 7-10, and 6, respectively. Loosely packed gels have coordination numbers lower than 6, and Fig. 10-5 shows that volume fraction and pore volume rapidly increases with decreasing coordination number. A coordination number of 3 corresponds to a volume fraction of 95% and a solid fraction of only 5%.

Figure 10-5 also shows that for a given packing geometry, defined by coordination number, the only way to increase the pore size of the gel is to increase the size of the primary particles, which, however, is accompanied by a loss of specific surface area. On the other hand, the specific surface area of gel can be maintained at a high value when the pore size is increased by changing the packing geometry, that is by reducing the coordination number. Opening up the gel structure in this manner will usually bring about a wider pore size distribution.

As will be described below, the degree of coalescence, packing, and particle size of a gel depends on carefully controlling the conditions of the various steps in the manufacturing of silica gels.

1.1.3. Manufacture of Silica Gel. Acid gelation of sodium silicate solutions is by far the most common and important general method for making silica gels, but they can also be made from silica sols or by hydrolysis of silicon alkoxides; see Chapter 2.

Sodium silicate, in particular of $SiO_2:Na_2O$ ratio 3.3, is the most commonly used source of silica and H_2SO_4, HCl, and CO_2 are the most commonly used acids in commercial manufacture of silica gels.

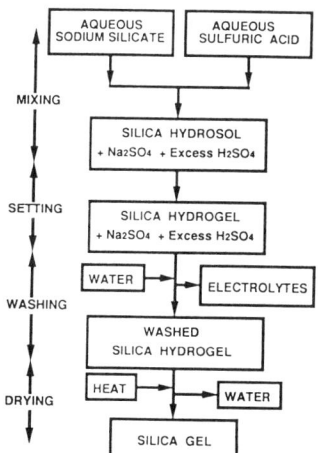

Fig. 10-6. The Patrick silica gel process. Patrick (1918).

There are many processes available for making silica gel but most of them are variations of the basic Patrick process, Patrick (1918), shown in Fig. 10-6. A laboratory procedure of making silica gel, which represents a further development of the original Patrick process was studied by Mattsson and Otterstedt (1995), is shown in Fig. 10-7a.

The sodium silicate and acid solutions are mixed, often into a heel of water with pH about 5 as in step 2 of Fig. 10-7a, in such proportions as to maintain an acid pH of about 5 in the reaction mixture. The temperature is preferably below about 50°C. At this stage the reaction mixture consists of a dispersion of very small silica particles, smaller than 4 nm, in an aqueous medium containing salt, usually Na_2SO_4 or NaCl, see Fig.10-6. The gel time of this sol depends on the

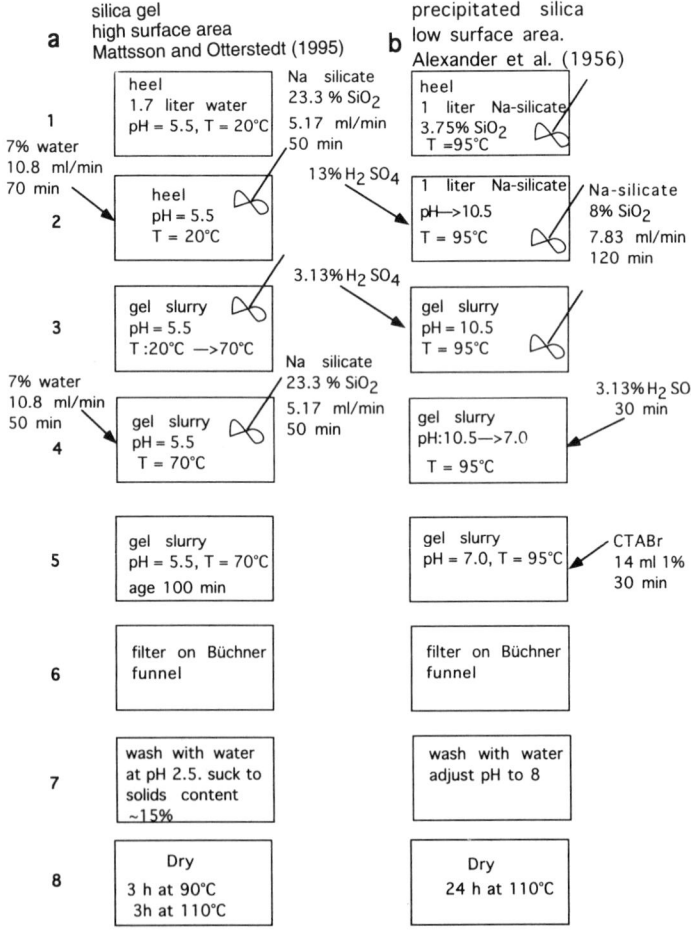

Fig. 10-7. Preparation of synthetic silica powders by wet methods.

silica concentration, temperature and concentration of electrolyte, but typically it falls in the range from 30 to 60 minutes.

In the setting stage, a coherent, three-dimensional network of contiguous silica particles is formed—see Fig. 10-8. Were the gel washed and dried at this point, the strong capillary forces acting during the drying step would completely collapse the open, highly porous gel to a dense material with very low porosity. It is therefore essential to reinforce the gel structure during the setting stage to make it so strong as to be able to withstand the capillary forces during drying.

This can be done in primarily two ways. One way, as used by Patrick, involves providing conditions of time, temperature and pH during the setting stage so as to increase the coalescence of the gel structure by the process of Ostwald ripening—that is silica dissolves from areas of positive curvature and is deposited at the contact points between the particles, which are areas of negative curvature; see Fig. 10-4. Iler (1953) and Alexander et al. (1956) strengthened the bond between particles by adding active silica to the gelling system of silica particles, see also step 4 in Fig. 10-7a. In step 5 of Fig. 10-7a the structure is further reinforced by heat-aging the wet gel.

In the washing stage, salts are washed out of the gel and the sodium content is reduced to a value typically in the range 0.1.- 0.2% Na_2O. It is important to realize that the seemingly simple step of washing is also an aging step. The pH of the wash water can affect the final properties of the gel in a major way.

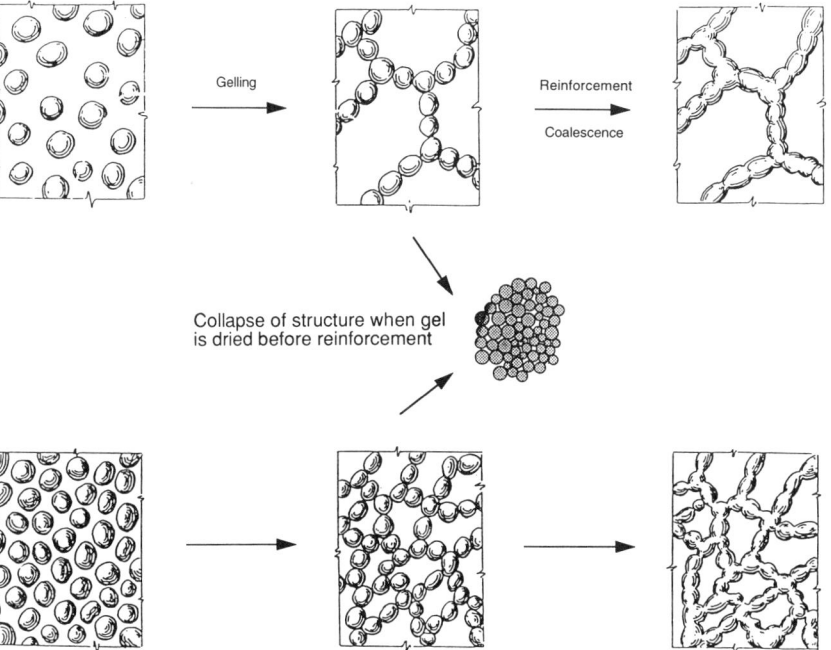

Fig. 10-8. Reinforcement of gel structure to reduce shrinkage on drying.

In the drying stage, liquid, usually water, is removed and the gel will shrink until its mechanical strength can withstand the compressive force exerted on the structure by the surface tension of the liquid around the boundary of the gel. Drying can be accomplished in several ways, and the particular method and the control of the drying conditions of that method will have a critical effect on the final properties of the gel.

Conventional drying, including drying under vacuum, is often done in two steps, first at a temperature somewhat below 100° C and then at a temperature somewhat above 100° C, and results in a gel called a *xerogel*. Mattsson and Otterstedt (1995) prepared silica gels according to the process outlined in Fig. 10-7a several times and obtained specific surface areas ranging from about 450 to about 650 m^2/g. Freeze drying causes less shrinkage than conventional drying, but is more expensive. Freeze-dried gels are called *cryogels*. The structure of the wet gel can be virtually preserved if the liquid is removed by supercritical drying. Gels dried in this manner are called aerogels— see below.

Schneider and Baiker (1995) discussed the drying of gels in terms of the following equation relating the capillary pressure P_c to the surface-to-volume ratio of a pore, S_p/V_p, the liquid-vapor interfacial energy of the solvent γ_{lv}, and the liquid-solid contact angle θ:

$$P_c = [\gamma_{lv} \cdot \cos \theta \cdot S_p]/V_p \qquad (10\text{-}1)$$

where

$$S_p/V_p = S \cdot \rho_b/(1-\rho) \qquad (10\text{-}2)$$

then

$$P_c = \gamma_{lv} \cdot \cos \theta \cdot [S_p\rho_b/(1-\rho)] \qquad (10\text{-}3)$$

where

ρ= relative density =ρ_b/ρ_s
ρ_b= bulk density of the solid network without liquid
ρ_s= skeletal density
S = specific surface area of the porous body

For a wet gel of a given structure and coalescence the capillary pressure can be reduced by substituting a solid-gas interface for the liquid-vapor interface as in freeze-drying, or by eliminating the surface separating the liquid and vapor phases as in supercritical drying. Supercritical drying means that the wet gel is heated at a constant rate in an autoclave to about 20° K above the critical point of the liquid. The critical point of water corresponds to a temperature of 647° K and a pressure of 22.1 MPa. Under these conditions silica is highly soluble in water and the gel structure will undergo radical changes. To avoid this problem, water can be replaced by an alcohol, e. g., methanol, with critical constants 513° K and 8.1 MPa, or an ether such as diethyl ether, with critical constants 467° K and 3.6 MPa. When thermal equilibrium has been reached, the pressure is isothermally released at a constant rate. To remove residual organic solvent the autoclave can be flushed with nitrogen and subsequently cooled to ambient temperature.

Silica gels can also be made in the shape of spherical beads, uniformly sized and with diameters up to several millimeters, by a process developed by Marisic and Dray (1945). Solutions of acid and silicate are mixed continuously in a flowing stream which is fed to a drop-forming device. From this device drops of appropriate size fall into a layer of petroleum oil, which is supported on a layer of water, before gelation takes place. As the drops slowly fall through the oil at a controlled temperature, they set to solid spheres. When the beads reach the interface they are hard enough to pass into the water layer without distortion and can be further processed without losing their form.

1.1.4. Aerogels. Known sometimes as "frozen smoke" for their ghost-like appearance, aerogels are among the lightest solids known with some types consisting of 99.8 % air. In addition, of all known solids, aerogels have the highest internal surface areas per unit weight of material due to their complicated, cross-linked internal molecular structure. An aerogel the size of a grape has the surface area of about two basketball courts. Probably the best known type of aerogel is a silica aerogel composed mainly of silicon dioxide. Today many other types including organic aerogels are known.

Although aerogels are not used as conventional adsorbents—an extremely promising use is water purification by electrically-induced adsorption of ionic impurities at low voltages with carbon aerogel electrodes, Farmer et al. (1996)—they are currently being studied with great interest because their extremely porous microstructure gives them unusual optical, thermal, and electrical properties and offers interesting opportunities in catalytic applications. They will therefore be briefly discussed here. For more information on aerogels we refer to Schneider and Baiker (1995), Fricke (1993), and Brinker and Scherer (1990).

Like silica sols, silica aerogels can be synthesized by hydrolysis of monomeric tetrafunctional alkoxide precursors, but using an acid, rather than a base, as in the case of sols, as catalyst; see Section B2 of Chapter 2. The ratio of water to silicon alkoxide in the synthesis of aerogels is much lower than in the synthesis of sol (more than 20:1); see Fig. 2-12. Under these conditions the formation of extended polymeric networks, rather than compact structures of the kind generally found in alkoxide-derived sols, is promoted.

Aerogels of many different metal oxides, e.g., Al_2O_3, Cr_2O_3, TiO_2, ZrO_2, V_2O_5, and MoO_2, can be prepared by similar methods, using the appropriate metal alkoxides as precursors.

As final products, aerogels are either loose and brittle solids in the shape of clumps, pellets or tiles, or very fine powders.

Silica aerogels are highly porous materials with porosities up to 99.8% and bulk densities ranging from 3 to 500 kg/m^3—the density of air is 1.29 kg/m^3. Due to the low bulk density and ultrafine pore size, the thermal conductivity is generally very low, ranging from 0.014 W m^{-1} K^{-1} at atmospheric pressure to 0.004 W m^{-1} K^{-1} at 1000 Pa for silica aerogel, Schneider and Baiker (1995).

Air of atmospheric pressure has a thermal conductivity of 0.2 W m^{-1} K^{-1}, that is, more than an order of magnitude higher than that of silica aerogel. The thermal conductivity of polystyrene, considered to be a good thermal insulator, is five times higher than that of silica aerogel.

Silica aerogels also have interesting acoustic properties. The speed of sound in air is 333 m/s whereas it is only about 20 m/s in silica aerogel with a bulk density of 5 kg/m^3.

The extremely low refractive index of silica aerogels, ranging from about 1.007 to 1.4, make these materials, in the shape of tiles, very useful in detectors of so called Cherenkov radiation, that is pions, muons and protons moving at velocities close to the speed of light.

1.1.5. Precipitated Silica. Precipitated silica is a material, which at a first glance resembles silica gel in several respects. Both materials are made by wet processes and consist of agglomerated small particles of silica. As prepared, their surface is fully hydroxylated and contains about 5 silanol groups per nm^2. However, the fact that precipitated silica is made at alkaline pH, whereas silica gel is prepared at acid pH, imparts several important differences in properties between the two silica materials; see below and also Table 10-2.

Generally, precipitated silica is made by adding an acid, usually H_2SO_4, or HCl, to a sodium silicate solution, as a rule of SiO_2:Na_2O ratio 3.3, whereas silica gel is made by adding the silicate solution to the acid. We have defined gels as structures that consist of branched chains of particles that fill the whole volume of the sol. Precipitated silica, on the other hand, consists of relatively close-packed clumps of particles. Fig. 10-9 shows the difference between a gel and a precipitate.

There are many different ways of making precipitated silica, but the flow-diagram shown in Fig. 10-7b is representative of a process developed by Alexander et al. (1956), which yields a material that is an excellent reinforcing filler in rubber; see section on plastics and polymers below. To a heel of 3.3 ratio sodium silicate solution at 95° C enough 13 wt. % H_2SO_4 is added at a rate of about 3.3 ml/min (~30 minutes) with vigorous agitation to lower the pH to 10.5. In this step, #2 in Fig. 10-7b, the small particles, about 2 nm, in the Na-silicate solution grow by Ostwald ripening to a diameter of about 7 nm. Next, Na-silicate solution, 8% in SiO_2, and 3.13% H_2SO_4, is fed to the sol with vigorous stirring simultaneously and at the same rate, 7.8 ml/min (120 minutes) in step 3.

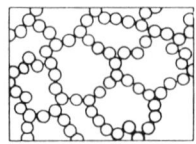 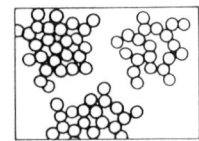

Fig. 10-9. Two-dimensional representation of the difference between a gel (left) and a precipitate (right). Iler (1965). Courtesy Cornell University Press

The amount of acid is adjusted to maintain the pH constant at 10.5. During this operation part of the silica is precipitated and the rest, in the form of active silica released by the reaction of sodium silicate with sulfuric acid, is accreted to the clumps of precipitated silica particles. The rate at which accretion can occur without forming secondary silica particles in the aqueous phase is related to the pH, the temperature, the specific surface area of silica in the aggregates, and the sodium ion concentration; see the "build-up process" in Chapter 2.

In step 4, more 3.13% H_2SO_4 is added with vigorous agitation over a period of 30 minutes to reduce the pH to 7. The pH is reduced slowly over a period of time so as to allow the remaining active silica to deposit on the aggregates, rather than to form secondary silica particles in the water phase. Cetyl trimethylammonium bromide in amount equal to 0.2% based on the weight of the silica is added as a dilute solution in order to coagulate the silica aggregates and act as a filter aid.

Alternatively, the slurry can be maintained at 95° C without agitation for about 4 hours in order to further coagulate the precipitate to aid in filtration. After filtration, the precipitate is washed with cold water of pH 8 and dried at 110° C for 24 hours. The surface area of precipitated silica, made by Mattsson and Otterstedt (1996) according to this process, was about 105 m^2/g.

1.1.6. Fumed Silica. Whereas silica gel and precipitated silica are made by wet processes, fumed or pyrogenic silica is manufactured by a dry process. Volatile silane compounds, usually $SiCl_4$, but $Si(OC_2H_5)_4$ can also be used, are hydrolyzed in an oxygen-hydrogen gas flame. The result is an extremely fine-particle size silica; see Table 10-2. A commercial fumed silica, Aerosil® 200 from Degussa, is reported to have 4.2 silanol groups/nm^2, Degussa (1982), a surprisingly high figure for a silica that has been treated at high temperatures, but work by Uytterhoeven et al. (1963) revealed that only 35% of the SiOH groups were on the surface, the remaining were internal groups.

Comparison between the properties of synthetic silicas. Typical properties of the synthetic silicas silica gel, precipitated silica, and fumed silica are shown in Table 10-2.

Comparison between fumed silica and silica made by wet processes. Table 10-2 shows that the loss on ignition is higher for silica gels and precipitated silicas than for fumed silica, reflecting the fact that the two silicas made by wet processes have a fully hydroxylated surface, i.e., about 5 SiOH groups/nm^2, whereas fumed silica, manufactured by a dry process at high temperatures, although water vapor is one of the reactants, does not. The compacted volume of fumed silica is higher than that of silica gel and precipitated silica since the former silica consist of chain-shaped aggregates, whereas the latter two silicas consist of corpuscular, 3-dimensional aggregates.

Comparison between silica gel and precipitated silica. The most striking difference between silica gels and precipitated silicas is the great contrast in appearance between their pore size distributions—see Fig. 10-10—reflecting

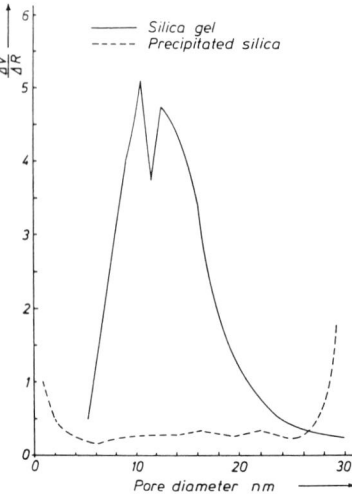

Fig. 10-10. Comparison of the pore volume distribution curves of a silica gel and a precipitated silica. V and R denote volume and radius, respectively. Ferch (1994). Courtesy American Chemical Society.

great dissimilarities in their pore structures, which in turn is a consequence of the differences in synthesis conditions, see Fig. 10-7. Silica gels have a fairly narrow pore size distribution, the width and mean pore size of which depend on the manufacturing conditions—the conditions of the drying step may thus have a great affect on the pore size distribution of the final product. Precipitated silicas,

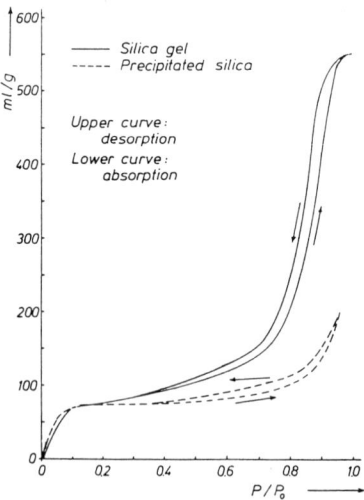

Fig. 10-11. Water adsorption isotherms of a silica gel and a precipitated silica at room temperature. P is the actual pressure and P_0 is the saturation pressure. Ferch (1994). Courtesy American Chemical Society.

on the other hand, have very wide pore size distributions with no discernible mean pore size.

This great difference in pore structure accounts for the dissimilarity in appearance between the adsorption isotherms for water of the two silicas; see Fig. 10-11. The adsorption isotherm of silica gel has the shape of a standing S, whereas that of precipitated silica has the shape of a lying S. During the adsorption of water vapor, silica gels approach the saturation point, whereas this is not the case for precipitated silicas.

1.1.7. Applications for Silica Gel Adsorbents. Silica gels are widely used as drying agents in various applications; seeTable 10-3.

2. Thickening, Suspending and Viscosity Control

The viscosity of dispersions of small particles depends not only on concentration but also, in some cases very strongly, on particle shape and charge, and pH. Small particles can be used as effective thickening, suspending and viscosity control agents.

Table 10-3. Applications for silica gel as drying agent. Courtesy W.R. Grace.

Application	Remarks
Ultrasonic cleaning equipment	Indicator gel for visual monitoring of the water loading.
Refrigerant drying	Silica gel with extremely low water loading for filter blocks.
Tank vent breathers	As tank empties moist air flows into the void. Substances sensitive to moisture can be protected when the incoming air is dried with silica gel.
Transformer breathers	The use of a breather filled with silica gel prevents penetration of moisture into the transformer, and thus maintains its functional efficiency.
Packaging protection	For protection of goods sensitive to moisture during storage and transportation.
Double glazing-Thermopane windows	The adsorbent prevents the condensation of Thermopane windows moisture in the space between the window panes.
Drying of flowers	For drying and preservation of flowers and plants.
Drying of substances in a desiccator	The change in color of blue silica gel (from blue to pink) enables a visual monitoring of the state of activity of silica gel.

2.1. General Considerations

It is known from experience that the viscosity of a suspension in general increases with concentration. In concentrated suspensions the particles bump into each other, and increase the frictional forces, with a resultant increase in viscosity. In dilute dispersions, where there are few collisions, the suspended particles will also contribute to the viscosity. Since the layers of the liquid phase must flow past the suspended particles, the smoth flow of the liquid is disturbed by the particles and an increase in viscosity occurs. For the case of a dilute suspension of spherical particles Einstein (1906) could—by assuming that the spheres were rigid, that their size was large compared with the size of the liquid molecules (but small compared with the dimensions of the measuring apparatus so that inertial effects could be neglected), that there was no slippage between spheres and liquid, and that the concentration was so low that the disturbance of the flow around a given particle was not affected by disturbances around neighboring particles—derive the famous equation, the "Einstein" equation:

$$\eta_{rel} = 1 + 2.5V \tag{10-4}$$

where V is the volume concentration of the suspended spheres and η_{rel} is the relative viscosity.

Einstein's method has been extended to ellipsoids, rods, discs, dumbbells, and non-rigid spheres, by, for instance, Alfrey (1948) and Frisch and Simha (1956).

The Einstein equation has been extended to more concentrated solutions by, e.g., Vand (1948). He assumed that two spheres rolled around each other when they met and were then separated by the shearing action of the liquid, and was able to derive the following equation, the Vand equation:

$$\eta_{rel} = 1 + 2.5V + 7.35V^2 \tag{10-5}$$

Adding spheres to a dispersion of spheres of the same size increased the viscosity, according to Mooney (1951), in two ways: The added spheres occupied the liquid space not filled by the original spheres, and also reduced the volume of liquid available for the original spheres. From these considerations Mooney could develop the following equation, the Mooney equation:

$$\eta_{rel} = \exp\left(\frac{2.5V}{1+kV}\right) \tag{10-6}$$

where k is a self-crowding and hydrodynamic interaction factor, and V is the volume fraction of the spheres. As V approaches 0 the Mooney equation is reduced to the Einstein equation.

Mooney predicted that the most probable value of k would be 1.35, corresponding to hexagonal close packing of the spheres with a volume fraction

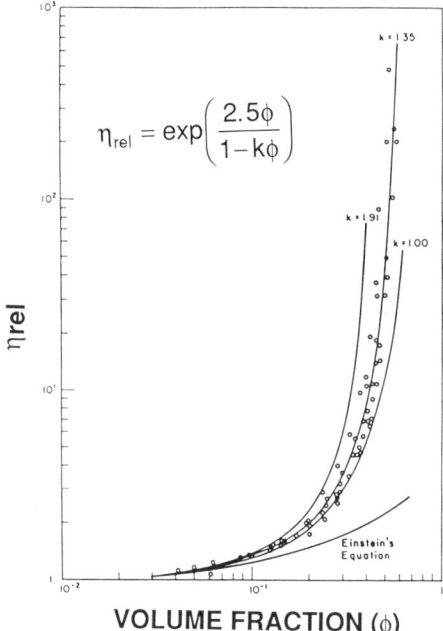

$$\eta_{rel} = \exp\left(\frac{2.5\phi}{1-k\phi}\right)$$

VOLUME FRACTION (ϕ)

Fig. 10-12. Mooney's equation plotted with various values of k. Brodnyan (1959). Courtesy the American Institute of Physics.

of 0.745; see also the Section on gels in this Chapter. Cubic close packing of the spheres, with a volume fraction of 0.524, is a looser packing that leaves the spheres just free to move and corresponds to a k-value of 1.91. Brodnyan (1959) applied experimentally obtained data to the Mooney equation and Fig. 10-12 shows that the data were approximated by k = 1.35 and fell between k = 1.91 - 1.00.

The Einstein equation is valid at volume fractions below about 6%. It is interesting to note that in a dispersion of spherical particles of volume fraction 0.07, the average distance between the particles is one particle diameter, independent of the diameter of the spheres. Above a volume fraction of 0.1-0.2 the relative viscosity increases very rapidly and approaches a thousand-fold increase at volume fraction 0.6.

2.2. Effect of pH and Charge on Viscosity

For most of the small particles dispersed in water and discussed in this book the charge on the particle surface depends on pH; see Fig. 6-30 for the case of silica particles, Fig. 10-13 for particles of fibrillar boehmite, and Fig. 10-14 for silica particles charge reversed with polycations of aluminum.

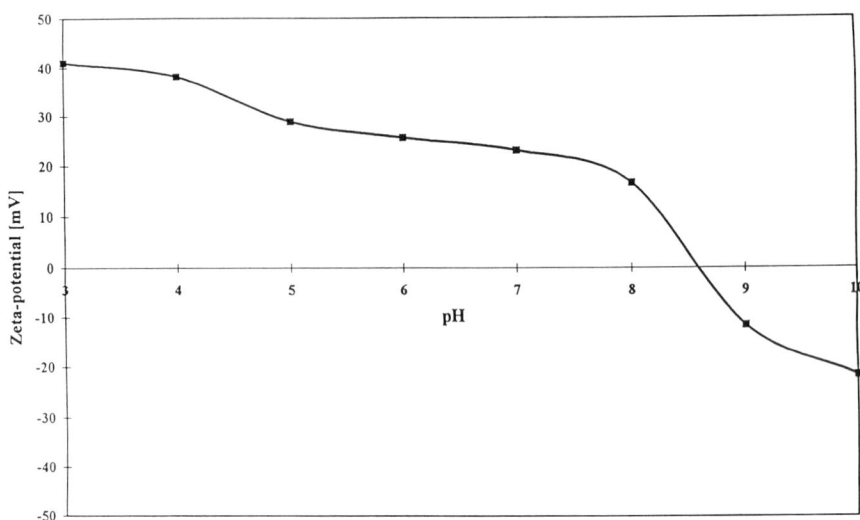

Fig. 10-13. Zeta potential of fibrillar boehmite versus pH. Mattsson and Ottersredt (1995).

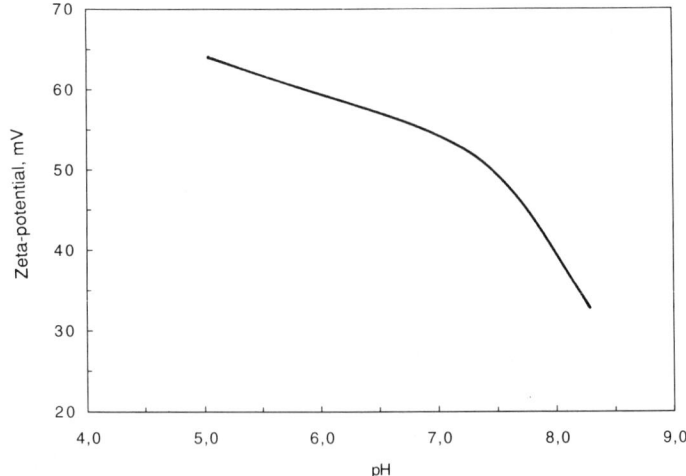

Fig. 10-14. ζ-potential of 100-nm silica sol particles charge reversed with $[Al_{13}O_4(OH)_{24}(H_2O)_{12}]^{7+}$ ions. Mattsson and Otterstedt (1995).

When the particle charge is high, the dispersion will consist of discrete particles kept apart by electrostatic repulsion and the viscosity is relatively low—it depends, of course, on the concentration. However, as the charge on the particles decreases, they can come closer together so as to interact with one another and form aggregate structures in which water will be trapped. The decrease in free water will cause the viscosity of the dispersion to increase. The viscosity of silica sols at pH of about 7 will increase with time and after a certain length of time, the sol will form a solid gel. At a given electrolyte concentration and temperature, the gel time decreases rapidly with the silica concentration.

Figure 10-15 shows that a dispersion of colloidal alumina, containing 4 wt. % fibrillar boehmite, has a minimum viscosity between pH 3 and 4, but that the viscosity rapidly increases with pH and reaches a fairly constant level at pH of about 6, about 1000 times higher than the viscosity at the minimum. The charge on the particles decreases from about 40 mV at the viscosity minimum to about 25 mV at pH 6; see Fig. 10-13. Particles having a platelike structure, as, for instance, some clays, may have oppositely charged edges and faces. In aqueous dispersions such particles may form "cardhouse structures", in which water is trapped, causing the viscosity to go up; see Chapter 4 and Fig. 4-9. High viscosities brought about by cardhouse structures are usually not strongly effected by changes in pH—unless the change in pH has been achieved by a dispersing agent such as a 3.3 ratio sodium silicate solution.

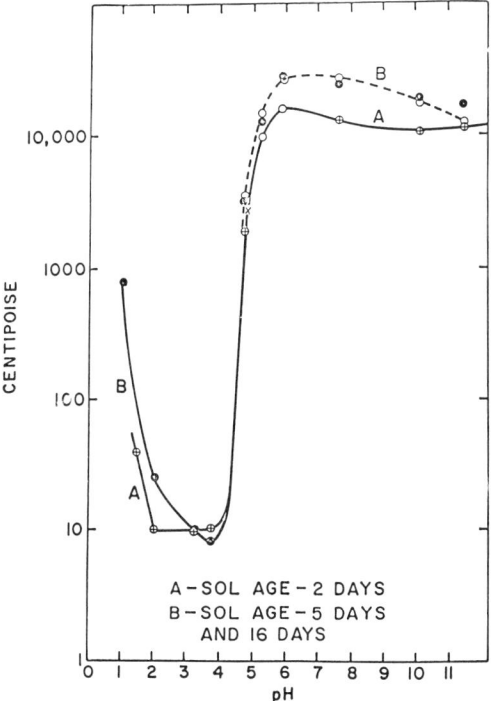

Fig. 10-15. Viscosity of fibrillar boehmite, 4% alumina at 25° C vs. pH. Courtesy E. I. Du Pont Co.

2.3. Effect of Particle Shape on Viscosity

The viscosity of dispersions of small particles is strongly affected by the particle shape and in this section the effect of some frequently occurring shapes will be discussed.

Spherical shape. Otterstedt and Brandreth (1996) used the information in Du Pont's technical brochures on Ludox® sols, to test whether the viscosity of such sols, consisting of spherical particles of quite uniform particle size, obey the Mooney equation, Eqn 10-6. Since the viscosities given in the Du Pont brochures are not relative viscosities the Mooney equation was transformed in the following way:

Assume silica sols 1 and 2 have viscosities η_1 and η_2, and volume fractions V_1 and V_2 then:

$$\ln \eta_{rel}^{sol1} = \ln \frac{\eta_1}{\eta_2} = \frac{2.5 V_1}{1 - kV_1} \tag{10-7}$$

$$\ln \eta_{rel}^{sol2} = \ln \frac{\eta_2}{\eta_0} = \frac{2.5 V_2}{1 - kV_2} \tag{10-8}$$

Subtracting Eqn. 8 from Eqn. 7 gives

$$\ln \frac{\eta_1}{\eta_2} = \frac{2.5 (V_1 - V_2)}{(1 - kV_2)(1 - kV_1)} \tag{10-9}$$

The pairs of sols in the Table 10-4, chosen from Table 2-12, were used to test the validity of Eqn. 10-9. A value of 1.6, which is the arithmetic mean of the k´s corresponding to cubic close packing, 1.91, and hexagonal close packing, 1.35, was chosen as the value of k in the Mooney equation. The left and right hand sides of Eqn. 10-9 were calculated from the data in Table 2-12 and the results are shown in Table 10-4.

Had the viscosities of the silica sols strictly obeyed the Mooney equation, the ratio in the last column of Table 10-4 should be 1. Although the ratio varies between 1.4 and 2.7, thus suggesting that the theory on which the Mooney equation is based does not give an accurate description of the viscosity of silica sols, this theory probably includes the most significant factors responsible for the viscosity of such sols.

Fibrillar shape. Attapulgite is an example of a clay having particles of fibrillar shape, about 500-1000 nm long and 20-60 nm wide. The viscosity of aqueous dispersions of attapulgite containing no dispersing agent increases rapidly with solids content and becomes intractable at solids contents above about 10 wt.%. By using dispersing agents such as Na-polyacrylates or 3.3 ratio Na-silicate, the solids content can be increased to about 25 wt.%; see also Fig. 4-10c.

Table 10-4. Test of the validity of the Mooney equation using pairs of silica sols. Otterstedt and Brandreth (1996).

Pairs of sols	η	V	$\ln\dfrac{\eta_1}{\eta_2}$	$\dfrac{2.5(V_1-V_2)}{(1-1.63V_1)(1-1.63V_2)}$	$\dfrac{\ln\dfrac{\eta_1}{\eta_2}}{\dfrac{2.5(V_1-V_2)}{(1-1.63V_1)(1-1.63V_2)}}$
Ludox® TM	4.0	0.36			
1:			2.0	1.1	1.8
Ludox® SM	5.5	0.25			
Ludox® TM	4.0	0.36			
2:			1.6	1.1	1.5
Ludox® LS	8.0	0.25			
Ludox® HS-40	16	0.31			
3:			1.1	0.52	2.1
Ludox® SM	5.5	0.25			
Ludox® HS-40	16	0.31			
4:			1.4	0.52	2.1
Ludox® HS-30	4.0	0.25			

Figure 10-16 shows that the viscosity of aqueous dispersion of fibrillar boehmite increases very rapidly with concentration at pH of 3.8. The sols are thixotropic and will thicken in time. An 8% sol is quite fluid when first prepared, but usually thickens in about one hour. The broken line in Fig. 10-16 indicates that more highly concentrated, fluid dispersions can be prepared if they contain barium. Sulfate ions have a strong destabilizing effect on positively charged alumina sols and they are almost always present, in small to moderate amounts, as impurities in alumina. The viscosity of alumina dispersions may be reduced by adding sufficient barium salt to react with the sulfate in the sol. The barium sulfate formed is very finely divided and will settle out only from dilute, very fluid sols.

Platelike shape. In a completely peptized dispersion of montmorillonite clay the ultimate particles are large in two dimensions, but only about 1 nm thick. With no dispersing agent, the viscosity of aqueous dispersions of montmorillonite increases rapidly with concentration and will be very high above about 5 wt.% clay. However, Fig. 4-10b shows that dispersions of much higher concentrations of montmorillonite, up to about 12 wt. %, can be made by using an anionic dispersant such as 3.3 ratio sodium silicate and Na-polyacrylate or cationic dispersing agents like basic aluminum salts.

Corpuscular shape. It is true that the ultimate structural element of kaolinite

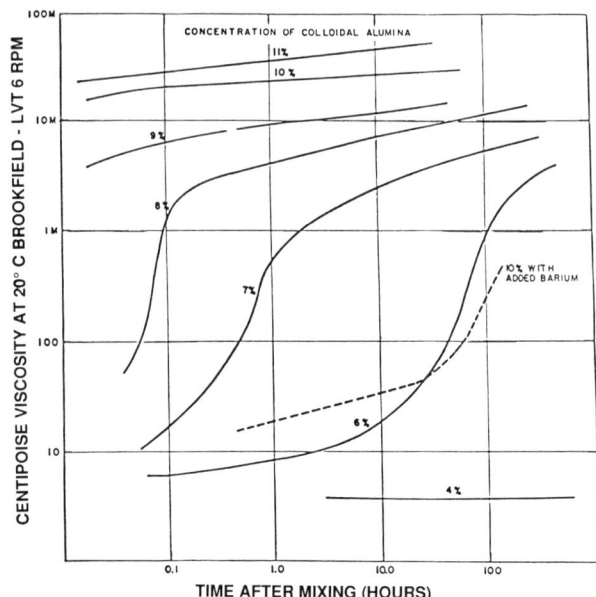

Fig. 10-16. Viscosity of dispersions of fibrillar boehmite of different concentrations as a function of time at pH 3.8. Courtesy of E. I. Du Pont Co.

is a silica-alumina double sheet, only about 0.7-0.8 nm thick, see Chapter 4, but these sheets are strongly aggregated into plates, so called "books", which in many types of kaolin clay are so thick that the clay particles are practically corpuscular in shape. The viscosity of aqueous dispersions of kaolin clay increase less rapidly with concentration than aqueous dispersions of attapulgite and montmorillonite and can go up to 30-40 wt.% before the viscosity becomes unmanageably high. By using dispersing agents, the solids content of kaolin can reach very high values, up to about 70 wt.%; see Fig. 4-10a.

The trick to obtaining dispersions of the highest possible solids content compatible with a reasonable ease of handling is therefore to prepare the dispersions at a suitable pH, using effective dispersants, and to choose particles of appropriate shape so as to achieve the densest possible packing of the particles in the dispersion. This is most readily accomplished with spherical and corpuscular particles.

Figure 10-12 shows that the viscosity of a dispersion of spherical particles increases very rapidly as the volume fraction of spheres approaches 0.524, corresponding to cubic close packing of the spheres, including the double layer around the particles. Table 10-5 shows that the highest solids content consistent with a manageable viscosity or stable dispersion is about 0.45 for colloidal silica and kaolin clay—i.e., corresponding to a packing of the colloidal particles that is only about 10% lower than cubically close packed particles. For dispersions of attapulgite and fibrillar alumina the volume fractions of the particles are only

Table 10-5. Maximum solids concentration of dispersions of small particles of different shpes		
Shape	Solids concentration	
	Weight %	Volume %
Spherical (colloidal silica, 100 nm)	65	44
Corpuscular (kaolin clay)	70	47
Fibrillar (attapulgite clay)	18	8
Sheetlike (Na-montmorillonite clay)	11	5

0.08 and 0.05, respectively—indicating that the particles are loosely packed in these dispersions. A good reference on the preparation of dispersions in liquids is Stein (1996).

2.4. Effect of Salt on Viscosity

Salts generally destabilize dispersions of small particles, often to the point where the particles will aggregate and form precipitates, in which case the salt is present in a concentration that exceeds the critical coagulation concentration, c.c.c., of the salt. The c.c.c depends on the charge of the ionic components of the salt—the higher the charge, the lower the c.c.c.—and the type of salt. Below the c.c.c., salts will in many cases increase the viscosity of sols of small particles. Cations will thicken anionic sols and anions will increase the viscosity of cationic dispersions. Thus, addition of 0.5 to 1% sulfate ion (based on particle solids) will thicken dispersions of fibrillar alumina and positively charged silica particles at pH of about 4. The effect is more marked in more concentrated sols. Acid aluminum phosphate has a similar effect and will also insolubilize films cast from dispersions of alumina particles or silica particles coated with alumina, e.g., in the form of basic aluminum chloride.

2.5. Thickening Effect of Small Particles

The question as to how to make dispersions of small particles of the highest possible solids content is very important in many technical applications, e.g., in the clay and ceramics industries, and has already been addressed.

The opposite problem, i.e., to thicken thin solutions and dispersions or to make dispersions of high viscosity but low solids content, is technologically equally important. Organic thickening agents such as cellulose derivatives, alginates, various gums and others, are thus widely used in, for instance, the food and paint industries. There are, however, many thickening and suspending applications where inorganic thickening agents would be very useful and desirable.

When using small particles as thickening agents the shape of the suspended particles obviously has a large effect on the viscosity of a dispersion. At the same solids content, platelike or fibrillar particles bring about much higher viscosity than spherical or corpuscular particles.

Also, if there are places of opposite electrical charge on the particles, for instance, the edges and faces of particles of kaolinite and montmorillonite, aggregate structures, e.g., "card house" structures, may form and thus sharply increase the viscosity. Small particles, in which these two effects are combined, will be particularly effective thickening agents. The thickening effect of small particles gives rise to thixotropic dispersions.

Thixotropy. The characteristic feature of thixotropic dispersions is that their viscosity decreases with increasing shear rate. The viscosity may be so high that the dispersions have the appearance of gels, as is the case with certain types of paints, so-called drip-free paints, yet when they are sheared, as for instance by the paint roller or brush, they become quite fluid.

When small particles in dilute colloidal dispersions are highly charged, e.g., colloidal silica at pH 9 and fibrillar alumina at pH 4, they repel and slide past one another. No bonds are formed between the particles under these conditions and the viscosity is Newtonian; that is it is independent of the rate of shear.

However, if the particles are able to touch one another and form linkages the viscosity will rapidly increase. The network formed is, on the other hand, usually not very strong and can readily be broken by shearing the dispersion, e.g., by stirring it. The more rapid the stirring, the greater the proportion of linkages broken at any one instant, the weaker the network at that instant, and the lower the viscosity. This phenomenon is called thixotropy.

Thixotropic behavior increases with:

• Increasing concentration of the small particles.

• Decreasing the charge on the small particles, which can be accomplished in the case of boehmite particles by raising the pH above about 4.5; see Fig. 10-13.

• Screening the charge of the particles by ions, preferably multivalent ions, of opposite charge. Thus sulfate, phosphate, citrate and tartrate ions effectively increase the thixotropic behavior of colloidal solutions of alumina.

Fibrillar alumina is one of the most effective inorganic thickeners known for aqueous systems and its use as a thickener will be demonstrated in three examples (from product bulletin on Baymal®, Du Pont's colloidal fibrillar alumina).

Synergistic thickening effects with clays. Swelling clays, such as hectorite and bentonite, are widely used as thickening agents. However, their thickening effect can be greatly enhanced if they are mixed with a sol of fibrillar boehmite.

For example, a 2% refined hectorite dispersion can be prepared and allowed to stand overnight. To 100 parts of this dispersion was added, with stirring, 100 parts of an aqueous dispersion containing appropriate amounts of

Table 10-6. Synergistic thickening of hectorite with fibrillar boehmite (courtesy E. I. DuPont)	
Composition	Viscosity, cps
A. 2% Hectorite (refined)	1600
B. 1% Hectorite	<50
C. 1% Hectorite + 0.02% Baymal®	350
D. 1% Hectorite + 0.05% Baymal®	2250
E. 1% Hectorite + 0.1% Baymal®	9300

Table 10-7. Synergistic thickening of bentonite (montmorillonite) with fibrillar boehmite (courtesy E. I. DuPont)	
Composition	Viscosity, cps
F. 4% Bentonite	5500
G. 2% Bentonite + 0.01% Baymal	13500

fibrillar boehmite to give the ratios shown in Table 10-6. Viscosities were measured with an LVT model Brookfield viscosimeter using a #2 spindle at 0.6 r p m.

Thus, one part of fibrillar boehmite can apparently replace approximately 10 to 20 times its weight of clay in the mixture. In similar tests with a bentonite dispersion the results shown in Table 10-7 were obtained.

It is important that the fibrillar boehmite and the clays are mixed in the form of individual dilute dispersions. Dry fibrillar boehmite powder added to a clay dispersion will not disperse and show its thickening effect, nor will a dry clay powder properly swell in a dispersion of fibrillar boehmite. Each of the components must be swelled and thoroughly dispersed separately, in water, before they can by brought together to develop maximum thickening effects.

Thickening of colloidal silica with fibrillar boehmite. Generally speaking, colloidal alumina and colloidal silica are not compatible since the two types of particles bear opposite charges. Thus, in dispersions containing more than a few percent each of silica and alumina, the particles link together to form gels or precipitates.

Dry fibrillar boehmite powder may be dispersed directly into colloidal silica with the proper precautions. The colloidal silica must first be acidified and then admixed with a small amount of aluminum salt, such as aluminum nitrate, so as to cover the silica surface with adsorbed aluminum ions. This mixture is considerably more compatible with colloidal alumina than unmodified silica sol.

For example, 90 grams of colloidal silica, e.g., Ludox® HS, see Table 2-12, can be acidified by adding about 1 cc of concentrated nitric acid to obtain a pH of 2.7 or at least less than 3.0. To the acidified sol is next added 0.33 cc of 2 M aluminum nitrate solution. As much as 10 grams of fibrillar boehmite powder may be added to the acidified silica sol, by adding the powder slowly to the sol as it is stirred vigorously, and then continuing high speed stirring for a few minutes.

If the pH is around 2 or lower, it is not even necessary to add the aluminum nitrate to the silica sol before adding the colloidal alumina powder. However, below pH 1.5 fibrillar boehmite will not swell or disperse properly. Mixtures containing 10% fibrillar boehmite are very viscous and thixotropic and may have the consistency of a soft grease. More fluid dispersions can be obtained by diluting the silica sol to half or quarter strength before adding the colloidal alumina. However, all of these mixtures tend to be viscous and thixotropic, even in the pH range of 2 to 3, if the concentrations of silica and alumina are higher than 3 to 4%. When the pH of these mixtures is raised to 5 or 6, they thicken or gel, unless the concentration of alumina is less than about 1%.

Thickening of anionic latices. Fibrillar boehmite may be used as a thickener for anionic latices, i.e., dispersions of negatively charged, spherical polymer particles in water, with which it is normally not compatible. This can be accomplished by raising the pH of the dispersion of fibrillar boehmite to about 8.5, at which pH it no longer bears a positive charge, and is therefore much more compatible with anionic dispersions.

Guidelines for successful results are that the fibrillar boehmite was thoroughly dispersed in water, that the added alkali be as dilute as possible, and that the mixture is thoroughly blended.

As an example, 2000 g of a 4% sol of fibrillar boehmite was placed in a blender and stirred at high speed. While stirring, 2000 g of water containing 12 g concentrated ammonia solution (28% NH_3) was added rapidly and stirring continued for about one minute. The pH of the freshly made mixture was 8.5, but may drop slightly on standing. A quantity of 100 g of this sol was placed in a mixer and 80 g of a styrene-butadiene type latex containing 48% solids was added slowly with good stirring. The resultant mixture was completely compatible.

3. Bonding and Reinforcing

Small particles may reinforce, e.g., polymers and plastics by interacting with the polymer chains in such a way, say, through hydrogen bonding, that mechanical properties such as modulus and energy at break are significantly improved. This reinforcing function increases with the specific surface area of the particles and will be described in more detail in the section "Plastics, Polymers and Resins" below. Small particles may also function as binders and reinforcing agents, either by themselves or in combination with other binders and fillers, for inorganic fibers and powders. This reinforcing function depends on the ability of the small particles to form coherent films free of cracks. Generally speaking, only those small particles or combination of small particles that can form strong, coherent films are good binders for inorganic fibers and powders.

3.1. Mechanism of Film Formation

To produce a dried, strong film or mass of material of small particles, the mass has to remain coherent, without cracking, as the water evaporates. Iler (1979) discussed the drying of silica sols on glass plates and concluded that it was very difficult, if not impossible, to obtain coherent films of silica if the particles were of uniform size. On the other hand, if the sols contained appropriate amounts of particles of different sizes so that the pores formed by the larger particles could be filled with successively smaller particles, strong coherent films could be formed; see Fig. 10-17.

Sols of platelike or fibrillar particles form coherent films more readily, i. e., they are better binders, than sols of spherical particles. For instance, as a sol of fibrillar boehmite dries on a substrate the concentration of fibrils in the film will increase and the sol will thicken and gel at a concentration between 10 and

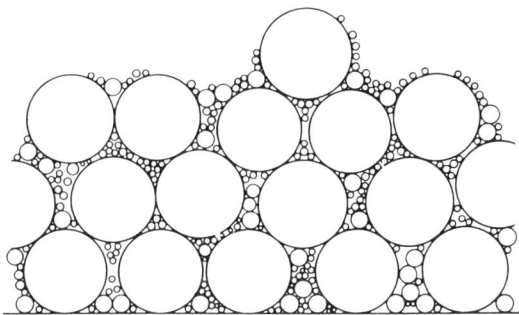

Fig. 10-17. Strong, dense gel is formed from a mixture of large and small particles. Iler (1979). Reprinted by permission of John Wiley & Sons, Inc.

20%. As drying continues, the gel shrinks, and this is the critical step in the film formation.

It is the shrinkage of the partially dried fibrils that causes cracking. Unless the gel is dried below a certain critical rate, depending on thickness, the film will crack due to uneven shrinkage. Assuming appropriate drying conditions, the gel film diminishes in thickness as the drying proceeds and becomes firm at about 50% solids. At this stage the film is transparent but becomes more opaque as part of the pores are dried out and then almost clear again and hard and brittle when all free water is gone and the pores are full of air. The film is a microporous mat of submicroscopic fibrils; see also Figs. 7-4 and 7-13. It is quite permeable to air, water, organic solvents, but not to many high polymers.

Table 10-8 shows a list of inorganic film-formers.

Table 10-8. Commercially available clays and minerals with film forming properties similar to fibrillar boehmite		
Product	Trade name (examples)	Uses —functions (see also Chapter 4)
Magnesium montmorillonite	Ben-a-gel®	thickening, gelling, stabilizing agent for latex paints, thixotropic gelling, particle suspension, ceramics
Magnesium aluminum silicate	Veegum®	retard pigment settling, impart body, stabilize emulsions, inorganic film former, ceramics
Bentonite	Volclay®	bonding of minerals, clarifying ceramics
Attapulgite	Attagel®	binder, thixotropic thickener, foam and emulsion stabilizer, adhesives, cosmetics
Alumina - boehmite	Disperal®	binder, catalysts, viscosity control, thixotropic thickener

3.2. Small Particles as Binders in Mats and Felts

Small particles can bind coarser fibers together into sheets, which will be stronger when the fiber mixture contain fibers not too much coarser than the size of the small particles. Table 10-9 shows the thickness of some typical fibers.

Figure 10-18 shows schematically that the film of small particles bonding together and reinforcing a fibrous mat covers all fiber surfaces but is concentrated at the fiber contacts.

Finer fibers are more readily bonded by small particles than are coarser fibers for the same reason that one can bond a mass of hair into a mat with glue whereas glue does not bond coarse rope together well. Something is needed to bridge the gap in size. That is why a mat bonded with small particles will be stronger if the distribution of fiber diameters is wide.

Using fibrillar boehmite as a binder for fibrous, and also for platelike, materials, there is no critical ratio of fibrils for best film-forming behavior. Compositions containing more fibers or plates than fibrils are generally referred to as fibers or plates, bonded with fibrils, e. g., aluminosilicate fibrous thermal insulation. On the other hand, when the fibrils make up half or more of the total solids, one generally considers the fibrous or platelike material, such as mica, asbestos or clay, to be reinforcing filler for the fibrils. Since these mineral fillers generally have a density of about 2.5 g/cc, whereas a film of fibrillar alumina alone has a density not greater than 1.25 g/cc, a 50-50 volume distribution will require 1 part by weight of fibrils to 2 parts by weight of these materials.

Table 10-9. Typical thickness of some fibers		
	Diameter	
Fiber	nm	mils
Fibrillar boehmite	5	0.0002
Chrysotile asbestos and Attapulgite (ultimate fibrils)	25	0.001
Glass fibers (ultrafine)	500	0.02
Many common fibers	5000	0.2

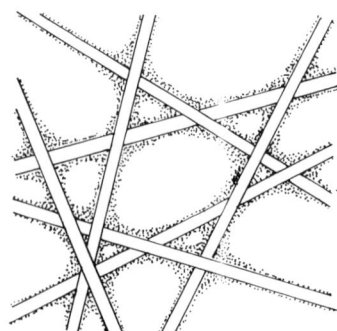

Fig. 10-18. Small binder particles cover all fiber surfaces, but are concentrated at fiber contacts.

Fibers form films of more open structure than platelets—compare the "pick-up-sticks" structure shown in Figs. 7-4 and 7-13 with a random pile of playing cards. Thus these two materials will require different amounts of fibrils, or any other type of film-forming small particles, to fill the pores between the fibers or the platelets. For greatest strength, it is desirable to have the most binder (small particles) that can be included without interfering with the closest packing of material being bonded. For fibrous materials such as asbestos and aluminosilicate fibers, 1 part of small particles, e.g., fibrillar boehmite, to two parts of asbestos or aluminosilicate fibers gives the greatest strength. For plate-like materials such as mica, up to 1 part of small particle binder may be used for 3 to 4 parts of platelets.

Conversely, chrysotile asbestos can be used as reinforcing filler in films of fibrillar boehmite. As little as 5 to 10% of the fibrous filler will give a marked improvement in film-forming behavior and strength. When using fibrillar boehmite as a binder, bonds are formed on drying at low temperatures, but much stronger bonds are obtained by heating at 400-500° C and preferably at 1000-1200° C.

Colloidal silicas are excellent binders in fiber composite systems for refractory fiber insulation. Aluminosilicate fibers are generally used in such composites, but other inorganic fibers such as mineral wool, glass, quartz, alumina, zirconia, mullite, and asbestos fibers can also be used. The fiber is slurried in water and colloidal silica, e.g., Ludox® HS-40, SM and AS, is added. The fiber insulation is vacuum formed on a screen mold and dried at about 120°C. Strength is developed after firing at about 1150° C for 2 hours.

3.3. Small Particles as Binders for Spherical or Rounded Particles

The most effective bonding of rounded particles, e.g., alpha alumina powder, is accomplished if the particle size distribution is such that the particles can pack to as dense a structure as possible—see Fig. 10-19—and fill the remaining small pores with a dispersion of film-forming small particles. On drying, these particles will bond the larger particles strongly together at their necks and points of contact.

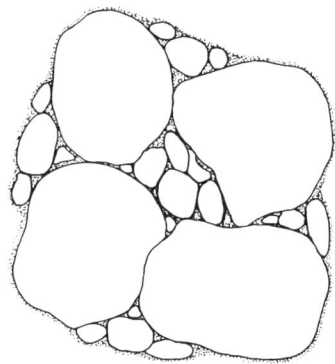

Fig. 10-19. Closer packing - Greater bond strength.

Thus, there are alpha aluminas available which require only 20-30 cc of water to make a fluid slip with 100 g of powder. If the water used to form the slip contains, e. g., fibrillar alumina—in general the ratio of fibrils to alpha alumina should be in the range 1:25 to 1:4—the dried composition will form a strong structure when it is fired at about 1200° C.

Many alumina powders and many inorganic pigments or fine dusts, consist of porous aggregates, which are not very strong and tend to shrink upon drying. Small particle binders do not bind such powders well to form coatings and films.

Porous powders can be identified by measuring the minimum volume of water required to make a fluid slurry, see also impregnation to incipient wetness in Chapter 7. Many powders of high specific surface area, indicating a porous structure, require from 50 to 100 cc of water to make a paste with 100 g of the powder. With porous particles of this kind, the reason for the low bond strength obtained with film-forming small particles is that much of the binder is drawn into the internal pores of the particles and only a small fraction remains at the points of contact between the particles to form a mechanical bond.

Such porous particles, when added as a dry powder, will thicken sols of binder particles by withdrawing water. If the pores are submicroscopic—as is usually the case if the specific surface is over 50 to 100 m^2 per gram—the water is drawn into the pores, but the small particles remain on the outside as a gelatinous sheath. Mixing is facilitated by first moistening the powder with water before mixing with the binder sol.

In ceramic investment casting, which simplifies the manufacture of complex metal parts, silica sols are used as binders for ceramic powders, e. g., zircon or fused silica. Detailed parts of complex design can be cast in a wide variety of metals and alloys. The technique is described in Du Pont's brochure "Ceramic Shell Investment Casting with Du Pont Ludox®" and in a series of patents, Moore (1973). In conventional precision investment casting, also called the "lost-wax" or disposable pattern technique, a disposable pattern, which is a replica of the part to be cast, is dipped into a refractory slurry consisting of a suspension of fine refractory grain, often zircon, with approximately 75% of the particles passing a 325-mesh sieve, in a bonding sol, often a silica sol. The disposable pattern is usually made of wax or plastic. After dipping, the excess slurry is drained from the coated pattern and, while the coating is still wet, it is stuccoed with a second slurry consisting of coarser refractory particles, also often zircon, but with particles in range from 80 to 140 mesh.

The stuccoing can be carried out by dipping the pattern coated with the first slurry into a fluidized bed of the second slurry or by sprinkling this slurry onto the pattern. The process of dipping and stuccoing is repeated until a refractory shell of sufficient thickness to resist stresses incurred in subsequent casting operations is built around the pattern. A common thickness of the shell is from one-eighth to one-half inch, but thinner or thicker shells may be made. The finished shell around the pattern is usually dried under ambient conditions for 24 hours, after which time the disposable pattern is removed from the shell mold in flash dewaxing furnaces, steam autoclaves, or boiling solvent baths: The refractory shell mold is then fired between 925 to 1050° C to prepare it for metal casting.

To increase the rate of buildup of the shell, the pattern can alternately be dipped in reagents containing oppositely charged materials such as a slurry of positively charged particles, in which positively charged alumina coated colloidal silica is the binder, followed by a slurry of negatively charged particles, in which regular, negatively charged colloidal silica is the bonding agent. This modification of the conventional precision casting process does not require drying after each spray-coating or dip since the slurries of oppositely charged particles coagulate and gel each other.

Iler (1979) gives several examples of inorganic binders, stiffeners, and molded refractory bodies, in which rounded particles are bonded together by conventional, anionic colloidal silica or surface modified colloidal silica.

4. Anchoring Agents, Mordants, Surface Modifiers

In Section D of Chapter 6 it was described not only how the surface of small particles could be modified in different ways, but also how the surface of various materials could be modified by treating them with small particles. Thus, microcoatings of small particles may be deposited on various inorganic and organic substances to modify surface characteristics. Such extremely thin coatings can be accurately controlled in thickness from monoparticle to multiparticle layers, see also those sections of Chapter 7 dealing with catalyst preparation.

Since the surface of most materials, natural and man-made, is negatively charged, it can most readily be modified or anchored to other materials by treatment with positively charged small particles such as particles of alumina, fibrillar or corpuscular, and of silica, charge-reversed by adsorption of polycations of aluminum or cationic polymers.

Very thin films of such positively charged particles, which have been exhausted from dilute sols onto various substrates, may be employed to anchor additional coatings to achieve other surface effects. Long chain fatty acids and different substituted acids, e.g., stearic or perfluorooctanoic acids, may thus be adsorbed from dilute solutions to confer water and grease repellancy. Acid polymers, for example, acrylic and methacrylic copolymers, can also be applied in this way. Acidic and direct dyestuffs, siliceous and other negatively charged inorganic pigments or colloids, and anionic polymer latices may be deposited in thin coatings on surfaces, charge-reversed by the adsorption of cationic, small particles.

Finely divided materials such as inorganic and organic powders and fibers can be surface coated with cationic small particles so as to alter their properties. The treated materials can be further modified by reaction with compounds that form strong chemical bonds with the small, positively charged particles. Such particles can thus be used in methods for altering surface charge and chemistry, modifying electrical and thermal properties, and allowing reactions with compounds normally not combining with the original material. The strong positive charge carried by the small particles in aqueous dispersions permits rapid exhaustion of the particles onto negative surfaces. Total exhaustion of the particles would be employed in beater treatment of organic and inorganic papers. Dipping and rinsing to remove excess particles would be useful in treating smooth surfaces like glass, porcelain, and many polymer films, and metals.

The thickness of the coating of cationic particles can be controlled within certain limits, which depend on the chemical nature, surface area and surface charge of the substrate as well as on the nature of the solution. Exhaustion of small particles is also sensitive to concentration and pH. In the case of alumina particles, fibrillar as well as corpuscular, and silica particles coated with alumina (polycations of aluminum), monolayer deposition occurs from very dilute sols at pH below about 7.0. Multilayer adsorption occurs at higher pH, where the particles may aggregate and precipitate onto the substrate surface. Silica particles coated with cationic polymers may deposit as monolayers at pH above 7 since they exist as stable sols at higher pH than particles of alumina or particles coated with alumina.

When mordanting negatively charged substrate surfaces with cationic small particles, and thereby reversing the charge on the surface from negative to positive, it is important that the system does not contain phosphate or sulfate ions, or organic polyvalent ions like tartrate or citrate, which will immediately precipitate the small particles and make them inactive. In the same way, anionic detergents, organic sulfonates or soaps will precipitate positively charged small particles.

In Chapter 4 it was described how filler particles of clays such as kaolin, attapulgite and bentonite could be dispersed to high solid slurries by adsorbing polycations of aluminum $Al_{13}O_{14}(OH)_{24}(H_2O)_{12}^{7+}$, onto the surface of the particles.

B. Applications

1. Ceramics

Ceramic materials have always been very important to mankind and the development of civilization. Already 20,000 years ago man made figurines, pottery and building materials of clay. Four to five thousand years ago the Egyptians had developed the technology to the point where they could fire their pottery in kilns at temperatures above 800°C. Traditional ceramics play an important role in modern society and in the last 25 years advanced ceramics have established themselves as outstanding engineering materials in many demanding and sophisticated applications.

Small particles are the starting material in ceramics processing, and here we will bring out the small particle aspects of ceramics technology.

For a more comprehensive coverage of ceramics and ceramics technology we refer to the works by Reed (1995), Richerson (1992), Pugh & Bergström (1994), and Yanagida et al. (1997).

1.1. Definition

The broad term "ceramic" is defined as an inorganic, non-metallic material. This wide definition includes natural stone, and concrete among ceramic materials. According to a more narrow definition ceramics are inorganic, non-metallic materials processsed or consolidated at high temperatures (> 600°C). That is, concrete and natural stone such as limestone and sandstone would not be included among ceramics according to the more narrow definition, but natural stone such as granite and gneiss would. The broad definition would include ice, used as building material in very cold climates, among ceramic materials.

With the wider definition any material can thus be classified as either a ceramic, a metal or a polymer/plastic. Table 10-10 compares the properties of ceramics, metals and polymers.

Compared with metals, and above all, polymers, ceramics have an outstanding ability to withstand high temperatures, see also Table 10-11. In many applications ceramics compete with metals, over which they have several advantages.

- Some ceramics are extremely hard. The hardest materials known, such as diamond, cubic boron nitride, boron carbide, and silicon carbide, are ceramics.
- Most ceramics are highly resistant to oxidation and other chemical attack, as well as to erosion.
- As a rule, ceramics are lighter than metals, sometimes as much as up to 40% lighter. This is of primary importance in aircraft, missile and spacecraft applications, but also in automotive applications, where reduced weight conserves fuel. In a gas turbine, a lightweight ceramic rotor accelerates more rapidly than a heavier metallic rotor because it has less inertia.
- Most ceramics are made from abundant raw materials. Silicon and aluminum are among the most common elements used in ceramics and they account for 27% and 8%, respectively, of the earth's crust.
- High-tech, advanced, ceramics have the potential of being less expensive than super alloys, although they are not nearly as cheap as might be expected from the abundant and inexpensive raw materials. For example, the silica and

Material	Density	Tempera-ture Stability	Corrosion Resistance	Erosion Resistance	Brittleness	Strength	Forming Work-ability	Price
Metals	1	2	1	1	2-3	3	3	2
Plastics	3	1	2	2	3	2	2-3	3
Ceramics	2	3	3	3	1(2)	1-2(3)	1(2)	1(2)

Table 10-10. Comparison between properties of metals, plastics, and structural ceramics. (Courtesy of Professor R. Carlsson, Swedish Ceramic Institute, Gothenburg, Sweden)

Key: 1 = disadvantageous; 2 = not quite advantageous; 3 = advantageous; numbers in parentheses indicate potential of new structural ceramics

and alumina used in making advanced ceramics are not cheap, technical-grade materials. They must undergo a sequence of costly purification and processing steps before they can be used to produce high-performance ceramics. Also, advanced ceramics are today still fairly low-volume products. In the years ahead, if advanced ceramics are made in large quantities, they will be much less expensive than super alloys.

- Because of their low coefficient of friction, high compressive strength, and wear resistance, some ceramics can be used in bearings and other moving mechanical parts without requiring lubrication.

1.2. Classification

The classification of ceramics into the four groups shown in Fig. 10-20 is based mainly on differences in the methods of manufacturing or forming of these ceramic materials.

As the name suggests, melt based ceramics are made from a melt of the raw materials. Glasses may be the most well-known materials in this group. Inorganic glasses can be made from melts of many compositions in the general areas of silicates, phosphates, aluminates, borates, halides, and chalcogenides. Common glass, i.e., the glass used in window panes, bottles and jars is made of lime, soda, and quartz sand. Pure SiO_2-glass, quartz glass, is used in crucibles and furnace windows due to its high acid resistance and low thermal expansion coefficient. Hair-thin fibers of super-pure glass can carry thousands of times more information than a conventional copper cable.

A glass melt will solidify to a continuous and rigid structure, which, unfortunately, is brittle and susceptible to mechanical failure. When a fracture is introduced it can, despite the great strength of the covalent Si-O bond, propagate through a large mass of glass. At a sharp crack tip there is virtually no plastic deformation, which results in an enormous increase of stress.

However, by crystallizing glass, new high-performance materials called glass-ceramics can be produced. Nucleating agents, in the form of small particles of, e.g., noble metal, titanate, zirconate, and fluoride, are homogeneously incorporated into the glass melt. When the solidified glass is reheated, they precipitate very efficiently within the body of the glass since they are only a few hundred angstroms apart. The bulk of the material, usually a silicate, crystallizes on these tiny nuclei. A silicate crystal on one of these nuclei will grow until it bumps into neighboring silicate crystals, creating a highly crystalline material with a very small amount of residual amorphous glass. Where the crystals meet, structural discontinuities or grain boundaries are created that may cause deflection, branching or splintering of cracks. Several glass-ceramics, e.g., lithium aluminum silicate, and cordierite, have very low thermal expansion, making them very useful as stove-tops and catalyst supports in catalytic converters for cars. Glass ceramic rods of nepheline, a sodium aluminosilicate, can be strengthened by ion-exchange-induced surface compression to obtain modulus of rupture values of 300,000 psi and a strength-to-weight ratio greater than that of steel.

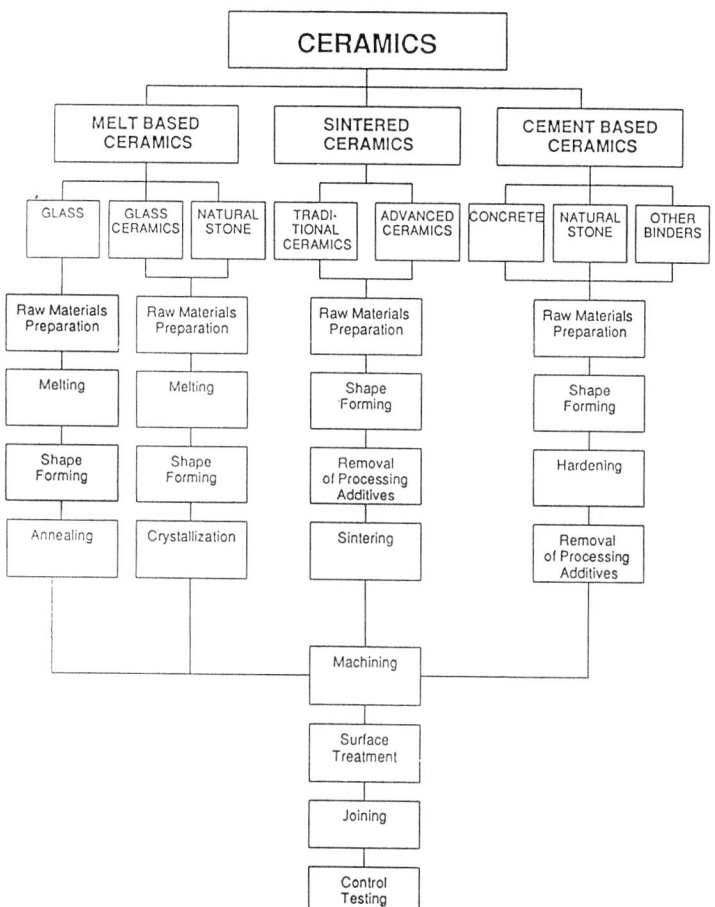

Fig. 10-20. Classification of ceramics. (Courtesy Prof. R. Carlsson, Swedish Ceramic Institute, Gothenburg, Sweden).

Natural stone, such as granite and gneiss, is formed by solidification of magma from the interior of the earth. Natural stone of the granite family is composed of quartz and feldspar and also contains minor accessory minerals such as biotite, muscovite, hornblende, and pyroxene. Gneiss has the mineral composition of the granite family, but has a texture that shows distinct planar or linear properties due to the parallel alignment of grains.

Sintered ceramics include advanced ceramics, which will be discussed separately below, and traditional ceramics. The starting material, a powder blend, is compacted and given its geometrical shape at room temperature. The shaped, unfired, and, as the case may be, dried component is porous with a pore volume of about 50%. In the firing-sintering step it obtains its final form and strength size. The sintered material is preferably compacted to density with no pores at all.

All traditional ceramics contain clay minerals—see Chapter 4—and products made from them include floor and wall tile brick, domestic tableware, sanitary ware, refractories and electrical insulators. The most widely used clay mineral is kaolin. Traditional ceramics also contain feldspars, usually the three silicate minerals $KAlSi_3O_8$, $NaAlSi_3O_8$, and $CaAl_2Si_2O_8$, and fillers, often quartz or α-alumina. Obviously, an elegant teacup of chinaware looks quite different from a brick, but the raw material, usually clay, and the underlying chemistry in their production are quite similar. The clay, usually kaolin in the form of a thick slurry of small particles as the starting material, loses chemically bound water in the firing step and the particles adhere into a rigid solid; see thermal reactions of kaolin in Chapter 4.

The third group of ceramics, cement-based ceramics, comprises concrete, natural stone, and other binders. They are formed at ambient temperature, e.g., by casting, and rely on chemical reactions at low temperature, and not on melting or sintering at high temperatures, to obtain their final strength. Since shrinkage is low, cement-based ceramics are always quite porous. Cement chemistry allowed Etruscan and Roman builders, as it allows us today, to readily convert limestone rock to durable bodies of many different shapes for many uses and application in the building industry and road construction. Limestone, for cement production in the form of finely ground powder, decomposes at temperatures above 600° C—in modern cement production temperatures between 1200 and 1500° C are used—into lime and CO_2 according to reaction.

$$CaCO_3 = CaO + CO_2$$

Lime reacts with water to give slaked lime,

$$CaO + H_2O = Ca(OH)_2$$

which can be cast into the desired shape. The product or component of slaked lime, hardens by reacting with atmospheric carbon dioxide to form $CaCO_3$, i.e., limestone

$$Ca(OH)_2 + CO_2 = CaCO_3 + H_2O$$

Modern cement, e.g., Portland cement contains more components, such as Ca compounds of silicon, aluminum and iron oxides, than the early lime-based materials, but the basic chemistry leading to particles bonded together at ambient temperatures into strong, rigid structures, is similar. Another important difference between Portland cement and the primitive cement is that the former will set and harden under water. The term mortar refers to a slurry of cement powder in water. Concrete is made of mortar containing stone, i.e., mineral particles exceeding about 2 mm in diameter—if the particles are smaller in size they are referred to as sand.

Examples of cement-based natural stone are lime stone and sandstone.

Inorganic binders used in cement-based ceramics are alkali-silicates, silica sols, and tetraethylorthosilicate; see Chapter 2.

Ceramic coatings constitute the fourth group of ceramics and will be described in a separate section.

1.3. Advanced Ceramics

Small particle technology plays a very important role in ceramics technology in general, but is especially important to the technology of advanced ceramics. Such ceramics will therefore be covered in somewhat more detail than traditional ceramics.

1.3.1. Classification and Definition.

Advanced ceramics can be classified as structural ceramics, electronic ceramics, and coating ceramics.

Typical structural ceramics are made of such materials as silicon nitride, S_3N_4, silicon carbide, SiC, zirconia, ZrO_2, boron carbide, B_4C, alumina, Al_2O_3, and sialon, SiAlON. Structural ceramics are used as cutting tools, bearings, wear components, components in combustion engines, and heat exchangers. The high value-in-use of structural ceramics in these applications depends on their high hardness, outstanding corrosion and chemical resistance, ability to withstand high temperatures and low-density.

Electronic ceramics include such materials as barium titanate, $BaTiO_3$, zinc oxide, ZnO, lead zirconate titanate, $Pb(Zr_xTi_{1-y})O_3$, lithium niobate, $LiNbO_3$, aluminum nitride, AlN, and alumina, Al_2O_3. They are not valued so much for outstanding mechanical properties and durability as for electrical properties such as relative permittivity and permeability, and electrical and thermal conductivity for power dissipation. Applications of electronic ceramics include insulators, capacitors, integrated circuit systems, magnetic ferrites, piezoelectric materials, semiconductors and superconductors.

Coating ceramics will be covered in a separate section.

1.3.2. Raw Materials.

Sintered ceramics are traditionally made from powders via solid state reactions at high temperatures and during rather long times. The properties of the small particles making up the powders for advanced ceramics must meet very stringent specifications. Thus the particles should have a uniform shape (spherical), be smaller than 1 μm, and have a narrow particle size distribution. The nature of the particle surface plays an important role in all the manufacturing steps, although it is not yet completely known in what way and how the final properties are affected; see also the subsection "ultrastructure processing" below.

Chemical purity and, of course, the cost, are important. Powders meeting the strictest specifications can be made today, but at a cost that is too high for the great majority of applications.

More than 90% of the powders used for the manufacture of ceramics today are made by grinding. Using special ball mills, the particle size can be reduced to about 1μm, Becker (1987).

Ceramic powders can also be produced by wet chemical methods of the type described in Chapter 3 and in Section D of Chapter 6. Thus, powders of zirconia, and highly pure barium titanate, mullite, and lead zirconium titanate are produced in this way; see also the section on sol-gel synthesis below.

Very fine, highly pure powders of SiO_2 and TiO_2 are produced commercially in large quantities by flame hydrolysis of $SiCl_4$ and $TiCl_4$, respectively. High quality powders of many ceramic materials can also be made by gas phase reactions, but the cost is still too high for commercial applications.

Mechanical and physical properties of some structural ceramics are shown in Table 10-11 and compared with those of hard metal and stainless steel.

The outstanding hardness of ceramics stems from the nature of their interatomic bonding, which is partially covalent and therefore directional. Thus, slipping of one crystal plane over another in a ceramic is difficult to bring about, and ceramics cannot readily be deformed. Metals, on the other hand, are malleable because their bonding is non-directional. One can hammer a chunk of iron into shape, but not a piece of ceramic.

The nature of their bonding is also the reason why ceramics are brittle materials, although some are less brittle than others. Thus, yttrium stabilized zirconia is not much less brittle than hard metal.

Ceramics are usually thought of as being electrically non-conducting materials, that is insulators, but Table 10-11 shows that the resistivity of titanium diboride is about 1000 times lower than that of stainless steel.

Usually ceramic are insulators; thus heat conduction must be accomplished by phonons. The more perfect the crystalline structure of the ceramics, the higher the thermal conductivity since there are fewer imperfections to scatter the travel of phonons through the material. The almost perfect crystal structure of gem-quality diamonds makes them conduct heat well. Similarly, sapphire, a single crystal of alumina, is a good heat conductor, whereas brick, a conglomerate mass of many small crystals of alumina (corundum), is an insulator.

1.3.3. Processing. Figure 10-20 shows that many steps in the processing of melt-based, sintered, and cement-based ceramics are similar, at least in their objective if not in design and execution. We are mainly concerned with advanced ceramics and focus on the steps of shape forming and sintering.

Shape forming. There are three principle methods of shape forming based on the techniques of pressing, plastic forming, and casting, respectively.

A dry powder, which may or may not be plastic, is normally used in shape forming by pressing, e.g., isostatic pressing.

Plastic forming utilizes a plastically deformable powder. Plasticity may be achieved by making a paste with water or by adding plastic, often polymeric, process aids to the powder. Extruding and injection molding are examples of methods of plastic forming.

Table 10-11. Mechanical and physical properties of some common structural ceramics. (Courtesy Prof. R. Carlsson, Swedish Ceramic Institute, Gothenburg, Sweden)

Property	Si_3N_4	ZrO_2 (Y_2O_3)	SiC	Al_2TiO_5	TiB_2	AlN	Al_2O_3 Aluminum oxide	Al_2O_3 porcelain	Hard metal WC/Co	Stainless steel
Flexural strength, MPa	800	1000	400	30	600	350	350	185	2500	200 (σ_s) 490 (σ_B)
Compressive Strength, MPa	3000	2200	2200	-	-	-	4100	950	5000	200 (σ_s) 490 (σ_B)
Young's Moduls, GPa	300	210	400	20	570	310	390	125	600	200
Hardness, Hv	1600	1250	2500	-	2200	-	1800	800	1400	200
Fracture toughness, MPa m$^{0.5}$	6.5	10.5	4.0	-	-	-	4.5	1.4	12	~50
Resistivity, ohm- cm	10^{14}	25×10^{12}	10^{-1}	-	1.5×10^{-7}	10^{14}	10^{14}	10^{14}	20×10^{-6}	10^{-4}
Thermal expansion, 10^{-6} K^{-1}	2.5	10.5	3.5	1.0	7.2	4.5	7.5	5.0	5.0	16.8
Thermal conductivity, W m^{-1}k^{-1}	25	1.5	100	1.4	110	150	30	2.6	95	15
Thermal shock resistance, ΔT(°C)	700	200	200	1200	110	190	80		625	-
Maximum temperature with retained strength, °C	1400	300	1500	-	-	-	1000	-	-	-
Density g/cm^3	3.2	5.9	3.2	3.4	4.4	3.3	3.95	2.85	15	7.9

The third principle method of shape forming is casting, which requires that the ceramic powder is dispersed in water to a high solids slurry, or slip. Thus, fine-grained alumina can be slip-cast at almost 60 vol. % solids, and plastic fireclay can be cast at nearly 50 vol. % solids. Casting can also be used to shapeform melt based ceramics.

Sintering. Sintering involves heating the formed body so as to compact it to a ceramic product with good mechanical and physical properties. In reaction sintering a precursor of the ceramic material is made to react and form the ceramic during the sintering process. Thus, a shaped body of silicon can be reacted with nitrogen to silicon nitride. Reaction-sintered ceramics usually have some remaining porosity or contain incompletely reacted material. This method is therefore primarily used to make cheaper ceramics, of which fully developed mechanical properties are not expected. By applying pressure, shaped bodies can be sintered to completely dense ceramics. Hot pressing and hot isostatic pressing are examples of sintering techniques which can yield ceramics of the highest quality.

1.3.4. A Ceramic Car Engine - A Dream? In an evaluation of the prospects for new ceramics, Alexander (1976) used the Carnot efficiency factor as the starting point for his analysis. The efficiency of heat engines, i.e., steam, internal combustion and turbine engines, increases with the difference between the temperature inside the engine and the temperature of the environment. In engines made of metal, engineers had worked hard and long to increase this temperature difference by testing different metals and compositions of metals. However, even the most temperature resistant alloys began to soften and corrode at temperatures in the range from 1600 to 1900°F.

On the other hand, engineers were well aware that ceramic materials could withstand much higher temperatures than metals and that ancient man routinely fired his pots in flames hotter than 2000°F. At the time many companies were looking into techniques for mechanical engineering with ceramics, Westinghouse, for instance, was working on ceramic stator vanes for a 40000-horsepower turbine that could be used either for propelling ships or for generating electric power. Vanes of silicon nitride were found to easily survive tests at 2500°F, whereas metal vanes could not withstand more than 1650°F in this application. Automotive engineers were dreaming of the "pottery engine", which, if it could be built to operate at temperatures of 3000°F, would reduce the fuel consumption by half.

This was at the time of the energy crisis and the federal Energy Research and Development Administration, ERDA, decided that the quickest way to save energy was to improve gas mileage. Together with NASA, which had much experience with high temperature research, ERDA proposed a 350 million dollar auto-engine development program. The objective was to develop a ceramic turbine engine capable of operating at 2500°F and have it ready for use by 1985, which was to be followed by a 3000°F engine sometime later. NASA Studies indicated that the 3000°F engine would be equal in performance to the eight-cylinder engines of the day and get about fifty-one miles per gallon of gasoline.

Today, more than 20 years later, no such engine is available, although many car engines contain light-weight, durable ceramic components such as valves and vanes. Although it is possible to make a small number of engines under such controlled conditions that they will perform well in a car, it is not possible to cost-effectively make a large number of ceramic engines with reliable performance. The reason is that it is very difficult to make large ceramic components, the structures of which are sufficiently free of defects. Defects in materials in general, and in brittle materials as ceramics in particular, will cause failure under stresses far below those that can be estimated from theory for the material.

1.3.5. Failure of Materials. The fact that ceramics are inherently brittle, which is a consequence of their incomparable hardness, is not the same thing as saying that they are weak. However, cracks initiated at a defect, propagate much more readily through a ceramic than a metal. Applied stresses will concentrate at the tip of a flaw, and, due to the unyielding nature of ceramics, wedge and propagate it right through the material. By contrast, the ductility of most metals lets the crack tip blunt itself and dissipate the forces.

It is well known by mechanical engineers that there is a difference in strength between large and small components or constructions of the same material. Already Gallileo, analyzing this problem, noted: "One cannot reason from the small to the large, because many mechanical devices succeed on a small scale that cannot exist in great size", Drake (1974). In a discussion on how materials break, Marder and Fineberg (1996) observe "that the strength of solids calculated from an excessively idealized starting point comes out completely wrong; it is not determined by performance under ideal circumstances, but instead by the survival of the most vulnerable spot under the most adverse conditions".

Using a very simple model, Marder and Fineberg (1996) estimated the stresses at which materials would break. A material in the form of a rod, of height h and cross-section A, is pulled by a force F. The rod increases in length by an amount δh, and breaks as a single unit when δh exceeds a critical value; that is when the atoms of the material are pulled beyond the breaking point. The stress, σ, on a material is related to its extension, δh, through the equation

$$\sigma = \frac{F}{A} = \frac{\delta h}{h} Y \qquad (10\text{-}10)$$

where Y is Young's modulus.

Marder and Fineberg (1996) assumed that the rod broke when the atomic bonds were stretched by 20% of their original length.
The stress at break, σ_B, is then

$$\sigma_B = Y/5 \qquad (10\text{-}11)$$

Table 10-12. Practical and theoretical strengths of materials. Marder and Fineberg (1996). Courtesy American Institute of Physics					
Material	Young's Modulus y 10^{11} dyne/cm^2	y/s 10^{11} dyne/cm^2	Theoretical strength 10^{11} dyne/cm^2	Practical strength 10^{11} dyne/cm^2	Practical stregth/ theoretical strength x 100
Iron	16	3	3	0.085	2.8
Copper	19	4	3	0.049	1.6
Silicon	18	4	3	0.062	2.0
Glass	7	1	4	0.002	0.05

Table 10-12 shows that the theoretical strength, as estimated from Eqn. 10-11, for iron, copper, and silicon is about fifty times larger than the practical strength.

For glass, which is a ceramic, the theoretical strength is about 2000 times larger than the practical strength. Marder and Fineberg maintained that these large discrepancies were not due to the crude approximations used to obtain Eqn. 10-11 since much more sophisticated quantum mechanical calculations would give similar results; but that there was something else involved, namely, defects or flaws. Since stresses on a material concentrate on defects, the practical strength of materials is determined by the presence and the size of flaws. Inglis studied the distribution of stress in a large plate with a defect in the form of an elliptical hole of length, L, and radius of curvature, ρ, at the narrow end. He found that the stress, which was uniformly applied to the material far from the hole, was increased by a factor $2(L/\rho)^{1/2}$ near the narrow end of the hole.

Flaws may, therefore, cause a material to fail by increasing a stress on the material to the destructive limit, the break stress, at the tip of the flaw, whereas the bulk of the material safely remains below this limit. Slit-like defects of a given radius of curvature at the narrow end, say 1 nm, and of lengths 0.1, 1, and 10 μm would increase a stress on the material by a factor of 20, 63, and 200, respectively, at the tip of the defect. A very tiny scratch of length 100 μm and radius of curvature 0.1 nm on the surface of a glass rod would increase the stress on the rod at the narrow end of the defect by a factor of 2000, which is also the difference between the theoretical and practical strength of glass, Table 10-12. See too § 6 in Applications for a discussion of small particle strengthening of metals.

1.3.6. Principles for Making High-Performance Ceramics. Since the performance of ceramics depends so critically on microstructure, they must be manufactured under conditions that will guarantee that the structure does not contain defects or irregularities of a size above a critical value, which certainly should not be above 1 μm. This can be accomplished by what is called ultrastructure processing.

Ultrastructure processing. According to Ulrich (1984) "ultrastructure processing refers to the manipulation and control of surfaces and interfaces during the earliest stages of formation at scales of 50-1000 Å through chemical processes". The result will be high performance ceramics with predictable properties.

Figure 10-21 indicates that poor control of powder quality will result in a structure that, even after sophisticated processing involving high temperatures and pressures, will contain many defects in the form of pores. Controlling the powder geometry, that is, using powders of particles of uniform size and shape (spherical), for instance, made by the methods described in Chapters 2 and 3, will improve the packing density of the powder. Dispersing the powder well in a liquid medium prior to shaping and sintering gives much better control of the density and microstructure of green and sintered compacts than dry pressing of the powder, Hirata (1997) and Russel (1991).

Best performance, that is, ceramics with the densest and most homogenous structures, are obtained by also controlling the powder chemistry. The most obvious and direct way to accomplish this is to add special additives to the slurry of well-dispersed small particles of uniform size and shape. Another approach involves modifying the surface of the powder particles, e.g., by the techniques described in Ch. 6. Still another way of controlling the chemistry of the particle surface is to employ the technique of mechanical alloying, Matteazzi et al. (1997) and De Barbadillo (1993). Grinding can be used, not only to reduce the size of particles of different materials, but also to synthesize advanced materials in the form of small particles of composite structure by the process of mechanical alloying. A large variety of materials with a diversity of structures can be synthesized by ball milling mixtures of particles (mechanosynthesis). See also Ch. 1, § 7.1 on fine grinding.

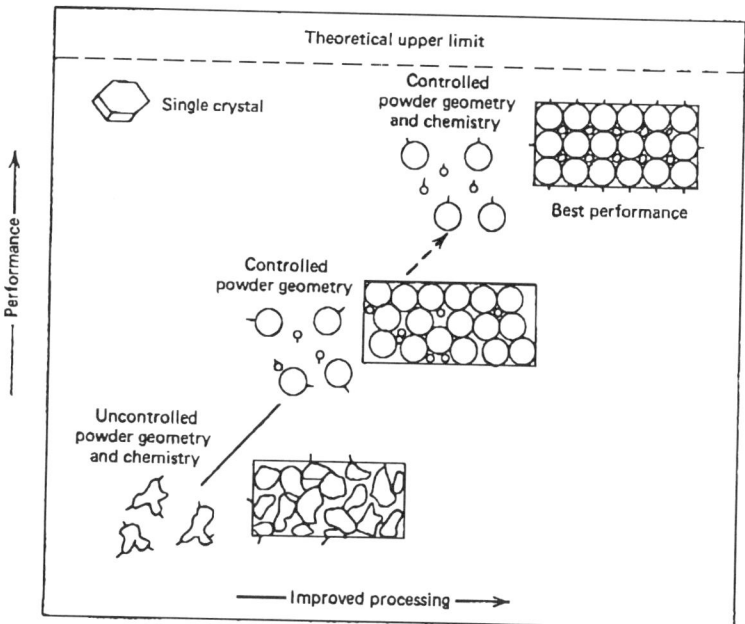

Fig. 10-21. Importance of particle geometry and chemistry in ultrastructure processing of ceramics. Ulrich (1984). Reprinted by permission of John Wiley & Sons, Inc.

Barringer et al. (1984), discussing ultrastructure processing of monosized powders, pointed out that the ideal process of making high performance ceramics requires: (a) an unagglomerated powder of monosized particles, preferably 0.1-1.0 μm in diameter; (b) a green microstructure with a uniform distribution of pores and the particles as densely packed as possible; and (c) choice of variables of temperature, pressure and composition such that densification rather than particle coarsening occurs.

In some very demanding applications even ultrastructure processing may not bring the ceramic sufficiently close to its theoretical strength limit, and in others the cost of ultrastructure processing may simply be too high. A second reinforcing method, transformation toughening, can then be used to supplement or replace ultrastructure processing.

Reinforcement by transformation toughening. Partially stabilized zirconium dioxide, PSZ, containing yttrium oxide is one of the strongest ceramics commercially available for applications in the temperature range up to 300° C; see Table 10-11. It consists of a matrix of tetragonal ZrO_2 containing from about 20 to 50 volume % of metastable tetragonal ZrO_2 particles. When cubic ZrO_2, the starting powder, is heated, Y_2O_3 promotes the transformation of some of the cubic crystals to tetragonal ZrO_2 crystals, Garvie (1975). Magnesium and calcium oxides can also act as promoters of the tranformation of cubic to tetragonal zirconia.

The improved crack resistance of PSZ is associated with the 3 to 5 percent increase in volume that occurs when stresses at a crack tip induce transformation of tetragonal to monoclinic zirconia. Thus, an advancing crack approaches tetragonal grains embedded in a monoclinic zirconia matrix and causes them to expand. The ensuing compression of the adjacent matrix stops the crack by pinching it shut.

Alumina ceramics can also be transformation toughened by incorporating into the material about 15 volume % of fine particles of tetragonal zirconia, preferably smaller than 1μm in diameter, Claussen (1976). The mechanism of reinforcement is the same as for PSZ. Tetragonal zirconia can also be used to strengthen other ceramics such as spinel, $MgAl_2O_4$, and mullite, $3Al_2O_4 \cdot 2SiO_2$.

Strengthening by microcracks. The analysis of why materials fail, showing that stresses at the tip of a crack increases by a factor $(L/\rho)^{1/2}$—see above—suggest an elegant method of toughening by introducing microcracks into the structure. When the tip of a crack reaches a microcrack, it is immediately blunted; that is, the tip runs into a small hole and its radius of curvature increases abruptly, causing a sharp decrease in the stress-increasing factor $(L/\rho)^{1/2}$. The microcracks, essentially small cavities with a diameter preferably smaller than 1 μm, are deliberately dispersed throughout the material in the course of its processing.

Fiber reinforcement. Ultrastructure-processed ceramics, transformation-toughened ceramics, and ceramics strengthened by microcracks come much closer to their theoretical strength than other ceramics, but they are still brittle materials. Another way to strengthen ceramics, which also significantly increases their resistance to brittle failure, is to interlace fine ceramic fibers in them. The thickness of the fibers falls typically between 5 to 10 μm, and their length can vary from a few millimeters to long continuous fibers. Whiskers are also used for reinforcing ceramics, although they do not increase the resistance to brittle failure so much as longer fibers do. As in a polymer-fiber composite, the ceramic fibers span cracks and thus prevent them from widening and growing.

1.3.7. Sol-Gel Science and Technology. Compared with conventional processing, in which ceramics are generally molded from plastic masses of powders, often wetted by a liquid or dispersed in a liquid, sol-gel science and technology brings more chemistry into ceramics processing. A molecular precursor or a mixture of molecular precursors in solution in converted by a chemical reaction to a sol or a gel, which by different routes can be processed to a dense ceramic.

Sol-gel technology was briefly discussed earlier in the text—preparation of Stöber silica sols in Chapter 2, and of monosized particles of other metal oxides in Chapter 3, and of aerogels, primarily of silica, in Section 1 of this chapter. For detailed information on the many aspects of sol-gel science and technology we refer to the excellent review by Brinker and Scherer (1990).

One can distinguish between two principally different uses of sol-gel science and technology in processing advanced ceramics. In the first route, organic compounds of metals, usually alkoxides of such metals as aluminum, silicon, iron, zirconium and titanium, are hydrolyzed in solution to very small particles, usually only 3 to 4 nanometers in size, of the hydrous oxides. In sufficient concentration, these small particles will link together in chains and form three dimensional networks that fill the liquid phase as a gel. The gel is then dried and heated to a dense ceramic.

One advantage of sol-gel processing is that sol-gels fire to dense ceramics at lower temperature than conventional ceramics. Another advantage is that very intimate and uniform mixtures of different colloidal oxides can form gels that are molecularly dispersed. Also, the gel can readily be molded, cast, or spun into the final desired object.

Disadvantages of the sol-gel method are high cost of raw materials, large shrinkage during processing, residual carbon and long processing times. However, this principal application of sol-gel processing has great potential for making thin films, coatings and fibers.

The other principal use of sol-gel methods for making advanced ceramics involves the preparation of colloidal particles of uniform size and shape (spherical). These particles are then used to form a green body with minimum pore volume and uniform pore diameter, which can be sintered to density. The starting material is again alkoxides of metals, which are hydrolyzed under conditions that promote the formation of particles rather than gel; see Section B of Chapter 2.

Thus, Barringer and coworkers (1984) at MIT developed a process for making finely dispersed titanium dioxide by hydrolyzing titanium tetraethoxide,

$Ti(OC_2H_5)_4$; see also Section C of Chapter 3. This method yields TiO_2 powder of spherical monosized particles in the size range from about 0.3 to 0.6 μm, which has been used to produce ceramics with more than 99% of the theoretical density. Superpure fine powders of the oxides of silicon, zinc, zirconium, and aluminum can also be prepared by this method.

Thick slurries of these finely divided metal oxide particles can also be prepared with solids contents ranging from 30 to 50 vol. %, which can be cast into various shapes by different methods, e.g., slip casting. This alternative use of sol-gel methods has many of the advantages of the first alternative, but somewhat fewer disadvantages. Thus, multicomponent materials can readily be made but shrinkage during processing is much less. The cost of raw materials is still high. However, particles of the same high quality can probably be made by purely inorganic reactions, i.e., hydrolysis of solutions of inorganic salts, similar to those described in Section B of Chapter 3, at considerably lower cost.

2. Coatings

The coating process modifies the surface of a material, providing a gradual difference in composition or property between the surface and the bulk. The properties of an inexpensive material can be upgraded by applying to its surface a thin coat of an expensive material. The classification of coatings in Fig. 10-22 is by no means generally accepted, but is used by the authors to bring out the usefulness and importance of small particle technology to the vitally important technology of coating materials.

For a comprehensive and detailed coverage of coating technology we refer to the modern and excellent works by Bunshah (1994) and Ohring (1992).

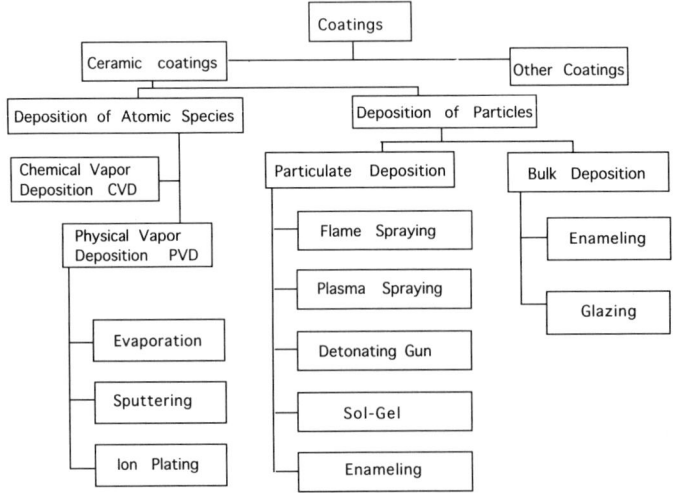

Fig. 10-22. Classification of coatings.

2.1. Ceramics Coatings

By definition, a ceramic coating involves the deposition of a ceramic material onto the surface of a substrate and heat treating at temperature above 600°C.

Ceramic coatings are extremely important in such uses as wear reduction (Al_2O_3, B_4C, SiC, TiC, WC and diamond-like carbon, DLC), friction reduction (MoS_2 and BN), corrosion reduction (Al_2O_3, Si_3N_4, and SiO_2) and thermal protection (ZrO_2, Ca_2Si_4, and $MgAl_2O_4$). They can also provide electrical conductivity (In_2O_3/SnO_2), semiconductivity (Si and GaAs), electrical insulation (SiO_2), and ferroelectricity ($Bi_4Ti_2O_{12}$).

In optical applications they can bring about selective optical transmission and reflectivity (CeO, CdS, BaF_2/ZnS, and SnO_2), and serve as optical waveguides (SiO_2), and sensors (SiO_2, SnO_2, and ZrO_2).

There is a wide range of methods that can be used to deposit ceramic coatings. Most of them fall under the categories of chemical vapor depositon, CVD, physical vapor deposition, PVD, and wet chemical techniques using liquid precursors such as slurries and sol-gels.

2.1.1. Deposition of Atomic Species.
Coatings made by deposition of atomic species do not involve much small particle technology in the sense that the term is used in this text, but they are technically and commercially very important and will therefore be briefly described.

Chemical vapor deposition, CVD, and physical vapor deposition, PVD, involves deposition of a solid from a vapor phase. CVD is typically carried out by passing a mixture of gases across the heated surface of the material to be coated. The gases will react or decompose when they come into contact with the hot surface. Thermal decomposition most often involves the use of organometallic compounds. Thus, a mixture of CH_3SiCl_3 vapor and hydrogen react in the temperature range from 1000 to 1400° C to form a coating of SiC on the heated substrate. The deposition rate is moderately low, i.e., from about 20 to 250 nm per minute. Ceramic coatings with a thickness of the order of micrometers can be applied to various substrates to provide abrasion resistance, (Al_2O_3, TiC), oxidation resistance (Al_2O_3), and friction reduction (TiN).

Physical vapor deposition includes the methods of evaporation, sputtering and ion plating and involves vaporizing material from one surface and depositing it on another. All three methods are vacuum processes and can be either of the direct type or the reactive type. In the direct type the evaporant is the ceramic material itself whereas in the reactive type a metal or a compound in a low valence state reacts with a reactive gas to form a compound deposit. The deposition rates of evaporation and sputtering can be very high, up to 75 and 25 μm/min, respectively, whereas it is low, < 20 nm min^{-1}, for ion plating. Evaporation is widely used for applying coatings having such electrical properties as conductivity, including superconductivity, and resistivity.

2.1.2. Deposition of Particles.
Deposition of particles to form a coating on a substrate can be accomplished either by particulate deposition or bulk deposition. Some techniques such as enameling and sol-gel can be found in either category.

Particulate deposition. Particulate deposition implies that discrete particles of ceramic materials are applied to a substrate surface, often by a spraying technique at high temperature.

In flame spraying, a fine powder of relatively uniform size in the range from 10 to 40 µm is carried in a gas stream and passed through a hot flame, where it becomes at least partially molten. The rapid heating of the gas causes it to expand and spray the molten particles onto the substrate where they solidify.

Detonating gun coating is a variant of flame spraying in which a controlled amount of powder and a mixture of oxygen and acetylene are injected into a gun. Ignition of the gas mixture heats the particles and accelerates them to high velocities before they strike the substrate surface. The procedure is repeated several times a second.

In plasma spraying particles of uniform size are injected into a plasma jet produced by heating an inert gas in an electric arc. The particles melt rapidly in the plasma and are accelerated towards the substrate surface, where they are rapidly quenched and consolidated on impact. The deposition rates of flame spraying and plasma spraying methods are very high, several tens of microns per minute. In general the densities of flame and plasma coatings vary with the particle size of the powder.

Using fine powders and high impact velocities, it is possible to achieve densities of more than 95% of theoretical. Typical coating thicknesses range from 50 to 500 µm.

Enamel is a glassy coating on various metals or a decorative coating on glass. A powder of alkali borosilicate glass is electrostatically applied to the substrate surface and consolidated by a fusion process. The glass is made by fusing the component oxides, e.g., Na_2O or K_2O from the corresponding carbonates, B_2O_3 from borax, Al_2O_3 from alumina, and SiO_2 from quartz, with adhesion-promoting oxides of such metals as cobalt, nickel, and copper between 1100 and 1400° C. The glass is ball-milled to a fineness of 95% of particles smaller than about 75 µm (<200 mesh) for the ground coat and to a fineness of 98% of the particles smaller than about 45 µm (< 325 mesh) for the cover coat.

The particles of the ground glass, the frit, are surface treated with a silicone before they are delivered to spray guns, in which each particle is provided with a negative electric charge. The thickness of the coating that can be applied increases with its resistivity. Surface-treated frit particles may give coatings with a resistivity as high as 10^{15} ohm cm corresponding to a potential thickness of at least 100 to 200 µm. Without surface treatment the coatings may have a resistivity of only about 10^6 ohm cm.

The dry frit is applied to the substrate surface, usually surfaces of steel, cast iron, and aluminum, in one or two applications, i.e., in the latter case one ground coat and one cover coat. In two-coat applications of dry frit, the cover coat can be applied directly on the ground coat. A firing step between the two applications is not necessary; compare two-coat applications by bulk deposition below. The enamel powder is consolidated to a mature coating by firing to 750 to 850° C for about 3 to 6 minutes in a continuous furnace.

Bulk deposition Bulk coating. We shall only consider bulk coating by such wetting processes in which coating material is applied in liquid form by methods such as rolling, dipping, brushing, and spraying and becomes solid by solvent (water) evaporation.

Enameling. Enamels can also be applied to a substrate, usually a metal surface, as a dispersion of frit particles in water, the slip, by a two-coat application. In the first application, the part dipped into or sprayed with the ground slip. The ground coat is dried, and sometimes also fired, before the part is coated with the cover slip. Firing is done at between 750 and 850° C for about 3 to 6 minutes.

Sol-gel coating. Stable sols of hydrous oxides, see sections on sol-gel processing in this Chapter, Chapter 2 and Chapter 3, can be applied to substrate materials by for instance dipping and spin coating. After application, the sol forms a gel which after drying and densification yields a coating characterized by a high degree of chemical homogeneity.

2.2. Other Coatings

Thin, discrete coatings of small inorganic particles can be applied to the surface of many substrates by adsorption or mordanting or by bulk depositions methods. The composition of the particles is such, e.g., SiO_2, Al_2O_3, Al-silicates, etc., that they fall in the category of ceramic materials, but the coatings made of them are not true ceramic materials since they have not been fired and densified at temperatures above 600° C. Also, in some coatings small particles may serve as binders for other small particles, e.g., pigments, and the coating formulations are similar to paints. Table 10-13 shows eight examples of such coatings.

Table 10-13. Fibrillar alumina as binder in coating mixtures. (Courtesy E. I. DuPont Co.)			
Composition number	Pigment or particle component	Film air dried or baked at 100°C. not water resistant*	Film baked 5-15 min., 450°C water resistant
1	graphite	black, adherent, slippery, conductive	harder and more adherent
2	boron nitride	white, adherent, conductive	harder and more adherent
3	Teflon® TFE fluoro-carbon resin	translucent, adherent, slippery	partly decomposed
4	clay	translucent, somewhat soft	harder and more adherent
5	mica— wet ground	translucent to opaque, adherent	harder and more adherent
6	carbon black	dull black, easily polished, high electrical resistance	fair water resistance
7	titanium dioxide mica powder	white flat paint	fair water resistance and adhesion in thin coatings
8	alpha alumina	white, opaque, soft film	adherent and still soft but hardens at 1200° C or higher. Resistant to temperatures up to 1500°C.

* Films of mixtures which contain aluminum acid phosphate have fair water resistance when well dried at 100°C.

3. Paints, Lacquers, and Printing Inks

Paints, Lacquers, and Printing Inks are coating materials that are applied to the surface of various substrates for protective decorative, informative, or some other functional purpose.

Paints and lacquers form protective and decorative coatings on wood, plastics, metals, concrete and many other materials.

Inks are used to print text, pictures and decorative patterns on paper, textile, plastics, and other substrates. Most of these materials are today organic solvent-based, but for reasons of environment and health water-based products are rapidly gaining ground and will in the near future overtake solvent based counterparts. For this reason and because small particles play a larger role in these materials, only water-based paints, lacquers, and inks will be treated here.

Although it is highly desirable to reduce emissions of volatile organic compounds into the environment and exposure of workers and trades people to them, water as a solvent in these products also poses several problems, particularly in applications. Thus, water has a higher freezing point and surface tension, and lower rate of evaporation compared with organic solvents. Also, water-based systems are more difficult to formulate and they are quite prone to attack by microorganisms.

For a more comprehensive and detailed treatise of water-based, and solvent-based, paints, lacquers, and printing inks we refer to Paul (1995), Karsa & Davies (1995), and Boxall & von Fraunhofer (1980).

3.1. Components

In Table 10-14 the components of paints, lacquers, stains and inks are separated into the groups: binders-resins, pigments, extender-fillers, and additives. Several of the components consist of small particles. The order of mixing the raw materials and how they are mixed together play a very important role for obtaining the finished product as a homogeneous mixture of evenly distributed ingredients.

Table 10-14. Components of paints, lacquers, and printing inks			
Component	Paints	Lacquers (clear)	Printing Inks
Binder	√	√	√
Pigment/colorant	√	-	√
Extender-filler	optional	-	-
Additives:			
Defoaming	√	√	√
Wetting agent	√	√	√
Dispersant	√	-	√
Thickener	√	√	√
Surface flow control	√	√	-
Matting agent	optional	optional	optional
Wax	optional	optional	optional

3.1.1. Binders. The most important types of binders for water-based systems are **a)** aqueous dispersions of polymers, so-called latices, **b)** water-soluble or water-reducible polymers, and **c)** alkali silicates.

a. Polymer dispersions - latices. A latex is a stable dispersion of spherical, small particles of diameters in the range from about 50 to 1000 nm in an aqueous medium. The solids content is usually quite high, ranging from about 30 to 60 wt. %, which corresponds to a volume fraction of polymer between 0.40 and 0.70. The aqueous phase contains small amounts of substances such as electrolytes, surfactants, and initiator residues.

Latices are by far the most important binders in water-based systems and are usually prepared by emulsion polymerization of various monomers. Suspension polymerization may also be used, but it produces larger particles, $\geq$ 500 nm, than emulsion polymerization, $\leq$ 500 nm. The reaction can take place by addition, condensation, or a combination of addition and condensation, and is carried out by emulsifying the monomer or a mixture of monomers in water with a surfactant, e.g., sodium lauryl sulfate. In a free-radical addition polymerization, the water-soluble initiator, e.g., ammonium persulfate, decomposes above a certain temperature and produces free radicals which initiate polymerization in the micelles of monomer and surfactant.

When the reaction is complete, these micelles become polymer particles with a stabilizing layer of surfactant around them. Any type of latex can be functionalized, as well as additionally stabilized, by including a small amount of monomer units, no more than 5 mol %, containing functional groups such as carboxylic acid, hydroxyl, sulfonic acid, amine or amide groups. Most latices for industrial applications are anionic, i.e., they are stabilized by anionic surfactants or anionic groups on the polymer chain, but cationic latices can also be prepared by using cationic surfactants in the emulsion polymerization.

The film forming properties of a latex are very important when the latex is used as a binder in paints, lacquers, or inks. When a latex is coated onto a substrate and the water evaporates, a transparent homogeneous film is formed by a sequence of stages. In the first stage, the loss of water crowds the particles into close contact with one another. Next, the void volume of the film decreases because surface and osmotic forces deform the particles into a very close packing.

In the last stage, providing the temperature is above the minimum film forming temperature, MFT, of the polymer, the particles coalesce into a homogeneous, void-free film. However, the mechanical properties of the film continue to improve long after most of the water has evaporated by an aging process involving polymer diffusion across particle-particle interfaces in the films. The film formation process of latices can be promoted by addition of small amounts of coalescing aids, which usually are relatively nonvolatile organic solvents such as ethylene glycol and diethylene glycol derivatives; see section on additives below. Such coalescing aids lower the MFT at which the latex will form a film upon drying by acting as plasticizers for the polymer and hence aid diffusion of polymer across particle-particle interfaces in the film. Since the MFT decreases with the glass transition temperature of the polymer, Tg, a plasticizer will also lower Tg.

Acrylic latices are based on homo- and copolymers of esters of acrylic and methacrylic acids. The softer acrylic ester monomers can readily be copolymerized with the harder methacrylates, styrene, acrylonitrile, and vinyl acetate to yield products that range from soft rubbers to hard nonfilm-forming polymers. Due to their clarity, toughness, UV-stability, and chemical inertness acrylic latices are prime binders in all types of paint and lacquer formulations: for interior and exterior use; in flat, semiglossy and glossy products; and in primers and topcoats. Table 10-15 shows properties and suggested applications of several types latices made by Hoechst-Perstorp, Sweden, having the trade name Mowilith®: acrylic, styrene-acrylic, vinyl acetate, vinyl acetate-ethene, vinyl acetate-ethene-vinyl chloride, vinyl acetate-ethene-vinyl chloride, and vinyl acetate-dibutyl maleinate.

Polyurethane latices are mainly based on dispersions of condensation or addition-condensation polymers and are widely used as binders because of their outstanding flexibility and adhesion and excellent chemical solvent, abrasion and scratch resistance. Polyether and polyester prepolymers can be reacted with aliphatic or aromatic isocyanates to yield products for a wide range of binders applications. Compared with aromatic isocyanates, aliphatic isocyanates impart improved outdoor stability. Polyether-based polyurethanes have better mechanical properties and hydrolytic stability than polyester-based polyurethanes, but the latter show higher gloss retention under accelerated weathering tests.

Typical products are NeoRez® polyurethane dispersions from Zeneca, UK. Thus Neorez® R-940 is an aromatic polyether polyurethane latex with a solids content of 31 wt. % for coatings on wood and plastics. NeoRez® R-970 is an aliphatic polyester polyurethane dispersion with a solids content of 39 wt. % for coatings on flexible substrates. Daotan® is the trade name of Hoechst polyurethane dispersion and Daotan® 1233, for instance, is an aliphatic polyester-based latex with a solids content of 32 wt.% for coatings on wood.

Recently, dispersions of hybrid polymers, containing combinations of a radical copolymer and a condensation polymer, such as urethane-acrylates have been developed. Such latices offer a good balance of properties and a high value-in-use. Daotan® VTW 1267 is an aliphatic, polyester-based self-crosslinking urethane-acrylate hybrid latex from Hoechst with a solids content of 35 wt.% recommended for clear and pigmented coatings on wood, metal, and plastic. NeoPac® E-106 is an aromatic urethane-acrylic dispersion from Zeneca with a solids content of 33 wt. % suggested for use in parquet lacquers and industrial coatings.

b. Water soluble or water reducible polymers. Water soluble or water reducible binders are polymers of relatively low molecular weight, below 20000, containing functional groups such as carboxyl groups in polyacrylics. By neutralizing the carboxyl groups with a base, e.g., ammonia or NaOH, the polymers will become soluble in water. Representative examples of such binders are the Joncryl® series of products from S. C.Johnson & Son, Inc., USA.

Table 10-15. Binder dispersions for paints and lacquers. Polyvinylacetate- homopolymers (Courtesy Hoechst-Perstorp, Perstorp, Sweden)

Product	Polymer type	Stabilizing system	Solids content	Viscosity Brookfield RTV (Pa-s)	spindle	pH	Particle size (μm)	MFT (°C)	T_g (°C)	Film hardness (K-s)	Applications (paints unless noted otherwise)
Mowilith D 025	VA	PVA1	54	18-42	6	3-4	0.3-5	0	2	8	high pigment interior
Mowilith DC 20F	VA	C/T	60	5-12	5	4-5	0.5-3	0	11	6	fire retardant interior
Copolymers											
Mowilith FE 166S	VA/E	C/T	55	1-5		4-5	0.1-0.5	0	10	12	organic solvent free interior
Mowilith FE 174S	VA/E	C/T	55	1-5		4-5	0.1-0.5	0	10	14	organic solvent free interior
Mowilith LDM 1870	VA/E	PVA1/T	53	3-7		4-5	0.2-0.6	0	15	14	organic solvent free interior
Mowilith 9120 V	VA/E/VI	C/T	56	2-6		6-8.5	0.1-0.5	0	8	12	organic solvent free interior
Mowilith DM 123 S	VA/E/VI	C/T	55	4-7		4-5	0.1-.05	16	16	40	interior
Mowilith 1975 S	VA/E/VI	C/T	55	5-9		4-5	0.1-0.5	7	10	22	exterior
Mowilith DM 2	VA/DBM	PVA1	55	3.5-8.5		3.5-4.5	0.3-1.5	7	10	41	exterior
Mowilith DM 2 HB	VA/DBM	PVA1	53	1.6-4		3.5-4.5	0.3-2	0	-2	18	exterior
Mowilith DM 21	VA/VeoVa	T	50	9-15		3.5-5	0.1-1	10	15	45	exterior
Styrene-acrylates											
Mowilith DM 60	S/A	T	50	0.1-0.7		7.5-9.5	0.02-0.2	8	10	7	interior & exterior
Mowilith 6050½	S/A	T	50	0.05-0.2		6-7	0.08-0.12	16	27	15	high pigment interior
Mowilith FA 162 S	S/A	T	46	1-4		7-8	0.08-0.15	0	10	45	interior clear lacquer
Mowilith FA 163 S	S/A	T	46	1-4		7-8	0.08-0.15	3	10	70	interior clear lacquer
Mowilith DM 760 S	S/A	T	34	0.01-0.05		7.5-8.5	0.02-0.2	0	1	11	primer and stains
Mowilith DM 765 S	S/A	T	50	4-11		8-9	0.10-0.2	0	-16	7	exterior roof
Bonotex BT 805	S/A	T	44	0.1-0.2		8.5-9.5	0.070.12	10	10	30	roof tiles, alkaline substrates primer
Acrylates											
Mowilith DM 772	A	T	46	4-10		8-9	0.05-0.15	12	25	70	exterior wood
Mowilith DM 772 S	A	T	50	2.5-6		8-8.5	0.05-0.15	12	25	70	exterior wood
Mowilith BT 774	A	T	46	4-10		8-9	0.05-0.15	20	28	110	interior wood varnish
Bonotex BT 511	A	C/T	59	0.4-1.5		7.5-8.5	0.3-0.5	14	16	40	exterior wood & component in silicate paint
Bonotex BT 514	A	T	40	0.02-0.08		8.5-9.5	0.04-0.1	22	20	90	alkaline substrates paint
Bonotex BT 516	A	T	40	0.01-0.05		8.5-9.5	0.07-0.12	53	50	175	alkaline substrates paint
Bonotex BT 5131	A	T	40	0.02-0.1		9.3-9.7	0.08-0.12	10	18	65	alkaline substrates paint
	A	T	40	0.01-0.05		8.5-9.5	0.07-0.12	45	47	130	alkaline substrates paint

Key: C cellulose derivatives E ethylene T surfactant A acrylate DBM dibutyl maleinate NFEO nonylphenol ethoxilate TCEP trichlorophosphate S styrene DBP dibutyl phthalate PVAl polyvinyl alcohol VA vinyl acetate VCl vinyl chloride VeVa vinyl ester of neodecane acid

c. Alkali silicates - water glasses. Alkali silicates, primarily potassium silicate, are used as binders in weather-resistant and wash-resistant coatings on such mineral substrates as concrete, mortar, limestone, sandy limestone, and others, in both exterior and interior application. Water glass is soluble in water, but a water glass-based paint hardens and becomes water insoluble after drying by chemical reactions. Potassium silicate, for instance, reacts with atmospheric CO_2 and forms an insoluble silica gel:

$$K_2O \cdot mSiO_2 \cdot nH_2O + CO_2 = mSiO_2 \cdot nH_2O + K_2CO_3$$

On a mineral substrate, e.g., concrete, the water glass can also react with calcium hydroxide to form an insoluble mixture of calcium silicate and silica gel. The water will eventually evaporate and leave a hard, somewhat porous coating permeable to moisture. The potassium carbonate forms a transparent layer on the coating, which is slowly dissolved and removed by moisture. Potassium silicate is preferred to sodium silicate since sodium carbonate forms efflorescences on the coating.

For more information on the composition, nature, and properties of alkali silicates see Section C of Chapter 2.

3.1.2. Pigments. Pigments, organic and inorganic, colored, white, nacreous and transparent pigments, were discussed in Chapter 8. They are used in coatings primarily to hide the surface and provide color, but also to protect the binder from degradation by UV radiation and sometimes, as in the case of zinc chromate, as an anticorrosive.

Titanium dioxide is the most important pigment that furnishes the hiding power in white coatings.

Colored coatings can be produced by using either inorganic or organic pigments. Inorganic pigments have excellent lightfastness and weather resistance, are insoluble in water and organic solvents, and are resistant to alkalis but they have less color strength and brilliance than organic pigments.

Iron oxide pigments are the most widely used inorganic pigments and the color range includes red (hematite, α-Fe_2O_3), black (magnetite, Fe_3O_4), yellow (goethite, α-FeOOH), and orange (lepicrodite, γ-FeOOH). Micronized pigments, the surface of which is coated with metal oxides, e.g., Al_2O_3, have good dispersion properties and show high stability towards flocculation.

Chromium oxide pigments, Cr_2O_3, crystallized in a corundum lattice, compete successfully against more brilliant green organic pigments due to their outstanding color fastness and insolubility in solvents, acid, and alkali.

Prussian blue, ferric ferrous-ferrocyanide $Fe^{III}[Fe^{III}Fe^{II}(CN)_6]_3$, and Turnbull's blue, ferrous ferrous-ferrocyanide, $Fe^{II}[Fe^{III}Fe^{II}(CN)_6]_2$, are important pigments, which have very high tinctorial strength. They are primarily used in printing ink formulations.

Compared with inorganic pigments, organic pigments have high light absorption and low scattering power and are therefore used in applications where brilliance, color purity, transparency and tinting strength are important, e.g., printing inks, stains, and tinted lacquers. The most important types of organic pigments are azo pigments, phthalocyanines, and polycyclic compounds such as anthraquinones, quinacridones, and perylenes.

The pigment volume concentration of a coating is the ratio of pigment volume to total volume of nonvolatiles, i.e.,

$$PVC = \frac{\text{pigment volume}}{\text{pigment volume} + \text{binder volume}}$$

The PVC at which there is sufficient binder to barely cover the pigment and fill the voids between the particles is called the critical pigment volume concentration, CPVC.

At PVC's below CPVC coatings have high gloss, low porosity, low dirt-pick up, and low modulus at break, are flexible and have low water permeability. Above CPVC they have low gloss, high porosity, high dirt-pick-up, and high modulus at break, are brittle and have high water permeability, see Fig. 10-23.

The CPVC depends on particle size and is generally lower for pigments of smaller particle size, since these have a higher volume proportion of adsorbed binder than larger particle size pigments. It also depends on the particle size distribution and increases with the width of the distribution because the pigment particles will then pack more efficiently.

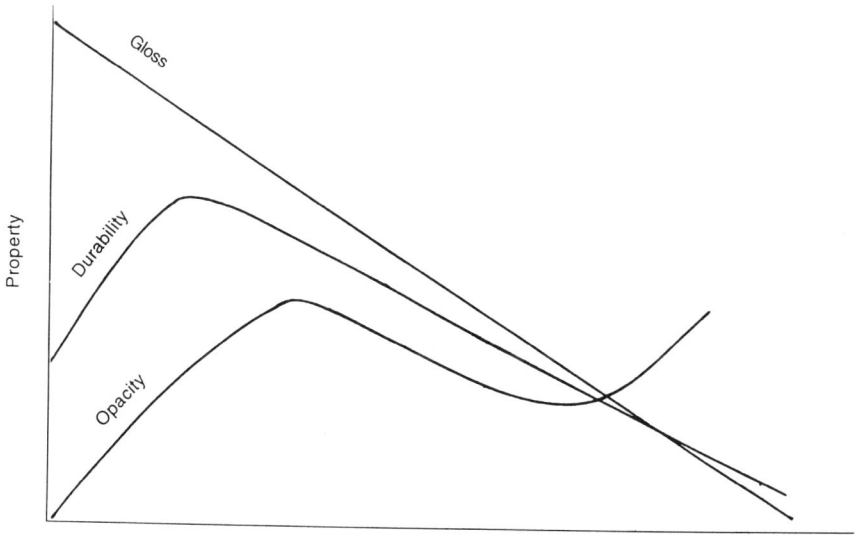

Fig. 10-23. General dependence of some paint film properties on the pigment volume concentration, PVC.

3.1.3. Extenders - Fillers. Extenders are small particles of minerals such as clays—see Chapter 4—and carbonates and sulfates such as calcite, chalk, and dolomite, and barite, and gypsum, respectively; see Chapter 9. They can also be man-made materials such as precipitated calcium carbonate, and sulfate—see Chapter 9—and precipitated and fumed silica; see Section A of this Chapter.

Extenders are usually white in appearance, but their refractive index is not much higher than that of most binders, i.e., below 1.7, so their hiding power is much less than that of such white pigments, as TiO_2, ZnO, and ZnS with refractive indices 2.75, 2.06, and 2.37, respectively.

Originally, they were used to extend or fill out, the coating composition so as to reduce the cost of the end product, e.g., a paint. Today, however, special processing, e.g., micronization and precipitation, and surface treatment, e.g., coating with stearic acid, will yield high quality extenders that improve the properties of or add valuable properties to a coating.

3.1.4. Additives. Additives are compounds that even in small amounts, usually less than 1%, significantly modify and improve the properties of a coating formulation so as to give a high quality coating free of surface defects. In Table 10-14 the most important additives in coating formulations are classified by function. However, several additives, added to a formulation to provide apparently different functions, actually do the same thing, e.g., lower the surface tension of the liquid medium. Thus wetting and dispersing agents lower the surface tension and prevent or reduce crawling, see below, as do such surface flow control agents as leveling agents and flow modifiers.

Defoamers and antifoaming agent. There is actually a distinction between the two terms. A defoamer acts by rapidly destroying a foam whereas an antifoaming agent prevents the formation of it. However, the same type of compounds are used both for antifoaming and defoaming.

Generally, defoamers are liquids of low surface tension and low solubility in the medium to be defoamed. Defoamers destabilize foams by penetrating into the foam lamella and spreading out on its surface. Defoamers are often sold as emulsions of the active component, which can be a hydrocarbon, a polyether, a silicone, or a fluorocarbon in water. Thus, Foamex® 805 from TEGO, Germany, is an emulsion with a solids content of up to 25 wt. % of a polysiloxane-polyether copolymer.

Wetting and dispersion agents. Wetting agents are surfactants of relatively low molecular weight, which lower the surface tension of a coating formulation so that it properly wets the substrate onto which it is applied. They also promote wetting of the surface of pigment and extender particles by the binder so as to ensure a very fine dispersion of these particles in the coating formulation.

A typical additive for improving wetting of substrate surfaces is represented by the BYK-300 series of wetting agents from BYK Chemie in Germany. They are solutions of polyether modified dimethylpolysiloxane with a solids content of about 50 wt. %.

Dispersing agents stabilize pigment particles in water-borne systems toward aggregation by charge repulsion. They are usually anionic surfactants, but cationics can also be used, and provide each pigment particle with an electric charge by adsorbing on the particle surface.

Thickeners. Thickeners are used to modify and adjust the viscosity of paints and other coating formulations so as to meet the requirements of different application methods. In drip-free paints the viscosity is so high that they have the appearance of a gel in the can, but decreases sharply when the paint is sheared by the paint roller or brush.

The two principal types of thickeners are non-associative and associative thickeners. Non-associative thickeners have been used for a long time and are water-soluble cellulosics, e.g., hydroxyethyl cellulose and hydroxypropylmethyl cellulose, and alkali-swellable or soluble emulsions.

Clays, such as montmorillonite and hectorite, which consist of platelike particles, and attapulgite and hectorite, which consist of rod-like particles—see Chapter 4—can also be used as non-associative thickeners, but only in coatings where clarity and transparency are not required.

Non-associative thickeners increase the viscosity by a hydrodynamic mechanism involving restricting the volume of free water in the system. Sometimes these thickeners show a very strong shear-thinning behavior, which in some application could be disadvantageous.

Associative thickeners are newer developments than non-associative thickeners and are hydrophobe-modified ethoxylate urethane, hydroxyethyl cellulose, or alkali swellable emulsions, or modified polyacrylics. They are commonly supplied as aqueous solutions with a solids content of between 30 and 50 wt. %. These thickeners increase viscosity by adsorbing on, associating with, binder, pigment and other particles in the formulation, thus forming a network between the particles of the system.

Surface flow control additives. A coating of a paint or a lacquer serves to provide protection and decorative appearance to a substrate. Surface defects may result from poor application techniques or local differences in surface tension of the coating material. Thus, application by brush gives brush marks which are not related to the bristles of the brush, but instead to the fact that the film of wet coating material, e.g., paint, is split between the substrate and the brush. If the brush marks remains in the wet film, the wet film can not flow properly so as to fill in the grooves and smooth out the ridges. Besides brush marks, the following defects are caused by surface tension gradients:

Bénard cells are hexagonal patterns with well marked centers caused by circulatory flows in thin films.

Craters are small round depressions, often having tiny mounds of material at their centers and raised circular edges, in a coating.

Crawling is dewetting caused by applying a coating with a high surface tension over a substrate of lower surface tension.

Floating is an uneven distribution of pigments at the film surface causing color streaks and mottled areas.

Orange peel is a surface bumpiness or waviness that resembles the skin of an orange.

These surface defects are caused by uneven flow-patterns in the film, which are the result of differences in surface tension. Thus, too little flow results in brush marks, roller striations and orange peel. Any low tension material in the form of small particles of oil, dirt, or dust will result in flow away from that material, leaving a depression such as a crater or a fish eye. Orange peel may develop as a result of spraying a coating onto a substrate. Some droplets may have traveled for a longer time than others before they strike the surface of the wet film. Since they have lost more solvent they will have higher surface tension and when they land on the wet film surface, material of lower surface tension will flow to cover them, causing bumps.

Flow modifiers or leveling agents are used to reduce or eliminate surface defects. Surface defect formation and the mode of action of flow modifiers has been discussed recently by Grolitzer and Erickson (1995). Most leveling agents act by reducing surface tension. Some of them, such as surfactants, fluorinated alkyl esters, polysiloxanes, and solvents, are highly surface active and markedly reduce the surface tension of coating material even when used at ppm levels.

Polyacrylates, on the other hand, have a rather weak effect on bulk surface tension although they are effective leveling agents. They act by forming a monomolecular layer at the film surface, resulting in an even and fairly constant surface tension across the surface. Modaflow® AQ-3000 from Monsanto represents a flow modifier of the polyacrylic type.

Matting agents. It is desirable to be able to control the luster of coatings on the surface of such substrates as furniture, floors, walls etc. from high gloss to matte appearance. This can be achieved by adding a matting agent to the coating formulation. Natural silicas, micronized wax grades and plastic materials, and fillers such as talc can be used as matting agents, but the most effective materials are micronized xerogels of silica, typical representatives of which are the Syloid® products from W. R. Grace, USA, and micronized fumed silica, such as the Acematt® series of products from Degussa, Germany.

The matting efficiency is primarily determined by the pore volume, particle size and particle size distribution, and surface treatment of the matting agent. The efficiency increases with the porosity of the material, and the particles of an efficient matting agent have a sponge-like appearance.

The particle size must be closely matched with the dry film thickness of the coating for optimum matting efficiency. The particle size distribution must therefore be carefully controlled so as to achieve optimum matting for coatings of different film thicknesses. Typical average particle sizes range from 30 to 75 μm with a particle size distribution of about ± 35% of the average particle size. Surface treatment of the particles, e.g., by waxes, will enhance the matting efficiency in some systems.

Waxes. Waxes are added to coating formulations primarily to reduce the coefficient of friction and block resistance and improve the release properties of a paint or lacquer. They can also improve the chemical resistance of a coating,

Table10-16. Molecular weight and melting point of important waxes		
Wax type	Melting point range, °C	Molecular weight
Polyethylene	105-140	60 - 12000
Polytetrafluoroethylene	318-323	> 100000
Microcrystalline	60-93	600-900
Paraffin	46-68	400-700
Carnauba	82-86	~700

e.g., resistance to salt spray, water and alkali.

Molecular weights and melting points of important types of waxes are shown in Table 10-16.

3.2. Applications - Products

In this section we will briefly describe composition and uses of some typical water-borne paints, lacquers, and printing inks. Compositions will be illustrated by guide formulations, most of which have been graciously supplied by the manufacturers of one or several of the components in the formulation. In these formulations, the amounts of the different materials are given as parts of component as delivered by the manufacturer per hundred parts of the finished coating composition. In several cases the component as delivered does not consist of 100% active material, but we have tried to indicate the content of active substrate in the formulation.

3.2.1. Paints and Lacquers. Paint is defined as "any liquid, liquefiable, or mastic composition designed for application to a substrate in a thin layer which is converted to an opaque solid film after application", Le Sota (1978).

In an analogous manner lacquer, or varnish, may be defined as "any liquid composition for application to a substrate in a thin layer which is converted to a transparent or translucent film after application".

The primary functions of paints and lacquers are protection and decoration, but they can also provide other functions such as information and identification (primarily paint), insulation, diffusion barrier and skid-prevention.
Paints. Architectural and industrial paints dominate among the various types of paints available on the market, and water-borne products play an important role in such coatings.

Paints contain pigments and usually also extenders. The manufacture of paint involves mixing, grinding, color matching, viscosity control and other adjustments. Table 10-17 shows that a paint formulation consists of a grind portion and a let-down portion. In the grinding operation using various kinds of grinding and milling equipment, pigment and extender are dispersed with dispersants, wetting agents, and other ingredients to yield a paste of suitable viscosity with as fine a particle size as possible, preferably below 1 μm.

Good dispersion requires that each individual particle is wetted thoroughly with vehicle, which is usually water and a small amount of some type of glycol, and that aggregates are broken up. After the grinding operation, the ingredients

Table 10-17. A general paint formulation		
Raw Materials - ingredients	Weight %	Volume %
Grind portion - Pigment Paste		
Water	5 - 20	5 - 20
Thickener	0 - 0.30	0 - 0.27
Wetting agent	0 - 0.30	0.20 - 0.30
Dispersant	0 - 0.30	0.08 - 0.24
Antifoam	0.10 - 0.40	0.11 - 0.44
Pigment (TiO$_2$)	20 - 30	5 - 7
Extender	0 - 20	0 - 7
Let-down portion - Paint		
Pigment paste	35 - 60	~(30 - 45)
Latex binder (~45%)	20 - 50	18 - 45
Film former	0.50 - 2.0	0.5 - 2.0
Antifoam	0.05 - 0.20	0.06 - 0.22
Thickener	0.20 - 0.40	0.18 - 0.36
Aqueous ammonia (pH control)	~0.10	~0.10
Water	3 - 7	3 - 7

PVC: 50 - 75% for interior flat paints
45 - 50 for exterior flat paints
20 - 25 for interior glossy paints
15 - 20 for exterior glossy paints

of the letdown portion are added to the grind portion with good stirring and blending. Many pigments, and also extenders, can be delivered very well dispersed in the form of pastes or thick slurries. The grinding step can then be eliminated and the manufacture of paint is simplified to blending the different ingredients in effective mixing equipment.

Gloss is not just a matter of appearance, but glossy coatings are lower in PVC (Pigment Volume Concentration) and have better mechanical properties— they can better withstand wear and abuse, e.g., washing and scrubbing—than flat coatings. The gloss of a coating is controlled by the PVC. High gloss, semi-gloss, satin, and flat coatings have PVC's of about 20, 30, 40, and from 50 to 70, respectively. Flat wall paint accounts for the largest volume of interior latex paints and a typical composition is shown in Table 10-18. Binders for such paints are polymers of acrylic and acid esters and polyvinyl acetate, with a predominance of the less expensive vinyl polymer. Interior flat paints have PVC near the CPVC, and in some applications, such as ceiling paints, paints with a PVC higher than CPVC, give good performance. The PVC of the flat wall paint in Table 10-18 is 74%.

Exterior house paints are also dominated by latex paints. The advantages of latex paints over oil and alkyd paints in exterior applications include superior durability and coverage, rapid drying, absence of odor, water cleanability, and excellent color retention (particularly in white). One hundred percent acrylic latices are widely recognized as the type of binders that give the most durable paints for exterior use. The composition of a typical semi-flat latex paint for exterior use is shown in Table 10-18. Exterior house paints are usually formulated at PVC well below CPVC for improved durability. For a given gloss

Table 10-18. Typical formulations for flat and semi-flat paints.
(Courtesy Hoechst-Perstorp, Perstorp, Sweden)

Raw materials - ingredients	Flat interior paint Weight %	Semi-flat exterior paint Weight %
Grind portion - Pigment Paste		
Pigment: rutile TiO$_2$	6.0	20.0
Extender: Omyacarb ® 10 GU	40.0	-
Kaolin speswhite	4.0	4.0
Microtalk ® AT1	-	4.0
Preservative: Metatin ® 55-31	0.2	1.0
Dispersant: Dispex® N 40	0.3	1.0
Antifoam: Nopco® 8034	0.3	0.3
Thickener: Natrosol® 250 GR, 5%	-	20.0
Natrosol® 250 HR, 2%	28.0	3.5
Water	8.3	4.7
	87.1	57.9
Let-down portion - Paint		
Pigment paste:	87.1	57.9
Latex binder: Mowilith® FE 174S	12.7	-
Mowilith® 9742S	-	40.1
Film former: butyl diglycol acetate	-	1.5
Antifoam: Nopco® 8034	0.2	0.3
NH$_3$	-	0.2
	100.0	100.0

Pigment volume concentration, PVC, %   74   27
Pigment:binder ratio   7.1:1   1.2:1
Solids content, %   58   54
Gloss, Gardner 60%, scale units   2   20

designation, the PVC of an exterior latex paint is lower than that of an interior latex paint. Thus, in Table 10-18 the PVC of the exterior paint is 27% whereas it is 74% for the interior paint.

A zinc-rich maintenance coating is one example of an industrial paint that can be water-based. Thus, an aqueous solution of alkali silicate with a high content of zinc dust, is extremely resistant to moisture and brine and gives cathodic protection to steel.

3.1.2. Lacquers. Since lacquers are defined as transparent or translucent coatings on a substrate they are usually not pigmented. Just as in the case of paints, lacquers for exterior use contain durable binders such as 100% acrylic latices, whereas in lacquers for interior application cheaper binders such as vinyl-acetate latices can be used. Gloss can be controlled by addition of matting agents. For high gloss lacquers all-acrylic, polyurethane, or urethane-acrylic latices are preferred.

Table 10-19 shows formulations for clear lacquers on wood panels and parquet, respective. The latter application requires a very hard and strong

Table 10-19. Clear coatings on wood panels and parquet floors. (Courtesy Hoechst-Perstorp, Perstorp, Sweden)

Clear glossy lacquers on wood panels		Clear, glossy lacquers on parquet floors	
Ingredient	Weight %	Ingredient	Weight %
Binder: Mowilith® LDM 7460 (46% TS)	74.5	Binder: Neorez®R-2001 (35% TS)	92.3
Water	17.0	Film former: Dowanol® DPnB	2.2
Film former: trimethyl pentane diolmono-isobutyrate (Texanol®) Methoxybutanol	1.0 2.5	Antifoam: Dehydran® 1513	0.8
Antifoam: Agitan® 295	0.02	Wetting agent: Byk® 344	0.3
Thickener: Tafigel® PUR 40	0.04	Flow control: propylene glycol	4.4
Flow control: 1,2-propylene glycol	1.0		Total 100.0 TS=34%
Wax: Südranol® 230	3.0		
Preservative	0.02		
pH control: AMP 90	0.02 Total 100.0 TS=37%		

lacquer and a latex of a self-crosslinking urethane-acrylate copolymer is therefore used as binder.

3.2.2 Printing Inks. Printing inks are colored formulations that are transferred to a substrate by a printing process. Based on the type of printing process, the inks used in printing are classified as typographic and lithographic inks, which have a pasty consistency, and flexographic and rotogravure inks, which are liquids.

The general composition of printing inks is similar to that of paints, see Table 10-17; i.e., a pigment, and sometimes also, an extender, is dispersed in a binder containing various additives. Most inks still use solvent-based binders, but there is a strong shift toward water-bases systems, particularly in flexographic and rotogravure inks—roughly 25% of the printing inks today are water-based. Since the thickness of the ink film on the substrate, e.g., paper or textile, is very thin, about 3-15 µm, the ink manufacturer formulates the ink for maximum color strength, which implies high loadings of pigments having high color strength. Furthermore, printing inks are usually required to give prints of brilliant, clear colors, and often also of high gloss appearance.

In Chapter 8, it was pointed out that organic pigments have much higher color strength, although, in general, lower light fastness than inorganic pigments. Organic pigments are, therefore, the pigments of choice for manufacturers of printing inks. In spite of the high color strength, the loading of organic pigments in a formulation is quite high, typically about 15-20 wt. % compared with about 3-5 wt. % (colored) pigment in a wall paint (a colored wall paint contains, in addition to colored pigment, also a white pigment, usually TiO_2, which may be present in amounts between 10 and 20 wt. %). Moreover, organic pigments give much brighter and clearer colors than inorganic pigments since they, due to their relatively lower refractive index, scatter light less than inorganic pigments, see Section 3.3, "Optical Properties", of Chapter 8.

Glossy prints of intense, brilliant colors can therefore be obtained by using printing inks containing organic pigments of very fine particle size, certainly smaller than 1 μm and preferably smaller than 0.3 μm (see Fig. 8-6) dispersed in a suitable binder at a PVC well below the CPVC. Suitable binders are mixtures of acrylate-based copolymer dispersions and water reducible polymers, also acrylate-based and containing carboxyl functional groups, typical examples of which are Joncryl® R.C. emulsions and Joncryl® alkali soluble resins, respectively, from Johnson & Son, USA. The water soluble polymers acts as a dispersant for the pigment as well as a binder that enhances the gloss of the print.

4. Fillers in Plastics

We have discussed small particle fillers in paper in Chapter 9. Several such fillers are clays, which were treated in Chapter 4. Moreover, we have already partially covered fillers in polymeric materials since the extenders used in paint formulations serve as fillers in the resulting polymer coating.

Most of the small particle fillers used in plastics are the same type of materials used in paper and paints, although the requirements on particle size, particle size distribution, particle morphology and surface modification may be different.

Important fillers in plastics, especially reinforcing fillers, consist of short fibers or ribbon-like particles, but are not really small particles in the sense of this book since the particle size is too large. Such fillers, and also hollow or solid microspheres of, e.g., glass, which are also used as fillers in plastics, will only be treated in a cursory manner.

For a comprehensive coverage of fillers in plastics we refer to the works by Katz and Milewski (1987).

4.1. Types of Fillers

Table 10-20 shows some of the more important fillers and their primary functions in plastics and elastomers.

Carbon black and calcium carbonate are by far the most widely used fillers. About 95% of the total amount of carbon black produced is used as fillers in rubber and only a small amount, less than 5%, is used as filler in plastics. The reverse is true of calcium carbonate, of which the bulk is being used as filler in plastics and only a relatively small amount in elastomers.

Kaolin is the largest filler in the world, but most of it is being used in the production of coated and uncoated paper and less than 10% of the world production of kaolin is used as filler in plastics and elastomers. About four times as much kaolin is used in elastomers as in plastics.

Alumina trihydrate, natural silicas and talc are intermediate to small volume products, primarily used as filler in plastics.

Barite, mica and synthetic silicas are small volume fillers. Barite is mainly used in plastics whereas fumed and precipitated silicas are chiefly used in elastomers.

Wollastonite (not in Table 10-20), calcium metasilicate with density 2.9-3.1 g/cm^3, is a small but growing reinforcing filler in plastics, replacing asbestos and more expensive glass fiber. Fibers of glass and aromatic polyamides (aramids) are used as reinforcing fillers in thermoplastics and thermoset plastics.

Solid spheres of primarily glass, but also of organic polymers, metals, and fly ash, are used in thermoplastics and thermoset plastics to improve processability and, in the case of fiber reinforced plastics, distribution of fibers in

Table 10-20. Fillers in plastics and rubbers and their primary functions

Filler	Specific Gravity	Plastics[1]	Primary Function	Rubbers[2]	Primary Function
Alumina trihydrate	2.4	many different	flame retardance	NR, CR, SR	flame retardance
Barium sulfate	4.3	thermoset	brightness, chemical resistance	NR, CR, SR	density, sound proofing
Carbon black	1.0 - 2.3	many different	color, cost, electrical & thermal conductivity, UV stability, reinforcement	NR, SBR, IR	reinforcement, UV stabilty, cost
Calcium carbonate	2.6 - 2.9	many different	cost, impact strength, weathering	NR, SR	cost, color, processability
Kaolin	2.6	many different	bloom prevention	NR, SBR, EPDM	cost, electrical properties, processability
Mica	2.8	PVC, PP, PE. PA	chemical, heat, UV resistance	NR, SR	
Silica, synthetic Fumed silica	2.2	FRP, PVC, epoxy	viscosity, flow	silicone rubber	thickeniny, thixotropy
Silica gel	varied	PVC, LDPE	matting	-	-
Precipitated silica	1.9 - 1.2	PVC, PE, EVA	thixotropy	IIR, CR, NBR, NR	reinforcement
Silica, natural Quartz (sand)	2.7	many different	cost, mechanical	NR	surface effect
Diatomaceous silica	2.7	LDPE	antiblocking	NR, IIR	oil absorbance
Talc	2.7 - 2.8	PP, HDPE, TPE, PVC	reinforcement	NR, CR, IIR	electrical properties, antiblocking

[1] HDPE, high-density polyethylene; LDPE, low-density polyethylene; FRP, fiber-reinforced plastic; EVA, ethylene vinyl alcohol; TPE, thermoplastic elastomer.
[2] NR, natural rubber; CR, chloroprene; SR, synthetic rubbers; SBR, styrene-butadiene; EPDM, ethylene propylene diene; IIR, isobutylene-isoprene; NBR, nitrile-butadiene.

the material. Hollow spheres of glass, in particular borosilicate glass, give in addition to lower density, the same benefits as solid spheres in thermoset plastics, e.g., polyester and epoxy resins.

4.2. Filler Particle Properties and their Effect on Resin Properties

The function of a filler in a resin depends on such factors as particle morphology, particle size, packing of particles, and the nature of the particle surface.

4.2.1. Morphology.

Most fillers have been processed by different methods resulting in particles of very variable shapes. Nevertheless, it is meaningful to classify filler particles in certain idealized shape classes. Thus, the shapes of particles of calcium carbonate, barite, and natural silica vary from irregular to prismatic or tabular. Particles of calcium carbonate, e.g., precipitated calcium carbonate, can have a cubic shape, whereas those of fumed silica are spherical. Kaolin, mica, talc, and alumina trihydrate have flaky or platy particles. The particles of wollastonite are fibrous or ribbon-like in appearance.

Since the variety of filler particle shapes is so large, it is difficult to unambiguously express particle size as a particle dimension such as length, diameter, or thickness. Instead, the particle size of many fillers is expressed as the equivalent spherical diameter or the diameter of a sphere having the same volume as the particle, which is a purely theoretical dimension.

The aspect ratio, which is the ratio of the smallest dimension of the particle to the largest, e.g., plate width to thickness and diameter or width to length for platy and fibrous particles, respectively, is also a measure of the irregularity of the particle shape, i.e., deviation from a spherical shape. For prismatic, platy, and ribbon-like particles the aspect ratios are 1-5, 4-15, and 5-100, respectively.

4.2.2. Particle Size, Particle Size Distribution and Packing.

Particle sizes in the range from 10 to about 45 μm are usually determined by the Coulter-type electronic particle counter, whereas finer particle sizes are evaluated by sedimentation methods. Both techniques yield particle size distributions which are represented as cumulative weight percent of particles finer than a given size versus the logarithm of the equivalent spherical diameter; see Figs. 4-6 and 4-7.

A mean particle size of the filler can be estimated from BET measurement of the surface area and using Eqn. 1-23. Two fillers may have the same mean particle size but quite different particle size distributions. For instance, one filler with more coarse fraction but a broader particle size distribution can have the same mean particle size as another filler with less coarse fraction and narrower particle size distribution.

The effect of a filler on the properties of a given resin will depend on the packing of the filler particles in the resin matrix. The packing of small particles was discussed in Section A of this chapter. Packing density depends strongly on the morphology, i.e., the aspect ratio, of the particles. The closer to unity the aspect ratio is, i.e., the more spherical the particles are, the closer the packing.

Inversely, the higher the aspect ratio, i.e., the more the particles deviate from spherical shape, the looser the packing, Thus, fillers with spherical particles will give the closest packing. Furthermore, multisized spheres with certain well-defined ratios between the diameters will pack more closely than monosized spheres, see Figs. 10-17 and 10-19. Dense packing of a filler in a resin matrix is therefore promoted by spherical or nearly spherical particles with broad particle size distribution.

The surface area of a filler is closely related to the mean particle size as determined by the BET method and, indirectly, to the particle size distribution. The nature of the surface and its area are two very important properties of a filler which determine the degree of wetting of the filler surface by polymer, surfactants and other components of the plastic or elastomer formulation, how much filler can be added without adversely affecting, e.g., the rheological properties of the formulation, etc.

The oil absorption of a filler refers the amount of linseed-oil that is required to wet 100 g filler to form a stiff puttylike paste that does not break or separate (ASTM D 281). It depends on the nature of the surface and the specific surface area of the filler and gives information on the flow and dispersion characteristics. Thus, fillers with low oil absorption,10-25 g oil per 100 g filler, are easily dispersed to high loadings with acceptable flow, whereas those with high oil absorption, 40-80 g oil per 100 g filler, may be difficult to disperse. Physical properties of some selected fillers are shown in Table 10-21.

4.2.3. Effect of Filler Properties on Filled Polymers. Fillers affect most of the properties of polymers and are used both for cost reduction and performance reasons. Filler loadings can be expressed in different ways. Manufacturers of rubber and vinyl plastics prefer *p h r*, parts of filler per 100 parts of resin. Others use weight percent filler or volume percent filler. It is

Table 10-21. Properties of selected fillers in plastics and elastomers.				
Filler Type	Specific Gravity	Mohs Hardness	Mean Particle Size, μm	Specific Surface Area, m^2/g
Barium sulfate	4.3	3.0 - 4.0	0.8 - 12.0	0.1 - 2.0
Calcium carbonate				
Ground mineral	2.6 - 2.9	3.0 - 4.0	1.0 - 7.5	0.3-2.0
Precipitated	2.9	3.5 - 4.0	0.20-2.0	1 - 10
Kaolin	2.6	2.0 - 4.0	0.3 - 5.0	5 - 30
Mica	2.5 - 3.2	2.5 - 4.0		
Silica, synthetic				
Fumed silica	2.2			50 - 600
Silica gel	varied		0.001 - 0.02	250 - 1000
Precipitated silica	1.9 - 2.2		0.001 - 0.04	30 - 800
Silica, natural				
Quartz (sand)	2.7	6.0 - 7.0	>100	<0.02
Diatomaceous silica	2.7	6.0 - 7.0		
Talc	2.7-2.8	1.0 - 2.0	>1	<2
Wollastonite	2.9 - 3.1	4.5 - 5.0	3.0 - 8.0	0.3 - 1.0

difficult to give typical loading values of different fillers in various polymers since they depend strongly on particle morphology, compability with polymer, and, most importantly, what the manufacturer wants to accomplish by using fillers. However, for a particulate materials to be referred to as a filler it must be added to the polymer in amounts exceeding 5 vol. %, and most filled polymer products have loadings in the range from 10 to 50 vol. % fillers.

- **Density.** Fillers, can increase the density, i.e., barium sulfate, or lower it, i.e., hollow microspheres.

- **Flame retardance.** Fillers that are inorganic in nature are non-combustible and provide some fire resistance by reducing the amount of combustible matter in the filled polymer. Alumina trihydrate will further enhance fire resistance by giving off water when it is heated.

The effect of fillers on the mechanical properties of polymers can best be discussed by referring to the stress-strain curve in Fig. 10-24, which shows the behaviour of a material under tension.

The effect of particle size on the mechanical properties will be discussed in more detail in Section 4 below.

- **Tensile strength at break** increases with decreasing particle size and, up to a maximum, with loading. It also increases with the aspect ratio of the filler, i.e., rod-like particles are usually strongly rein-forcing, and with increasing interaction-bonding-between the polymer and the filler particle surface.

- **Modulus (Stiffness)** is the slope of stress-strain curve and increases with loading and decreasing particle size. Reinforcing fillers yield higher modulus than inert fillers (extenders) at equal loading.

- **Elongation at break** is almost always reduced regardless of the type of filler used.

- **Hardness** is affected by the same factors and in the same manner as modulus.

- **Attrition resistance** increases with energy at break and decreasing friction coefficient of the surface of the material; see *Tack* below. The energy at break corresponds to the area under the stress-strain curve and is strongly affected by particle size (see Section 4 below), shape, morphology and the interaction between the fillers particle surface and polymer.

- **Tack** is a surface stickiness which can be reduced by fine-particle spherical fillers such as silica and flaky fillers such as mica. Reduced tack means lower coefficient of friction.

- **Tear strength and impact strength** are both favored by low loading of fillers with high aspect ratio, e.g., fibers or rod-like particles, good bonding between polymer and particle surface, and narrow particle size distribution.

- **Compressive strength** is promoted by high loading of particles of low aspect ratio, i.e., spherical or corpuscular particles, broad particle size distribution, and strong bonding between polymer and filler particle surface.

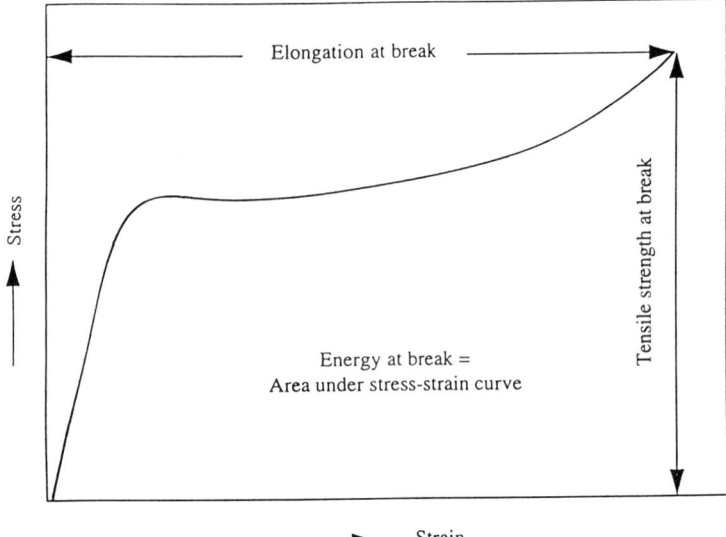

Fig. 10-24. Typical stress-strain curve of a polymeric material.

Important mechanical properties of polymers relating to the stress-strain curve are further illustrated in Fig. 10-25.

Description of Polymer	Characteristics of Stress-Strain Curve			
	Modulus	Yield Stress	Ultimate Strength	Elongation at Break
Soft, weak	Low	Low	Low	Moderate
Soft, tough	Low	Low	(Yield Stress)	High
Hard, strong	High	High	High	Moderate
Hard, tough	High	High	High	High
Hard, brittle	High	None	Moderate	Low

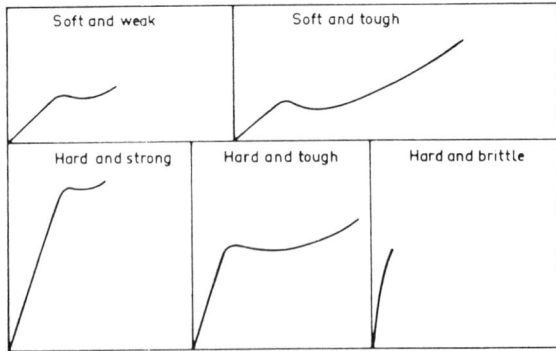

Fig. 10-25. Stress-strain curve and important mechanical properties. Winding and Hiatt (1961). Courtesy of McGraw-Hill.

4.3. Modifying the Filler Particle Surface: Coupling Agents

Most of the mechanical properties of filled polymers, with the exception of elongation at break, are improved by strong interaction-bonding between filler particles and polymer. The chemical natures of the filler particle surface and the polymer must thus be at least such that the particle surface is readily wetted by the polymer and, preferably, also can form bonds with it. If the polymer wets the particle surface poorly or not at all, the filler particles will be distributed throughout the polymer matrix, but will be separated from it—the particles fit loosely into small holes in the polymer matrix. This discontinuity between filler and polymer will create defects or weak spots in the material, at which cracks may form.

The problem of poor adhesion between two dissimilar surfaces can be overcome by chemical treatment of the surfaces. Inorganic surfaces receptive to chemical treatment are characterized by the presence of hydroxyl groups attached to such elements as silicon, aluminum, calcium and magnesium (in the case of filler particles). These surface hydroxyl groups can form chemical bonds with certain substances which either are readily wetted by the polymer or can form chemical bonds also with the polymer, i.e., they can couple the polymer to the filler particle surface.

Surface stearates. Stearic acid is the most commonly used compound for improving wetting between fillers and polymers. It can form an ester with the hydroxyl groups on the filler surface:

$$Filler\text{-}OH + HOOC(CH_2)_{16}CH_3 = Filler\text{-}OOC(CH_3)_{16}CH_3 + H_2O$$

However, most of the fatty acid will not form an ester on the surface, but will instead be adsorbed on it. Kirkland (1992) describes a stearate-treated calcium carbonate filler with an average particle size of 2 μm.

Silane coupling agents. Silane coupling agents are organosilicon compounds of the general formula:

$$R\text{-}SiX_3$$

where R is an organofunctional group attached to silicon by a hydrolytically stable bond, and X designates groups which can be converted to silanol groups, SiOH, by hydrolysis.

Usually, R consists of a reactive group R^1, which may be very simple, e.g., Cl, or contain several functional groups. X is usually a methoxy group and in this case the general formula is:

$$R^1CH_2CH_2CH_2Si(OCH_3)_3$$

After hydrolysis of the methoxy groups to silanol groups, the silane compound can be attached to the filler through bonds formed by reaction between the silanol groups and hydroxyl groups on the filler surface. The filler can next

Table 10-22. Silane coupling agents. (courtesy Dow Corning Corp.)

Designation	Chemical Type	Chemical Name	Applications
Z-6020	Diamino	N-(2-aminoethyly)-3-amino-propyltrimethoxy silane	epoxies, phenolics, malamines, nylons, PVC, acrylics, polyolefins, polyurethanes, nitrile rubbers
Z-6030	Methacrylate	3-methacryloxypropyl trimethoxysilane	free-radical crosslinked polyester, rubber, polyolefins, styrenics, acrylics
Z-6032	Styrylamine cationic	N-[2-(vinylbenzyl amino)-ethyl]-3-aminopropyl-trimethoxysilane	most thermoset and thermoplastic resins
Z-6040	Epoxy	3-glycidoxy-propyltrimethoxy silane	epoxies, urethane acrylic and polysulfide sealants
Z-6075	Vinyl	vinyltriacetoxy silane	polyesters, polyolefins, EPDM, EPM (peroxide cured)
Z-6076	Chloroalkyl	3-chloropropyltrimethoxy silane	epoxy, styrenics, nylon

be coupled to the polymer through bonds formed by reaction of R^1 with reactive groups on the polymer. Silane coupling agents available from Dow Corning are shown in Table 10-22.

Plueddemann (1991) gives an excellent account of silane coupling agents.

Other coupling agents. Volan® is a chrome-complex coupling agent supplied by Du Pont.

Titanates, e.g., products supplied under the trade name Tyzor® from Du Pont and several products from Keurich Chemical Co (USA), are also used as coupling agents.

$$(CH_2=\overset{\overset{\displaystyle CH_3}{|}}{C}-COO)_3 \, TiOCH(CH_3)_2 \qquad \text{tris(methacryl)isopropyl titanate}$$

The use of coupling agents and surface modifying agents requires that the filler surface is receptive to treatment with such compounds. The receptiveness, i.e., the reactivity, of a given filler depends on the concentration of hydroxyl groups on the surface. The concentration, and also the reactivity, of such groups can be increased on the surface of any filler by coating the filler particles with a thin layer of silica, a few nanometers would be enough, according to the methods described in Section D of Chapter 6.

Several benefits can be gained by using coupling agents or surface modifying agents.

- Improved mechanical properties such as tensile, flexural and compressive strength.
- Good retention of these properties under adverse conditions of use.
- Improved retention of clarity in certain composites. Formation of cracks and fractures, which tend to develop opacity, is prevented.
- Improved dispersion of filler in polymer matrix.
- Loading of inexpensive filler can be increased.

4.4. Effect of Filler Particle Size on Mechanical Properties

At a given loading of certain filler in a polymer the reinforcing effect on the mechanical properties of the material increases with increasing filler surface area that is exposed to and interacts with the polymer. Since surface area is closely related to particle size, e.g., through Eqn. 1-23, it is understandable why particle size is the property of a filler that has the largest influence on the reinforcing properties of a filler.

For most fillers and polymers, Table 10-23 can be used as a rule of thumb for the reinforcing effect of filler.

The reinforcing effect of a filler on the mechanical properties of polymers will, of course, also depend on the specific interaction-bonding between the filler surface and the polymer. Figure 10-26 is an interesting illustration of the difference in effect on the mechanical properties of a polymer, in this case natural rubber, of two fillers, carbon black and barite, $BaSO_4$, which differ greatly in particle size and their interaction with the polymer. Carbon black is a very fine particle organophilic filler, i.e., it has a high specific surface area, which interacts strongly with natural rubber.

Barite, on the other hand, is a coarse particle hydrophilic filler, i.e., it has a low specific surface area, which interacts only weakly with the polymer. The tensile strength at break increases with loading up to a maximum at about 20 vol. % carbon black, whereas the elongation at break decreases with loading; Fig. 10-26a. With barite as filler, tensile strength and elongation at break both decrease with loading. In Fig. 10-26b, the energy at rupture, which is a measure of the strength and toughness of the material, is plotted versus filler loading.

Table 10-23. Particle size and reinforcing effect of fillers.		
Filler	Mean diameter μm	Effect on properties
Ultrafine	< 0.1	Reinforcing
Fine	0.1-1	Semi-reinforcing
Medium	1 - 10	Extender
Coarse	> 10	Impairing

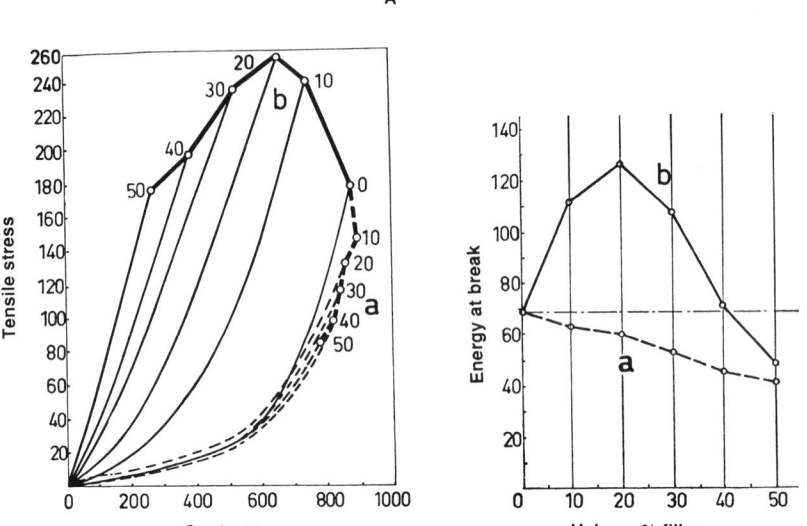

Fig. 10-26. A Stress-strain curves and B: Energy at break of natural rubber containing barite (a) and carbon black (b) as filler. Van Rossem (1958). Courtesy N. V. Servire

Carbon black is a strongly reinforcing filler at loadings of about 20 vol. %, and reinforcing up to loadings of about 40 vol. %, whereas barite is a weakening filler, and the weakening effect increases with loading. J. M. Huber corporation claims that grinding their dry-ground calcium carbonate filler Optifil® T from a mean particle size 1.5-2.0 μm to 1.4-1.5 μm with top particle sizes of 8 μm and 5 μm respectively, a seemingly modest reduction in particle size, improves impact and abrasion resistance, Schut (1995).

Iler (1979) studied the effect of particle size on the mechanical properties of rubber using silica sol particles, the surface of which had been esterified with n-butyl alcohol, as fillers; see also Section C of Chapter 6. Tensile strength and energy at break increased by 25% and 33%, respectively, when the particle size was reduced from 25 nm to 10 nm. His data are schematically plotted in Fig. 10-27 and it appears that the mechanical properties increases faster than linearly with decreasing particle size. It would be interesting to know what would happen if the particle size was reduced further. Would the increase in mechanical properties accelerate or would it level off?

In an effort to answer these questions Otterstedt et al. (1987) prepared organosols of silica with particle sizes ranging from 1.4 to 15 nm and used them as fillers in polyester-and polyether-based polyurethanes. The organosols, consisting of silica particles dispersed in tetrahydrofuran, THF, were prepared by the methods described in Section C of Chapter 6. Polymer films were made by dissolving the polyurethane in THF, adding appropiate amounts of organosol to the polymer solutions so as to give the desired loading of silica particles of a

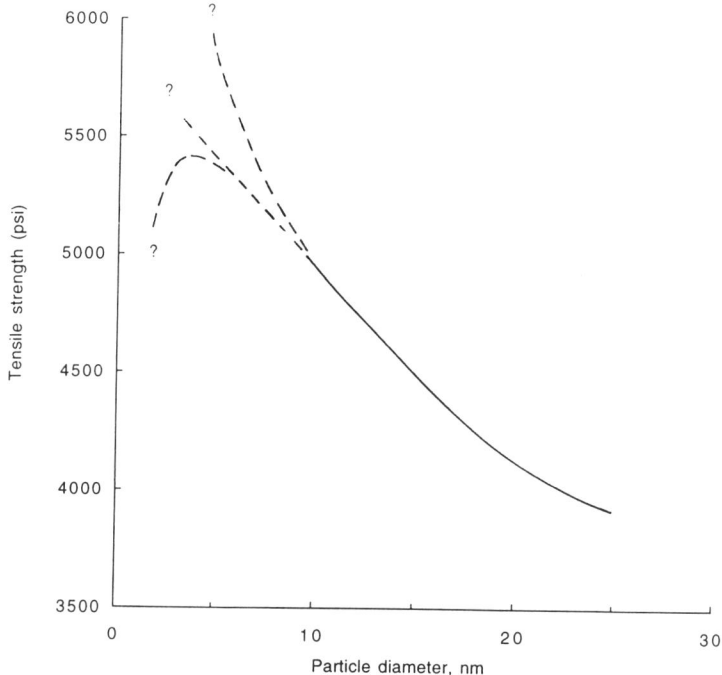

Fig. 10-27. Reinforcement of rubber with surface-esterified silica particles. Iler (1979).

given size, casting films of the solutions of polymer-silica in THF on glass plates, and evaporating the solvent.

From stress-strain curves, the curves in Fig. 10-28 show the relative area under the stress-strain curve versus particle size at a loading of 15 phr SiO_2, were constructed. The relative area is the ratio between the area under the stress-strain curve at 350% strain of the film containing SiO_2 to the same area of the pure polyurethane film. The areas go through a maximum at about 2 nm for the polyether-based polyurethane, Estane® 5714, Fig. 10-28a, and at about 5 nm for the polyester-based polyurethane, Fig. 10-28b. The reinforcement effect at the maxima is remarkably high, about 250% for both types of polyurethane, but decreased when the film samples were cycled between 0% and 350% strain fourteen times. The decrease was greater for the polyester-based than for the polyether-based polyurethane. The structure of polyurethane elastomers consists of hard and soft chain segments with the hard segments acting as bridges between the unordered soft segments. The hard segments make up about 10-20% of the molecular weight and are typically in the size range from 2 to 4 nm. The mean free path of the soft segments are in the range from 10 to 30 nm. Fifteen wt. % SiO_2 in the polymer corresponds to 7 volume %.

Assuming that the silica particles are uniformly dispersed in the polymer matrix, 7 vol. % corresponds to a distance between the particles of one particle

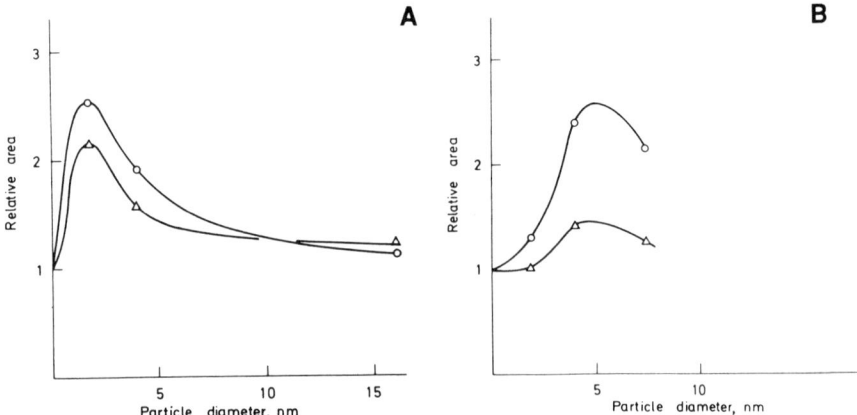

Fig. 10-28. Relative area of (A) Estane® 5714 and (B) Estane 5714 F1 containing 15 phr SiO₂ as a function of particle size. (o) first cycle; (Δ) 14th cycle. Otterstedt et al. (1987). Reprinted by permission of John Wiley & Sons, Inc.

diameter. The mean free path of the particles is thus one particle diameter.

The pronounced maxima of the area under the stress-strain curves, and also of the tensile strength, for particles in the range 2 to 5 nm indicate some type of geometrical and chemical matching of the particles to the hard segments. One possibility is close association by hydrogen bonding between the silanol groups of the particle surface and the hard segments. An additional reinforcing mechanism is bridging of hard segments by the particles. The large deformations in the first hysteresis loop will weaken the interaction and alter the geometrical arrangement of the particles. The plastic deformation after completed cycling is practically the same for polymer with and without silica particles, suggesting that the particles are about one order of magnitude smaller than the size of the soft chain segments.

5.Textiles

Fatty, organic compounds have long been used on synthetic fibers for altering the hand, providing lubrication, and preventing static. It is less well known, however, that small inorganic particles, including the extremely small particles of silica present in alkali silicate solutions, can be used to improve the properties of textile fibers and fabrics.

- Increase crispness and impart silky feel to synthetic fibers and blends. Fiber frictionizing.
- Increase the bulk of fabrics which may have too little body.
- Control of pilling and minimizing loss of pill resistance.
- Reduce soiling and staining.
- Control static and reduce static pick-up.
- Primer for fabric surfaces to improve adhesion of plastic

coatings or laminations.
* Can be combined with standard sizes, such as starch and polyvinyl alcohol, to modify properties, e.g., such as improving adhesion, modifying, modification of surface charge , etc.
* Improve cutting efficiency by reducing fiber slippage
* Dye acceptor on synthetics and blends
* Mordant for dyeing. Improves leveling of color in blended fabrics and uniformity of color in package dyeing.
* Insolubilization of dyes, especially acid dyes.
* Improve flame resistance.

5.1. Treatment with Alkalisilicates

Buckwalter (1942) treated cotton fibers with 3.7 ratio sodium silicate solution containing resin and claimed that their strength increased by 10 to 15 percent or more.The absorbancy of moisture by fibers in textiles can be increased by impregnating them with a sodium silicate solution and then treating the fibers with acid to precipitate it, Rodwell et al. (1937).

Cotton cloth can be made "flame-retardent" by impregnating it with a sodium silicate solution and then treating it with a $TiOCl_2$-$SbCl_3$ complex, Gulledge and Seidel (1950), Jacobsen et al. (1951), Lane and Dills (1951).

Boving (1933) found that sodium silicates in combination with selenium or selenium salts act as binder as well as flame inhibitor. The flameproofing effect of selenium is due to the formation of a deoxidizing blanket of vapor which supplements the protective action of the silicate coating.

5.2. Treatment with Silica Sols

Treatment of fibers, fabrics and yarns with colloidal silica, e.g.,140 m^2/g and 70 m^2/g sols such as Ludox® TM and Syton® W, increases fiber friction, leading to reduced yarn slippage during processing and improved weave setting, properties in fabrics. Colloidal silica, aluminum modified 220 m^2/g sols such as Ludox ® AM are particularly suitable—can be used to modify sizes, binders, and coatings to achieve special effects such as delustering and improved soil resistance. Usually about 0.1 to 2.0 wt. % based on the dry weight of the fabric is sufficient to achieve these effects with delustering requiring the least, and anti-soiling the most, silica.

Colloidal silica can be used in rug shampoos, upholstery cleaners and general cleaning compounds to make textile materials soil resistant. Such cleaning compounds contain nonionic and anionic detergents which remove soil and clean the interior surfaces, i.e., the fiber surfaces, of the textile. The silica particles of colloidal silica are deposited onto the surface during the cleaning process and protect the cleaned surface against soiling by occupying the sites where ordinary dirt and soil usually cling.

The strength of cotton can be increased by impregnating cotton fabrics with an organosol containing silica particles, on the surface of which quaternary ammonium ion have been adsorbed, Vossos (1972). The starting aqueous silica sol should preferably have a particle diameter in the range from 16 to 20 nm and an average surface area from 150 to 190 m²/g. Such a silica sol is added to a solution of the quaternary ammonium compound, e.g., lauryl trimethyl quaternary ammonium chloride, in a hydrophilic solvent such as isopropyl alcohol. A non-polar organic liquid such as a hydrocarbon oil is added to this mixture. After the mixture has been stirred for about 5 to 60 minutes two phases are obtained, the bottom layer generally being the organosol containing quaternary coated silica. The product layer is heated at 60 to 80° C for 2 to 20 minutes to drive off the hydrophilic solvent.

The strength of cotton can be increased by 10-20 % when the textile is treated with silica organosol corresponding to from about 0.05 to about 0.6 % SiO_2 based on the weight of the fibers.

5.3. Treatment with Alumina Particles

The adsorption of Baymal®, which was Du Pont's trademark for its fibrous colloidal alumina, on nylon, Dacron® polyester, Orlon® polyacrylic, and cotton fibers has been studied at the Sales Technical Laboratory of Du Pont's Industrial and Biochemicals department. The fibers were treated by the following general procedure.

1. Weigh fabric sample dry.
2. Immerse fabric in a Baymal® sol of such concentration that the fabric to sol ratio is in the range from 15:1 to 40:1 and contact for 10 minutes on a platform shaker.
3. Adjust pH with 1-M NaOH.
4. Wring out excess sol by passing twice through a hand wringer.
5. Weigh sample for wet pick-up.
6. Dry at 110° C for 30 minutes.

Figure 10-29 shows that the amount of fibrillar alumina adsorbed on the four types of fabric is small but measurable below pH 6 but increases rapidly with pH above pH 6. The shape of the adsorption curves depends on how the electrical charge on the fibers and the alumina particles vary with pH. The negative charge on the surface of the fibers increases with pH—the surface of cotton fibers, for instance, contains COOH-groups which becames increasingly charged at higher pH.

Figure 10-13, on the other hand, shows that although the positive charge on alumina particles decreases with pH, it is still appreciable at pH above 6—at least after about 10 minutes after the pH adjustment. In the adsorption experiments the fibers and the alumina particles were brought together at fairly low pH, about 4-5. When the pH subsequently was raised, say to 9, the negative charge on the fiber surface—e.g., the ionization of COOH-groups on the surface of cotton fibers—increased rapidly. The positive charge on the surface of the alumina particles, on the other hand, decreased much more slowly and was still

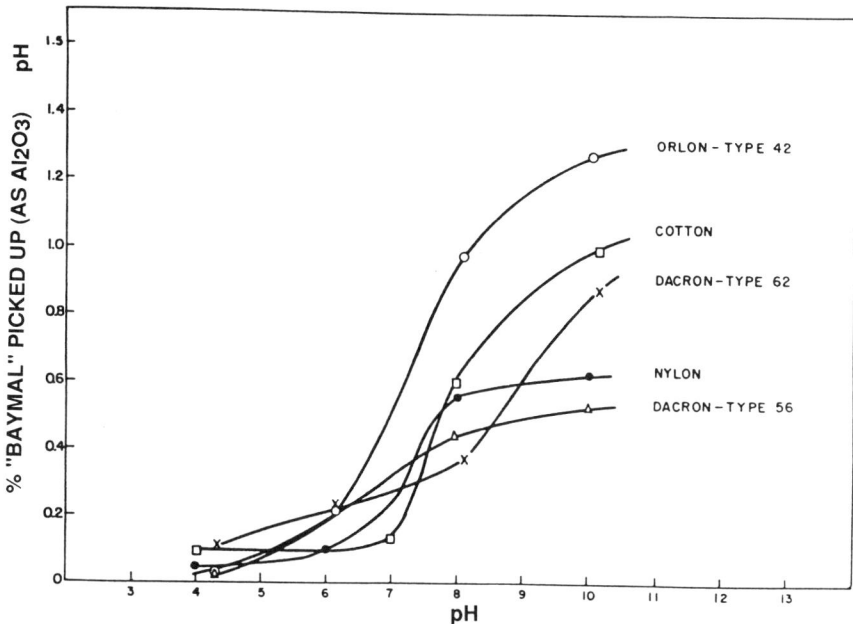

Fig. 10-29. Substantive pick-up of fibrillar boehmite (Baymal®) on various textiles vs. pH. Courtesy E.I. DuPont.

sufficiently high for the alumina particles to adhere strongly to the negatively charged textile fibers.

Figure 10-30 shows that adjustment of the pH of the Baymal dispersion before immersing the fabric in the bath did not affect the pick-up at pH below about 7 to 8. However, above pH 8 the pick-up decreased rapidly with pH. The rate of decrease of the positive charge on the alumina particles increases with pH, but is so slow at pH below 8 that the positive charge remains sufficiently high for strong adsorption on the fibers to occur during the time scale of the experiment. Above pH 8, however, the positive charge on the particles drops so fast before the fabric is immersed in the dispersion that the adsorption of the particles on the fiber surface is reduced and the pick-up falls.

Although the pick-up of Al_2O_3 increases with pH, Fig. 10-29, the alumina appears to be more loosely bound to the fibers at higher pH than at lower pH. Fig 10-31 shows that the amount of alumina that can be rinsed out increases with pH. Above pH 8, some of the alumina is not strongly adsorbed on the fiber surface, but is instead precipitated onto the fibers. The fraction of alumina that is precipitated onto the fibers increases with pH and can readily be rinsed out.

Du Pont (1961) also compared the effect of fibrous alumina particles, about 100 nm long and 5 nm in diameter, with that of spherical alumina particles, about 20 nm in diameter, on the hand of finished textile goods (Orlon®). The fibrous alumina, Baymal®, gave a much softer hand to the fabric than the spherical alumina.

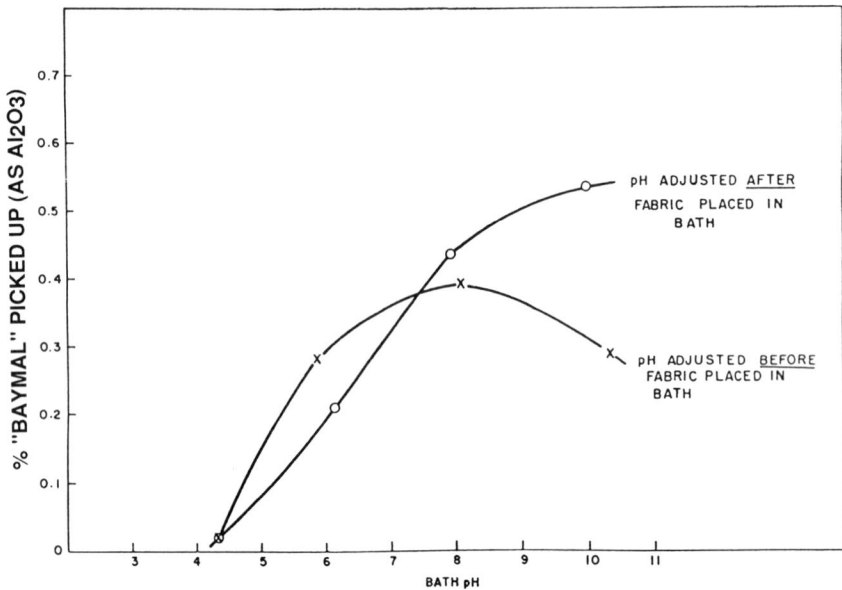

Fig. 10-30. Adjustment of pH before and after fabric is added to bath of colloidal alumina. Courtesy of E. I. Du Pont Co..

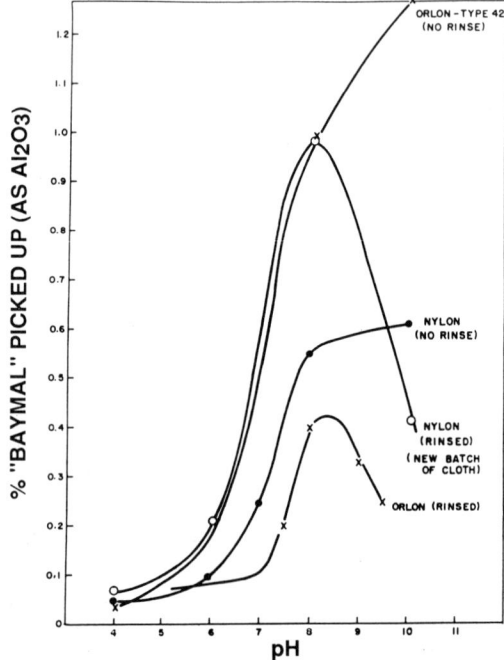

Fig. 10-31. Rinse-out of alumina particles ves. pH. Courtesy E. I. Du Pont Co.

Many of the properties of textiles and textile fibers listed at the beginning of this Section can be improved by application of 0.1 to 0.5 wt. % alumina, based on the weight of the textile goods, in the form of small particles. If the particles are applied to the textile at pH below 8 they will not be removed by washing.

6. Particle Strengthening of Metals and Alloys

A very important application of small particle technology occurs in the strengthening of metals and alloys by the formation of the particles *in situ* by precipitation from a supersaturated solid solution or by mechanical alloying or by internal oxidation. This is another application where a very small amount of fine-particle material has a very marked effect on the properties of the bulk material.

In the first decade of this century Wilm (1911) discovered that the hardness of some aluminum-copper-magnesium alloys quenched from elevated temperatures increased as they aged at room temperature. The probable size of the precipitates that strengthened Wilm's alloys was likely below 10 nm—too small to be seen with optical microscopes.

At that time there were no theories to explain the phenomenon now known as *age hardening*. In 1934 Orowan, Polyanyi, and Taylor independently reported the concept of dislocations in metals. Since then the theory has developed into a complex and well understood domain of metallurgy which has been quite successful in creating many superalloys of great commercial importance. Here we do little more than mention the topic due to its relevance, but in view of its very specialized nature, refer the reader to the excellent treatment given by Nembach (1997).

Essentially the mechanism of hardening is based on the small particles being obstacles to the movement of dislocations in the metal structure. Other inhomogeneities that can act as hardeners are grain boundaries, single dissolved atoms, and other dislocations brought about by cold-working, among others.

The fundamental importance of this concept of metal or alloy strengthening is that, for most uses, structural materials need to combine strength with a certain amount of ductility because otherwise catastrophic failure can occur if momentary stresses exceed the design limitations. Both strength and ductility are imparted by the mobility of dislocations. Thus any obstacles to the mobility of the dislocations cause profound effects.

Nembach (1997) cites a convincing example with the nickel-base superalloy NIMONIC PE16 whose main constituents are nickel, iron, chromium, aluminum, and titanium. Above the temperature 1150°K it is a homogeneous solid solution. If this state is preserved by quenching, the critical resolved shear stress, CRSS, is 69 MPa at room temperature. By heat treating this alloy at 1029°K, precipitation of fine particles of the γ'-phase, rich in nickel, aluminum, and titanium, occurs. Upon annealing for 330 h at 1029°K, the particles' volume fraction and radii are 0.09 and 25 nm, respectively, and in this state at room temperature the CRSS is 170 MPa—a very significant improvement.

Some of the many commercially important particle-strengthened materials are (basic metal noted first):

1. Aluminum-copper-magnesium alloys
2. Copper-beryllium alloys.
3. Nickel-base superalloys
4. Maraging steels
5. Titanium-rare earth oxides

In addition to the most important method of creating small particles in place from a supersaturated solid solution, two other methods—mechanical alloying and internal oxidation—utilize the physical addition of suitable small particles (generally below about 0.5μm)to achieve the objective.

In the 1960's Benjamin (1970, 1989) developed the mechanical alloying method for uniformly dispersing oxide particles in nickel-base superalloys. Typically yttria (Y_2O_3) was used. The process involved: (1) mixing powders of pure metals or of crushed master alloys with the oxides; (2) ball milling the mixture; compacting the ball-milled product by extrusion or isostatic pressing; (3) direct recrystallization; (4) precipitation of γ'-particles by a final heat treatment, Hack (1987).

A third method of introducing particles which cause hardening is by the internal oxidation process, which also has the objective of homogeneously distributing oxide particles in the matrix. In this method the starting materials are not powders, but dilute solid solutions whose solute is highly reactive with oxygen present during a heat treatment. This method has the limitations that only thin specimens can be treated within practical times since the rate of internal oxidation is diffusion controlled and the size distribution of the oxide particles is spatially nonhomogeneous. In addition, in complex systems it may not possible to limit the oxidation to those elements meant to be oxidized, Benjamin (1970), Böhm and Kahlweit (1964).

7. Small Particle Contamination

Before the advent of space technology this topic might have been called simply "dirt removal". As vacuum tubes were replaced with much smaller devices such as transistors and then printed circuit boards with large-scale integration of components, the importance of small particles in causing malfunctions by short circuiting, poisoning mechanisms in semi-conductors, combustion at hot spots, and other nefarious ways, became a critical limitation to miniaturization. In fact, without the contamination control technology now used, modern high-speed computers would be impractical due to unacceptably low reliability.

Today particles in the 0.2-0.5 μm range are a limitation to further speed and reliability increases. Although this size is at the high end of the range we are principally concerned with in this book, it is significant that in terms of particle removal, the important fact is that, at these sizes and below, a different regime is reached in terms of the relative adhesive force between particle and surface.

Most ordinary cleaning problems, such as particulate soil removal in washing clothes, generally involve a larger size range though many of the principles discussed in other chapters do apply in such cases.

Manufacturing areas for sensitive electronic and optical parts have to be designed so as to reduce particulate levels well below those ordinarily experienced in the workplace, i.e., to *clean room* specifications. In addition, an entire technology developed involving specialty solvents, largely based on CCl_2FCClF_2 and its azeotropes with active solvents such as acetone, methylene chloride, and lower alcohols to remove particulates and adsorbed layers of oils, greases, and ionic compounds from surfaces of printed circuit boards and other electronic components. Johnson and Brandreth (1977, 1978, 1979) studied the fundamentals of particulate and surface film removal. Here we summarize their principal findings.

The recognition of ozone depletion by chlorofluorocarbons as a serious world-wide hazard has markedly changed the situation in using 1,1,2-trifluorotrichloroethane as the key component in precision cleaning. Because that compound has a high ozone depletion potential, many other non-flammable, relatively non-toxic, relatively non-aggressive solvents have been looked at as replacements for 1,1,2-TFTC, but to date nothing has been found that is nearly so universally effective, cheap, and safe. Instead, such cleaning problems have been segmented into many different types, some of which can be solved with isopropanol, some with water + surfactants, some with terpenes, etc. Efforts continue in an attempt to find a really good and universal replacement for 1,1,2-TFTC.

What was generally not realized is that 1,1,2-TFTC (often known as Freon®113, R-113, or Freon® TF for the cleaning solvent grade) is basically a carrier for the active ingredients (normally in the range of 1-10% depending on the system—usually azeotropic mixtures with 1,1,2-TFTC) such as, methanol, ethanol, water (non-azeotroping) and acetone in the various formulations that were sold for microelectronics, optical, and other precision cleaning processes. It is difficult to find a better solvent in terms of the desirable moderate boiling point (47.6°C), non-flammability, and relatively high TLV that R-113 exhibits.

Microcircuit production costs depend strongly on the yield of usable circuits, i.e., the fraction of the total circuits free of defects. Defects are caused by a number of sources, among which are electrical short circuiting and pinholing in photomasks. Whereas the minimum average dimension in integrated circuits was on the order of 40 μm in 1960, it was about 6 μm in 1979 and it is under 1 μm today and is still dropping.

Particle-Surface Adhesion

It is convenient to divide particulates into three size regimes to discuss adhesion of particles to surfaces. These demarcations are roughly: over 1000μm, 1000 μm- 0.1 μm and below 0.1 μm. Gravitational, intermolecular, and electrostatic forces attract particles independently of their size, but it is the ratio of these forces which is relevant. Thus for a 1000-µm particle resting on a surface, the ratio of the gravitational attraction to the intermolecular forces (always attractive) is very large. Gravity is a long-range force wherein the entire bulk of the particle helps determine the strength of the interaction, whereas the intermolecular forces are of such short range that, in effect, only the tiny fraction of the particle in immediate contact with the surface enters into the calculation. As the particle size decreases, the relative amount of particle mass within the effective range of the intermolecular forces increases. At a particle diameter of 0.1 µm the intermolecular forces are vastly larger than the gravitational forces.

Of course, the total force of adhesion for a larger particle is much greater than for a small particle, but in practical terms the removal of the larger particle is much easier because larger inertial forces can readily be applied to larger particles.

Forces of attraction due to static electricity are also of importance in some cases; but with the R-113 azeotropes involving alcohols, rapid charge dissipation occurs and those forces are thereby overcome. Table 10-24 shows the relative adhesive force in air versus the particle size for four particle sizes from 500 µm to 0.5 µm. It is apparent that particles in the 1- µm range are held so tenaciously that mechanical cleaning cannot dislodge them. When a solvent replaces air as the medium, the attractive force is diminished. The resultant attraction depends very much on the chemical nature of the solvent.

The strong type of interaction known as hydrogen bonding, which occurs in alcohols and water, allows those components of the cleaning composition to interact strongly with the surface and thereby diminish the forces that hold the particulates to the surface. This adhesion force can be reduced by several orders of magnitude. Water and alcohols are the most common molecules which form strong hydrogen bonds, and it is these relatively simple, inexpensive substances which are of great importance in freeing surfaces of particulates.

However, water itself often must be excluded from consideration as a precision cleaning agent because it promotes corrosion. Alcohols and ketones are usually too flammable or too aggressive towards some components to be used by themselves though they are often excellent as cleaning agents.

A common misconception is that solubility, *per se*, is the determining criterion for a good cleaning solvent. In cases which involve gross amounts of soluble soils, high solvency for the soil is important, but such cases are not of particular relevance in cleaning microelectronic components. In removing particulates, the solvent serves two functions in varying degrees: (1) it adsorbs on the substrate by physical adsdorption or chemical adsorption, and (2) it provides

Table 10-24. Relative adhesive force (intermolecular forces) in air for various sizes of particles. Brandreth and Johnson (1979).	
Particle size (µm)	Relative adhesive force in air (g's)
500	2
50	200
5	20, 000
0.5	2, 000, 000

a different medium in which the interaction of the particles and the surface is changed.

When solvent molecules adsorb on both substrate and particulate, they create a different outermost layer which has its own (different) attraction for the soil particulate. But, in addition to this change in the attractive force between particle and surface, there is another change. The solvent acts as a different medium, not air or vacuum. To fully understand this "medium effect", one must resort to the physics of adhesion, which demands considerable development of definition of terms and notation, Hamaker (1937) and Krupp (1967).

To capture the essence of the idea, however, is to understand that the ultimate attraction of the substrate for the particle is influenced by the nature of the surrounding medium, which changes the relative values of the interactions between particle and surface, particle and solvent, and solvent and surface. A good solvent cleans by both adsorption on the surfaces and by the "medium effect", which diminishes the attraction between the surface and the particulates. With the addition of a solute, such as a surface-active-agent, the adsorption role can be enhanced still further.

It should further be pointed out that, whereas a particle not completely surrounded by liquid, experiences forces due to capillarity which tend to cause the particle to adhere to the surface, Bhattacharya and Mittal (1978), the disappearance of the gas/solid interface erases the capillary forces and yields the case of the totally immersed particle.

In one type of experiment, which directly measures the effect of solvents on the adhesion of particles to surfaces, diamond dust was deposited on a clean specimen. The particles were sized and counted after deposition. The system was then centrifuged while immersed in the test solvent, and the particles were again sized and counted. In some experiments where only low forces were studied, the system was merely inverted. The better solvents decrease the adhesive forces enough to enable the centrifugal force to release many of the particles. Although simple in concept, in practice meticulous care is required in these experiments.

The ability of the hydroxyl group to form hydrogen bonds was illustrated when cyclohexane and cyclohexanol were compared. Figure 10-32 shows the particle size distribution for diamond particles which adhere to glass after inverting the system in the presence of the solvent. There is pronounced superior particle removal by the cyclohexanol due to the stronger interaction of the hydroxyl group with the glass surface. Similar effects were found when low molecular weight alcohols such as methanol and ethanol were used to make azeotropes with R-113. These azeotropes, with the alcohol component in the 4-6 wt. % range, are effective cleaning agents for particulate matter because they combine the non-flammability, high specific gravity of R-113, and the relatively high solvent safety with the good adsorption characteristics of the alcohol component and the overall favorable "medium effect" of the solvent.

It was additionally found that the alcohol components of the azeotropes, being hygroscopic, were even more effective than very dry alcohols. This turns out to be fortuitous because the equilibrium amounts of absorbed water in the

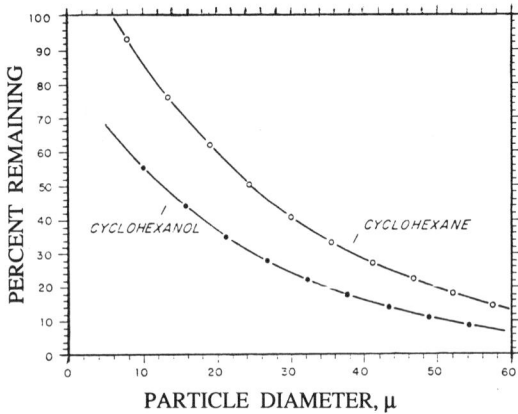

Fig. 10-32. Adhesion of diamond particles to glass in cyclohexane and in cyclohexanol at 1 g force. Brandreth and Johnson, (1979).

alcohol components at ambient conditions favors adsorption and consequently better particle removal.

Although the current non-availability of these R-113-based cleaning systems now makes some of these results academic, the principles still hold and should provide a basis for understanding and development of alternative cleaning compositions.

References

Alexander, G. B., Broge, E. C., and Iler, R. K., U.S. Patent 2,765,242 (1945).

Alexander, T., Fortune, April 1976, 153 (1976).

Alfrey, T., *Mechanical Behavior of High Polymers*, Interscience Publishers, New York (1948).

Barringer, E. A., Jubb, N., Fegley, B., Pober, R. L., and Bowen, H. K., *Ultrastructure Processing of Ceramics, Glasses, and Composites*, eds. L. L. Hench and D. R. Ulrich, John Wiley & Sons, New York, 315 (1984).

Barringer, E. A., Brook, R., and Bowen, H. K., Mat. Sci. Res., **16**, 1 (1984).

Becker, J. E., Interceram, **3**, 55 (1987).

Benjamin, J. S., Met. Trans., **1**, 2943 (1970).

Benjamin, J. S., *New Materials by Mechanical Alloying Techniques*, ed. E. Artz and L. Schultz, 3, DGM Informationsgesellschaft Verlag Oberursel, (1989).

Bergna, H. E., and Roberts, W. O., *Ullmann's Encyclopedia of Industrial Chemistry*, **A23**, 614 (1993).

Bhattacharya, S., and Mittal, K. L., Surface Technol., **7**, 413 (1978).

Böhm, G., and Kahlweit, M., Acta Metall., **12**, 641 (1964).

Boving, H., U.S. Patent 1,897,629 (1933).

Boxall, J., and von Fraunhofer, J. A., *Paint Formulation*, George Goodwin, Ltd., London (1980).

Brandreth, D. A., J. Testing Evaluation, 5, No. 1, ASTM (1977).

Brandreth, D. A., and Johnson, R. E., *Surface Contamination*, Vol. 1, ed. K. L. Mittal, Plenum Publishing, 83 (1979).

Brandreth, D. A., and Johnson, R. E., J. Opthalmic Dispensing, Jan., 109 (1979).

Brinker, C. J., and Scherer, G. W., *Sol-Gel Science*, Academic Press, Boston, (1990).

Brodnyan, J. G., Trans. Soc. Rheology, **3**, 61 (1959).

Buckwalter, H. M., U.S. Patent 2,297, 536 (1942).

Bunshah, R. F., *Handbook of Deposition Technologies for Films and Coatings*, Noyes Publications, Park Ridge, N.J. (1994).

Claussen, N., J. Am. Ceram. Soc. **59**, 49 and 179 (1976).

De Barbadillo, J. J., Key Eng. Mater., **77-78**, 187 (1993).

Degussa Co., Technical Bulletin Pigments No 11: Basic Characteristics (1982).

Drake, S., *Two New Sciences, Including Centers of Gravity and Force of Percussion*, U. of Wisconsin Press, Madison, WI (1974).

Du Pont "Orlon" Acrylic Fiber, Bulletin OR-109 (1961).

Einstein, A., Ann. Physik (4), 549 (1905).

Farmer, J. C., Fix, D. V., Mack, G. V., Pekala, R. W., and Poco, J. F., J. Electrochem. Soc., **143**, 159 (1996).

Ferch, H. K., *The Colloid Chemistry of Silica*, Ed. H. E. Bergna, Advances in Chemistry Series 234, American Chemical Society, 481 (1994).

Fricke, J., New Scientist, 30 January, 31, (1993).

Frisch, H. L., Simha, R., *Rheology*, Vol. 1, ed. F. Eirich, Academic Press, NY (1956).

Garvie, R. C., Hannink, R. H. and Pascoe, R. T., Nature (London), **258**, 703 (1975).

Grolitzer, M. A. and Erickson, D. E., Journal of Coatings Technology, **67**, No. 845, 89 (1995).

Gulledge, H. C., and Seidel, G. R., Ind. Eng. Chem. **42**,440, (1950).

Hack, G. A., *Advanced Materials and Processing Techniques for Structural Applications*, Ed. T. Khan and A. Lasalmonie, Office National d'Etudes et de Recherces Aerospatiales, Chatillon, 310 (1987).

Hamaker, H. C., Physica, **IV**, 1058 (1937).

Hirata, Y., Ceramics International, **23**, 93 (1997).

Iler, R. K. *The Chemistry of Silica*, John Wiley & Sons, NY (1979).

Iler, R. K., *The Colloid Chemistry of Silica and Silicates*, Cornell University Press, Ithaca, N.Y., 133 (1955).

Iler, R. K., U.S. Patent 2,657,149 (1953).

Inglis, G. E., Trans. Inst. Nav. Archit. **55**, 219 (1913).

Jacobsen, A. E., Sullivan, W. F. and Panik, I. M., Am. Dyestuff Reptr., **40**, (14), 439 (1995).

Karsa, D. R., and Davies, W. D., *Waterborne Coatings and Additives*, The Royal Society of Chemistry, Cambridge (1995).

Katz, H. S., Milewski, J. V., *Handbook of Fillers for Plastics*, Van Nostrand Reinhold Co., NY, (1987).

Kirkland, C., Plastics World, March 1992, 36 (1992).

Krupp, H., Adv. Colloid Interface Sci., **1**, 111 (1967).

Lane, F. W., and Dills, W. L., U.S. Patent 2,570,566 (1951).

Le Sota, S., *Paint/Coatings Dictionary*, Federation of Societies for Coatings Technology, Blue Bell, PA (1978).

Marder, M., and Fineberg, J., Physics Today. September 1996, 24 (1996).

Marisic, M. M., and Dray, S., U.S. Patent 2,386,810 (1945).

Matteazzi, P., Le Caër, G., Mocellin, A., Ceramics International, **23**, 39 (1997).

Mattsson, M., and Otterstedt, J-E., unpublished work (1995).

Mooney, M., J. Colloid Sci. **6**, 162 (1951).

Moore, E. P., U.S. Patents 3,748,157; 3,754,945; 3,764,355 (1973).

Nembach, E., *Particle Strengthening of Metals and Alloys*, John Wiley & Sons, NY (1997).

Ohring, M., The Materials Science of Thin Films, Academic Press, Boston, (1992).

Orowan, E., Z. Phys., **89**, 634 (1934).

Otterstedt, J-E., and Brandreth, D. A., unpublished work (1996).

Otterstedt, O. E., Otterstedt, J-E., Ekdahl, J., Backman, J., and Andersson, C. H., J. Applied Polymer Science, 34, 2575 (1987).

Patrick, W. A., U.S. Patent 1,297,724 (1918)

Paul, S., *Surface Coatings*, John Wiley & Sons, NY (1995).

Plueddemann, E. P., *Silane Coupling Agents*, Plenum Press, NY (1992).

Polanyi, M., Z. Phys., **89**, 660 (1934).

Pugh, R. J., and Bergstrom, L., *Surface and Colloid Chemistry in Advanced Ceramics Processing,* Marcel Dekker, Inc. (1994).

Reed, J. S.,. *Introduction to the Principles of Ceramics Processing*, John Wiley & Sons, NY (1995).

Richerson, D. W., *Modern Ceramic Engineering—Properties, Processing, and Use in Design*, Marcel Dekker, Inc., NY (1992).

Rodwell, A. G., and Barber, S. G., British Patent 462, 230 (1937).

Russel, W. B., Mater. Res. Bull. **XVI**, 27 (1991).

Schneider, M., and Baiker, A., Catal. Rev.- Sci. Eng. **37**(4), 515 (1995).

Schut, J. H., Plastics World, November, 61 (1995).

Stein, H. N., The Preparation of Dispersions in Liquids, Marcel Dekker, Inc., NY, (1996).

Taylor, G. I., Proc. R. Soc. London A, **145**, 362 (1934).

Ulrich, D. R., *Ultrastructure Processing of Ceramics, Glasses, and* Ultrastructure Processing of Ceramics, Glasses, and Composites, eds. L. L. Hench and D. R. Ulrich, University Press, Ithaca, 133 (1955).

Uytterhoeven, J. Hellinckx, E., and Fripiat, J. J., Silic. Ind. **28**, 241 (1963).

Vail, J. G. *Soluble Silicates*, Volumes 1 and 2. Reinhold Publishing Corp., NY (1952).

Van Rossem, A., *Rubber*, N. V. Servire, The Hague, 136 (1958).

Vand, V., J. Phys. and Colloid Chem. **52**, 277, 300, 314 (1948).

Vossos, P., U.S. Patent 3,652,329 (1972).

Wilm, A., Metallurgie, **8**, 225 (1911).

Winding, C. C., and Hiatt, G. D., *Polymeric Materials*, McGraw-Hill, NY (1961).

Yanagida, H., Koumoto, K., and Miyayama, M., *The Chemistry of Ceramics*, John Wiley and Sons, NY (1997).

Index